T0324134

Handbook of Machining with Grinding Wheels

Second Edition

Handbook of Machining with Grinding Wheels

Second Edition

Ioan D. Marinescu • Mike P. Hitchiner
Eckart Uhlmann • W. Brian Rowe • Ichiro Inasaki

CRC Press
Taylor & Francis Group
Boca Raton London New York

CRC Press is an imprint of the
Taylor & Francis Group, an **informa** business

CRC Press
Taylor & Francis Group
6000 Broken Sound Parkway NW, Suite 300
Boca Raton, FL 33487-2742

First issued in paperback 2019

© 2016 by Taylor & Francis Group, LLC
CRC Press is an imprint of Taylor & Francis Group, an Informa business

No claim to original U.S. Government works

ISBN-13: 978-1-4822-0668-5 (hbk)
ISBN-13: 978-0-367-86870-3 (pbk)

Library of Congress Cataloging-in-Publication Data

Handbook of machining with grinding wheels / Ioan D. Marinescu [and four others]. -- Second edition.
    pages cm
"A CRC title."
Includes bibliographical references and index.
ISBN 978-1-4822-0668-5 (alk. paper)
   1. Grinding wheels--Handbooks, manuals, etc. 2. Grinding and polishing--Handbooks, manuals, etc. 3. Machine-tools--Handbooks, manuals, etc. I. Marinescu, Ioan D.

TJ1280.H425 2016
671.3'5--dc23

2015028915

Visit the Taylor & Francis Web site at
http://www.taylorandfrancis.com

and the CRC Press Web site at
http://www.crcpress.com

Contents

Section II Application of Grinding Processes

Preface to First Edition

Grinding, once considered primarily a finishing operation involving low rates of removal, has evolved as a major competitor to cutting, as the term "abrasive machining" suggests. This is what Milton Shaw was saying about 10 years ago, the man who is considered the great pioneer and father of American grinding. Milton led the development of grinding in the United States over the past 50 years.

We named this book *Handbook of Machining with Grinding Wheels* because the borders between grinding and other operations like superfinishing, lapping, polishing, and flat honing are no longer distinct. Machining with grinding wheels extends from high-removal rate processes into the domains of ultrahigh accuracy and superfinishing. This book aims to explore some of the new "transition operations," and for this reason we chose this title.

This book presents a wide range of abrasive machining technology in fundamental and application terms. The emphasis is on why things happen as they do, rather than a how-to-do-it approach. The topics detailed in this book cover a range of abrasive machining processes with grinding wheels, making this probably the most complete book regarding all kinds of grinding operations.

The aim of this monograph is to present a unified approach to machining with grinding wheels that will be useful in solving new grinding problems of the future. It should be of value to engineers and technicians involved in solving problems in industry and to those doing research on machining with grinding wheels in universities and research organizations.

The team of authors is composed of famous researchers who have devoted their entire lives to doing research in this field and who are still actively contributing to new research and development. The authors represent a large region of the world where abrasive machining with grinding wheels are most advanced: the United States, Great Britain, Japan, and Germany. I would like to take this opportunity to thank my coauthors for taking time from their busy activities to write and review this book over a period of two years.

All the coauthors are my long time friends, and with some of them I have previously published, or we are still in the process of finishing other books. I would like to make a short presentation of them:

Professor Brian Rowe is considered the world's Father of Centerless Grinding in addition to other notable research concerning grinding aspects: thermal aspects, dynamic aspects, fluid-film bearings, etc. He established a great laboratory and school in manufacturing processes at Liverpool John Moores University. As an emeritus professor, Brian is busier than before retirement. Being a native English speaker he spent a lot of time polishing our English in order to have a unitary book. I thank him for similar great work on our previous book: *Tribology of Abrasive Machining Processes*.

Professor Ichiro Inasaki is the leading figure in grinding in Japan. As dean of the Graduate School of Science and Technology at Keio University, he developed a great laboratory with outstanding research activities. His "intelligent grinding wheel" is featured in the Noritake Museum and represents one of his best accomplishments and contributions. He led CIRP in 2004/2005 as the president and was granted several awards including an SME award. With Ichiro-san I have already written two books: *Handbook of Ceramic Grinding and Polishing* and *Tribology of Abrasive Machining Processes*.

Professor Eckart Uhlmann is professor and director of the Institute for Machine-Tools and Management at Technical University of Berlin. Dr. Uhlmann earned this chaired professorship after a very successful industrial career with Hermes Abrasive in Germany. His main research is on one of these transition processes: grinding with lapping kinematics. As the head of his Institute, one of the largest in Germany, he holds the leading position in research on all aspects of abrasive machining with grinding wheels. A future book with Dr. Uhlmann will be published this year: *Handbook of Lapping and Polishing/CMP*.

Dr. Mike Hitchiner is manager of precision technology at Saint Gobain Abrasives Co., the largest grinding wheel company in the world. Mike has devoted all his life to research and development of grinding processes. He started this activity during his PhD at Oxford University in England, and today he is considered "Mr. CBN Grinding" by the automotive industry. He has brought an important industrial perspective to this book, as well as hundreds of applications.

As leading author, my own experience in abrasive machining research complements and extends the experience of the other authors, widely across industrial and fundamental areas of investigation. My research has focused on new and challenging techniques of abrasive machining particularly for new materials. I have been fortunate to have studied the latest technologies developed in countries across the world first hand and have contributed to developing new techniques for application in industry and in research.

The main aim of this book is to present abrasive machining processes as a science more than an art. Research and development on abrasive machining processes have greatly increased the level of science compared to 25 years ago, when many aspects of abrasive machining processes still depended largely on the expertise of individual technicians, engineers, and scientists.

The book has two sections: The Basic Process of Grinding and Application of Grinding Processes. This structure allows us to present more about *understanding of grinding behavior* in Section I and more about *industrial application* in Section II.

Ioan D. Marinescu
January 2006
Toledo, Ohio

Preface to Second Edition

Today many machining operations are considered "old fashioned" processes. The last decade has brought new fashionable research topics including nanotechnology, alternative energy, and the latest is "additive manufacturing."

Even additive manufacturing requires a finishing process as the dimensions and surface quality of the parts are not final due to the limitation of the process itself. This means that abrasive processes, and especially grinding, are still necessary for these new processes. At the same time the bulk of manufacturing processes for automotive industry and aerospace industry have not changed too much during the past 10 years.

Writing the second edition of the book, *Handbook of Machining with Grinding Wheels*, has been a challenge due to the fact that I insisted on keeping the same team of authors. Some of them retired, some received even more demanding responsibilities, and I am thankful that they agreed to work on the second edition of the book.

As you can see from the Preface to the First Edition, the team is well-known internationally and one with a reputation difficult to match. They brought to this book experience from their laboratories and from their work with industry, and this is a combination that makes this book useful for industry and academia.

Regarding the contents of the second edition, there is a significant new work on Abrasives in Chapter 5, on Bonds in Chapter 6, and Dressing in Chapter 7. In Chapter 8, Dynamics, there is a new figure showing classic stability lobes for grinding. Also a new method for tracking dynamic instability in centerless grinding is presented in Chapter 19. Chapter 20, on Ultrasonic-Assisted Grinding, includes a new section that contains recent work on modelling of the process. In Chapter 11, Process Monitoring, new material showing experimental results for in-process feedback to the grinding process was added. There are also changes in some other chapters. Some work on fluid cooling was added to Chapter 10, Coolants. Chapter 15, Grinding Machine Technology, presents many new examples, particularly for dressing. We appreciate very much the work Dr. Mike Hitchiner did on these chapters while facing conditions of serious illness.

I am particularly thankful to Dr. Brian Rowe who agreed in his retirement to put together and review all the chapters using his tremendous experience. Brian worked with the team for many years and he was also instrumental in putting together *Tribology of Abrasive Machining Processes*, which was published in 2013 (second edition).

The first edition of the *Handbook of Machining with Grinding Wheels* was reviewed by the Abrasive Engineering Society and was considered the best comprehensive book on grinding after the Milton Shaw 1996 book. We are very proud of this review, and the second edition aims to be a better and more complete book.

Again, I would like to thank my coauthors and their families for taking the time to work on this second edition, at the time when all of us are getting older and having different priorities in our lives. But writing a new book makes each of us feel younger and useful to our profession and we hope to leave a good legacy.

Ioan D. Marinescu
July 2015
White Lake, Michigan

Section I

The Basic Process of Grinding

1

Introduction

1.1 From Craft to Science

Grinding has been employed in manufacture for more than a hundred years, although earliest practice can be traced back to Neolithic times (Woodbury 1959). The lack of machine tool technology meant that primitive operations were mostly limited to simple hand-held operations. An early device for dressing a sandstone grinding wheel was patented by Altzschner in 1860 (Woodbury 1959).

The twentieth century saw the burgeoning of grinding as a modern process. Seminal publications by Alden and Guest started the process of bringing the art of grinding onto a scientific basis (Alden 1914; Guest 1915).

Grinding is a machining process that employs an abrasive grinding wheel rotating at high speed to remove material from a softer material. In modern industry, grinding technology is highly developed according to particular product and process requirements. Modern machine tools may be inexpensive machines with a simple reciprocating table or they may be expensive machines. Many grinding machines combine computer-controlled feed drives and slideway motions, allowing complex shapes to be manufactured free from manual intervention. Modern systems will usually incorporate algorithms to compensate for wheel and dressing tool wear processes. Programmable controls may also allow fast push-button setup. Monitoring sensors and intelligent control introduce the potential for a degree of self-optimization (Rowe et al. 1994, 1999).

Faster grinding wheel speeds and improved grinding wheel technology have allowed greatly increased removal rates. Grinding wheel speeds have increased by 2–10 times over the last century. Removal rates have increased by a similar factor and in some cases by even more. Removal rates of 30 mm^3/mm s were considered fast 50 years ago, whereas today, specific removal rates of 300 mm^3/mm s, are increasingly reported for easy-to-grind materials. In some cases, removal rates exceed 1000 mm^3/mm s. Depths of cut have increased by up to 1000 times values possible 50 years ago. This was achieved through the introduction of creep-feed and high-efficiency deep grinding technology.

Advances in productivity have relied on increasing sophistication in the application of abrasives. The range of abrasives employed in grinding wheels has increased with the introduction of new ceramic abrasives based on sol–gel technology, the development of superabrasive cubic boron nitride (CBN), and diamond abrasives based on natural and synthetic diamond.

New grinding fluids and methods of delivering grinding fluid have also been an essential part in achieving higher removal rates, while maintaining quality. Developments include high-velocity jets, shoe nozzles, factory-centralized delivery systems, neat mineral oils, synthetic oils, vegetable ester oils, and new additives. Minimum quantity lubrication provides an alternative to flood and jet delivery aimed at environment-friendly manufacture.

Grinding is not a process without its share of problems. Problems experienced may include thermal damage, rough surfaces, vibrations, chatter, wheel glazing, and rapid wheel wear. Overcoming these problems quickly and efficiently is helped by a correct understanding of the interplay of factors in grinding. Commonly encountered problems are analyzed in succeeding chapters to show how parameters can be optimized and grinding quality can be improved.

Grinding dynamics and the sources of vibration problems are explained, and different approaches to avoiding vibrations are explored. Some of the techniques described may be surprising to some practitioners. For example, it is shown that increased flexibility of the grinding wheel can be an advantage for vibration suppression.

Attitudes to costs have changed over the years. Buying the cheapest grinding wheels has given way to evaluation of system costs including labor, equipment, and nonproductive time. Examples are included in Chapters 12 and 19 to show how systematic analysis can greatly increase productivity and quality while reducing cost per part. Often the key to reducing costs is to reduce nonproductive time.

1.2 Basic Uses of Grinding

Grinding is a key technology for production of advanced products and surfaces in a wide range of industries. Grinding is usually employed where one or more of the following factors apply.

1.2.1 High-Accuracy Required

Grinding processes are mostly used to produce high-quality parts to high accuracy and to close tolerances. Examples range from the very large such as machine tool slideways to the small such as contact lenses, needles, electronic components, silicon wafers, and rolling bearings.

1.2.2 High Removal Rate Required

Grinding processes are also used for high removal rate. A typical example is high removal rate grinding for the flutes of hardened twist drills. The flutes are ground into solid round bars in one fast operation. Twist drills are produced in very large quantities at high speeds explaining why grinding is a key process for low costs, high production rates, and high quality.

1.2.3 Machining of Hard Materials

While accuracy and surface texture requirements are common reasons for selecting abrasive processes, there is another common reason. Abrasive processes are the natural choice for machining and finishing very hard materials and hardened surfaces. In many cases, grinding is the only practical way of machining some hard materials. The ability to machine hard material has become increasingly important with the increasing application of brittle ceramics and other hard materials such as those used in aerospace engines.

1.3 Elements of the Grinding System

1.3.1 The Basic Grinding Process

Figure 1.1 illustrates a surface-grinding operation. Six basic elements are involved: the grinding machine, the grinding wheel, the workpiece, the grinding fluid, the atmosphere, and the grinding swarf. In addition, there is the need for a dressing device to prepare the grinding wheel shape and cutting surface.

The grinding wheel removes material from the workpiece, although inevitably the workpiece wears the surface of the grinding wheel. An important aspect of grinding is to get the balance right between high removal rate from the workpiece and moderate wear of the grinding wheel.

Grinding swarf is produced from the workpiece material and is mixed with a residue of grinding fluid and worn particles from the abrasive grains of the wheel. The swarf is not necessarily valueless but has to be disposed of or recycled.

The grinding fluid is required to lubricate the process to reduce friction and wear of the grinding wheel. It is also required to cool the process, the workpiece, and the machine preventing thermal damage to the workpiece and improving accuracy by limiting thermal expansion of both workpiece and machine. The grinding fluid also transports swarf away from the grinding zone.

The atmosphere plays an important role in grinding most metals by reducing friction. Newly formed metal surfaces at high temperature are highly reactive leading to oxides that can help to lubricate the process. It is usual to emphasize physical aspects of grinding but chemical and thermal aspects play an extremely important role that is easily overlooked.

The machine tool provides static and dynamic constraint on displacements between the tool and the workpiece. The machine tool stiffness is, therefore, vital for productivity and for achievement of tolerances for geometry, size, roughness, and waviness. Vibration behavior of the machine also affects fracture and wear behavior of the abrasive grains.

To summarize, the main elements of an abrasive machining system are (Marinescu et al. 2013):

1. The workpiece material, shape, hardness, speed, stiffness, thermal, and chemical properties.
2. The abrasive tool, structure, hardness, speed, stiffness, thermal and chemical properties, grain size, and bonding.

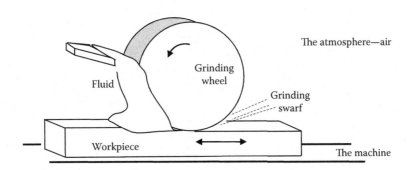

FIGURE 1.1
The six basic elements involved in surface grinding.

3. The geometry and motions governing the engagement between the abrasive tool and the workpiece (kinematics).

4. The process fluid, flowrate, velocity, pressure, physical, chemical, and thermal properties.

5. The atmospheric environment.

6. The machine, accuracy, stiffness, temperature stability, and vibrations.

1.3.2 Four Basic Grinding Operations

Four basic grinding processes are illustrated in Figure 1.2. The figure shows examples of peripheral grinding of flat surfaces and cylindrical surfaces. The figure also shows examples of face grinding of nonrotational flat surfaces and face grinding of rotational flat surfaces. Face grinding of rotational flat surfaces can be carried out on a cylindrical grinding machine and may therefore be simply termed cylindrical face grinding.

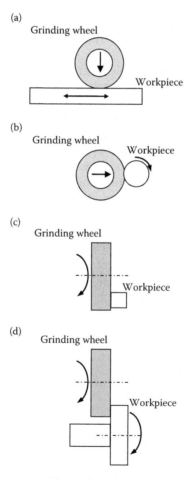

FIGURE 1.2
Examples of four basic grinding operations using straight wheels. (a) Peripheral surface grinding, (b) peripheral cylindrical grinding, (c) face surface grinding, and (d) face cylindrical grinding.

Figure 1.2 introduces common terms with four basic operations. A distinction is drawn between grinding with the face of the grinding wheel known as face grinding and grinding with the periphery of the wheel known as peripheral grinding. Surface grinding usually refers to grinding flat or profiled surface with a linear motion. Cylindrical grinding refers to grinding a rotating workpiece. Cylindrical grinding may be performed internally or externally. A full description of grinding operations commonly employed is necessarily rather more complex and is described in other chapters.

In practice, the range of possible grinding processes is large and includes a number of profile generating operations, profile copying operations, slitting, and grooving. Profiling processes include grinding of spiral flutes, screw threads, spur gears, and helical gears using methods similar to gear cutting, shaping, planing, or hobbing with cutting tools. There are other processes suitable for grinding crankshafts, cam plates, rotary cams, and ball joints. Terminology for these different processes can be confusing. The International College of Production Research (CIRP) has published a number of terms and definitions (CIRP 2005). Details of CIRP publications can be found on the Internet at www.cirp.net. Further details of process classification are given in Chapter 3 and later chapters dealing with applications.

1.4 The Importance of the Abrasive

The importance of the abrasive cannot be overemphasized. The enormous differences in typical hardness values of abrasive grains are illustrated in Table 1.1 (after De Beers). A value for a typical M2 tool steel is given for comparison. The values given are approximate, since variations can arise due to the particular form, composition, and directionality of the abrasive.

In grinding, it is essential that the abrasive grain is harder than the workpiece at the point of interaction. This means that the grain must be harder than the workpiece at the temperature of the interaction. Since these temperatures of short duration can be very high, the abrasive grains must retain hot hardness. This is true in all abrasive processes, without exception, since if the workpiece is harder than the grain, it is the grain that suffers most wear.

The hardness of the abrasive is substantially reduced at typical contact temperatures between a grain and a workpiece. At 1000°C, the hardness of most abrasives is approximately halved. CBN retains its hardness better than most abrasives, which makes CBN a wear-resistant material. Fortunately, the hardness of the workpiece is also reduced. As can

TABLE 1.1

Typical Hardness of Abrasive Grain Materials at Ambient Temperatures

	Units (GPa)
Diamond	56–102
Cubic boron nitride	42–46
Silicon carbide	~24
Aluminum oxide	~21

be seen from Table 1.1, the abrasive grains are at least one order of magnitude harder than hardened steel.

The behavior of an abrasive depends not only on hardness but on wear mode. Depending on whether wear progresses by attritious wear, microfracture or macrofracture, will determine whether the process remains stable or whether problems will progressively develop through wheel blunting or wheel breakdown. This range of alternatives means that productivity is improved when grinding wheels are best suited for the particular grinding purpose.

1.5 Grinding Wheels for a Purpose

Grinding wheels vary enormously in design according to the purpose for which the wheel is to be used. Apart from the variety of abrasives already mentioned, there is a variety of bonds employed including plastic, resinoid, vitrified, metal bonds, and plated wheels.

Within each class of bond, there is scope for engineering bond properties to achieve strength and wear behavior suited to the particular abrasive. The bond must hold the abrasive until wear makes the abrasive too inefficient as a cutting tool. In addition, the porosity of the wheel must be sufficient for fluid transport and chip clearance. However, porosity affects grit retention strength and so the wheel must be correctly engineered for the workpiece material and the removal rate regime.

A grinding wheel is bonded and engineered according to the particular process requirement. A general-purpose wheel will give greatly inferior removal rates and economics compared to an optimized wheel for the particular product. This may be relatively unimportant in a toolroom dealing with various tools of similar material. However, wheel selection and optimization become critical for large-scale repeated batches of aerospace and automotive parts. In such cases, the process engineer should adopt a systematic approach to problem solving and work closely with the grinding wheel and machine tool manufacturers.

1.6 Problem Solving

Few readers have time and fortitude to read a handbook from beginning to end. Although much could be learned from such an approach, readers are encouraged to cherry-pick their way through the most appropriate chapters. Readers are mostly busy people who want to solve a problem. The handbook is therefore structured to allow individual areas of interest to be pursued without necessarily reading chapters consecutively.

Part I: The 12 chapters in Part I cover the principles of grinding. This part includes all aspects that relate to grinding generally. Topics include basic grinding parameters, grinding wheels and grinding wheel structure, and wheel dressing processes used for preparing wheels for grinding and restoring grinding efficiency. Further chapters include vibrations, wheel wear mechanisms, coolants, process monitoring, and grinding costs. Principles are explained as directly as possible and references are given to further sources of information. For example, some readers may wish to explore the science and tribology of grinding more deeply (Marinescu et al. 2013). Tribology is the science of friction, lubrication, and wear (HMSO 1966). The tribology of abrasive machining processes brings together the branches of science at the core of grinding and grinding wheel behavior.

Part II: The eight chapters in Part II explore applications of grinding. This part covers grinding of conventional ductile materials, grinding of ceramics, grinding machine technology and rotary dressers, surface grinding, external cylindrical grinding, internal cylindrical grinding, centerless grinding, and ultrasonically assisted grinding. A particular emphasis is placed on developments in technology that can lead to improved part quality, higher productivity, and lower costs.

The authors draw on industrial and research experience, giving numerous references to scientific publications and trade brochures where appropriate. Readers will find the references to the various manufacturers of machine tools, auxiliary equipment, and abrasives, a useful starting point for sourcing suppliers. The references to scientific publications provide an indication of the wide scope of research and development in this field around the world.

The second edition introduces additional material where this is considered helpful and updates chapters with more recent technological advances.

References

Alden, G. I., 1914, Operation of grinding wheels in machine grinding. *Transactions of the American Society of Mechanical Engineers*, 36, 451–460.

CIRP, 2005, *Dictionary of Production Engineering II—Material Removal Processes*. Springer, Heidelberg.

Guest, J. J., 1915, *Grinding Machinery*. Edward Arnold, London.

HMSO, 1966, Lubrication (tribology) education and research. DES (Jost) Report, London.

Marinescu, I. D., Rowe, W. B., Dimitrov, B., Ohmori, H., 2013, *Tribology of Abrasive Machining Processes*, 2nd edition. Elsevier/William Andrew Publishing, Norwich, NY.

Rowe, W. B., Li, Y., Inasaki, I., Malkin, S., 1994, Applications of artificial intelligence in grinding. *Annals of the CIRP, Keynote Paper*, 43, 2, 521–532.

Rowe, W. B., Statham, C., Liverton, J., Moruzzi, J., 1999, An open CNC interface for grinding machines. *International Journal of Manufacturing Science and Technology*, 1, 1, 17–23.

Woodbury, R. S., 1959, *History of the Grinding Machine*. The Technology Press, MIT, USA.

2

Grinding Parameters

2.1 Introduction

Grinding, in comparison to turning or milling, is often considered somewhat of a "black art," where wheel life and cycle times cannot be determined from standard tables and charts. Certainly precision grinding, being a finishing process with chip formation at submicron dimensions occurring by extrusion created at cutting edges with extreme negative rake angles, is prone to process variability such as chatter, system instability, coolant inconsistency, etc. Nevertheless, with grinding equipment in a competent state of repair, performance can be controlled and predicted within an acceptable range. As important, rules and guidelines are readily available to the end user to modify a process to allow for system changes. It is also essential to ensure surface quality of the parts produced. These objectives are balanced through an analysis of costs as described in subsequent chapters on economics and on centerless grinding. The importance of the grinding parameters presented below is to provide an understanding of how process adjustments change wheel performance, cycle time, and part quality.

Probably the best way for an end user to ensure a reliable and predictable process is to develop it with the machine tool builder, wheel maker, and other tooling suppliers at the time of the machine purchase using actual production parts. This then combines the best of the benefits from controlled laboratory testing with real components without production pressures, resulting in a baseline against which all future development work or process deterioration can be monitored.

The number of grinding parameters that an end user needs to understand is actually quite limited. The key factors are generally associated with either wheel life, cycle time, or part quality. The purpose of this discussion is to define various parameters that relate to wheel life, cycle time, and part quality and to demonstrate how these parameters may be used to understand and improve the grinding process. In most cases, the author has avoided the derivations of the formulae providing instead the final equation. Derivations and more detailed discussion can be found in publications such as Marinescu et al. (2013), Rowe (2014), or Malkin (1989).

2.1.1 Wheel Life

The statement that a process can be controlled "within an acceptable range" requires some definition. A recent study by Hitchiner and McSpadden (2005) investigated the process variability of various vitrified cubic boron nitride (CBN) processes as part of a larger program to develop improved wheel technology. They showed that under "ideal" conditions

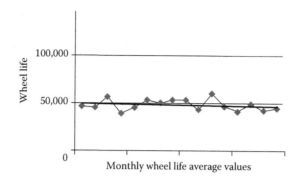

FIGURE 2.1
Monthly average wheel life values for high-production internal grinding operation.

repeatability of wheel life within ±15% or better could be achieved. However, variability from one wheel to another associated with just wheel grade (±1% porosity), all within the standard limits of a commercial specification, made the process less repeatable, and increased the variability to ±25%. In the field, for example, in a high-production internal grinding operation with 20 machines, the average monthly wheel life was tightly maintained within ±5%. However, these average values obscured an actual individual wheel life variability of ±100%. Of these, wheels with very low or zero life were associated with setup problems, while the large variability at the high end of wheel life was associated with machine to machine variables, such as coolant pressure, spindle condition, or gauging errors. A process apparently in control based on monthly usage numbers was actually quite the opposite (Figures 2.1 and 2.2).

Wheel makers and machine tool builders are usually in the best position to make predictions as to wheel performance. Predictions are based on either laboratory tests or past experience on comparable applications. Laboratory tests tend to reproduce ideal conditions but can make little allowance for a deficiency in fixturing or coolant, etc. The author well remembers a situation, where the laboratory results and the actual field wheel life differed by a factor of 40. The loss of wheel life in the field was caused by vibration from poor part clamping, and wheel bond erosion from excessively high coolant pressure. Laboratory data were able to inform the end user that there was a major problem and provide evidence to search for the solution.

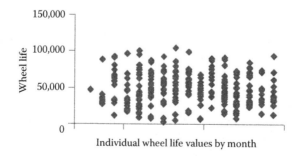

FIGURE 2.2
Individual wheel life values over the same period as Figure 2.1.

2.1.2 Redress Life

In practice, the end user seeks to reduce cycle time for part production as a route to reducing costs and increasing production throughput. The number of parts produced per dress is critical for economic production (Rowe et al. 2004). For parts produced in large batches, redress life can be given as the number of parts per dress n_d. If redress has to take place for every part produced, the cost of grinding is greatly increased. Long redress life depends on having the correct grinding wheel for the grinding conditions and also on the dressing process. Dressing parameters are discussed further in the chapter on dressing.

2.1.3 Cycle Time

Cycle time is usually defined as the average total time to grind a part. For a batch of n_b parts produced in a total time t_b, the cycle time is

$$t_c = \frac{t_b}{n_b}$$

The cycle time, therefore, depends on the dressing time as well as the grinding time, and the loading and unloading time.

2.2 Process Parameters

2.2.1 Uncut Chip Thickness or Grain Penetration Depth

The starting point for any discussion on grinding parameters is "uncut chip thickness," h_{cu}, as this provides the basis for predictions of roughness, power, and wear (Shaw 1996). Uncut chip calculations are typically based on representations of the material removed in the grind process as a long slender triangular shape with a mean thickness h_{cu}. However, a more practical way of looking at this parameter is to think of h_{cu} as representing the depth of abrasive grit penetration into the work material. In fact, this parameter is often termed the grain penetration depth. The magnitude of h_{cu} may be calculated from the various standard parameters for grinding and the surface morphology of the wheel.

$$h_{cu} = \sqrt{\frac{v_w}{v_s} \cdot \frac{1}{C \cdot r} \sqrt{\frac{a_e}{d_e}}} \quad h_{cu} \ll a_e$$

where v_s is the wheel speed, v_w is the work speed, a_e is the depth of cut, d_e is the equivalent wheel diameter, C is the active grit density, and r is the grit cutting point shape factor.

Other useful measures of grain penetration include equivalent chip thickness $h_{eq} = a_e \cdot v_w/v_s$. However, equivalent chip thickness takes no account of the spacing of the grains in the wheel surface.

2.2.2 Wheel Speed

Wheel speed v_s is given in either meter/second (m/s) or surface feet per minute (sfpm). To convert the former to the latter, a rule of thumb multiplication factor is 200 (or 196.7 to be precise).

2.2.3 Work Speed

Work speed v_w is a term most typically applied to cylindrical grinding; equivalent terms for surface grinding are either traverse speed or table speed.

2.2.4 Depth of Cut

Depth of cut a_e is the actual depth of work material removed per revolution or table pass after taking account of machine deflection. It is not the same as the programmed or set depth of cut a_p.

2.2.5 Equivalent Wheel Diameter

Equivalent wheel diameter d_e is a parameter that takes into account the conformity of the wheel and the workpiece in cylindrical grinding and gives the equivalent wheel diameter for the same contact length in a surface grinding application (i.e., $d_e \rightarrow d_s$ as $d_w \rightarrow \infty$). The plus sign is for external cylindrical grinding, while the negative sign is for internal cylindrical grinding:

$$d_e = \frac{d_s \times d_w}{d_s \pm d_w}$$

where d_w is the workpiece part diameter and d_s the wheel diameter.

2.2.6 Active Grit Density

Active grit density C is the number of active cutting points per unit area on the wheel surface. For convenience in predicting the effect of process changes on wheel behavior, it is often assumed that C is constant. While this is a reasonable assumption for single-layer superabrasive wheels, it is difficult to justify for vitrified and elastic bond wheels. An example of measurements on a vitrified alumina wheel by Darafon et al. (2013) is shown in Figure 2.3. The factor C varies greatly with grain penetration depth, increasing to a maximum as more grains are brought into active contact with the workpiece. The increase in C with depth of grain penetration is further greatly increased by local wheel deflection. This is particularly the case for vitrified and elastic bonds where the normal grinding force compresses the wheel surface bringing more grains into active grinding contact (Cai and Rowe 2004; Rowe 2014).

2.2.7 Grit Shape Factor

Grit shape factor r is the ratio of chip width to chip thickness. In most discussions of precision grinding, the product $C \cdot r$ is considered as a single factor that can be somewhat

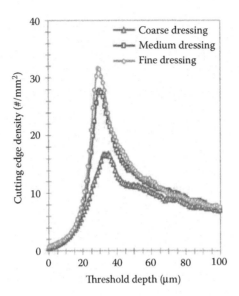

FIGURE 2.3
Cutting-edge density of a 60 grit alumina wheel measured by Darafon et al. (2013).

affected by dress conditions. Under stable grinding conditions, with a fixed or limited range of dress conditions, r is usually considered as a constant for a given wheel specification. The experimental support for this assumption is poor, but it proves useful technique for considering the direct effect of a change in process conditions.

The several key parameters that research has shown to be directly dependent on h_{cu} are as follows.

2.2.8 Force per Grit

A basic model of the force on the grit f_g gives

$$f_g \propto h_{cu}^{1.7} \propto \left\{ \frac{v_w}{v_s} \cdot \frac{1}{C \cdot r} \sqrt{\frac{a_e}{d_e}} \right\}^{0.85}$$

Grit retention is directly related to the forces experienced by the grit and these forces increase with uncut chip thickness. It can be seen that for a constant stock removal rate per unit width (Q') equal to $a_e \cdot v_w$, forces are lower at large depth of cut and low table speed. Hence, a softer grade might be used for creep feed rather than for reciprocal surface grinding. A softer grade has a better self-sharpening action and reduces grinding forces.

Wheel wear can accelerate as the wheel diameter gets smaller and force/grit increases.

2.2.9 Specific Grinding Energy

Specific grinding energy e_c (or u in older publications) is the energy expended to remove a unit volume of workpiece material. The units are usually J/mm^3 or lbf in/in^3; conversion

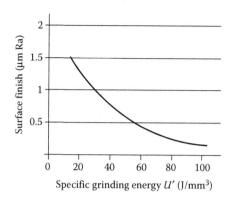

FIGURE 2.4
Example of the relationship between surface roughness and specific grinding energy for a fixed Q'.

from metric to English requires a multiplication factor of 1.45×10^5. Analysis of the energy to create chips leads to the following relationship between e_c and h_{cu}:

$$e_c \propto \frac{1}{h_{cu}^n} \propto \sqrt{\frac{v_s}{v_w}} C \cdot r \cdot \sqrt{\frac{d_e}{a_e}}$$

where $n = 1$ for precision grinding. The relationship is logical so far as it takes more energy to make smaller chips but is valid only as long as chip formation is the dominant source. In general terms, for precision grinding of hardened steel, the surface roughness will follow a trend rather like that is shown in Figure 2.4 as a function of specific energy (see below).

Hahn (1962) and also Malkin (1989) showed that in most cases, especially in fine grinding or low metal removal rates, significant energy is consumed by rubbing and ploughing. Under these circumstances, specific energy e_c varies with removal rate Q' as illustrated in Figure 2.4.

2.2.10 Specific Removal Rate

Specific removal rate Q' or Q'_w is defined as the metal removal rate of the workpiece per unit width of wheel contact, $Q' = a_e \cdot v_w$. The units are either mm³/mms or in³/in min. To convert from the former to the latter requires a rule of thumb multiplication factor of approximately 0.1 (or 0.1075 to be precise).

For very low values of Q', rubbing and ploughing dominate but as Q' increases so does the proportion of energy consumed in chip formation. More to the point, the energy consumed by rubbing and ploughing remain constant thereby becoming a smaller proportion of the total energy consumed as stock removal rates increase. Precision grinding for the steels illustrated by Figure 2.5 gives specific energy values of 60–30 J/mm³ of which about 20 J/mm³ is associated with chip formation.

Chip formation dominates in high removal rate precision applications such as camlobe grinding or peel grinding with vitrified CBN or rough grinding with plated CBN. Under these circumstances, $e_c \propto 1/h_{cu}$ is a good predictor of performance.

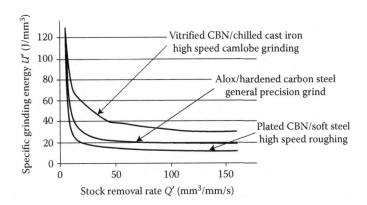

FIGURE 2.5
Examples of specific grinding energy U' trends versus stock removal rate Q'.

2.2.11 Grinding Power

Grinding power P can be estimated from the specific grinding energy e_c using the equation:

$$P = e_c \cdot Q' \cdot b_w$$

where b_w is the width of grind.

2.2.12 Tangential Grinding Force

Tangential grinding force F_t may then be calculated from

$$F_t = \frac{P}{v_s} = \frac{e_c \cdot Q' \cdot b_w}{v_s}$$

2.2.13 Normal Grinding Force

Normal grinding force F_n is related to the tangential grinding force by the coefficient of grinding; a parameter defined in a similar way to friction coefficient.

2.2.14 Coefficient of Grinding

Coefficient of grinding is μ, where

$$\mu = \frac{F_t}{F_n}$$

The value for μ can vary from as little as 0.2 for low-stock removal applications, grinding hard steels and ceramics, to as high as 0.8 in very high-stock removal applications such as peel grinding, or grinding soft steels or gray cast iron. Coolant can also have a major impact on the value as a result of the hydrodynamic pressure created by high wheel speeds. Strictly, the force created by the coolant should be subtracted from the measured normal force to establish a realistic value of normal grinding force. The effect is

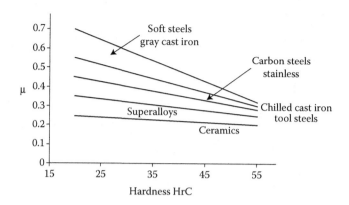

FIGURE 2.6
μ (F_t/F_n) for major material types in precision grinding.

particularly noticeable with high viscosity straight oils. Typical precision grinding applications on steels have values of μ in the range of 0.25–0.5.

Since tangential force can be readily calculated from power but not from normal force, knowledge of μ is particularly useful to calculate required system stiffness, work holding requirements, chuck stiffness, etc. Figure 2.6 plots general values for μ as a function of material classes and hardness. For most precision production grinding processes with hardened steel or cast iron, it can be seen that μ tends to be a value of about 0.3. Note, however, that these numbers are for flat profile wheels in a straight plunge mode. If a profile is added to a wheel or the angle of approach is changed from 90° then allowance must be made for increased normal forces and for side forces.

2.2.15 Surface Roughness

Surface roughness, not surprisingly, is closely related to uncut chip thickness.

$$R_t \propto \frac{h_{cu}^{4/3}}{a_e^{1/3}}$$

$$\approx \left(\frac{v_w}{v_s} \cdot \frac{1}{C \cdot r \cdot \sqrt{d_e}} \right)^{2/3}$$

2.2.16 R_t Roughness

R_t roughness is the SI parameter for maximum surface roughness; the maximum difference between peak height and valley depth within the sampling length. As a first approximation R_t is independent of the depth of cut but is dependent on v_w, v_s, $C \cdot r$, and d_e. The relationship between surface roughness and specific grinding energy can also be readily obtained by direct substitution.

R_t is but one of several measures of surface roughness. Two other common roughness standards are discussed below.

2.2.17 R_a Roughness

R_a roughness is the arithmetic average of all profile ordinates from a mean line within a sampling length after filtering out form deviations.

2.2.18 R_z Roughness

R_z roughness is the arithmetic average of maximum peak to valley readings over five adjacent individual sampling lengths. R_t and R_z values are much larger than R_a roughness values for measurements from the same surface.

Two other parameters related to surfaces, especially those used for rubbing contact, are defined as follows.

2.2.19 Material or Bearing Ratio

Material or bearing ratio t_p is the proportion of bearing surface at a depth p below the highest peak.

2.2.20 Peak Count

Peak count P_c is the number of local peaks which project through a given band height. t_p is less for grinding than for other operations such as honing, although it can be improved to some extent by a two-stage rough and finish grind with wheels of very different grit size. P_c can be controlled somewhat by adjusting dress parameters.

2.2.21 Comparison of Roughness Classes

Comparison of various international surface roughness systems is given in Table 2.1.

2.2.22 Factors That Affect Roughness Measurements

Relative values between different roughness systems will vary by up to 20% depending on the metal-cutting process by which they were generated. Even when considering just grinding, the abrasive type can alter the ratio of R_z to R_a; CBN often giving a higher ratio to alumina. This difference also shows up cosmetically when looking visually at surfaces ground with alumina or CBN. When changing from lapping to fine grinding, the change in appearance of the finish can be dramatic going from a matt-pitted surface to a shiny but scratched surface, both of which have comparable surface roughness values. The type of grinding process will also affect the appearance in terms of the grind line pattern. For example, in face grinding of a shoulder using, for example, a 2A2 or 6A2 wheel, the grind gives a cross-hatch appearance as in Figure 2.7a. In angle approach grinding, the face is produced with line contact and the lines are concentric with the journal diameter as in Figure 2.7b.

2.2.23 Roughness Specifications on Drawings

Common roughness specifications (marks) on part drawings are shown in Figure 2.8. This gives both the current standard practice, especially in Europe, and the older machining marks still seen especially on Japanese drawings. 1 (∇) or 2 ($\nabla\nabla$) marks are indicative of a

TABLE 2.1

Guideline Comparisons of International Surface Finish Systems

R_a (μm)	R_t (μm)	R_z (μm)	RMS (μin)	CLA (μin)	PVA (μin)	Roughness Class France (μm)	Renault R France (μm)	Citreon β France (μm)	Citreon V France (μm)	Roughness Class China	Quality Class Russia
0.025	0.2	0.16	1.12	1	6.3	12C	0.13	0.08	0.15	N1	12
0.05	0.3	0.32	2.2	2	12	11C	0.25	0.15	0.3	N2	11
0.06	0.5	0.38	2.7	2.4	16	11B	0.3	0.18	0.36	N2	11
0.08	0.6	0.5	3.6	3.2	20	11A	0.4	0.24	0.48	N2/N3	11
0.1	0.8	0.63	4.5	4	25	10C	0.5	0.3	0.6	N3	10
0.12	1	0.75	5.3	5	32	10B	0.6	0.37	0.73	N3	10
0.16	1.25	1	7.1	6.3	40	10A	0.8	0.48	0.97	N3/N4	10
0.2	1.5	1.25	9	8	50	9C	1	0.61	1.22	N4	9
0.25	2	1.6	11.2	10	63	9B	1.25	0.76	1.52	N4	9
0.31	2.5	2	14	12.5	80	9A	1.6	0.95	1.8	N4/N5	9
0.4	3.2	2.5	18	16	100	8C	2	1.2	2.4	N5	8
0.5	4	3.2	22.4	20	125	8B	2.5	1.5	3	N5	8
0.63	5	4	28	25	160	8A	3.2	1.9	3.8	N5/N6	8
0.8	6.3	5	35.5	31.5	200	7C	4	2.4	4.8	N6	7
1	8	6.3	45	40	250	7B	5	3	6	N6	7
1.25	10	8	56	50	320	7A	6.3	3.8	7.6	N6/N7	7
1.6	12.5	10	71	63	400	6C	8	4.7	9.4	N7	7

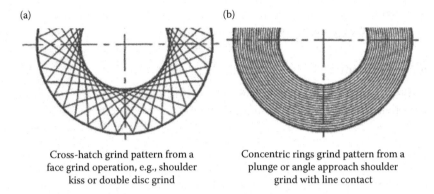

(a) (b)

Cross-hatch grind pattern from a
face grind operation, e.g., shoulder
kiss or double disc grind

Concentric rings grind pattern from a
plunge or angle approach shoulder
grind with line contact

FIGURE 2.7
Comparison of grind patterns from (a) face and (b) angle approach line contact grinding.

turning or milling operation, but 3 ($\nabla\nabla\nabla$) or 4 ($\nabla\nabla\nabla\nabla$) marks are indicative of the requirement to grind or even lap.

Two other force-related factors are of particular interest to end users with low stiffness systems such as internal grinding.

2.2.24 Stock Removal Parameter

The first factor is the stock removal parameter Λ, which is defined as the ratio between stock removal rate and normal grinding force:

$$\Lambda = \frac{Q'}{F_n}$$

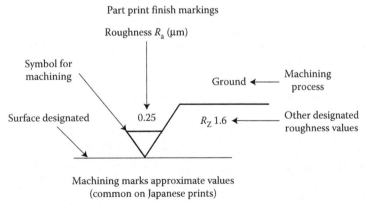

Part print finish markings

Roughness R_a (μm)

Symbol for
machining

Ground ← Machining
process

Surface designated   0.25 R_Z 1.6 ← Other designated
roughness values

Machining marks approximate values
(common on Japanese prints)

∇	25 R_a (μm)
∇ ∇	3.2 R_a (μm)
∇ ∇ ∇	0.8 R_a (μm)
∇ ∇ ∇ ∇	≤0.16 R_a (μm)

FIGURE 2.8
Print markings for surface finish.

Λ is an indicator of the sharpness of the wheel, but is limited by the fact that it must be defined for each wheel speed and removal rate.

2.2.25 Decay Constant τ

When the infeed reaches its final feed point, the grinding force F will change with time t as the system relaxes according to the equation:

$$F = F_0 e^{-t/\tau}$$

F_t and power are directly related; therefore, the decay time constant τ can be determined from a log plot of the decay in power during spark out. After 3τ, virtually all grinding has ceased preventing any improvement in part tolerance, while roughness, as shown above, will not improve further. Consequently, spark out times in internal grinding should be limited to no more than 3τ.

2.2.26 G Ratio

G ratio is used as the primary measure of wheel wear. This is defined as

$$G \text{ ratio} = \frac{\text{Volume of material ground per unit wheel width}}{\text{Volume of wheel worn per unit wheel width}}$$

G ratio is dimensionless with values that can vary from <1 for some soft alox creep feed vitrified wheels to as high as 100,000 for vitrified CBN wheels. G ratio will fall linearly with increases in Q' accelerating to an exponential drop as the maximum metal removal rate for the wheel structure is reached.

2.2.27 *P* Ratio

P ratio is a closely related index that has started to be used as an alternative to G ratio for plated superabrasive wheels.

$$P \text{ ratio} = \text{Volume of metal ground per unit area of wheel surface}$$

This allows for the fact that it is hard to define a wear depth on a plated wheel. P ratio usually has the dimensions of (mm^3/mm^2). For high-speed, high-stock removal applications in oil-cooled grinding, for example, crankshafts, P ratio values have reached 25,000 mm^3/mm^2. Since the usable layer depth on a plated wheel is only at most about 0.1 mm, a P value of 25,000 mm corresponds to a G ratio greater than or equal to 250,000.

2.2.28 Contact Length

Contact length I_c is the length of the grinding contact zone. Abrasive grits in full active contact with the workpiece remain in contact over the whole contact length. A very long contact length correlates with rubbing wear of the grits according to Archards law of wear. Contact length is also the length over which the heat input to the workpiece is spread.

The contact length is approximately equal to the geometric contact length for rigid metal bond wheels but is greatly increased for flexible and vitreous bond wheels.

2.2.29 Geometric Contact Length

Geometrical contact length applies for the theoretical case of ideally rigid wheels and rigid workpieces. The geometric contact length l_g is related to real depth of cut a_e and equivalent diameter d_e of the grinding wheel contact.

$$l_g = \sqrt{a_e \cdot d_e}$$

2.2.30 Real Contact Length

The real contact length l_c is typically twice this value or greater for more elastic vitrified wheels, Rowe et al. (1993), Marinescu et al. (2013), and Rowe (2014) who showed that allowing for elastic deflections:

$$l_c = \sqrt{l_g^2 + l_f^2}$$

where $l_f^2 = 8 \cdot R_r^2 \cdot F_n' \cdot d_e / \pi \cdot E^*$ gives the contribution to the contact length due to elastic deflection between the abrasive and the workpiece under the influence of the normal grinding force. This deflection is increased for rough surfaces such as an abrasive wheel, because the normal force is balanced on a smaller proportion of the apparent contact area. A typical value for the roughness factor is $R_r \approx 5$. The combined elastic modulus for the workpiece and abrasive materials is given by

$$\frac{1}{E^*} = \frac{1 - \upsilon_1^2}{E_1} + \frac{1 - \upsilon_2^2}{E_2}$$

A number of independent grinding experiments have confirmed these findings. An interesting and novel experimental investigation of the acoustic emission spikes in typical examples of surface grinding showed contact length varying from three times geometric contact length at small depths of cut and two times geometric contact length at large depths of cut (Babel et al. 2013).

2.3 Grinding Temperatures

2.3.1 Surface Temperature, *T*

Prediction of grinding temperatures and the avoidance of burn are critical to grinding quality. Numerous calculations modeling the partition of heat between the elements in the grinding zone have been developed over the last 50 years. Maximum temperature of a workpiece in early work was predicted by Jaeger (1942) for small depths of cut based on the principles of sliding heat sources described by Carslaw and Jaeger (1959). Temperature solutions were found by Rowe and Jin (2001) for deep cuts.

Heat partitioning is described in detail by Marinescu et al. (2013) and Rowe (2014). The following simple version suffices to illustrate key factors governing maximum surface temperature.

2.3.2 Maximum Workpiece Surface Temperature

The maximum surface temperature depends on the grinding power ($F_t' \cdot v_s$), the grinding speeds, and material parameters.

$$T_{max} = C_{max} \cdot R_w \cdot \frac{F_t' \cdot v_s}{\beta_w} \cdot \sqrt{\frac{1}{v_w \cdot l_c}}$$

The thermal parameters that affect grinding temperature are discussed in the following sections.

2.3.3 The C_{max} Factor

This is a constant that allows for the kinematic conditions and gives the maximum temperature. The value is approximately equal to 1 for conventional grinding at small depths of cut and high work speeds. The value is reduced for deep grinding. Rowe and Jin (2001) investigated deep cuts both at low and high work speed. Charts of C values were derived for maximum temperature and for finish surface temperature, thus covering the whole range of grinding conditions. Values are given by Rowe (2014).

2.3.4 The Thermal Property for Moving Heat Sources β_w

The thermal property of the workpiece material for moving heat sources is given by

$$\beta_w = \sqrt{k \cdot \rho \cdot c}$$

where k is the thermal conductivity, ρ is the density, and c is the heat capacity.

2.3.5 Workpiece Partition Ratio R_w

Workpiece partition ratio is the proportion of the total grinding energy in the grinding contact zone that is conducted into the workpiece. The work partition ratio is a complex function of the wheel grain conductivity and sharpness and of the workpiece thermal property. Ignoring for the present, coolant convection, and convection by the grinding chips, R_w approximates to R_{ws}. Hahn (1962) modeled heat transfer between a sliding grain and a workpiece. It can be shown for the steady state that

$$R_{ws} = \left(1 + \frac{k_g}{\beta_w \cdot \sqrt{r_0 \cdot v_s}} \right)^{-1}$$

where k_g is the thermal conductivity of the abrasive grain and r_0 is the contact radius of the grain. R_{ws} is relatively insensitive to variations of r_0 except for extremely sharp grains.

Typically, R_{ws} for conventional grinding varies between 0.7 and 0.9 for vitrified wheels and between 0.4 and 0.6 for CBN wheels. A simple modification of this formula provides a solution for steady state and nonsteady states (Rowe 2014).

2.3.6 Effect of Grinding Variables on Temperature

The temperature equation can be shown very approximately to vary with process conditions for a given wheel/work/machine configuration according to:

$$T_{max} \propto \sqrt{a_e \cdot v_s \cdot C \cdot r}$$

The relationship applies roughly for conventional precision grinding and is found by taking account of specific energy and power variation as expressed in the above paragraphs. $C \cdot r$ represents the active grain density in the wheel surface and the active grain width to depth ratio. It follows that increasing the wheel speed, increasing the depth of cut, or increasing the number of active cutting edges (e.g., by fine dressing) will increase the surface temperatures. Further discussion of temperatures generated when grinding at very high wheel speeds is made in a later chapter.

2.3.7 Heat Convection by Coolant and Chips

A note of caution should be sounded for deep grinding where the long contact length allows substantial convective cooling from the grinding coolant. Also in high-rate grinding with low specific energy, the heat taken away by the grinding chips reduces maximum temperature very substantially (Rowe and Jin 2001).

Allowance can be made for convective cooling by subtracting the heat taken away by the coolant and chips as described by Rowe (2014). Allowance for convective cooling is essential for creep grinding as shown by Andrew et al. (1985). It has also been found important for other high-efficiency deep grinding processes as employed for drill flute grinding, crankshaft grinding, and cutoff grinding. If allowance is not made for convective cooling, the temperatures are greatly overestimated.

The maximum temperature equation modified to allow for convective cooling according to Rowe (2014) has the form:

$$T_{max} = \frac{R_{ws}(F_t' v_s - \rho \cdot c \cdot T_{mp} \cdot a_e \cdot v_w)}{\left(\beta_w \sqrt{v_w l_c} \,/ C_{max}\right) + h_f \cdot l_c}$$

where T_{mp} is a temperature approaching the melting point of the workpiece material. For steels, the material is very soft at 1400°C and this temperature gives a reasonable estimate for the chip convection term.

h_f is the coolant convection coefficient that applies as long as the maximum temperature does not cause the fluid to burn out in the grinding zone. If burnout occurs, the convection coefficient is assumed to be zero. Burnout is a common condition in grinding but should be avoided for low-stress grinding and to achieve low temperatures in creep grinding. Values estimated for convection coefficient when grinding with efficient fluid delivery and low grinding temperatures at 50 m/s range from 84,000 to 144,000 W/m²K for emulsions and 26,000–78,000 W/m²K for oil based on using the above temperature model and results

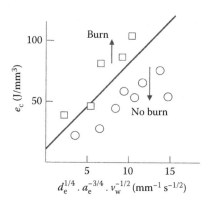

FIGURE 2.9
Specific energy values below the threshold avoid burn. Specific energy values above the threshold cause burn.

presented by Jin and Stephenson (2008). Other experimental values are given by Rowe (2014) and alternative models for convection are given by Zhang et al. (2013) and by Rowe (2014). A short summary of convection theory as applied to grinding is given in Chapter 10: Coolant.

2.3.8 Control of Thermal Damage

An increasingly popular semiempirical approach to control thermal damage has been developed by Malkin (1989) with literature examples of its application in industry by General Motors on cast iron (Meyer 2001) with Bell Helicopter on hardened steel and by Stephenson et al. (2001) on Inconel to impose a limit on grinding temperatures. Malkin (1989) provides an empirical method to monitor maximum allowable specific grinding energy for a given maximum temperature rise as

$$e_c = A + C \cdot T_{max}(d_e^{1/4} \cdot a_e^{-3/4} \cdot v_w^{-1/2})$$

A and C are constants based on the thermal conductivity and diffusivity properties of the workpiece and wheel. A series of tests are made for different values of a_e, v_w, and d_e and the workpieces analyzed for burn. Plotting these on a graph of e_c against $d_e^{1/4}a_e^{-3/4}v_w^{-1/2}$ establishes the slope CT_{max} and intercept A.

The method is illustrated schematically in Figure 2.9. In an industrial situation, a power meter is used to monitor specific energy values. If the specific energy values exceed the threshold level for burn, it is necessary to take corrective action to the process. This can mean redressing the wheel or making some other process change such as reducing the depth of cut, increasing the work speed, or using a different grinding wheel (Figure 2.9).

Appendix: Drawing Form and Profile Tolerancing

Examples of drawing form and profile tolerance are shown in Figure 2A.1.

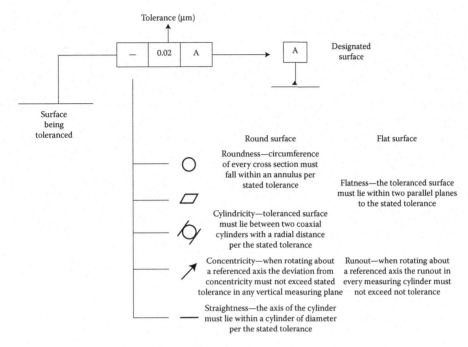

FIGURE 2A.1
Examples of form and position tolerances.

References

Andrew, C., Howes, T. D., Pearce, T. R. A., 1985, *Creep Feed Grinding.* Holt, Rinehart and Winston, London, UK.

Babel, R., Koshy, P., Weiss, M., 2013, Acoustic emission spikes at workpiece edges in grinding. *International Journal of Machine Tools and Manufacture,* 64, 96–101.

Cai, R., Rowe, W. B., 2004, Assessment of vitrified CBN wheels for precision grinding. *International Journal of Machine Tools and Manufacture,* 44, 12/13, 1391–1402.

Carslaw, H. S., Jaeger, J. C., 1959, *Conduction of Heat in Solids.* Oxford Science Publications, OUP.

Darafon, A., Warkentin, A., Bauer, R., 2013, Characterization of grinding wheel topography using a white chromatic sensor. *International Journal of Machine Tools and Manufacture,* 70, 22–31.

Hahn, R. S., 1962, On the nature of the grinding process. *Proceedings of the 3rd Machine Tool Design & Research Conference.* Pergamon Press, Oxford, UK, 129–154.

Hitchiner, M. F., McSpadden, S. B., 2005, Evaluation of factors controlling CBN abrasive selection for vitrified bonded wheels. *Annals of the CIRP,* 54, 1, G3.

Jaeger, J. C., 1942, Moving sources of heat and the temperature at sliding contacts. *Proceedings of the Royal Society of New South Wales,* 76, p. 203.

Jin, T., Stephenson, D. J., 2008, A study of heat transfer coefficients of grinding fluids. *Annals of the CIRP,* 57, 1, 367–370.

Malkin, S., 1989, *Grinding Technology.* Ellis Horwood Publications Chichester, UK and John Wiley & Sons, NY.

Marinescu, I. D., Rowe, W. B., Dimitrov, B., Ohmori, H., 2013, *Tribology of Abrasive Machining Processes,* 2nd edition. Elsevier Press, Amsterdam.

Meyer, J. E., 2001, Specific grinding energy causing thermal damage in helicopter gear steels. *SME 4th International Machining & Grinding Conference,* USA. 2001.

Rowe, W. B., 2014, *Principles of Modern Grinding Technology*, 2nd edition. Elsevier Press, William Andrew Imprint, Oxford, UK.

Rowe, W. B., Ebbrell, S., Morgan, M. N., 2004, Process requirements for precision grinding. *Annals of the CIRP*, 44, 1, 12–13.

Rowe, W. B., Jin, T., 2001, Temperatures in high efficiency deep grinding (HEDG). *Annals of the CIRP*, 50, 1, 205–208.

Rowe, W. B., Morgan, M. N., Qi, H. S., Zheng, H. W., 1993, The effect of deformation on the contact area in grinding. *Annals of the CIRP*, 42, 1, 409–412.

Shaw, M. C., 1996, *Principles of Abrasives Processing*. Oxford Science Series. Clarendon Press, Oxford.

Stephenson, D. J. et al., 2001, Burn threshold studies for superabrasive grinding using electroplated CBN wheels. *SME 4th International Machining & Grinding Conference*, USA.

Zhang, L., Rowe, W. B., Morgan, M. N., 2013, An improved convection solution in conventional grinding. *Proceedings of the Institution of Mechanical Engineers, Part B, Journal of Engineering Manufacture*, 227, 6, 332–338.

3

Material Removal Mechanisms

3.1 Significance

3.1.1 Introduction

Knowledge of the basic principles of a process is a prerequisite for its effective improvement and optimization. During grinding, surface formation is one of the basic mechanisms. In the case of cutting with geometrically defined cutting edges, a singular engagement of the cutting edge defines the removal mechanism. The consequent removal mechanisms can be directly observed by means of modern investigation methods.

In the case of grinding, the investigation of removal mechanisms is complicated due to many different factors. The first challenge is posed by the specification of the tool. The abrasive grains are three-dimensional and statistically distributed in the volume of the grinding wheel. The geometry of the cutting edges is complex. Moreover, there is a partially simultaneous engagement of the cutting edges involved in the process. The surface formation is the sum of these interdependent cutting-edge engagements, which are distributed stochastically. Furthermore, the chip formation during grinding takes place within a range of a few microns. The small chip sizes make the observation even more difficult.

3.1.2 Defining Basic Behavior

In spite of the complexity, some statements can be made on the removal mechanisms, surface formation, and the wear behavior during grinding. Analogy tests and theoretical considerations on the basis of the results of physical and chemical investigations are used for this purpose. In the past few years, chip and surface formation has been modeled with the help of high-performance computers and enhanced simulation processes.

- *Indentation tests*—In analogy tests, the engagement of the cutting edge in the material surface is investigated first. The advantage of this method is that single cutting edges can be investigated before and after the process, and their geometry is known. With the help of the so-called indentation tests with singular cutting edges, the material behavior to a static stress can be observed without the influence of the movement components which are typical for grinding. On the basis of these indentation tests, elastic and plastic behavior as well as crack formation can be observed in the case of brittle–hard materials at the moment the cutting edge penetrates the material.

- *Scratch tests*—A further method is the investigation of the removal mechanisms during scratching with single cutting edges, which allows the accurate examination of the geometry and the wear of the cutting edges. Contrary to the indentation tests, there is chip formation during this test method. Furthermore, the influence of different cooling lubricants can be investigated.

- *Cutting-edge geometry*—A further prerequisite of a comprehensive understanding of the material removal during grinding is the geometrical specification of the single cutting edges. This mainly takes place in analogy to the geometrical relations at geometrically defined cutting edges.

- *Thermal and mechanical properties*—The thermal and mechanical characteristics of the active partners of the grinding process also have a significant influence. Heat is generated in the working zone through friction. This contact zone temperature influences both the mechanical characteristics of the workpiece and the characteristics of the tool.

- *Surface modification*—As a result of the removal mechanisms, the subsurface of the machined workpiece is influenced by the grinding wheel due to mechanical stress. Residual stresses occur depending on the specification of the machined workpiece material. These stresses can have a positive effect on component characteristics; hence, they are in some cases specifically induced. Due to the mechanical stresses, cracks, structural, and/or phase changes that can occur on the subsurface of the material may have a negative effect on the component characteristics.

This shows the complexity of the mechanisms of surface formation during grinding. The better the surface formation is known, the more specifically and accurately the process parameters, the tool specification, and the choice of an appropriate cooling lubricant can be optimized.

3.2 Grinding Wheel Topography

3.2.1 Introduction

In the case of grinding, the cutting process is the sum of singular microscopic cutting processes, whose temporal and local superposition leads to a macroscopic material removal. As a consequence, the cause-and-effect principle of grinding can only be described on the basis of the cutting behavior of the individual abrasive grains (Sawluk 1964). The most important parameter is the number of the currently engaged cutting edges (Kassen 1969). An exact determination of the geometrical engagement conditions of the single cutting edges, however, is not possible for manufacturing processes like grinding or honing. Due to the stochastic distribution of the geometrically not defined cutting edges, their position and shape cannot be exactly determined.

Therefore, the position, number, and shape of the abrasive grains are analyzed statistically and related to the process kinematics and geometry to achieve a specification of the engagement conditions of the abrasive grain. Thus, grinding results can be related to events at the effective area of grinding contact for particular input values of machine and workpiece parameters and other specifications of the process. The main cutting parameters of the removal process are the theoretical chip thickness, chip length, and engagement angle. Knowing the overall relations between input values, cutting and chip values, as well as process output values, the behavior of the process can be cohesively described and used to improve the setup of the machining process. This implies a wheel specification

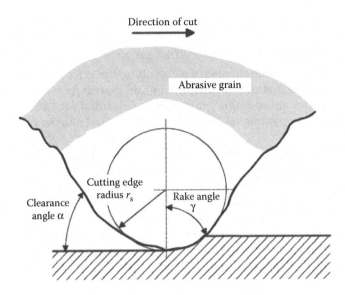

FIGURE 3.1
Average shape and analytic description of a cutting edge. (From König, W., Klocke, F., 1996, *Fertigungsverfahren Band 2. Schleifen, Honen, Läppen*. 3. Auflage, VDI-Verlag GmbH, Düsseldorf. With permission.)

suitable for the grinding situation and the choice of parameters leading to an economical grinding process.

Different authors have described the material removal mechanisms of diverse grinding processes. Thereby, a distinction is made between topography, uncut chip thickness, grinding force, grinding energy, surface, and temperature models in relation to different basic models (Kurrein 1927; Pahlitzsch and Helmerdig 1943; Reichenbach et al. 1956; Kassen 1969; Werner 1971; Inasaki et al. 1989; Lierath et al. 1990; Malyshev et al. 1990; Paulmann 1990; Toenshoff et al. 1992; Marinescu et al. 2013; Rowe 2014).

3.2.2 Specification of Single Cutting Edges

The geometry of single cutting edges may be described statistically by measurement of cutting-edge profiles. The depiction of the form of a cutting edge of an abrasive grain represents the average of all measured geometries. The main characteristic of the cutting edges acting during grinding is the clearly negative rake angle (Figure 3.1).

3.3 Determination of Grinding Wheel Topography

3.3.1 Characterization Methods

The determination of the grinding wheel topography can be divided into static, kinematic, and dynamic methods (Brücher 1996):

- *Static methods*—All abrasive grains on the surface of the grinding tool are considered. The kinematics of the grinding process are not taken into account.

- *Dynamic methods*—Using this kind of method, the number of actual abrasive grain engagements are measured. The number of active cutting edges is the total number of cutting edges involved in the cutting process.

- *Kinematic methods*—Kinematic methods combine the effects of the kinematics of the process with the statically determined grain distribution for the specification of microkinematics at the single grain that are essential for the determination of microscopic cutting parameters like chip thickness and chip length.

If static assessment methods are used, basically all cutting edges in the cutting area are included in the topography analysis. There is no distinction whether a cutting edge of the grinding process is actively involved or not in the cutting process. Hence, static methods are independent of the grinding conditions (Shaw and Komanduri 1977; Verkerk 1977).

Figure 3.2 shows a schematic section of a cutting surface of a grinding wheel. All abrasive grains protruding from the bond have cutting edges—the so-called static cutting edges. Since grains usually have more than one cutting edge, the distance between the static cutting edges does not correspond to the average statistical grain separation on the grinding wheel. Therefore, instead of the distance between the static cutting edges, the number of cutting edges is specified per unit length, which is the static cutting-edge number S_{stat}, the cutting-edge density per surface unit N_{stat}, or the number per unit volume of the cutting area C_{stat} (Daude 1966; Lortz 1975; Kaiser 1977; Shaw and Komanduri 1977; Verkerk 1977; Rohde 1985; Treffert 1995; Marinescu et al. 2013; Rowe 2014).

In contrast to static methods, dynamic methods depend on the grinding conditions. Taking the grinding process, the process parameters, and the geometrical engagement conditions into account, information is obtained about which areas of the effective surface of the grinding wheel are actually involved in the process (Verkerk 1977; Gaertner 1982).

Figure 3.3 shows some of the existing cutting edges actively involved in the process. The number and density of the active cutting edges S_{act} and C_{act} are therefore smaller than those of the static cutting edges S_{stat} and C_{stat}. Their value is mainly determined by the geometric and kinematic parameters. As a result of constant wheel wear and of the consequent topography changes of the wheel, the cutting-edge density has continually changing values during the grinding process.

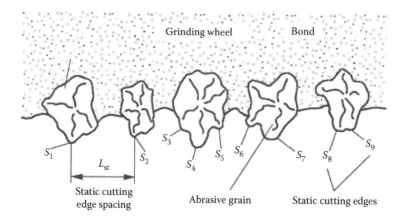

FIGURE 3.2
Static cutting edges. (From König, W., Klocke, F., 1996, *Fertigungsverfahren Band 2. Schleifen, Honen, Läppen. 3. Auflage*, VDI-Verlag GmbH, Düsseldorf. With permission.)

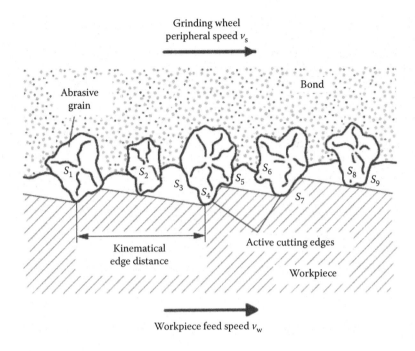

FIGURE 3.3
Kinematic cutting edges. (From König, W., Klocke, F., 1996, *Fertigungsverfahren Band 2. Schleifen, Honen, Läppen.* 3. Auflage, VDI-Verlag GmbH, Düsseldorf. With permission.)

In the kinematic approach, besides the statically determined grinding wheel topography, the process kinematics are additionally taken into account for the specification of the effective kinematic topography and cutting parameters. On the basis of the kinematically determined cutting-edge number, the trajectories of the single grains are reproduced when considering the grinding process, the setting parameters, and the geometrical engagement conditions (Lortz 1975; Gärtner 1982; Steffens 1983; Bouzakis and Karachaliou 1988; Stuckenholz 1988; Treffert 1995).

3.3.2 Measurement Technologies for Grinding Wheel Topography

Figure 3.4 summarizes different measuring technologies for the topography determination of grinding wheels and also indicates which topography describing parameters can be derived from which measurement technology. Most of these measurement technologies are carried out after the grinding process, while some can be applied "in-process."

3.3.2.1 Roughness Measures

While roughness measures in the past often were carried out with tactile devices like profilometers, the advance of measurement technology in the present leads to high precision optical measurement devices for topography and roughness determination. The advantage of optical methods in comparison to tactile measurement methods is a more accurate reproduction of the real topography, due to the absence of an adulterating caliper interface. Also, the measuring times with optical measurement systems are usually much

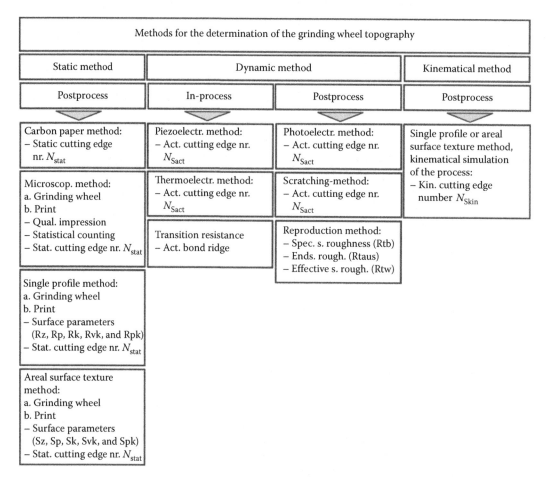

FIGURE 3.4
Methods for characterizing topography of grinding wheels. (From Brücher, T., 1996, Kühlschmierung beim Schleifen keramischer Werkstoffe. PhD thesis, Technische Universität, Berlin. With permission.)

lower than with tactile devices, especially when 3D surfaces and not only 2D single profiles are observed. On the other hand, the measurement of grinding wheel topographies with optical measurement systems is challenging and demands a well-adjusted measurement setup, since the various abrasive and bond materials have different and sometimes hard to measure reflection properties (Hübert et al. 2009).

Various surface parameters, such as Rz, Rk, Rvk, and Rpk, are suitable for the characterization of grinding wheel topographies and, in particular, the maximum profile height, Rp, is useful due to its chip space describing character (Schleich 1982; Werner and Kentner 1987; Warnecke and Spiegel 1990; Uhlmann 1994; Bohlheim 1995). Besides these parameters, which are based on a profilometer profile, a more accurate description of the grinding wheel topography is possible by the use of areal surface texture-describing parameters, which are defined by ISO 25178. In accordance to the above named R values, an areal description of the grinding wheel topography is possible by Sz, Sk, Svk, Spk, and Sp (ISO 25178).

A further value derived from static methods is the static cutting-edge number per length or surface unit S_{stat} or N_{stat} (Daude 1966; Lortz 1975; Kaiser 1977; Shaw and Komanduri 1977;

Verkerk 1977; Rohde 1985; Treffert 1995). The cutting edges are determined on the basis of the envelope curve of the effective area of the grinding wheels, which is defined by the external cutting edges. With increasing depth, the cutting edges penetrate equidistant intersection surfaces or lines. Similar to a material ratio curve, a frequency curve of the static cutting-edge numbers is formed depending on the cutting-edge depth z_S. In Figure 3.5a, grinding wheel topography mapped with a chromatic white light sensor and the derivation of topography describing parameters are shown.

3.3.2.2 Qualitative Assessment

SEM, optical, and stereomicroscopic images are used for the qualitative evaluation of wear processes (Schleich 1982; Stuckenholz 1988; Dennis and Schmieden 1989; Warnecke and Spiegel 1990; Wobker 1992; Uhlmann 1994; Marinescu et al. 2013; Rowe 2014).

3.3.2.3 Counting Methods

Microscopic processes can, however, also be used to make quantitative statements on the state of the grinding wheel effective area. For this purpose, grains of different wear characteristics were statically counted (Büttner 1968; Bohlheim 1995). The measurements are either carried out directly on the grinding wheel surface or indirectly on a print of the surface. In the case of the carbon paper method, white paper and carbon paper are put between the grinding wheel and a slightly conical, polished plastic ring. The grinding wheel topography is reproduced on the white paper by rolling the grinding wheel on the plastic ring. Due to the conical shape of the ring, the measurement

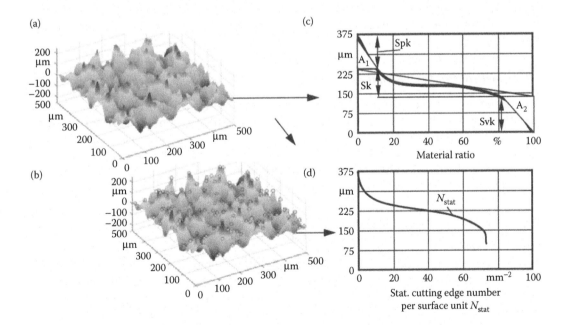

FIGURE 3.5
Grinding wheel topography (a), from which the single cutting-edge identification (b) and the determination of Abbott curve values (c) as well as cutting-edge distribution (d) is performed. (From Hübert, C., 2011, Schleifen von hartmetall- und vollkeramik-schaftfräsern. PhD thesis, Technische Universität, Berlin. With permission.)

of the cutting-edge distribution is expected to be dependent on the cutting area depth (Nakayama 1973).

3.3.2.4 Piezo and Thermoelectric Measurements

Piezoelectric and thermoelectric processes for the determination of the active cutting-edge number are based on the measurement of force signals or temperature peaks of single active cutting edges (Daude 1966; Kaiser 1977; Shaw and Komanduri 1977; Verkerk 1977; Damlos 1985). The piezoelectric process is only applicable for small contact surfaces, since it has to be ensured that no more than one cutting edge is engaged in the contact zone at any time. Small samples of a width of circa 0.3 mm (Kaiser 1977) are ground with relatively small feed. In the thermoelectric process, a thermocouple wire of a small diameter is divided by a thin insulating layer from the surrounding material within the workpiece. Every active cutting edge or bond ridge destroys the insulating layer and creates a thermocouple junction through plastic deformation. This leads to the emission of a measurable thermoelectric voltage signal from which a corresponding temperature can be derived. Therefore, the material has to be electrically conductive and sufficiently ductile. Hence, this process cannot be applied to ceramics (Shaw and Komanduri 1977). A contact resistance measurement can be applied for diamond and cubic boron nitride grinding wheels with an electrically conducting bond for the purpose of distinguishing between active cutting edges and active bond ridges (Kaiser 1977).

3.3.2.5 Photoelectric Method

Photoelectric methods according to the scattered light principle are based on light reflected by the cutting area and collected using a radiation detector. In this method, the scattered light distribution and the duration and number of light impulses in the direction of the regular direct reflection are evaluated (Werner 1994).

3.3.2.6 Mirror Workpiece Method

The topography of the grinding wheel can be depicted in a control workpiece. For this purpose, a mirroring workpiece angled diagonally to the grinding direction is ground once. Counting the scratch marks, a conclusion can be drawn to the number of active cutting edges per unit surface area. Since the scratch marks of successive cutting edges might overlap each other, it is not possible to determine the overall number of engaged cutting edges (Shaw and Komanduri 1977; Verkerk 1977).

3.3.2.7 Workpiece Penetration Method

A further method is the penetration of a thin steel plate or a stationary workpiece by the effective area of the grinding wheel. The roughness profile of the ground test workpiece results from the overlapping of the profiles of the active cutting edges in the removal process. The roughnesses of the test workpieces are called specific surface roughness Rtb, end surface roughness Rtaus, or effective surface roughness Rtw (Karatzoglu 1973; Saljé 1975; Frühling 1976; Weinert 1976; Jacobs 1980; Gärtner 1982; Rohde 1985; Stukenholz 1988). These processes are suitable for a comparative assessment of the cutting-edge topographies. It is, however, not possible to make any predication on the shape and number of cutting edges (Lortz 1975; Werner 1994).

3.4 Kinematics of the Cutting-Edge Engagement

To compare a variety of grinding processes, it is necessary to define general and comparable process parameters. With these parameters, different processes can be compared with each other, allowing an efficient optimization of the process. The most important grinding parameters are the geometric contact length l_g, chip length l_{cu}, and the chip thickness h_{cu}.

The kinematics and the contact conditions of different grinding processes are represented in Figure 3.6. They constitute the basis of many process parameters.

Neglecting the elastic deformation of the active partners of the grinding process, the grinding wheel penetrates the workpiece with the real depth of cut a_e. The arc contact length is defined by the geometric contact length l_g

$$l_g = \sqrt{a_e \cdot d_{eq}} \tag{3.1}$$

For the same depth of cut values, different contact lengths result in the case of cylindrical and surface grinding. The equivalent wheel diameter d_{eq} is a calculation method of representing geometric contact length independent of the grinding process, where

$$d_{eq} = \frac{d_w \cdot d_s}{d_w + d_s} \tag{3.2}$$

applies to the equivalent grinding wheel diameter in external cylindrical grinding. In the case of internal cylindrical grinding, the equivalent grinding wheel diameter is

$$d_{eq} = \frac{d_w \cdot d_s}{d_w - d_s} \tag{3.3}$$

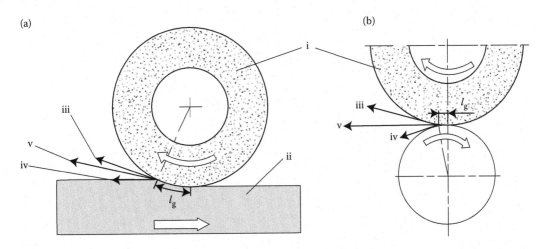

FIGURE 3.6
Contact conditions for different peripheral grinding processes. (a) Surface grinding. (b) Cylindrical grinding—(i) grinding wheel; (ii) workpiece; (iii) cutting speed; (iv) feed speed; and (v) resulting effective speed.

The equivalent grinding wheel diameter d_{eq} indicates the diameter of the grinding wheel, which has the same contact length in surface grinding. Thus, the equivalent grinding wheel diameter d_{eq} corresponds to the actual grinding wheel diameter in surface grinding.

The movement of the wheel relative to the workpiece is put in relation to the speed quotient q. In upgrinding, it is negative

$$q = \pm \frac{v_s}{v_f} \tag{3.4}$$

In peripheral grinding with rotating grinding tools, the cutting edges move on orthocycloidal paths due to the interference of the speed components.

In Figure 3.7, the paths of two successive cuttings edges are demonstrated. Both points have the same radial distance from the wheel center. The path the center travels between the two engagements of the wheel results from the feed movement and the time required.

The cutting-edge engagement and the resulting uncut chip parameters depend on the statistical average of the cutting edges distributed on the grinding tool. The equation

$$h_{cu} = k \left[\frac{l}{C_1} \right]^\alpha \left[\frac{v_f}{v_c} \right]^\beta \left[\frac{a_e}{d_{eq}} \right]^\gamma \tag{3.5}$$

relates the maximum uncut chip thickness h_{cu} to the cutting-edge distribution, the grinding parameters, and the geometric values (Kassen 1969). The mean maximum uncut chip cross-sectional area $\bar{Q}_{max}$ is a further characteristic parameter of the grinding process. Like the maximum uncut chip thickness h_{cu}, it depends on the parameters, the cutting-edge distribution, and the geometry of the active partners of the grinding process. The mean maximum uncut chip cross-sectional area $\bar{Q}_{max}$ is calculated from

$$\bar{Q}_{max} = \frac{2}{A_N} (C_1)^{-\beta} \left(\frac{v_f}{v_c} \right)^{1-\alpha} \left(\frac{a_e}{d_{eq}} \right)^{1-\alpha/2} \tag{3.6}$$

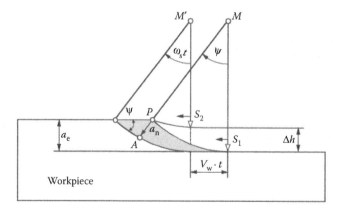

FIGURE 3.7
Contact conditions for two cutting edges. M = Center of grinding wheel; S_1 = leading cutting edge; and S_2 = trailing cutting edge.

On the basis of these values, a theoretical assessment can be made of the grinding process. There is a direct relation between the cutting parameters and the resulting surface quality. Equations 3.5 and 3.6 lead to the conclusion that the surface quality improves with increasing cutting velocity v_c and equivalent grinding wheel diameter d_{eq}. With increasing feed speed v_f and higher depth of cut a_e, however, the surface quality decreases.

The value of the uncut chip parameters, however, is only applicable to a real grinding process to a limited extent, since the kinematic relations were only derived for idealized engagement conditions (Marinescu et al. 2013; Rowe 2014). The cutting process is not a simple geometric process; there are also plastic and elastoplastic processes in real grinding so that chip formation is different from the geometric theory. For that reason, experiments are indispensable to gain knowledge and understanding about the material removal mechanisms.

3.5 Fundamental Removal Mechanisms

The removal process during the engagement of an abrasive cutting edge on the surface of a workpiece mainly depends on the physical properties between the active partners. A basic distinction can be made between three different mechanisms: microplowing, microchipping, and microbreaking (Figure 3.8).

In microplowing, there is a continual plastic, or elastoplastic, material deformation toward the trace border with negligible material loss. In real processes, the simultaneous impact of several abrasive particles or the repeated impact of one abrasive particle leads to material failure at the border of the traces. Ideal microchipping provokes chip formation. The chip volume equals the volume of the evolving trace. Microplowing and microchipping mainly occur during the machining of ductile materials. The relation between microplowing and microcutting basically depends on the prevailing conditions such as the matching of the active partners of the grinding process, grinding parameters, and cutting-edge geometry.

Microbreaking occurs in case of crack formation and spreading. The volume of a chip removed can be several times higher than the volume of the trace. Microbreaking mostly occurs during the machining of brittle–hard materials such as glass, ceramics, and silicon.

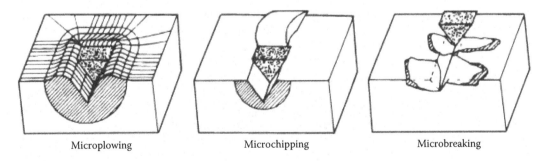

Microplowing　　　　　　Microchipping　　　　　　Microbreaking

FIGURE 3.8
Physical interaction between abrasive particles and the workpiece surface. (From Zum Gar, K. H., 1987, Grundlagen des Verschleißes. *VDI Berichte Nr. 600.3: Metallische und Nichtmetallische Werkstoffe und ihre Verarbeitungsverfahren im Vergleich*, VDI-Verlag, Düsseldorf. With permission.)

Hence, the mechanisms of surface formation during grinding consist of these three basic processes. Which of them predominates strongly depends on the workpiece material. Therefore, material removal mechanisms will be presented in Section 3.6 for ductile materials on one hand, and for brittle–hard materials on the other.

3.6 Material Removal in Grinding of Ductile Materials

During grinding, the cutting edge of the grain penetrates the workpiece on a very flat path causing plastic flow of the material after a very short phase of elastic deformation. Since the angle between cutting-edge contour and workpiece surface is very small due to the cutting-edge rounding, no chip is formed initially. The workpiece material is only thrust aside, forms material outbursts or side ridges, and flows to the flank underneath the cutting edge (König and Klocke 1996). In Figure 3.9, the chip formation during grinding of ductile materials is shown.

Only when the cutting edge penetrates the workpiece to a depth at which the undeformed chip thickness h_{cu}, equals the so-called critical cutting depth T_μ, the actual chip formation begins. Since displacement processes and chip formation occur simultaneously in the further process, it is crucial for the efficiency of the material removal how much of the uncut chip thickness h_{cu} is actually removed as chip, and what the effective chip thickness h_{cueff} is.

Grof (1977) has further differentiated the chip formation process during the machining of ductile materials with high cutting velocities. On the basis of experiments, he determined altogether six phases of singular chip formation during grinding (Figure 3.10). In the first quarter of the contact length (phase 1), the engaging abrasive grain first makes a groove, causing plastic and elastic deformation in the material, which is then thrust aside

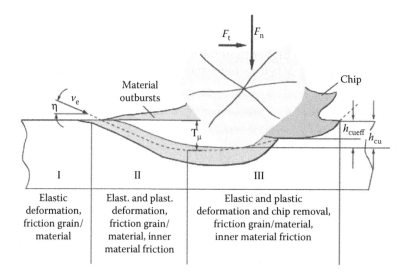

FIGURE 3.9
Removal process during the machining of ductile materials. (From König, W., Klocke, F., 1996, *Fertigungsverfahren Band 2. Schleifen, Honen, Läppen*. 3. Auflage, VDI-Verlag GmbH, Düsseldorf. With permission.)

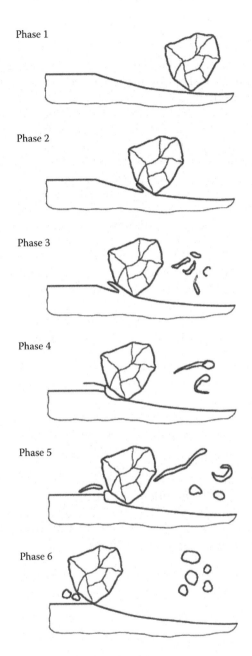

Phase 1

Phase 2

Phase 3

Phase 4

Phase 5

Phase 6

FIGURE 3.10
Removal process during the machining with high cutting speeds. (From Grof, H. E., 1977, Beitrag zur Klärung des Trennvorgangs beim Schleifen von Metallen. PhD thesis, Technische Universität, München. With permission.)

from the groove. The surface is presumed to consolidate during this contact phase (Werner 1971) without chip formation.

Through the further advance of the cutting edge (phase 2), a flow chip is removed with a nearly parallelogram-shaped cross section. This chip is compressed and bent in dependence of the pore space. Due to the large point angle of a cutting edge, the chip is very flat and offers a big surface for heat discharge through radiation and convection. In case of small infeeds or feeds, the working contact ends in the third phase, forming almost exclusively thread-shaped chips.

In case of large infeeds and feeds, the cutting edge penetrates the material deeper. This leads to a distinctive shear zone at approximately three-fourths of the maximum engagement length resulting in strong heat development (phase 3). The strong material accumulation through the high pressure causes an increase of the contact zone temperature. Due to the small effective surface of the cooling lubricant, this heat cannot be discharged. This leads to a melting of the formed chip above the plastic state.

If the cutting-edge engagement is terminated in this phase, tadpole-shaped chips are formed (phase 4). If the engagement takes place along the entire contact length, the whole material is liquefied in the pore space after the thread-shaped chip falls off (phase 5).

Due to the surface tension, the molten chip becomes spherical after leaving the contact surface. This takes place in a zone where there is none or only a small amount of cooling lubricant (phase 6). The fact of sphere formation has been observed by other researchers as well (Hughes 1974; Marinescu et al. 2013; Rowe 2014).

A further model of cutting-edge engagement has been presented by Steffens (1983). The cutting-edge engagement can be ideally considered as an even-yielding process. In this model, the contact zone is divided into layers that are arranged parallel to the movement axis. The material starts to yield in different directions at the cutting-edge engagement. Thus, the individual layers are thrust aside at the point of engagement of the first cutting edge. When successive cutting edges engage, these areas are removed or thrust aside anew. Through these processes, the number of kinematic cutting edges increases partially in a subarea. Since these changes are statistically distributed to the whole surface, there is overall no material accumulation.

The description shown in Figure 3.11 can serve as a model for the alternative description of a layer as an interaction of all displacement processes. A further prerequisite for the application of slip line theory is the knowledge of the rheological properties of the material, which is supposed to be inelastic ideal plastic.

On the basis of these theoretical considerations and experimental investigations, statements can be made on the friction conditions, which are decisively influenced by the lubrication (König et al. 1981; Steffens 1983; Vits 1985). If the friction is increased, the critical cutting depth decreases, which is additionally influenced by the radius of the cutting edge of the grain (König et al. 1981; Steffens 1983). Improved lubrication increases the plastic deformation toward a higher critical cutting depth. Thus, there is a reduction of friction between the active partners. With constant uncut chip thickness h_{cu}, the effective chip thickness h_{cueff} (thickness of the formed chip) decreases simultaneously with a reduction of friction (Steffens 1983; Vits 1985).

3.7 Surface Formation in Grinding of Brittle–Hard Materials

3.7.1 Indentation Tests

Fundamental mechanisms of crack formation and spreading in the case of brittle–hard materials were carried out by Lawn and Wilshaw (1975). The stressing of a ceramic surface with a cutting edge causes hydrostatic compression stress around a core area in the subsurface of the workpiece. This leads to a plastic deformation of the material (Figure 3.12a). If a

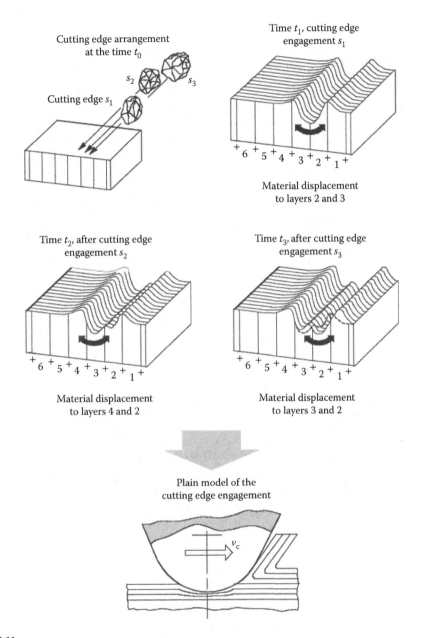

Cutting edge arrangement
at the time t_0

Cutting edge s_1

s_2 s_3

Time t_1, cutting edge
engagement s_1

$+ 6 + 5 + 4 + 3 + 2 + 1 +$

Material displacement
to layers 2 and 3

Time t_2, after cutting edge
engagement s_2

$+ 6 + 5 + 4 + 3 + 2 + 1 +$

Material displacement
to layers 4 and 2

Time t_3, after cutting edge
engagement s_3

$+ 6 + 5 + 4 + 3 + 2 + 1 +$

Material displacement
to layers 3 and 2

Plain model of the
cutting edge engagement

v_c

FIGURE 3.11
Idealization of the cutting-edge engagement through plain yielding. (From Steffens, K., 1983. Thermomechanik des Schleifens. *Fortschr.-Ber*, VDI, Reihe 2, Nr. 65, VDI-Verlag, Düsseldorf. With permission.)

certain boundary stress is exceeded, a radial crack develops below the plastically deformed zone (Figure 3.12b), which expands with increasing stress (Figure 3.12c).

After discharge, the radial crack closes in the initial phase (Figure 3.12d). During a further reduction of the stress state, axial stresses occur around the plastically deformed zone leading to lateral cracks below the surface (Figure 3.12e). The lateral cracks grow with decreasing stress. This growth can continue up to the surface of the material after

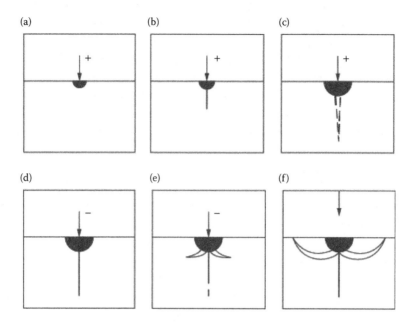

FIGURE 3.12
Mechanisms of crack spread in case of punctual stress. (From Lawn, B., Wilshaw, R., 1975, *Journal of Materials Science* 10(6), 1049–1081. With permission.)

complete discharge (Figure 3.12f) leading to the break off of material particles, which form a slab around the indentation zone (Lawn and Wilshaw 1975).

3.7.2 Scratch and Grinding Behavior of Brittle–Hard Materials

Despite the low ductility, elastoplastic deformations occur as removal mechanisms alongside brittle fracture during the grinding of brittle–hard materials (Figure 3.13). Thereby, material removal through brittle fracture is based on the induction of microcracks. Ductile

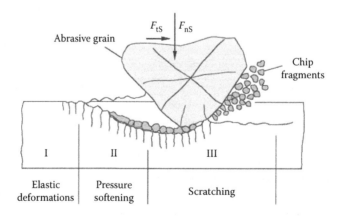

FIGURE 3.13
Material removal process in machining of brittle–hard materials. (From Saljé, E., Möhlen, H., 1987, Prozessoptimierung beim Schleifen keramischer Werkstoffe, Industrie Diamanten Rundschau. *IDR*, 21, 4. With permission.)

behavior of brittle–hard materials during grinding can be derived from the presumption that, below a threshold of boundary chip thickness defined by a critical stress, the converted energy is insufficient for crack formation and the material is plastically deformed (Bifano et al. 1987; Komanduri and Ramamohan 1994). It is supposed, however, that cracks that do not reach the surface are also formed under these conditions.

Thus, not only the amount of stress, defined by the uncut chip thickness, is responsible for the occurrence of plastic deformations, but also the above-mentioned hydrostatic compression stress below the cutting edge (Komanduri and Ramamohan 1994; Shaw 1995). The transition from mainly ductile material removal to brittle friction is decisively determined by uncut chip thickness at the single grain by grain shape and by material properties. Large cutting-edge radii promote plastic material behavior and shift the boundary of the commencing brittle friction to larger engagement depths. As a consequence, Hertzian contact stresses occur below the cutting edge, which cause hydrostatic stress countering the formation of cracks (Uhlmann 1994).

Roth (1995) investigated removal mechanisms during the grinding of aluminum oxide ceramics. He observed the influence of different grain sizes of the material as well as different diamond geometries on the material removal processes.

3.7.2.1 Fine-Grained Materials

When a scratching tool enters a fine-grained material, an entry section is formed by pure plastic deformation. The length of the entry section strongly depends on the corner radius of the diamond. If the material-specific shear stress is exceeded due to increasing scratching depth, a permanent deformation occurs thrusting aside the material and causing bulgings along the scratched groove. The base of the scratch and the flanks are even and cover a thin layer of plastically deformed material (Figure 3.14a).

Different scratches occur in the material depending on the critical scratch depth and the current boundary conditions. Along with lateral crack systems observed during penetration tests, a median crack occurs through a succession of semielliptical radial cracks at the base that runs vertically to the surface in the direction of the scratch. Similar scratch structures were observed during Vickers indentation tests. Also, V-shaped cracks are formed vertically to the surface and spread. The aperture angle is between 40° and 60° and grows with the increasing distance from the scratch. Obviously, they are crack structures that develop due to the shear stress of the tangential scratching force (shear scratches). If lateral cracks grow until the V-shaped cracks extend vertically to the surface, whole material particles break off on both sides of the scratch.

3.7.2.2 Coarse-Grained Materials

In the case of coarse-grained materials, the removal processes take place in a different way. A sharp diamond plastically divides the grains in the structure under high-energy input. With increasing infeed cracks develop mainly along the grain boundary. This is accompanied by intercrystalline failure, which leads to a break off of grains near to the surface (Figure 3.14b).

In the case of blunt scratching grains, however, strong plastic deformations occur deeper, since the maxima of the shear stresses grow. Thin, strongly deformed flakes occur on the workpiece surface. Through the extremely negative rake angle, the material is pushed forward and "extruded," not provoking ductile material removal. The stress increases with further growing infeed until the stability is exceeded at the grain boundaries and the top grain layer breaks off in the form of whole plates (Figure 3.14b).

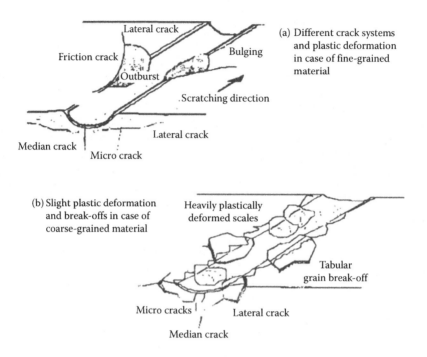

FIGURE 3.14
Material removal mechanisms during the scratching of aluminium oxide. (a) Different crack systems and plastic deformation of fine-grained material. (b) Slight plastic deformation and break-offs in coarse-grained material. (From Roth, P., 1995, *Abtrennmechanismen beim Schleifen von Aluminiumoxidkeramik. Fortschr.-Ber.* VDI, Reihe 2, Nr. 335, VDI-Verlag, Düsseldorf. With permission.)

On the basis of scratch tests and transmission electron microscopic structure analyses, Uhlmann assembles the mechanisms of surface formation during ceramic grinding (Uhlmann 1994).

Surface formation can be divided into three types of mechanisms:

- Primary mechanisms, which act during the penetration of the cutting edges into the material.
- Secondary mechanisms, which occur through the discharge by the first cutting edge and the multiple stress by successive cutting edges.
- Tertiary mechanisms, which prevent the spread of cracks in the case of materials with a high glass phase ratio.

The individual mechanisms are summarized in Table 3.1.

These material-removal processes also occur during the machining of glass. For the industrial application of optical glass, however, the conditions must allow a ductile surface formation. Thus, the development of an extensive crack system on the surface can be prevented. In Figure 3.15, the conditions for a model of ductile grinding of glass on the basis of scratch tests are summarized. According to this, a flattened blunt grain has to penetrate the material with a very small, single uncut chip thickness. The single uncut chip thickness has to be below the critical chip depth, allowing an almost crack-free surface of glass materials.

Monocrystal silicon is the most frequently applied substrate material in microsystems technology. The full range of physical properties of the material can only be exploited by

TABLE 3.1

Mechanisms of Surface Formation (Uhlmann 1994)

Surface Formation	
Primary mechanisms	
Mechanical stress	Thermal stress
Material condensation	Plastification or melting of the material or
Formation of obstructions	material phases
Crack induction	Ductile removal of material in front of the
Brittle break off of particles in front of and alongside the	cutting edge (chip formation)
cutting edge (particle formation)	
Secondary mechanisms	
Mechanical stress	Thermal stress
Break off of particles behind and alongside the cutting edge	Crack spread through thermal stresses due to
Induction of deep-lying cracks	temperature gradients
Spread of deep-lying cracks through multiple stress of	Induction of deep-lying cracks
successive cutting edges	Blistering of ceramic particles after crack
Break off of particles after crack spread toward the	spread toward the workpiece surface
workpiece surface	
Tertiary mechanisms	
Thermal stress	
Crack stopping effects at thermally plastified grain boundary phases	
Crack diversion at partially plastified grain boundary conditions in surface-near areas	

a high degree of purity, homogeneity, and crystal perfection. However, machining wafers induce damage in the subsurface and its monocrystal structure. Therefore, the subsurface damage has to be removed by polishing and etching prior to the processing of the electronic structures on the wafer front side. A concerted implementation of ductile material removal for the grinding process represents a possibility to improve the surface and subsurface quality as well as the economic efficiency of the entire process chain for wafer production (Holz 1994; Menz 1997; Kerstan et al. 1998; Tricard et al. 1998; Lehnicke 1999; Tönshoff and Lehnicke 1999; Klocke et al. 2000).

In grinding, surface formation mechanisms can be classified as ductile and brittle mechanisms. In brittle surface formation, microcracks and microcrack spread are generated in the subsurface by tensile stresses caused by the engagement of the abrasive grains of the grinding wheel into the surface of the wafer (Cook and Pharr 1990; Lawn 1993; Holz 1994; Menz 1997; Brinksmeier et al. 1998; Lehnicke 1999; Tönshoff and Lehnicke 1999). In ductile surface formation, the induced tensile stresses are insufficient to cause microcracks or crack propagation. The induced shear stress strength causes deformation and movement of dislocations that promote ductile material behavior (Figure 3.15). This leads to better surface qualities and smaller depths of subsurface damage (Lawn 1993; Menz 1997; Brinksmeier et al. 1998; Kerstan et al. 1998; Tönshoff and Lehnicke 1999). However, the basic mechanisms of ductile material removal are not known. It is further controversial whether chip formation is actually taking place or whether brittle-effective mechanisms continue to cause chips, through which the grooves on the surface are covered by plastified material. In this case, the surface seems to be formed in a ductile mode. The subsurface below would nevertheless be severely damaged by microcracks.

The mode, size, and shape of the subsurface microcrack system depend on the induced contact stresses during the engagement of the abrasive grain. The contact stress field is basically determined by the geometry of the indenter and the modulus of elasticity, hardness,

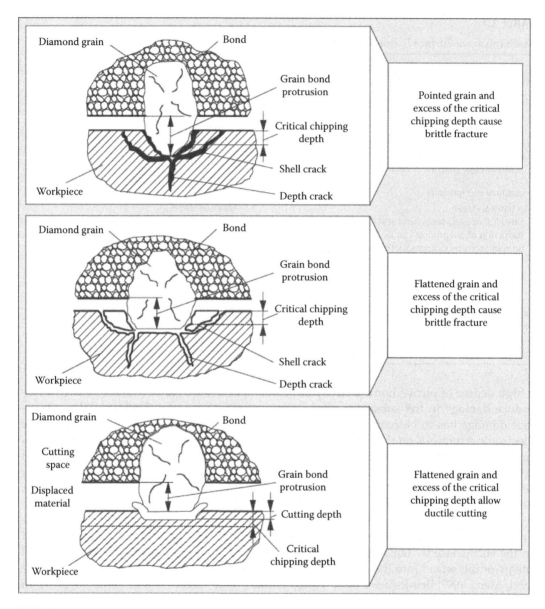

FIGURE 3.15
Theoretical prerequisites for ductile grinding of optical glass. (From Koch, E. N., 1991, Technologie zum Schleifen asphärischer optischer Linsen. Dissertation, PhD thesis, Rheinisch Westfälische Technische Hochschule Aachen. With permission.)

and fracture toughness of the indenter and the workpiece material (Lawn 1993). In the case of monocrystal silicon as workpiece material, the anisotropy of these properties, as well as the position of the slip system, is also decisive. The formation of such microcrack systems was frequently investigated in analogy tests with several brittle–hard materials such as silicon (Cook and Pharr 1990; Holz 1994).

The transition of elastoductile to mainly brittle surface formation can be classified into four different scratching morphologies (Figure 3.16). The individual areas of the scratches

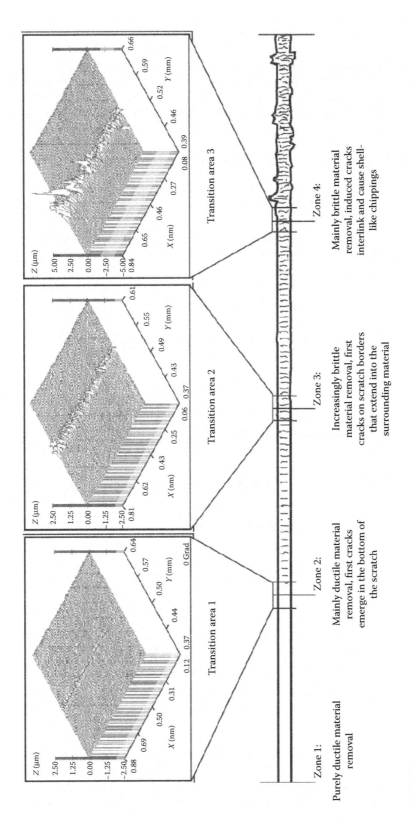

FIGURE 3.16

Classification of scratch marks in four morphological zones and three transition areas characterizing the change from purely ductile to mainly brittle surface formation.

were identified and marked using a microscope. To provide a three-dimensional image, the transition areas between the areas of different scratch morphologies were measured at a length of 1 mm. Then, the scratch depths were determined on the individual profiles.

Knowing the material removal mechanisms during the machining of silicon, it is possible to realize a low-damage process with the adequate settings and tools during the grinding of semiconductors. This low-damage grinding process is necessary to shorten the whole process chain and to decrease machining costs.

3.8 Energy Transformation

Mechanical energy is introduced into the grinding process by relative movement between the tool and the workpiece (Figure 3.17). This energy is mainly transformed into heat in an energy transformation process, leading to temperature increase in the contact zone. The transformation of mechanical energy into thermal energy takes place through friction and deformation processes (Grof 1977; Lowin 1980). External friction processes between abrasive grain and workpiece surface as well as between chip and abrasive grain are partly responsible for the heat development during grinding. However, heat also arises as a result of internal friction through displacement processes and plastic deformations (Grof 1977; Lowin 1980; Marinescu et al. 2013; Rowe 2014).

Heat development and heat flow during the grinding of brittle–hard materials differ decisively from the process in the machining of ductile materials (Figure 3.18). Heat development in the case of ceramics has been investigated in many studies. Due to the relatively poor heat conductivity of ceramics and, in contrast, to a very high heat conductivity of diamond as a grinding agent, a big percentage of the heat flow to the

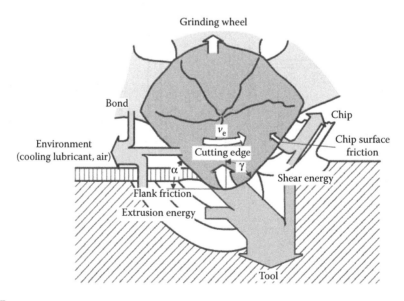

FIGURE 3.17
Heat flow during the grinding of metallic materials. (From König, W., Klocke, F., 1996, *Fertigungsverfahren Band 2. Schleifen, Honen, Läppen.* 3. Auflage, VDI-Verlag GmbH, Düsseldorf. With permission.)

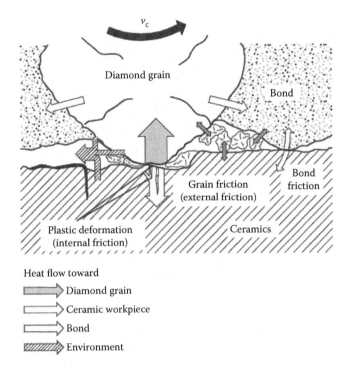

FIGURE 3.18
Heat flow during the grinding of ceramics. (From Uhlmann, E., 1994, Tiefschleifen hochfester keramischer Werkstoffe. *Produktionstech-Berlin, Forschungsber. für die Praxis*, Bd. 129, Hrsg.: Prof. Dr. H. C. Mult. Dr. Ing. G. Spur, München, Wien, Hanser. With permission.)

tool and a considerably smaller heat flow to the workpiece was observed (Wobker 1992; Uhlmann 1994).

The following energy transformation processes occur during the grinding of ceramics (Uhlmann 1994):

- Energy from retained dislocations (plastic surface areas) after particle removal
- Deformation energy at the workpiece surface (plastic scratch marks with a bulging at the edge)
- Elastic excess energy from the extension of existing microcracks during particle removal
- Elastic energy from microscopic surface areas returning in the initial position
- Friction work between diamond cutting edge and workpiece surface
- Friction work between ceramic particles (workpiece surface as well as ceramic particles) and diamond cutting edge
- Friction work between bond and workpiece surface

The use of cooling lubricant influences the heat development and the heat flow. The lubricant reduces the friction between the active partners entailing smaller heat development. The cooling mainly takes place through the amount of water in the cooling lubricant. Through the cooling effect, the percentage of conducted heat increases.

References

Bifano, T. et al., 1987, Precision machining of ceramic materials. *Intersociety Symposium on Machining of Advanced Ceramic Materials and Components*, Westerville, Ohio.

Bohlheim, W., 1995, Verfahren zur Charakterisierung der Topografie von Diamantschleifscheiben. *IDR* 29, 2.

Bouzakis, K. D., Karachaliou, C., 1988, Erfassung der Spanungsgeometrie und der Zerspankraftkomponenten beim Flachschleifen aufgrund einer Dreidimensionalen Beschreibung der Schleifscheibentopomorphie. *Fortschr.-Ber.* VDI Reihe 2 Nr. 165, VDI, Düsseldorf.

Brinksmeier, E., Preuß, W., Riemer, O., Malz, R., 1988, Ductile to brittle transition investigated by plunge-cut experiments in monocrystalline silicon. *Proceedings of the ASPE Spring Topical Meeting on Silicon Machining*.

Brücher, T., 1996, Kühlschmierung beim Schleifen keramischer Werkstoffe. PhD thesis, Technische Universität, Berlin.

Büttner, A., 1968, Das Schleifen sprödharter Werkstoffe mit Diamant-Topfscheiben unter besonderer Berücksichtigung des Tiefschleifens. PhD thesis, Technische Universität, Hannover.

Cook, R. F., Pharr, G. M., 1990, Direct observation and analysis of indentation cracking in glasses and ceramics. *J. Am. Ceram. Soc.* 73, 4.

Damlos, H. H., 1985, Prozessablauf und Schleifergebnisse beim Tief- und Pendelschleifen von Profilen. *Fortschr.-Ber. VDI*, Reihe 2, Nr. 88, VDI Verlag, Düsseldorf.

Daude, O., 1966, *Untersuchung des Schleifprozesses. PhD thesis, Rheinisch Westfälische Technische Hochschule Aachen.*

Dennis, P., van Schmieden, W., 1989, Abdruckverfahren zur Dokumentation von Verschleißvorgängen. *VDI-Z* 131(1, S), 72–75.

Frühling, R., 1976, Topographische Gestalt des Schleifscheiben-Schneidenraumes und Werkstückrauhtiefe beim Außenrund-Einstechschleifen. PhD thesis, Technische Universität, Braunschweig.

Gärtner, W., 1982, Untersuchungen zum Abrichten von Diamant- und Bornitridschleifscheiben. PhD thesis, Technische Universität, Hannover.

Grof, H. E., 1977, Beitrag zur Klärung des Trennvorgangs beim Schleifen von Metallen. PhD thesis, Technische Universität, München.

Holz, B., 1994, *Oberflächenqualität und Randzonenbeeinflussung beim Planschleifen einkristalliner Siliciumscheiben.* Produktionstechnik—Berlin, Forschungsberichte für die Praxis, Bd. 143, Hrsg.: Prof. Dr. H. C. Mult. Dr. Ing, G. Spur. München, Wien, Hanser.

Hübert, C., 2011, Schleifen von Hartmetall- und ollkeramik-Schaftfräsern. PhD thesis, Technische Universität, Berlin.

Hübert, C., Mauren, F., van der Meer, M., Hahmann, D., Rickens, K., Mutlugünes, Y., Hahmann, W. C., Pekarek, M., 2009, Charakterisierung von Schleifscheibentopographien aus fertigungstechnischer Sicht. *dihw* 1, 4.

Hughes, F. H., 1974, Wärme im Schleifprozess—ein Vergleich zwischen Diamant- und konventionellen Schleifmitteln. *IDR* 8, 2.

Inasaki, C., Chen, C., Jung, Y., 1989, Surface, cylindrical and internal grinding of advanced ceramics. Grinding fundamentals and applications. *Transactions of the ASME* 39(S), 201–211.

ISO 25178, 2012, Geometrical product specifications (GPS)—Surface texture: Areal. International Organization for Standardization.

Jacobs, U., 1980, Beitrag zum Einsatz von Schleifscheiben mit kubisch-kristallinem Bornitrid als Schneidstoff. PhD thesis, Technische Universität, Braunschweig.

Kaiser, M., 1977, *Tiefschleifen von Hartmetall.* Fertigungstechnische Berichte, Bd. 9, Hrsg.: H. K. Tönshoff, Gräfelfing, Resch.

Karatzoglou, K., 1973, Auswirkungen der Schneidflächenbeschaffenheit und der Einstellbedingungen auf das Schleifergebnis beim Flach-Einstechschleifen. Dr. Ing. dissertation, TU Braunschweig.

Kassen, G., 1969, Beschreibung der elementaren Kinematik des Schleifvorganges. PhD thesis, Rheinisch Westfälische Technische Hochschule Aachen.

Kerstan, M., Ehlert, A., Huber, A., Helmreich, D., Beinert, J., Döll, W., 1998, Ultraprecision grinding and single point diamond turning of silicon wafers and their characterisation. *Proceedings of the ASPE Spring Topical Meeting on Silicon Machining*, Camel-by-the-Sea, CA.

Klocke, F., Gerent, O., Pähler, D., 2000, Effiziente Prozesskette zur Waferfertigung. *ZwF*, 95, 3.

Koch, E. N., 1991, Technologie zum Schleifen asphärischer optischer Linsen. Dissertation, PhD thesis, Rheinisch Westfälische Technische Hochschule Aachen.

Komanduri, R., Ramamohan, T. R., 1994, On the mechanisms of material removal in fine grinding and polishing of advanced ceramics and glasses, in *Advancement of Intelligence Production. The Japan Society for Precision Engineering*, Elsevier Science, Amsterdam.

König, W., Klocke, F., 1996, *Fertigungsverfahren Band 2. Schleifen, Honen, Läppen*. 3. Auflage, VDI-Verlag GmbH, Düsseldorf.

König, W., Steffens, K., Yegenoglu, K., 1981, Modellversuche zur Erfassung der Wechselwirkung zwischen Reibbedingungen und Stofffluss. *Industrie Anzeiger* 103, 35.

Kurrein, M., 1927, Die Bearbeitbarkeit der Metalle im Zusammenhang mit der Festigkeitsprüfung. *Werkstattstechnik*. 21(S), 612–621.

Lawn, B., 1993, *Fracture of Brittle Solids*, 2nd edition. Cambridge University Press, New York.

Lawn, B., Wilshaw, R., 1975, Review indentation fracture: Principles and applications. *Journal of Materials Science* 10(6), 1049–1081.

Lehnicke, S., 1999, *Rotationsschleifen von Silizium-Wafern*. Fortschritt-Berichte VDI, Reihe 2, Nr. 534, VDI-Verlag, Düsseldorf.

Lierath, F., Jankowski, R., Schnekel, S., Bage, T., 1990, Prozessmodelle zur Qualitätssteigerung von Arbeitsabläufen in der Feinbearbeitung. Tagungsband zum 6, Intern. Braunschweiger Feinbearbeitungskolloquium.

Lortz, W., 1975, Schleifscheibentopographie und Spanbildungsmechanismus beim Schleifen. PhD thesis, Rheinisch Westfälische Technische Hochschule Aachen.

Lowin, R., 1980, Schleiftemperaturen und ihre Auswirkungen im Werkstück. PhD thesis, Rheinisch Westfälische Technische Hochschule Aachen.

Malyshev, V., Levin, B., Kovalev, A., 1990, *Grinding with Ultrasonic Cleaning and Dressing of Abrasive Wheels*. 61, 9, Stanki I Instrument, Moscow, 22–26.

Marinescu, I. D., Rowe, W. B., Dimitrov, B., Ohmori, H., 2013, *Tribology of Abrasive Machining Processes*. Elsevier (William Andrew imprint), Oxford.

Menz, C., 1997, *Randzonenanalyse bearbeiteter Siliziumoberflächen. Randzonenanalyse bearbeiteter Silizium-Oberflächen. Fortschritt-Berichte* VDI, Reihe 2, Nr. 431, VDI-Verlag, Düsseldorf.

Nakayama, K., 1973, Taper print method for the measurement of grinding wheel surface. *Bulletin of the Japan Society of Precision Engineering*. 7, 2.

Pahlitzsch, G., Helmerdig, H., 1943, Bestimmung und Bedeutung der Spandicke beim Schleifen. *Werkstattstechnik*. 11/12(S), 397–399.

Paulmann, R., 1990, Grundlagen zu einem Verfahrensvergleich. *Jahrbuch Schleifen, Honen, Läppen und Polieren*. 56, Ausgabe, Vulkann-Verlag, Essen.

Reichenbach, G. S., Mayer, I. E., Kalpakcioglu, S., Shaw, M. C., 1956, The role of chip thickness in grinding. *Transactions of the ASME* 18(S), 847–850.

Rohde, G., 1985, Beitrag zum Verhalten von keramisch-gebundenen Schleifscheiben im Abricht- und Schleifprozess. PhD thesis, Technische Universität, Braunschweig.

Roth, P., 1995, *Abtrennmechanismen beim Schleifen von Aluminiumoxidkeramik. Fortschr.-Ber.* VDI, Reihe 2, Nr. 335, VDI-Verlag, Düsseldorf.

Rowe W. B., 2014, *Principles of Modern Grinding Technology*, 2nd edition. Elsevier, Oxford.

Saljé, E., 1975, Die Wirkrauhtiefe als Kenngröße des Schleifprozesses. *Jahrbuch der Schleif-, Hon-, Läpp- und Poliertechnik und der Oberflächenbearbeitung*, 47, Ausgabe, Vulkan, Essen, S, 23–35.

Saljé, E., Möhlen, H., 1987, Prozessoptimierung beim Schleifen keramischer Werkstoffe, Industrie Diamanten Rundschau. *IDR*, 21, 4.

Sawluk, W., 1964, Flachschleifen von oxydkeramischen Werkstoffen mit Diamant-Topfscheiben. Dissertation, Technische Hochschule Braunschweig.

Schleich, H., 1982, Schärfen von Bornitridschleifscheiben. PhD thesis, Rheinisch Westfälische Technische Hochschule Aachen.

Shaw, M. C., 1995, Cutting and grinding of difficult materials. *Technical Paper Presented at the Abrasive Engineering Society, Ceramic Industry Manufacturing Conference and Exposition*, Pittsburgh, PA.

Shaw, M. C., Komanduri, R., 1977, The role of stylus curvature in grinding wheel surface characterization. *Annals of the CIRP* 25, 1.

Steffens, K., 1983. Thermomechanik des Schleifens. *Fortschr.-Ber*, VDI, Reihe 2, Nr. 65, VDI-Verlag, Düsseldorf.

Stuckenholz, B., 1988, Das Abrichten von CBN-Schleifscheiben mit kleinen Abrichtzustellungen. PhD thesis, Rheinisch Westfälische Technische Hochschule Aachen.

Tönshoff, H. K., Lehnicke, S., 1999, Subsurface damage reduction of ground silicon wafers. *Proceedings of Euspen*, Bremen, Germany.

Tönshoff, H. K., Peters, J., Inasaki, I., Paul, T., 1992, Modelling and simulation of grinding processes. *Annals of the CIRP* 41(2), 677–688.

Treffert, C., 1995. Hochgeschwindigkeitsschleifen mit galvanisch gebundenen CBN-Schleifscheiben. *Berichte aus der Produktionstechnik*, Bd. 4/95, Aachen, Shaker.

Tricard, M., Kassir, S., Herron, P., Pei, Z. J., 1998, New abrasive trends in manufacturing of silicon wafers. *Proceedings of the ASPE Annual Meeting*, Indianapolis, IN.

Uhlmann, E., 1994, Tiefschleifen hochfester keramischer Werkstoffe. *Produktionstech-Berlin, Forschungsber. für die Praxis*, Bd. 129, Hrsg.: Prof. Dr. H. C. Mult. Dr. Ing. G. Spur, München, Wien, Hanser.

Verkerk, J., 1977, Final report concerning CIRP cooperative work on the characterization of grinding wheel topography. *Annals of the CIRP* 26, 2.

Vits, R., 1985, Technologische Aspekte der Kühlschmierung beim Schleifen. PhD thesis, Rheinisch Westfälische Technische Hochschule Aachen.

Warnecke, G., Spiegel, P., 1990, Abrichten kunstharzgebundener CBN- und Diamantschleifscheiben. *IDR* 24(4, S), 229–235.

Weinert, K., 1976, Die zeitliche Änderung des Schleifscheibenzustandes beim Außenrund-Einstechschleifen. PhD thesis, Technische Universität Braunschweig.

Werner, F., 1994, *Hochgeschwindigkeitstriangulation zur Verschleißdiagnose an Schleifwerkzeugen. Fortschr.-Ber. VDI*, Reihe 8, Nr. 429, VDI-Verlag, Düsseldorf.

Werner, G., 1971, Kinematik und Mechanik des Schleifprozesses. PhD thesis, Rheinisch Westfälische Technische Hochschule Aachen.

Werner, G., Kenter, M., 1987, Work material removal and wheel wear mechanisms in grinding of polycrystalline diamond compacts. Vortrag anläßl. *Intersociety Symposium on Machining of Advanced Ceramic Materials and Components*, Westerville, Ohio.

Wobker, H. G., 1992, *Schleifen keramischer Schneidstoffe. Fortschr.-Ber. VDI*, Reihe 2, Nr. 237, VDI-Verlag, Düsseldorf.

Zum Gar, K. H., 1987, Grundlagen des Verschleißes. *VDI Berichte Nr. 600.3: Metallische und Nichtmetallische Werkstoffe und ihre Verarbeitungsverfahren im Vergleich*, VDI-Verlag, Düsseldorf.

4

Grinding Wheels

4.1 Introduction

4.1.1 Developments in Productivity

Huge increases in productivity have been achieved in recent decades due to advances in grinding wheel technology. These increases have only been possible by parallel developments in the machines and auxiliary equipment employed, since the greatest gains have been from not only the grinding wheel but also the performance of the total grinding system.

Grinding wheels operating at low wheel speeds employed in the early twentieth century have progressed to advanced conventional abrasives and superabrasives operating at high wheel speeds in the present era. Over this period, material removal rates have increased for some grinding processes by a staggering 10–100 times. The grinding wheel technology that made such advances possible primarily involved the development of new abrasives as described in this chapter. In addition, mechanical design aspects of wheel design are introduced that affect grinding quality, performance, and safety. Essential information is given on wheel design for high-speed operation including design of segmented wheels.

4.1.2 System Development

New abrasives require new ways of working that reflect in new designs of grinding wheel assembly, trueing, dressing and conditioning techniques, coolant delivery and coolant formulation, and finally new designs of machines capable of high wheel speeds and capable of delivering higher power to the grinding wheel. A variety of wheel designs have developed to cope with differing product geometries. However, two other considerations gave rise to a new approach to wheel design:

- High wheel speeds must be designed for much greater wheel strength.
- Expensive but hard-wearing diamond and cubic boron nitride (CBN) superabrasives only need thin layers of abrasive to achieve a long wheel life.

4.1.3 Conventional and Superabrasive Wheel Design

In the next few chapters, the distinction will be made repeatedly between operation with conventional abrasives such as alumina and silicon carbide and operation with superabrasives such as CBN and diamond. The wheel designs tend to be distinctly different.

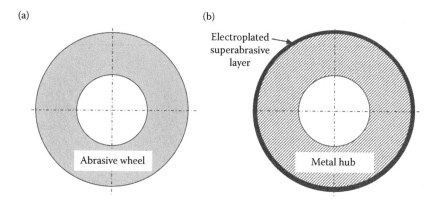

FIGURE 4.1

A conventional wheel and an EP superabrasive wheel. (a) Low- or medium-speed conventional abrasive. (b) High-speed EP superabrasive wheel.

One reason is the expense of the raw materials used for diamond and CBN superabrasives. Another reason is that these materials, especially CBN, tend to call for higher wheel speeds to take advantage of the potential for increased production rate and long wheel life. The higher speeds also drive the difference in wheel design.

The following sections provide essential information on basic dimensions and geometry of grinding wheels in these two categories. Figure 4.1 contrasts the difference in wheel design at the two extremes between a conventional wheel and an electroplated superabrasive wheel. In between these two extremes lie a range of wheel designs including high-speed segmented designs as described below.

4.2 Wheel Shape Specification

4.2.1 Basic Shapes

Grinding wheels come in a variety of shapes and sizes. Standard international wheel shapes and examples for conventional and superabrasive wheels are given in Tables 4.1 and 4.2. Wheel dimensions are usually expressed as diameter (D) × thickness (T) × hole (H). For superabrasives, the layer thickness (X) is added afterward. Conventional wheels are typically sold as standard stock sizes, although they can be cut to size and bushed in the bore to order. They can also be preprofiled for certain applications such as worm gear grinding. Many superabrasive wheels, especially resin bond, also come in standard stock sizes but many are custom built, often with complex premolded profiled layers. The cost of this is readily offset against the savings in abrasive and initial dress time.

4.2.2 Hole Tolerances

Wheel dimensional tolerancing is very dependent on application and supplier. Some typical manufacturer's guidelines are given below. Conventional cored wheels should not have a tight fit for fear of cracking due to thermal expansion differentials with the steel

TABLE 4.1

International Standard Shapes for Conventional Wheels

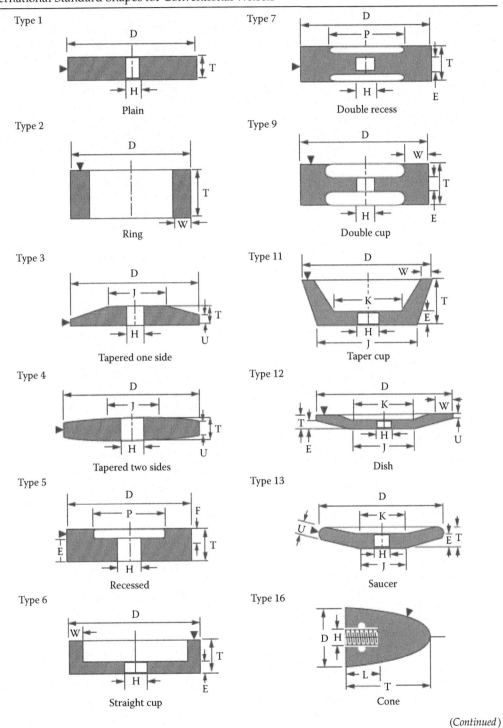

Type 1 — Plain

Type 2 — Ring

Type 3 — Tapered one side

Type 4 — Tapered two sides

Type 5 — Recessed

Type 6 — Straight cup

Type 7 — Double recess

Type 9 — Double cup

Type 11 — Taper cup

Type 12 — Dish

Type 13 — Saucer

Type 16 — Cone

(Continued)

TABLE 4.1 (*Continued*)

International Standard Shapes for Conventional Wheels

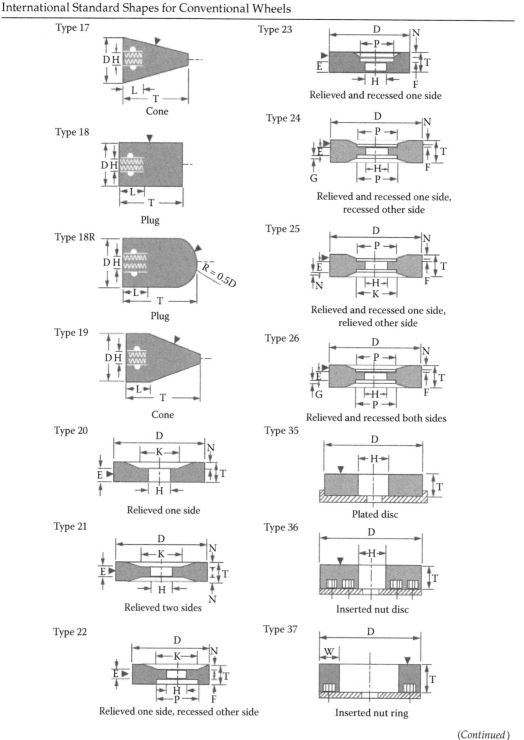

Type 17 — Cone

Type 18 — Plug

Type 18R — Plug

Type 19 — Cone

Type 20 — Relieved one side

Type 21 — Relieved two sides

Type 22 — Relieved one side, recessed other side

Type 23 — Relieved and recessed one side

Type 24 — Relieved and recessed one side, recessed other side

Type 25 — Relieved and recessed one side, relieved other side

Type 26 — Relieved and recessed both sides

Type 35 — Plated disc

Type 36 — Inserted nut disc

Type 37 — Inserted nut ring

(Continued)

TABLE 4.1 (*Continued*)

International Standard Shapes for Conventional Wheels

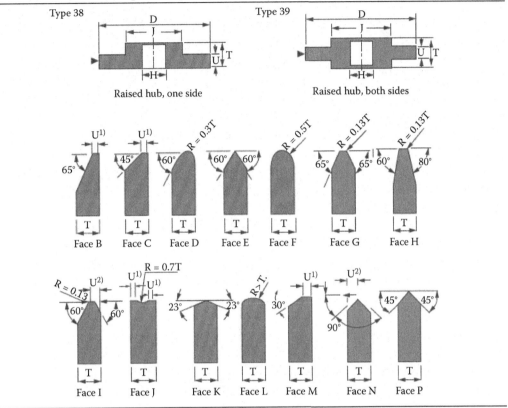

wheel mounts. On the other hand, steel-cored superabrasives for high speed require the best possible running truth to eliminate problems of chatter and vibration.

Conventional roughing wheels	H13 bore fit
Conventional precision wheels	H11 bore fit
Standard superabrasive wheels	H7 bore fit
High-speed CBN wheels	H6–H4 bore fits

Tolerances based on bore diameters are given in Table 4.3.

4.2.3 Side and Diameter Tolerances

Side and outer diameter (o.d.) run-out of wheels vary from one manufacturer to another and depend on application. For superabrasive wheels, the following tolerances are recommended (Table 4.4).

Outer diameter tolerances may be considerably more open, as long as running truth is maintained.

TABLE 4.2

Superabrasive Wheel Shape Examples

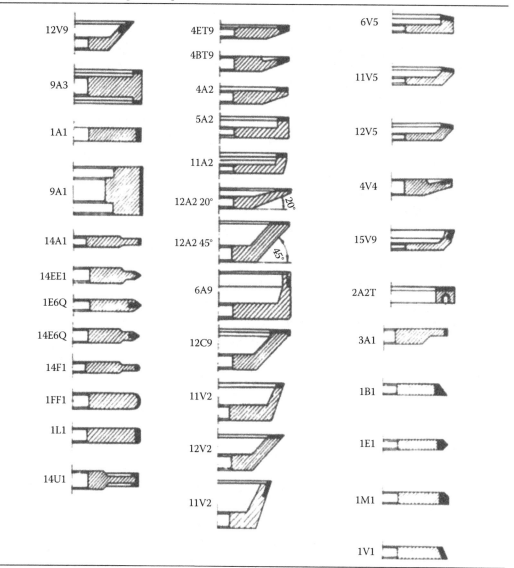

4.3 Wheel Balance

4.3.1 Introduction to Wheel Balance

Balance is closely associated with runout. As the degree of unbalance force increases, runout will also increase. Balance tolerances depend on application. The Japanese JIS B4131 code gives the following balance tolerances in terms of center of gravity displacement g (Table 4.5). For high speed, the balance requirements are significantly more stringent

TABLE 4.3

Standard Fits for Wheel Bore Tolerancing[a]

Bore Diameter (mm)	H4 Upper Limit (μm)	H5 Upper Limit (μm)	H6 Upper Limit (μm)	H7 Upper Limit (μm)	H11 Upper Limit (μm)	H13 Upper Limit (μm)
>3–6	4	5	8	12	75	180
>6–10	4	6	9	15	90	220
>10–18	5	8	11	18	110	270
>18–30	6	9	13	21	130	330
>30–40	7	11	16	25	160	390
>40–50	7	11	16	25	160	390
>50–65	8	13	19	30	190	460
>65–80	8	13	19	30	190	460
>80–100	10	15	22	35	220	540
>100–120	10	15	22	35	220	540
>120–140	12	18	25	40	250	630
>140–160	12	18	25	40	250	630
>160–180	12	18	25	40	250	630
>180–200	14	20	29	46	290	720
>200–225	14	20	29	46	290	720
>225–250	14	20	29	46	290	720
>250–280	16	23	32	52	320	810
>280–315	16	23	32	52	320	810
>315–355	18	25	36	57	360	890
>355–400	18	25	36	57	360	890
>400–450	20	27	40	63	400	970
>450–500	20	27	40	63	400	970

[a] Lower limit nominal (0).

TABLE 4.4

Side Runout Tolerances for Superabrasive Wheels

Wheel Diameter (mm)	Standard Superabrasive Wheels (μm)	High-Speed CBN (μm)
<250	20	5
250–400	30	10
400–600	50	15
600–750	70	20

than those shown in Table 4.5. For camshaft grinding, a high-speed CBN wheel of 14"
diameter weighing 25 lbs operating at 5000 rpm must be balanced to <0.015 oz in order
to prevent visual chatter even when used on a grinder with a high stiffness hydrostatic
spindle. This is equivalent to $g = 0.4$ which is almost an order of magnitude tighter than
current standards.

4.3.2 Static and Dynamic Unbalance

"Static" unbalance is the term employed for unbalance within a single plane. Balancing
of static unbalance may be performed either at zero speed or at running speed. Wheels

TABLE 4.5

Japanese JIS B4131 Specification for Wheel Balance Tolerances

Diameter (mm)	Max g
>75–150	0.5
>150–200	1
>200–300	2
>300–400	3
>400–500	5
>500–600	8
>600–700	12
>700–750	15

Center of gravity displacement (in.)

0.005
0.001
0.0005
0.0001
0.00005
0.00001
0.000004

15
6.3
2.5
0.1
0.04

100   500 1000   5000 10,000   50,000

RPM

are initially balanced statically off the machine but more frequently balanced at a fixed running speed. A confusion of terms is easily caused. Balancing at speed is often incorrectly called dynamic balancing, although strictly, the term "dynamic" balancing means balancing in two or more planes to avoid a conical gyration.

4.3.3 Automatic Wheel Balancers

Chatter is visible down to displacements of 10 micro-inches. Achieving displacements below this, even with the most sophisticated hydrostatic wheel bearings, is becoming increasingly more of a challenge as wheel speeds increase. For regular hydrodynamic or ball-bearing-based spindles, some form of dynamic balancer mounted to the machine is essential.

Automatic balancers are mounted on the spindle nose and function by adjusting the position of eccentric weights. Older systems actually pumped chlorofluorocarbon gas from one chamber to another but these have been phased out for environmental reasons. A separate sensor is employed to detect the level of vibration. The wheel guarding may need to be altered to allow a wheel balancer to be accommodated in an older machine. Wheel balancers are standard on the majority of new cylindrical grinders.

Figure 4.2 shows an automatic balancing device incorporated within a conventional wheel mounting arrangement.

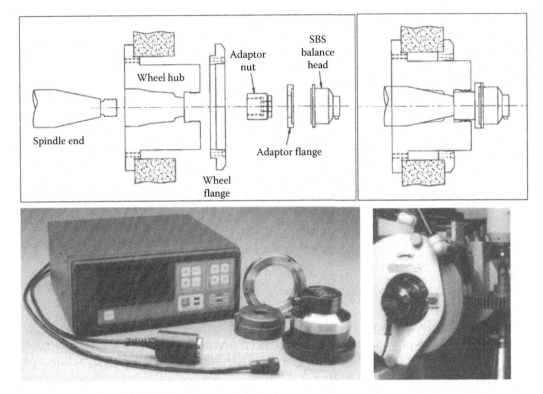

FIGURE 4.2
Automatic wheel balancer. (Courtesy of SBS, Schmidt Industries.)

The response times of automatic balancers have managed to keep pace with higher speed requirements and many can operate at 10,000 rpm or higher. The range has expanded to cover not only large cylindrical wheel applications but also smaller wheels down to 6 in. for use on multitasking machining centers. Others have the vibration sensor built into the balance head and can also be used as a crash protection and acoustic dress sensor.

4.3.4 Dynamic Balancing in Two Planes

Most recently, automatic balancers have been developed for dynamic balancing in two planes for compensation of long wheels such as for through-feed centerless grinding, or for complete wheel/spindle/motor assemblies.

4.3.5 Coolant Unbalance

Coolant is a key factor for maintaining balance.

A grinding wheel can absorb a considerable quantity of coolant (e.g., 220# WA 1A1 wheel can hold up to 16 wt%). When spun the coolant is not released instantaneously but may take several minutes or even hours depending on its viscosity. This can be seen in Figure 4.3 which shows the rate of loss of retained coolant in a 12½-in. WA 220# 1A1 wheel spun at 48 m/s. The primary problem arises when coolant is allowed to drip on a stationary wheel or a stationary wheel is allowed to drain vertically. This will throw the wheel into an unbalanced condition when next used. The effect is unlikely to cause

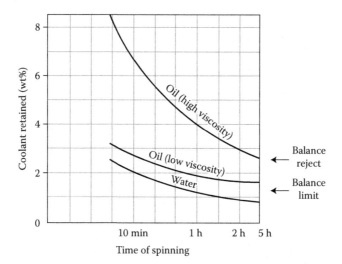

FIGURE 4.3
Coolant retention in a spinning grinding wheel as a function of time.

actual wheel failure, although it can happen. However, coolant unbalance will generate prolonged problems of vibration and chatter even when constantly dressing the wheel. It is also an issue with vitrified CBN wheels at high enough wheel speed (>60 m/s), even though the porous layer may only be a few millimeter thick.

4.4 Design of High-Speed Wheels

4.4.1 Trend Toward Higher Speeds

Vitrified CBN wheel speeds have risen significantly in the last 10 years. In 1980, 60 m/s was considered high speed, by 1990, 80 m/s was becoming common in production, by 1995, 120 m/s, and by 2000, 160 m/s. At the time of writing, several machines for vitrified wheels have been reported entering production for grinding cast iron at 200 m/s. Speeds of up to 500 m/s have been reported experimentally with plated CBN (Köenig and Ferlemann 1991). Such wheel speeds place increased safety demands on both the wheel maker and machine tool builder.

Conventional vitrified bonded wheels generally default to a maximum wheel speed of 23–35 m/s depending on bond strength and wheel shape. Certain exceptions exist; thread and flute grinding wheels tend to operate at 40–60 m/s and internal wheels up to 42 m/s. A full list is given in ANSIB7.1 2000 Table 23.

4.4.2 How Wheels Fail

To achieve higher speeds require an understanding of how wheels fail. Vitrified bonds are brittle, elastic materials that will fail catastrophically when the localized stresses exceed material strength. Stresses occur from clamping of the wheel, grinding forces, acceleration and deceleration forces on starting, stopping or changing speed, wheel unbalance, or thermal stresses. However, under normal and proper handling and use of the wheel, the greatest factor is the centrifugal stresses due to constant rotation at operating speed.

4.4.3 Hoop Stress and Radial Stress

The stresses and displacements created in a monolithic grinding wheel can be readily calculated from the classic equations for linear elasticity. The radial displacement U is given by

$$r^2 \cdot \frac{d^2U}{dr^2} + r \cdot \frac{du}{dr} - U + \frac{r}{h} \cdot \frac{dh}{dr} \cdot \left[r \cdot \frac{dU}{dr} + U \cdot v \right] = -\frac{1-v^2}{E} \cdot \rho \cdot \omega^2 \cdot r^2$$

This can be solved using finite difference approximations to give radial displacements at any radius of the wheel. The circumferential or hoop stress and radial stress equations are given by the following, where it is assumed the wheel o.d. is >10 times wheel thickness.

$$\sigma_{\theta\theta} = \frac{E}{(1-v^2)} \cdot \left[\frac{U}{r} + v \cdot \frac{dU}{dr} \right], \quad \sigma_{rr} = \frac{E}{(1-v^2)} \cdot \left[\frac{U}{r} + v \cdot \frac{dU}{dr} \right]$$

The solutions to these equations are given by Barlow and Rowe (1983) and Barlow et al. (1995, 1996) as:

$$U = \frac{r}{E} \left[(1-v) \cdot C_1 - (1+v) \cdot \frac{C_2}{r^2} - \frac{(1-v^2)}{8} \cdot \rho \cdot \omega^2 \cdot r^2 \right]$$

$$\sigma_{rr} = C_1 + \frac{C_2}{r^2} - \frac{3+v}{8} \cdot \rho \cdot \omega^2 \cdot r^2$$

$$\sigma_{\sigma\sigma} = C_1 - \frac{C_2}{r^r} - \frac{1+3v}{8} \cdot \rho \cdot \omega^2 \cdot r^2$$

The constants C_1 and C_2 are subject to the appropriate boundary conditions:

1. Radial stress at the periphery of the wheel is zero.
2. Free radial displacement at the bore.

The second boundary condition concerns the displacement of the bore and depends on the level of clamping of the wheel. It is usual in the design of wheels to assume the worst-case situation, which is free radial displacement, as this gives the highest level of circumferential stress. The maximum hoop stress occurs at the bore. In this case, the radial stress is zero.

Full constraint at the bore leads to zero displacement but the radial stress is now nonzero. Figure 4.4 gives an example of the stress distribution for the two extremes.

4.4.4 Reinforced Wheels

Wheel failure, in line with this analysis, occurs from cracks generated at or near the bore, where the stress is highest. The failure is catastrophic with conventional wheels. Typically

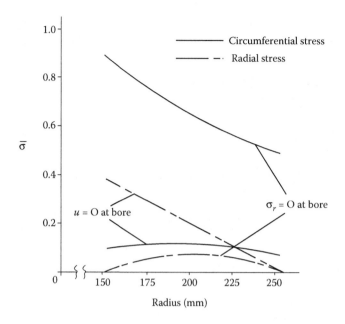

FIGURE 4.4
Normalized stress distribution across a rotating grinding wheel for both free spinning and constrained wheels. (From Barlow, N., Rowe, W. B., 1983, *International Journal of the Machine Tool Design and Research*, 23(2/3), 153–160.)

four or five large, highly dangerous pieces are flung out. To increase the burst speed either the overall strength of the bond must be increased (e.g., finer grit size, lower porosity, and better processing methods must be applied to eliminate large flaws) or the strength must be increased where the stress is highest. For conventional wheels, this has often been achieved with a two-component vitrified structure, where the inner portion is higher strength although not necessarily suitable for grinding.

For vitrified CBN wheels, higher speeds are achieved by substituting the inner section of the wheel with a higher strength material, such as aluminum, carbon fiber reinforced plastic (CFRP), and especially steel.

4.4.5 Segmented Wheels

Wheel manufacture of a high-speed segmented wheel consists of epoxy bonding or cementing a ring of vitrified CBN segments to the periphery of a steel core, as shown in Figure 4.5.

The segmented design serves several purposes. First, it produces a much more consistent product than a continuous or monolithic structure because of the limited movements required in pressing segments of such small volume. This is especially true when, as in the examples shown above, a conventional backing (white) layer is added behind the CBN to allow the use of the full layer of the abrasive layer. This gives both a better consistency in grinding and a higher Weibull number for strength consistency. Second, it allows a wheel to be repaired in the event of being damaged providing a considerable cost saving for an expensive CBN wheel. Third and very important, the segments provide stress relief, acting as "expansion joints" as wheel speed increases and the steel core expands due to centrifugal force.

FIGURE 4.5
Segmented vitrified CBN wheels and molded segment cross sections. (Courtesy of Saint-Gobain Abrasives.)

4.4.6 Segment Design

Trying to model segmented wheels using the traditional laws of elasticity has proved difficult because of complex effects within and around the adhesive layer. Finite element analysis (FEA)-based models are now more common with much of the groundwork having been done by Barlow et al. (1995, 1996).

Both hoop stresses and radial stresses can lead to wheel failure. Hoop stress is dependent on the expansion of the core and the segment length. Radial stress is dependent both on the expansion of the core and, more importantly, on the mass and therefore thickness of the segment. Figure 4.6 plots the principal stresses of a 20-in. diameter aluminum body wheel (12 in bore) as a function of segment number and abrasive layer thickness. σ_1 is the maximum principal stress, σ_2 is the minimum principal stress, and σ_3 is axial or out of plane stress. As can be seen, there is an optimum number of segments, 35 in the example below. Higher segment numbers give rise to additional stresses at the joint edges because as the wheel expands in a radial direction, it must contract in the axial direction.

4.4.7 Abrasive Layer Depth

For thin segments, the major stress is circumferential, but for thicker segments, the dominant stress shifts to radial. For this reason, an abrasive layer thickness of 10 mm maximum is typical for a high-speed wheel. This immediately places limitations on profile forms allowed. The key factor is the mass of the segment, and its impact on radial stress is also important when considering the effect of wheel radius on burst speed. As the wheel radius is reduced the centripetal force (mv^2/r) must increase which will directly increase the radial stress.

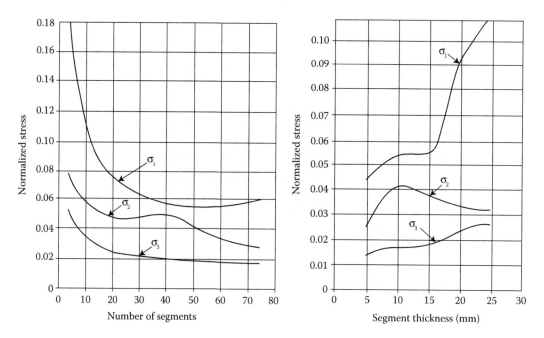

FIGURE 4.6
Maximum stress levels in a rotating segmented wheel as a function of segment number and abrasive thickness.

Figure 4.7 plots burst speed as a function of wheel radius for various abrasive layer depths. Ideally, the calculated burst speed should be at least twice the maximum recommended operating speed. For the 5 mm CBN layer, the burst speed levels off at 320 m/s because this was the calculated burst speed for the steel core used in the example. The most striking factor about this graph is that the burst speed drops rapidly as the wheel diameter is reduced. For a wheel diameter of 6 in., the maximum recommended wheel speed would be only 100 m/s based on the particular bond strength used in the example. The value employed is believed typical of current CBN technology. Not surprisingly, therefore, high-speed wheels operating in the range of 100–200 m/s tend to be >12 in. diameter with flat or shallow forms.

4.4.8 Recent Development of High-Speed Conventional Wheels

Segmental wheel research first began with conventional wheels in the 1970s as part of an effort to evaluate the effect of high speed (Yamamoto 1972; Anon 1979; Abdel-Alim et al. 1980). However, the labor-intensive manufacturing costs were not competitive for the economic gains in productivity possible at that time. However, the recent development of ultrahigh porosity specifications (and therefore low density) using extruded SG abrasives has allowed the development of wheels with thick layers of conventional abrasive capable of operation at up to 180 m/s. An example is illustrated in Figure 4.8.

4.4.9 Safety of Segmented Wheel Designs

The last and most important benefit of segmental designs is safety. The two high-speed photographs below compare the failure of a conventional wheel (Figure 4.9) with a high-speed segmental wheel (Figure 4.10). The failure in the latter can just be seen between segments 11 and 12.

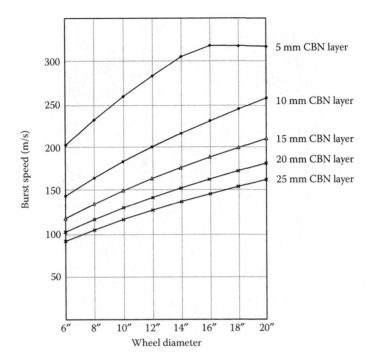

FIGURE 4.7
Segmented wheel burst speed as a function of wheel diameter and abrasive thickness (for defined vitrified bond strength).

FIGURE 4.8
Segmented vitrified Optimos™ Altos wheel containing needle-shaped ceramic (TG2) abrasive rated for 180 m/s. (Courtesy of Saint-Gobain Abrasives.)

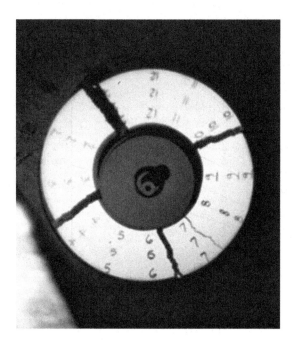

FIGURE 4.9
Solid wheel failure 90 m/s. Monolithic alox body.

FIGURE 4.10
Segmental wheel failure. 255 m/s steel core.

FIGURE 4.11
Wheel failure in a multiwheel cylindrical grinding operation.

The major difference in the two failures is the energy release. The results of a failure in a large conventional wheel can be extremely destructive for the machine tool as illustrated in the photograph (Figure 4.11) showing the aftermath of a multiwheel crankshaft journal wheel set failure. With this level of energy released, the situation is potentially very dangerous, costly, and time consuming even when well guarded. By comparison, the energy released by a segment failure, and the level of damage accompanying it, is very small. Any failure though is still unacceptable to the wheel maker or end user.

Figure 4.11 compares the energy release in wheel failure for a quadrant of a conventional wheel compared with the energy release for a segment of a segmented wheel.

4.4.10 Speed Rating of Grinding Wheels

Wheels are speed tested by overspinning the wheel by a factor prescribed by the appropriate safety code for the country of use. In the United States, ANSI B7.1 specifies that all wheels must be spin tested with an overspeed factor of 1.5 times operating speed. The theory behind this reverts back to conventional wheel research, where the 1.5 factor was proposed to detect preexisting flaws that might otherwise cause fatigue failure in the presence of moisture or water-based coolants during the expected life of the wheel (GWI 1983). In Germany, for the highest speed wheels, the DSA 104 code requires the wheel design be tested to withstand three times the operating stress. This gives an overspeed factor of $\sqrt{3}$ or 1.71. However, all production wheels must be tested at only 1.1 times operating speed. This code was due to the result of research in Germany that indicated that high overspeed factors could induce flaws that could themselves lead to failure. This discrepancy in the safety laws has a major impact on transfer of technology in the context of the global economy (Service 1991, 1993, 1996). The author has seen numerous examples of machine tools imported into the United States with incorrect spin test factors. In other countries with imported machine tools but no official safety code for wheel speed testing, the factor reverts back to the process as specified by the

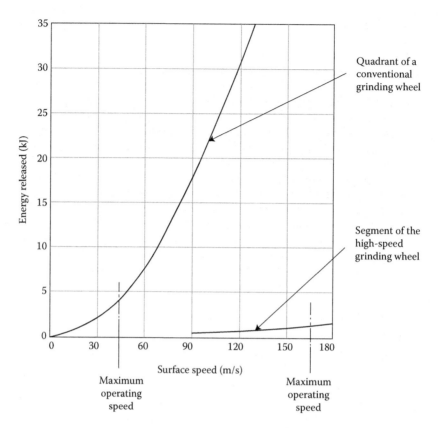

FIGURE 4.12
Energy release comparison for a conventional alox wheel with a steel-cored segmented wheel.

particular machine tool builder with the associated machine guarding on which the wheel is run.

Wheel makers must obey the safety codes of the appropriate country but they must also ensure above and beyond that the wheels are safe to the best of their abilities. For the Saint Gobain Abrasives (SGA) brands, the author has been associated with, the wheels are spin tested at 1.5× operating speed for the U.S. market per the ANSI standards but the wheels are designed and tested to >2× burst speed. For example, the wheel photographed in Figure 4.12 had a theoretical burst speed of 252 m/s. Seven wheels in total were tested and all failed at speeds between 255 and 273 m/s. As expected, failure occurred in the abrasive layer immediately adjacent to the epoxy bond, where the stress in the abrasive was highest (Hitchiner 1991).

4.5 Bond Life

Spin testing in itself, however, is not sufficient. Higher stresses actually occur in the epoxy bond than in the bond. Fortunately, the bond strength of epoxy is about 10 times greater than the abrasive. However, epoxy is prone to attack by moisture and coolant and will weaken over time. Efforts have been made to seal the bond from the coolant (Kunihito et al. 1991) but these are generally ineffective and the wheel maker must have life data for his particular bonding agent in coolant. Since wheels may have a life of several years on the machine with spares held in stock for a comparable time, this data take considerable time to accrue. Currently, Saint-Gobain Abrasives recommends a maximum life of 3 years on the machine or 5 years total including appropriate storage without respin testing.

4.6 Wheel Mount Design

Holding the abrasive section together on the wheel body has already been discussed. A second problem is how to hold the wheel body on the machine spindle.

Centrifugal forces cause the wheel to expand radially both on the o.d. and on the bore. It must also *contract* axially. The problem is, therefore, to prevent movement of the wheel on the hub either by minimizing bore expansion/contraction or by maintaining sufficient clamping pressure on the wheel to resist torsional slippage.

4.6.1 A Conventional Wheel Mount

An example of a conventional wheel mount (ANSI B7.1 1988, Figure 43) is shown in Figure 4.13.

4.6.2 Use of Blotters

Blotters are required for conventional wheels to equalize the variations in pressure due to the effects of microasperities in the grit structure of the vitrified body. Failure to incorporate

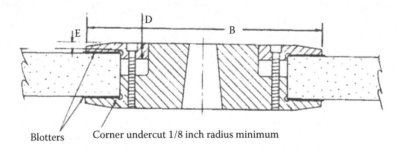

Blotters   Corner undercut 1/8 inch radius minimum

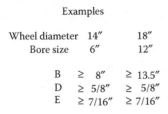

Examples

Wheel diameter	14″	18″
Bore size	6″	12″
B	≥ 8″	≥ 13.5″
D	≥ 5/8″	≥ 5/8″
E	≥ 7/16″	≥ 7/16″

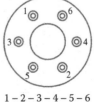

1 − 2 − 3 − 4 − 5 − 6

Bolt tightening sequence

FIGURE 4.13
Flange design for conventional wheels and bolt tightening sequence.

compressible blotters gives rise to local stress concentrations that are very dangerous. The blotters are made of either paper or plastic (polyester) with a thickness of typically 0.015 in.

4.6.3 Clamping Forces

The flanges are fixed together with a series of bolts, six in the example above, that are torqued to ≤20 ft lb in the sequence shown unless otherwise recommended. Overall clamping pressures must to be kept to <1000 psi and usually are considerably below this value.

Optimum torque values can lead to lower rotational stresses and higher burst speeds.

Barlow et al. (1995) carried out FEA analysis of clamping pressures and the effect on wheel stresses. An example showing the reduction in maximum radial stress is shown in Figure 4.14.

However, overtorquing causes distortion of the flange leading to a high stress peak that can readily exceed 2000 psi and lead to wheel failure (Meyer, R., 1996, Safe clamping of Cylindrical Grinding Wheels Consultant for ANSI B7.1 code Private communication 2/12/96). De Vicq (1979) recommended using tapered flange contact faces to compensate. Unfortunately, this is impractical except for dedicated machines and the accuracy of torquing methods is not always sufficient to ensure correct flange deflection.

In the absence of any significant axial contraction, it is relatively straightforward to calculate clamping forces required to prevent rotational slippage (Menard, J. C., 1983, Document WG6-4E, Calculations based on studies conducted in 1963 at the Technical High School of Hannover Germany. Private communication 5/3/1983).

1. Clamping force to compensate for the weight of the wheel:

$$F_{M} = \frac{M_{s}g}{\mu_{b}}$$

where M_s is the mass of the wheel, μ_b is the coefficient of friction for the blotter, paper blotter = 0.25, and plastic blotter = 0.15 (De Vicq 1979).

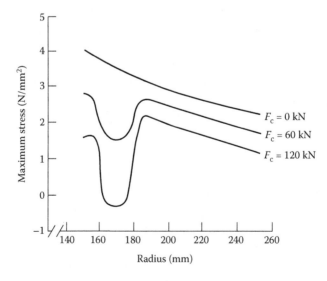

FIGURE 4.14
Effect of clamping force from a 175-mm radius flange on wheel stress.

2. Clamping force for unbalance of the wheel:

$$F_u = \frac{\delta v_s^2}{r_s^2 \mu_b}$$

where δ is the unbalance (force × distance), v_s is the wheel speed, and r_s = wheel radius.

3. Clamping force for motor power surge:

It is assumed that electric motors can develop a surge torque of 2.5 times their rated torque before stalling.

$$F_s = \frac{2.5 P_{sp} r_f}{v_s \mu_b r_s}$$

where r_f is the flange average radius and P_{sp} is the spindle motor power.

4. Clamping force for reaction of wheel to workpiece:

Again assume a motor surge capability of 2.5:

$$F_n = \frac{2.5 P_{sp}}{v_s \mu_b \mu_g}$$

where μ_g is the coefficient of grinding (0.4 typical).

In addition to these forces, there will be effects of accidental vibration and shocks, possible compression of the blotters, and the increased clamping force required as the wheel wears when holding constant surface footage. Practical experience leads to another factor 2 on forces. Total clamping force required becomes:

$$F_{total} = 2(F_m + F_u + F_s + F_n)$$

Knowing the number of bolts, tables are available giving torque/load values for the required clamping force. Calculations then need to be made to determine the flange deflection.

4.6.4 High-Speed Wheel Mounts

With high-speed, steel-cored wheels, the need for blotters is eliminated. Clamping is therefore steel on steel and not prone to the same brittle failure from stress raisers. Nevertheless, there is the uncertainty on wheel contraction and its effect on clamping. One solution is to eliminate the flanges entirely. Landis (Waynesboro, PA) developed a one-piece wheel hub, where the entire wheel body and tapered mount is a single piece of steel with the vitrified CBN segments bonded on to the periphery (Figure 4.15; Pflager 1997).

FEA analysis was carried out on a range of steel-cored wheel shapes as shown in Figure 4.16 based on 500 mm diameter × 20 mm o.d. face running at 6000 rpm (157 m/s).

4.6.5 The Single-Piece Wheel Hub

The straight one-piece hub was found to give the minimum level of bore expansion and is the design currently used in production (Table 4.6). The "turbine" or parabolic profile

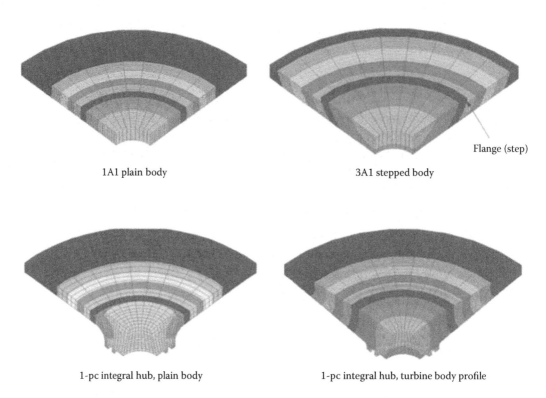

FIGURE 4.15
Optimization of hub design for minimal bore and diameter expansion.

minimizes o.d. expansion but at the expense of some additional bore (inner diamerter) expansion. It is also considerably more expensive to machine. This one-piece design concept has proved extremely successful in the crankshaft pin grinding and camshaft lobe grinding industries for speeds in the range of 60–120 m/s. It has eliminated the need for automatic balancers and allows fast change over times for lean manufacturing with minimal or no redress requirements.

4.6.6 Direct Mounting on the Spindle

Nevertheless, the design still has problems with bore expansion albeit now directed to a movement on a keyed tapered arbor. For the very highest wheel speeds, original equipment

TABLE 4.6

Dimensional Changes in a High-Speed Rotating Steel-Cored Wheel

Body Shape	Outer Diameter Expansion (mm)	Inner Diameter Expansion (mm)	Axial Contraction (mm)
1A1 plain	47	29	3
3A1 stepped body	164	13	3
One-piece hub plain	41	9	5
One-piece turbine profile	35	14	6

FIGURE 4.16
Example of one-piece integral hub design.

manufacturers and wheel makers are designing wheels to bolt directly to the motor spindle. Several examples for both vitrified and plated CBN wheels are shown below.

Figure 4.17 shows a typical steel-cored wheel for grinding camshafts at speeds up to 160 m/s. This particular example was made by TVMK for operation on a TMW camlobe grinder. The wheel has a small 40 mm hole governed partly by the motor spindle shaft size but also allows a bore tolerance of ±2.5 µm to be practicable. There are a large number of bolts (10) to hold the wheel thus allowing a high clamping force to be achieved. Since the wheel face is flat and flush to the spindle, there is no flange distortion. The bolt circle diameter is also small (68 mm) which minimizes problems from motor surge leading to slippage during start up. The body is twice as wide as the CBN section and tapered down toward the o.d. This reduces the bore expansion without creating some of the stress distortions seen with the 3A1 shape above. The vitrified CBN layer depth is 5 mm with a total layer of abrasive of about 6.5 mm.

4.6.7 CFRP Wheel Hubs

A 200 m/s is considered the current limit for steel-cored vitrified CBN wheels due to stresses in the abrasive layer from core expansion. This will vary somewhat depending on the particular bond and layer depth. However, for speeds of 160 m/s and greater, the steel is sometimes replaced with a material of comparable elastic modulus but one-third the density, namely CFRP or even titanium. This reduces wheel expansion by a factor of 3. Several wheel suppliers offer high-speed wheels with CFRP hubs in their literature. However, the cores are expensive if provided with the appropriate carbon fiber reinforcement level. CFRP hubs with lower CF content are available for lower wheel speeds that offer purely weight benefits. The primary problem with mounting a carbon fiber center is that the fibers are layered mats with a high Young's Modulus in the radial direction but a low compressive modulus axially. Either a steel flange ring is required for the bolt heads to lock against or steel inserts must be added into countersunk bolt holes.

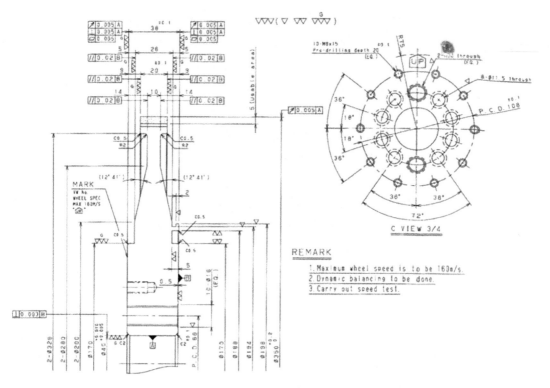

FIGURE 4.17
Steel-cored vitrified CBN wheel for frinding camshafts at 160 m/s.

4.6.8 Electroplated Wheels

Electroplated CBN wheels have been developed for considerably higher wheel speeds than vitrified CBN. The plated layer can withstand greater expansion of the hub. Research was reported as early as 1991 by Köenig and Ferlemann (1991) at 500 m/s using Winter wheels, while Tyrolit recently also offered a similar product design rated for 440 m/s in its literature.

The wheel design described by Köenig and Ferlemann has several novel features including the use of light-weight aluminum alloy for the hub material, a lack of a bore hole to further reduce radial stress, and an optimized wheel body profile based on turbine blade research to give the minimum wheel mass for uniform strength. Although this technology has been available for 10 years, there are only a limited number of machines in actual production running over 200 m/s. The radial expansion at 500 m/s is about 160 μm. A plated CBN layer can withstand the expansion at high speed, but the expansion is rarely perfectly uniform. Even a 1% difference due to any anisotropy in the hub material will lead to regenerative chatter and performance issues well before the expected life of the wheel (Figure 4.18).

4.6.9 Aluminum Hubs

Aircraft grade aluminum alloys are used as hub materials for some high-speed vitrified CBN. The obvious attraction is the lower density relative to steel. Various grades are

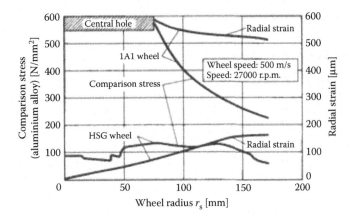

Stress and strain equalization of a 1A1 wheel compared to
an optimized wheel

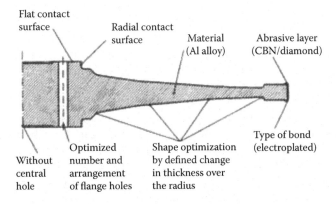

Main points of wheel optimization

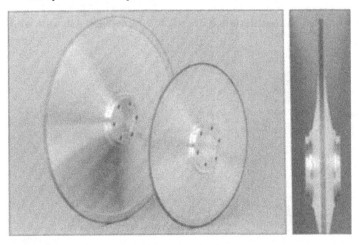

FIGURE 4.18
Optimized wheel shapes for high-speed plated CBN. (From Köenig, W., Ferlemann, F., 1991, CBN grinding at
five hundred m/s. *IDR*, 2, 72–79. With permission.)

available with tensile strengths of 120–140 kpsi. However, they have higher thermal expansion and appear to give more size and stability problems.

4.6.10 Junker Bayonet-Style Mounts

Other methods have been developed to compensate for bore expansion. Erwin Junker Maschinenfabrik developed a patented bayonet-style cam and follower three-point type mount. Various forms of the design are shown in Figure 4.19. The first consists of three-roller bearings inserted into the bore, and the second is the later, more common, version consisting of a hardened steel ring insert with three premolded raised areas as detailed. The design assures <1.25 µm runout repeatability at speeds up to 140 m/s (Junker 1992, n.d.).

4.6.11 HSK Hollow Taper Mount

Another method is the incorporation of the increasingly popular hollow taper shank (HSK) tool-holder shank. The HSK system was developed in the late 1980s at Aachen T.H., Germany as a hollow tapered shaft capable of handling high-speed machining. It became a shank standard in 1993 with the issuance in Germany of DIN 69893 for the hollow 1:10 taper shank together with DIN 69063 for the spindle receiver (Figures 4.20 and 4.21).

This system has seen a rapid growth, especially in Europe, as a replacement to the various 7:24 steep taper shanks known as the CAT or V-flange taper in the United States (ASME-B5.10 1994), the SK taper in Germany (DIN69871), and the BT taper in Japan (JISC-B-6339/BT.JISC). These tapers previously dominated the CNC milling industry.

The HSK system actually consists of a family of shank sizes from 25 to 160 mm flange diameter and designs from A through F, of which the HSK-A is the most common for grinding applications with flange diameters from 80 to 125 mm. Muller-Held reported that the HSK system could limit maximum position deviations to 0.3 µm compared with 2 µm for an SK taper (Muller-Held 1998). It is said to be three times more accurate in the X and Y planes and 400 times better in the Z axis. Bending stiffness was seven times better, while the short length of the taper allowed faster tool change times. The important detail though for this discussion is the fact that the design has a *hollow* taper that expands under centrifugal load to aid the maintenance of contact. However, Aoyama and Inasaki (2001)

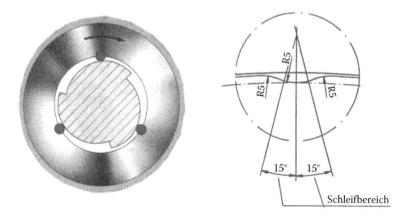

FIGURE 4.19
Junker patented bayonet-style mount systems for high-speed wheels.

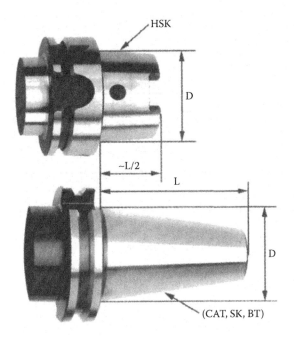

FIGURE 4.20
Comparison of the HSK mount with earlier CAT and other taper systems.

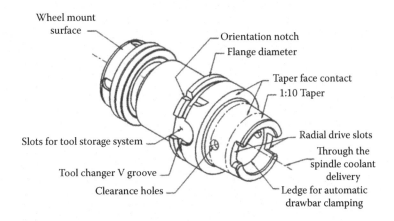

FIGURE 4.21
Elements of a typical HSK tool mount for automatic tool changing.

reported that radial stiffness reduces with increased rotational speed but is still far superior to any other taper mounting system.

The HSK wheel mount system has been widely adopted for hybrid grinding/metal cutting machines. Regular 1A1 style plated wheels have been routinely run at up to 140 m/s on HSK taper arbors. At least one research machine has been built using a one-piece hub design incorporating HSK mounting for use up to 250 m/s. Currently, the primary hesitation in broader adoption of this technique is the cost, availability, and delivery due to the limited number of capable high-precision manufacturing sources (Figure 4.22).

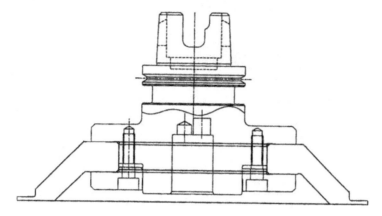

FIGURE 4.22
Plated CBN groove grinding wheel mounted on HSK adaptor.

4.6.12 Titanium Hub Design

As a final note, Ramesh (2001) reported using titanium flanges as a wheel mount option in a thesis on high-speed spindle design and grinding. This design is shown in Figure 4.23.

The wheels are made with steel cores and without a center hole being held by titanium flanges clamping to shoulders on the wheel. Titanium has comparable strength to high-tensile steel but has one-third the density. It is, therefore, expected that the wheel core will try to expand more than the flanges at high speeds and therefore the radial clamping will increase.

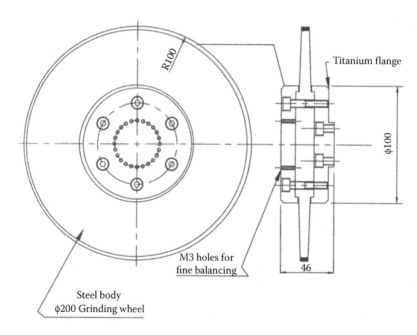

FIGURE 4.23
Mount method for high-speed wheels using titanium flanges. (From Ramesh, K., 2001, Towards grinding efficiency improvement using a new oil-air mist lubricated spindle. PhD thesis, Nanyang Technical University, Singapore. May. With permission.)

4.7 Wheel Design and Chatter Suppression

Chatter is an ever-present problem in grinding. Many claims have been made that the design of the wheel, especially regards the use of hub materials with high *damping* characteristics used in conjunction with superabrasive wheels, can suppress its occurrence (Broetz 2001; Tyrolit 2001). The reality is much more complicated and requires a brief discussion of the sources of vibration and chatter in grinding.

4.7.1 The Role of Damping

The basic equation for motion of a single degree of freedom system is given by:

$$x''(t) + 2\zeta\omega_n x'(t) + \omega_n^2 x(t) = F(t) \tag{4.1}$$

where ζ is the damping factor and ω_n is the natural frequency.

In the absence of damping, energy is exchanged without loss during the course of motion at particular natural frequencies and the amplitude of vibration will build over time depending on the rate of input of energy. Damping absorbs energy either through internal friction of the particular material or more often in joints and seams. Prediction of the damping of a machine is not possible, although damping in a particular mode of vibration may be determined empirically by use of a hammer test and measuring the decay rate. Damping is related to decay rate according to:

$$\delta = \frac{(2\pi\zeta)}{(1-\zeta^2)^{1/2}} = \text{logarithmic decrement} \tag{4.2}$$

4.7.2 Forced and Self-Excited Vibrations

The source of the energy that creates vibration can be either external leading to forced vibration or inherent in the instability of the grind process leading to self-excited vibration.

1. *Forced vibrations*: Forced vibrations can be eliminated in three ways. First is to eliminate the energy at its source. Wheels, motors, belts, and workpieces should all be balanced as should the three-phase power supply. Ultraprecision bearings and ballscrews should be used and properly maintained in work and wheel spindles and slides. Second, the grinder should be insulated from sources of vibration such as hydraulic and coolant pumps, and vibrations carried through the foundations. Third, where resonances cannot be eliminated, the machine dynamics must be modified. Where a particularly prominent frequency exists in, for example, a motor or cantilevered member, tuned mass dampers consisting of a weight with a damped spring can be fitted at the point, where the vibration needs to be reduced. This may often consist of a weight attached to the member via a rubber sheet sandwiched between. The sizes of the mass, spring, and damper are selected so that the mass oscillates out of phase with the driving frequency and hence dissipates energy. A relatively small tuned mass can have a large effect in reducing vibration amplitude.

2. *Self-excited vibration*: Self-excited chatter occurs only during grinding and the amplitude of vibration climbs with time. A small perturbation due to instability

in the system causes a regular variation in grinding forces that in turn creates an uneven level of wear around the wheel. The process is thus regenerative. There are several methods available for suppressing this form of chatter.

First, the system stiffness and damping can be increased. Second, the grinding conditions can be continuously varied by changing the workspeed, wheel speed, work support compliance, or by periodically disengaging the wheel from the workpiece. For example, Gallemaers et al. (1986) reported that by periodically varying the workspeed to prevent lobe build up on the wheel they could increase the G ratio by up to 40% and productivity (by extending the time between dresses) by up to 300%. Third, the stiffness of the contact area can be reduced to shift the state of the system more toward a stable grinding configuration (Snoeys 1968). Fourth is the use of various filter effects to reduce the wavelength to less than the contact width. The workspeed can be slowed to the point the chatter lines merge. Alternatively, and more interesting, the frequency of the chatter can be increased to the point that the grinding process itself acts as a filter to absorb the vibration energy. For this reason, the natural frequency of wheels is targeted at >500 Hz or ideally >1000 Hz.

4.7.3 Damped Wheel Designs and Wheel Compliance

It is interesting to note that most scientific studies on "damped" wheel designs are based on suppressing self-excited chatter. Furthermore, they all use hub materials that not only have good damping characteristics but are also considerably more compliant and light weight than "standard" hub materials such as steel. As the following examples illustrate, it is not only damping that is important for reducing vibrations. Reducing stiffness at the wheel contact has a similar effect. The analysis of chatter with added compliance at the wheel contact is given in Chapter 19 on centerless grinding. Sexton et al. reported excellent results in reducing chatter when grinding steel with resin CBN wheels by the use of a "Retimet" nickel foam hub material with a radial stiffness of 0.5 N/μm mm (Sexton et al. 1982). This was compared with values of 4–10 N/μm mm for standard phenolic (Bakelite) or aluminum filled phenolic hubs. McFarland et al. used polypropylene with a radial stiffness of 1.56 N/μm mm and natural frequency of 1169 Hz (McFarland et al. 1991). This was compared to a radial stiffness for an aluminum hub of 24 N/μm mm. Warnecke and Barth compared the performance of a resin-bonded diamond wheel on a flexible phenolic aluminum composite hub with a similar bond on an aluminum hub grinding SiN and demonstrated an improvement in life of over 70% (Warnecke and Barth 1999). FEA analysis of the contact zone revealed over twice the radial deflection with the flexible hub. It would appear that compliance in the hub can be transferred through a resin superabrasive layer and can significantly increase contact width (see also Zitt and Warnecke 1996).

In 1989, Frost carried out an internal study for Unicorn (Saint-Gobain Abrasives) to evaluate the impact of the higher stiffness of vitrified CBN bonds on the centerless grinding process (Frost 1989). The following radial contact stiffness values were obtained for conventional and CBN vitrified specifications:

47A100 L6YMRAA	0.06 N/μm mm
5B46 P50 VSS	0.78 N/μm mm
5B76 P50 VSS	0.31 N/μm mm

The stiffness of the CBN bonds was an order of magnitude greater than the conventional bond, and approached or exceeded that of the hub materials described above.

It would therefore be expected that as with the example earlier of resin-bonded diamond wheels, flexible hubs with radial stiffness values of the order of 0.5 N/μm mm could increase the contact width in the grind zone for wheels with a thin-vitrified CBN layer. Further analysis of the effect of wheel compliance on chatter in centerless grinding is given in Chapter 19.

4.7.4 Wheel Frequency and Chatter

The effect of wheel frequency on chatter was experienced firsthand by the author, while developing a process for grinding large diameter thin-walled casings with vitrified CBN. The project was initially prone to extreme chatter and noise. Maximizing the stiffness and nodal frequency of a steel-cored wheel by reducing the diameter by 30% and then doubling the body width increased wheel life by an order of magnitude and reduced noise by >20 dB. However, subsequently changing the hub material from steel to CFRP of comparable stiffness but one-third the density further doubled wheel life. CFRP would have provided some additional damping but also significantly increased the natural frequency of the wheel due to its low density.

4.7.5 Summary

In conclusion, light-weight flexible hubs can provide benefit in grinding by limiting self-excited chatter generation with superabrasive wheels. Damping may also be an issue, but hub compliance and frequency responses are more likely to be the controlling factors. The concept is unlikely to be effective where significant forced vibration is present, although it is sometimes difficult to differentiate the two.

Compliance is much higher in conventional wheels. The effect on contact width for suppressing chatter is particularly pronounced when using plastic bonds for camshaft grinding or shellac bonds for roll grinding. Some benefit is even seen using rubber inserts in the bores of vitrified alox wheels for roll grinding. Even with resin diamond wheels, Busch (1970) was able to show a 300% improvement in life merely placing a rubber sleeve between the wheel and flange to increase compliance.

Further research is likely to be focused on this aspect of wheel design as superabrasive technology targets applications such as roll and centerless grinding. It should be noted that efforts have been published regards commercial product introducing microelasticity into vitrified diamond and CBN bonds (Graf 1992). A more comprehensive review of the whole subject of grinding chatter excluding centerless grinding is given by Inasaki et al. (2001).

References

Abdel-Alim, A., Hannam, R. G., Hinduja, S., 1980, A feasibility analysis of a novel form of high speed grinding wheel. *21st International Machine Tool Design and Research, Conference. Swansea, Wales, UK.*

Anon, 1979, High-speed plunge grinding. *Manufacturing Engineering,* June, 67–69.

Aoyama, T., Inasaki, I., 2001, Performance of HSK tool interfaces under high rotational speed. *Annals of the CIRP,* 50, M09.

Barlow, N., Rowe, W. B., 1983, Discussion of stresses in plain and reinforced cylindrical grinding wheels. *International Journal of the Machine Tool Design and Research,* 23(2/3), 153–160.

Barlow, N., Jackson, M. J., Mills, B., Rowe, W. B., 1995, Optimum clamping of CBN and conventional vitreous-bonded cylindrical grinding wheels. *International Journal of the Machine Tools and Manufacture*, 35/1, 119–132.

Barlow, N., Jackson, M. J., Hitchiner, M. P., 1996, Mechanical design of high-speed vitrified CBN grinding wheels. In I.D. Marinescu (Ed.), *Manufacturing Engineering: IMEC Conference Proceedings*, 568–570, UK.

Broetz, A., 2001, Innovative grinding tools increase the productivity in mass production: Grinding of crankshafts and camshafts with Al2O3 and CBN grinding wheels. *Precision Grinding and Finishing in the Global Economy 2001, Oak Brook IL Conference Proceedings*, Gorham.

Busch, D. M., 1970, Machine vibrations and their effect on the diamond wheel. *IDR*, 30/360, 447–453.

De Vicq, A. N., 1979, *An Investigation of Some Important Factors Affecting the Clamping of Grinding Wheels under Loose Flanges*. Machine Tool Industry Research Association, Macclesfield, UK.

Frost, M., 1989, *An Evaluation of Two Experimental CBN Wheels for Use in Centerless Grinding*. University of Bristol Project report for Unicorn Industries, Bristol University, UK.

Gallemaers, J. P., Yegenoglu, K., Vatovez, C., 1986, Optimizing grinding efficiency with large diameter CBN wheels. *International Grinding Conference*, Philadelphia, USA SME86-644.

Graf, W., 1992, CBN- und Diamantschleifscheiben mit mikroelastischer Keramikbindung. (CBN and diamond grinding with micro-elastic ceramic bond.) *VSI-Z-Special*, Werzeuge, 72–74.

GWI, 1983, *Fatigue Proof Test Procedure for Vitrified Grinding Wheels*. Grinding Wheel Institute.

Hitchiner, M. P., 1991, Systems approach to production grinding with vitrified CBN. *Superabrasives' 91 Conference Proceedings, SME*.

Inasaki, I., Karpuschewski, B., Lee, H. S., 2001, Grinding chatter—Origin and suppression. *Annals of the CIRP Keynote Paper*, 50, 2.

Junker, n.d., Junker group international. *Commercial Company Presentation*.

Junker, M., 1992. A new era in the field of o.d. grinding. Trade brochure.

Köenig, W., Ferlemann, F., 1991, CBN grinding at five hundred m/s. *IDR*, 2, 72–79.

Kunihito et al., 1991, Segmented grinding wheel. *EP*, 433, 692 A2.

McFarland, D. M., Bailey, G. E., Howes, T. D., 1991, The design and analysis of a polypropylene hub CBN wheel to suppress grinding chatter. *Transactions of ASME*, 121, 28–31.

Muller-Held, B., 1998, Development of a repeatable tool-holder based on a statically deterministic coupling. MIT/RWTH Aachen project report.

Pflager, W. W., 1997, Finite element analysis of various wheel configurations at 160 m/sec. Landis Internal Report 9/23/97.

Ramesh, K., 2001, Towards grinding efficiency improvement using a new oil-air mist lubricated spindle. PhD thesis, Nanyang Technical University, Singapore. May.

Service, T., 1991, Safe at any speed. *Cutting Tool Engineering*, June, 1991, 99–101.

Service, T., 1993, Rethinking grinding wheel standards. *Cutting Tool Engineering*, December, 26–29.

Service, T., 1996, Superabrasive safety. *Cutting Tool Engineering*, June, 22–27.

Sexton, J., Howes, T. D., Stone, B. J., 1982, The use of increased wheel flexibility to improve chatter performance in grinding. *Proceedings of the Institution of Mechanical Engineers*, 1, 196, 291–300.

Snoeys, R., 1968, Cause and control of chatter vibration in grinding operations. *Technical Society for Tool and Manufacturing Engineers*.

Tyrolit Schleifmittel Swarovski, 2001, Grinding disk comprises an intermediate vibration damping ring which is made as a separate part of impregnated high-strength fibers, and is glued to the central carrier body and/or the grinding ring. Patent DE 20102684 4/19/01.

Warnecke, G., Barth, C., 1999. Optimization of the dynamic behavior of grinding wheels for grinding of hard and brittle materials using the finite element method. *Annals of the CIRP*, 48, 1, 261–264.

Yamamoto, A., 1972, A design of reinforced grinding wheels. *Bulletin of JSPE*, 6/4, 127–128.

Zitt, U., Warnecke, G, 1996, The influence of a new hub material concept on process behavior and work result in high performance grinding processes with CBN. *Abrasives Magazine*, April/May, 16–24.

5

The Nature of the Abrasive

5.1 Introduction

The abrasive grain is at the heart of the grinding process, where its shape, mechanical, thermal, and chemical properties are paramount to the wheel performance. Although an abrasive grain does not have the optimized orientation and geometry of a turning insert, there are many active grains on the face of a wheel and the average of these is surprisingly predictable; in many ways, more so than a turning operation where the ultimate performance of a single cutting edge is far less predictable being subject to random variations of production, where a single-edge chip can spell the end of the tool life.

Grain shape, for example, makes an enormous impact on grain strength, grinding performance, and packing characteristics that in turn impact wheel formulation and manufacture. Shape and size are interlinked especially for particles of indeterminate shape, that is, not perfectly spherical, cubic, etc. Synthetic diamond particles, for example, can exist in an infinite combination of particle shapes derived from the transition between octahedral, dodecahedron, and cubic forms. Moreover, crystal imperfections, and polycrystalline particles, further add to the variety of diamond forms (Figure 5.1).

Two key diametral dimensions are the major and minor diameters, d_a and d_b, respectively. From these two dimensions, the most common shape parameter is *aspect ratio* or *ellipticity*, and the ratio of the major to minor diameter is d_a/d_b. Additional factors include the *projection area*, the area enclosed by the boundary of its projection and *convexity*. A grain is convex if an idealized elastic membrane stretched across its projection leaves space between itself and the grain's surface. The higher the degree of convexity, the lower the mechanical integrity but the higher the abrasive aggressiveness. Convexity correlates closely with the notion of grain irregularity. Convexity C as a parameter is defined as:

$$C = \frac{A_f + A_p}{A_p}$$

where
  A_p = the projected area of the grain,
  A_f = the fill area between the grain projection and the idealized elastic membrane stretched across the projection. Finally, grain *sharpness* is a parameter that has been developed specifically for the characterization of abrasive grains based on chip formation modeling based on the shape of the projected area of the abrasive and depth penetrated into the workpiece (Figure 5.2).

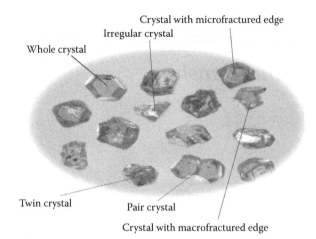

FIGURE 5.1
Diamond morphologies in typical commercial grain. (Courtesy of FACT Diamond.)

Various techniques for measuring grain shape parameters including aspect ratio, convexity, and sharpness are reported by De Pellegrin et al. (2002, 2004, 2008). Dullness which is the converse of sharpness is also considered by Rowe (2014).

The 2-dimensional (2D) projections of two most extreme samples (blocky → angular) are shown in Figure 5.2. The difference in shape is visually apparent, but the strength of the sharpness technique is that it is able to quantify much subtler differences in particle abrasiveness. Wear tests of grinding wheels fabricated from these particles are also included. Aspect ratio, convexity, and sharpness are increasingly sensitive predictors of

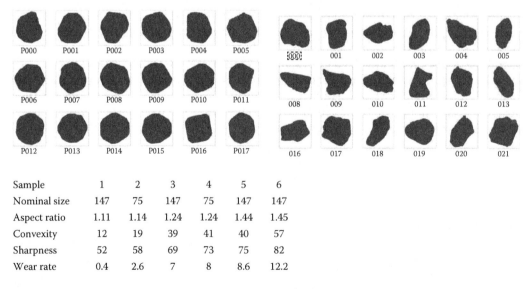

Sample	1	2	3	4	5	6
Nominal size	147	75	147	75	147	147
Aspect ratio	1.11	1.14	1.24	1.24	1.44	1.45
Convexity	12	19	39	41	40	57
Sharpness	52	58	69	73	75	82
Wear rate	0.4	2.6	7	8	8.6	12.2

FIGURE 5.2
Examples of grain shape analysis.

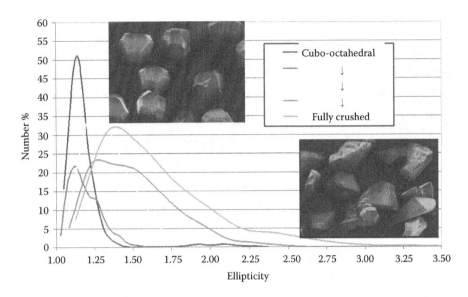

FIGURE 5.3
Influence of diamond morphology on ellipticity.

wear rate and grinding forces. In general, the tangential grinding force is proportional to the cross-sectional area of grain independent of material, grain type, or speed and is the basis of undeformed chip thickness models, but the grain shape has a significant impact on normal force that will impact threshold force and work removal parameter.

Various methods of crushing of grain have a dramatic effect on both the average value of aspect ratio and the range of aspect ratios within a sample (Figure 5.3). Crushing grain can dramatically increase cut rates but at the cost of increasing wheel wear.

Traditionally, attention to shape has been more focused on superabrasives, particularly synthetic diamond where aspect ratio ranges from about 1.1 to 1.5, but this is rapidly shifting with advances in engineered ceramic-shaped grains, where aspect ratios can be as high as 8:1. Shape parameters are being challenged in characterizing appropriately the engineered shapes of recent grain developments exemplified in Figure 5.4.

Grain is sized and measured in the grade range 46#–400# (400–40 μm), which covers the "precision" grinding portion of the market, by sieving techniques. Sieving consists of shaking the dried, free flowing, grain through a series of sieves of decreasing aperture sizes. The distribution is defined in terms of the mass portion retained on a particular sieve or as the cumulative mass portion retained on all sieves above a given sieve size. The method will categorize grain size based on the minimum cross section, that is, 2D shape which will lead to a much longer third dimension making it through a sieve for a high aspect ratio particle shape. To allow this, the approach taken by wheel makers to specifying a grit size in a wheel specification appears to give a grain size based on that expected for a given surface finish on the ground part and facilitate direct substitution in a factory environment.

Several National and International Standards define particle size distributions of abrasive grains. All are based on sizes by sieving in ranges typical of most regular grinding applications.

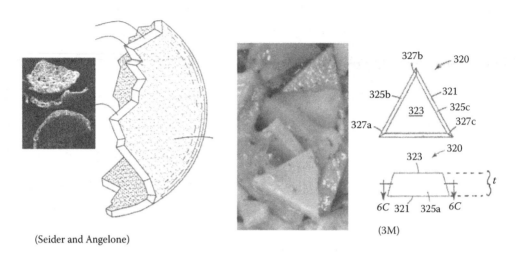

(Seider and Angelone)   (3M)

FIGURE 5.4
Extreme examples of engineered grain shapes.

The American National Standards Institute (ANSI) standard. In the case of the ANSI standard (B74.16 1995), mesh size is defined by a pair of numbers that correspond to sieves with particular mesh sizes. The lower number gives the number of meshes per linear inch through which the grain can only just fall, while staying on the surface of the sieve with the next highest number of meshes which is the higher number.

The Federation of the European Producers of Abrasives (FEPA) standard. The FEPA (ISO R 565-1990, also DIN 848-1988) gives the grit size in microns of the larger mesh hole size through which the grit will just pass.

Comparison of FEPA and ANSI standards. The FEPA and ANSI sizing standards are closely related. FEPA has a tighter limit for oversize and undersize (5%–12%) but no medium nominal particle size. ANSI has somewhat more open limits for oversize and undersize (8%–15%) but a targeted midpoint grit dimension. FEPA is more attuned to the superabrasive industry, especially in Europe, and may be further size controlled by the wheel maker. ANSI is more attuned to conventional wheels and in many cases may be further broadened by mixing two or three adjacent sizes. In tests, no discernable difference could be seen between wheels made using grain to the FEPA or ANSI size distribution (Hitchiner and McSpadden 2004). A major attraction of working with the FEPA system is that it provides a measure of the actual size of the grain in microns, whereas with the ANSI system, the mesh size increases with the numbers of wires in the sieve mesh and therefore becomes larger as the grain size becomes smaller. The grain size can be calculated very roughly from:

$$\text{Mesh size} = 0.6/\text{grain size(inch)}$$

U.S. grit size number. There is also a system called U.S. grit size number with a single number that does not quite correlate with either the upper or lower ANSI grit size number. This has created considerable confusion especially when using a single number in a specification. An FEPA grit size of 64 could be equivalent to a 280, 230, or 270 U.S. grit size depending on the wheel manufacturer's particular coding system. This can readily lead to error of one grit size when selecting wheel specifications unless the code system is well defined. Table 5.1 gives the nearest equivalents for each system.

TABLE 5.1

Grain Size Designations

	Sieve Sizes					
Size (mm)	U.S. Mesh	U.S. Grit Size	Europe FEPA	Japanese JIS	Chinese Standard	Russian Standard
5600						
4750	4					
4000	5					
3350	6					
2800	7					
2360	8	8/10				
2000	10	10/12				
1700	12					
1400	14	12/24				
1300		14/16				
1180	16	16/18	D1181		16/18	1250/1000
		16/20	D1182		16/20	
1000	20	18/20	D1001		18/20	1000/800
850	22	20/25	D851		20/25	
		20/30	D852	20	20/30	
710	24	25/30	D711		25/30	800/630
600	30	30/35	D601		30/35	630/500
		30/40	D602	30	30/40	630/400
500	36	35/40	D501		35/50	500/425
		35/40	D502			
425	40	40/45	D426		40/45	
		40/50	D427	40	40/50	
355	46	45/50	D356		45/50	400/315
300	54	50/60	D301	50	50/60	315/250
250	60	60/70	D251	60	60/70	250/200
		60/80	D252		60/80	250/160
212	70	70/80	D213		70/80	
180	80	80/100	D181	80	80/100	200/160
150	90	100/120	D151	100	100/120	160/125
125	100	120–240	D126	120	120/140	125/100
106	120	140/170	D107	140	140/170	
		170/200	D91	170	170/200	100/80
75	150	200–230	D76	200	200/230	80/63
63	180	230–270	D64	230	230/270	63/50
53	220	270/325	D54	270	270/325	50/40
45	240	325/400	D46	325	325/400	

(Contniued)

TABLE 5.1 (*Contniued*)

Grain Size Designations

	Micron Sizes			
Size (mm)	U.S. Mesh	U.S. Grit Size	Europe FEPA	Japanese JIS
50		400–500	40–60	400
45			40–50	500
38	280	500–600	25–50	
40			35–45	550
38		700–800	36–40	
35			30–40	600
29	320	600–700	22–36	700
25			20–30	800
20	400		15–25	1000
15	500		10–20	1200
10		1800	8–12	1800
8	600	2200	6–10	2000
7.5			5–10	2200
6	900	3000	4–8	3000
5			4–6	4000
4		6000	2–6	5000
3.5			2–5	7000
3		8000	2–4	8000
2			1–3	11000
1.5			0–3	12000
1.5			1–2	13000
1		15000	0–2	14000
0.5		25000	0–1	60000
0.25			0–0.5	100000
0.0125			0–0.25	200000

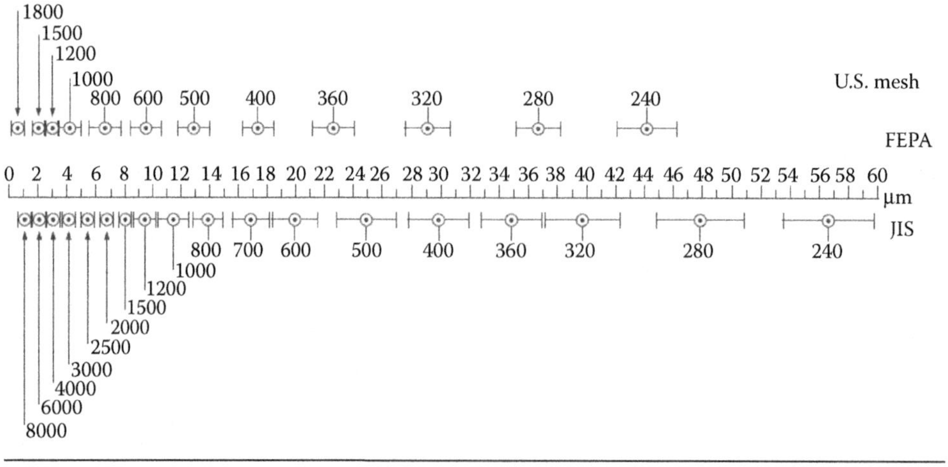

Other standards such as the Japanese Industrial Standards (JIS) tend to follow ANSI with only minor differences except at very fine micron sizes where the JIS system can give much higher mesh numbers than elsewhere again creating confusion when looking for local sourcing in a global market.

Standard grain distributions may be further subsieved and size sorted for specific application such as high precision electroplated superabrasives. It is also very common, especially for alox wheels to blend sizes to broaden the size distribution. This aids packing density to strengthen wheel structures, adjusts pore distribution and homogeneity, and increases cutting point density for improved finish. These blends will often include both different sizes and grain types. For example, Kitajima (1994) has shown that blending high toughness and medium toughness cubic boron nitride (CBN) in a suitable ratio could significantly improve ability to maintain surface finish. The size distribution may also be adjusted by using blends of grain two to three sizes apart, where the finer grain is of a "weaker" grade but is actually comparable in toughness allowing for size effects (grain toughness falls off with size as larger grains have larger or a higher probability of flaws).

Grain sizes in the micron or ultraprecision range, that is, <500# or <40 μm are produced by alternate separation techniques such as elutriation/sedimentation or dry cyclonic methods.

Sieving for measuring size distribution also gives way to alternate techniques such as sedimentation, that is, looking at the rate the particles fall through a liquid sometimes combined with centrifuging, or by laser diffraction or dynamic light scattering. Each technique relies upon different specific characteristics of the grain and will give slightly different results affected by factors such as turbidity, Brownian motion, and diffraction patterns. They therefore offer precision of measurement over accuracy.

Modern grinding abrasives mainly fall into one of two groups, namely:

1. *Conventional abrasives* based either on silicon carbide (SiC), alumina (Alox), or zirconia–alumina.
2. *Superabrasives* based either on diamond or on CBN.

The division into two groups is based on a dramatic difference in hardness of the grains leading to very different wheel wear characteristics and grinding strategies. The division is also based on cost; wheels made using superabrasives are typically 10–100 times more expensive. That range however has recently narrowed as advances in alox-based technologies in particular have outpaced new technologies in superabrasive grain.

5.2 Silicon Carbide

5.2.1 Development of SiC

SiC was first synthesized in 1891 by Dr. E. G. Acheson who gave it the trade name "Carborundum." It was initially produced in only small quantities and sold for $0.40/ct or $880/lb as a substitute for diamond powder for lapping precious stones. In its time, it might well have been described as the first synthetic "superabrasive"; certainly compared to the natural emery and corundum minerals then otherwise available. However, once a commercially viable process of manufacture was determined its price fell precipitously, and by 1938 it sold for $0.10/lb (Heywood 1938). Today the material costs between $0.80/lb and $3.00/lb depending on quality and particle size. Recent nongrinding uses of SiC such as for tank and body armor, and lapping in wafer manufacture, along with rising energy costs, have created major fluctuations in price and availability.

5.2.2 Manufacture of SiC

SiC abrasive is manufactured in an Acheson resistance heating furnace through the reaction of silica sand and coke at a temperature of around 2400°C. The overall reaction is described by the equation

$$SiO_2 + 3C \rightarrow SiC + 2CO$$

A large carbon resistor rod is placed on a bed of raw materials to which a heavy current is applied. The raw material also includes sawdust to add porosity to help release the CO, and salt to remove iron impurities. The whole process takes about 36 h and yields 10–50 tons of product. From the time, it is formed the SiC remains a solid as no melting occurs (SiC sublimates at 2700°C). After cooling the SiC is sorted by color; from green SiC which is 99% pure to black SiC which is 97% pure. Green dominates in fusions using virgin feedstock, while black dominates with recycled feedstock. It is then crushed and sized as described for alumina below.

5.2.3 Hardness of SiC

SiC has a Knoop hardness of about 2500–2800 and is very friable. The impurities within the black grade increase the toughness somewhat but the resulting grain is still significantly more friable than alumina. Above 750°C, SiC shows a chemical reactivity toward metals with an affinity for carbon, such as iron and nickel. This limits its use to grinding hard, nonferrous metals. SiC also reacts with boron oxide and sodium silicate, common constituents of vitrified wheel bonds (Viernekes 1987). SiC also shows reactivity with titanium in air through the routes:

$$SiC + O_2 \rightarrow SiO_2 + C$$

$$SiC + 2O_2 \rightarrow SiO_2 + CO_2$$

$$Ti + SiO_2 + (O_2) \rightarrow TiO_2 + SiO_2$$

$$Ti + SiC + (O_2) \rightarrow TiC + SiO_2$$

Despite this (or perhaps more by default because titanium reacts or shows adhesion at some level with every other abrasive!), and because SiC is so sharp/friable, it remains the preferred abrasive for titanium and its intermetallics such as γTiAl. It is also recommended on other nonferrous alloys as well as some cast irons and stainless steels in many cases blended with engineered ceramic grain.

5.3 Alumina (Alox)-Based Abrasives

Alumina-based abrasives are derived either from a traditional route of electrofusion, or more recently by chemical precipitation and/or sintering. Unlike SiC, alumina is available in a large range of grades because it allows substitution of other oxides in a solid solution, and defect content can be much more readily controlled. The crystallite size in the individual abrasive grains can be closely regulated from nanometer dimensions up to that of the grain itself. Shape, strength, fracture modes, and even fluid permeability can be "engineered" to a greater degree than with any other abrasive composition.

The following description of alumina-based abrasives is classified into electrofused alumina abrasives, chemically precipitated or sintered alumina abrasives, and agglomerates.

5.4 Electrofused Alumina Abrasives

5.4.1 Manufacture

The most common raw material for electrofused alumina is bauxite, which depending on source contains 85%–90% alumina, 2%–5% TiO_2, and up to 10% of iron oxide, silica and basic oxides. The bauxite is fused in an electric-arc furnace at 2600°C using a process demonstrated by Charles Jacobs in 1897 but first brought to commercial viability under the name "alundum" with the introduction of the Higgins furnace by Aldus C. Higgins of the Norton Company in 1904 (Tymeson 1953).

A Higgins furnace consists of a thin metal shell on a heavy metal hearth. A wall of water running over the outside of the shell is sufficient to maintain the shell integrity. A bed of crushed and calcined bauxite (mixed with some coke and iron to remove impurities) is poured into the bottom of the furnace and a carbon starter rod laid on it. Two or three large vertical carbon rods are then brought down to touch and a heavy current applied. The starter rod is rapidly consumed but the heat generated melts the bauxite, which then becomes an electrolyte. Bauxite is added continually over the next several hours to build up the volume of melt to as much as 20 tons. Current flow is controlled by adjusting the height of the electrodes which are eventually consumed in the process.

Perhaps, the most surprising feature of the process is the fact that a thin, water-cooled, steel shell is sufficient to contain the process. This is indicative of the low thermal conductivity of alumina, a factor that is also significant for its grinding performance as will be described in later sections. The alumina forms a solid insulating crust next to the steel. After the fusion is complete, the furnace is left to cool over the course of a day. During the solidification that starts from the shell advancing into the center, impurities are concentrated in the center and the material must generally be sorted prior to milling.

The resultant abrasive is called *brown-fused alumina* (BFA) and contains typically 3% TiO_2. Electrofused alumina is also made using low-soda Bayer process alumina that is >99% pure to produce *white-fused alumina* (WFA; Figure 5.5). As with BFA, when produced in a Higgins pot, the subsequent slow solidification process creates a range of WFA products throughout the ingot that must be sorted (Figure 5.6).

FIGURE 5.5
Higgins-type electric-arc furnace for fusion processes of alumina and zirconia. (Courtesy of Whiting Corp.)

Some modern furnaces in the United States and Canada pour the melt onto a water-cooled steel hearth to better control microstructure and reduce crystallite size. These furnaces are also especially prevalent in recent plant investments in China that now produce over 50% of the worlds fused alumina production. The microstructure and chemistry distribution are controlled by the size and depth of the pour pot (Figure 5.7).

Once cooled, the alumina is broken up and passed through a series of hammer, beater, crush, roller, and/or ball mills to reduce it to the required grain size. The type of crush process also controls the grain shape, producing either blocky or thin splintered grains. After milling, the product is sieved to the appropriate sizes down to about 40 μm (400#).

White-fused alumina (WFA) is the most popular grade for micron-sized abrasives in part because the crushing process concentrates impurities in the fines when processing other alumina grades. To produce micron sizes, the alumina is further ball milled or vibro milled after crushing and then traditionally separated into sizes using an elutriation process. This is achieved by passing a slurry of the abrasive and water through a series of vertical columns. The width of the columns is adjusted to produce a progressively slower vertical flow velocity from column to column. Heavier abrasive settles out in the faster flowing columns, while the lighter particles are carried over to the next. The process is effective down to about 5 μm and is also used for micron sizing SiC. More recently, air classification has also been adopted.

5.4.2 Brown-Fused Alumina

Brown-fused alumina has a Knoop hardness of 2090 and a medium friability. Increasing the TiO_2 content increases the toughness but reduces hardness. Although termed brown, the high-temperature furnacing in air required in subsequent vitrified wheel manufacture turns the brown alumina grains a gray-blue color due to further oxidation of the TiO_2.

5.4.3 White-Fused Alumina

White-fused alumina is one of the hardest, but most friable, of the alumina abrasive family providing a cool-cutting action especially suitable for precision grinding in vitrified bonds. Also, its low sodium content deters wheel breakdown from coolant attack when used in resin bonds.

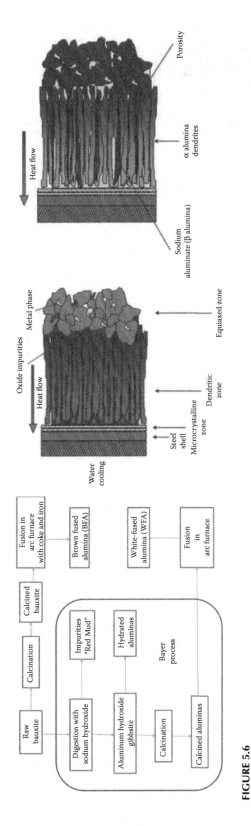

FIGURE 5.6
Fusion processes for white and brown-fused aluminas. (Courtesy of Saint Gobain Abrasives.)

FIGURE 5.7
Pouring of molten alumina. (Courtesy of Whiting Corp.)

Silicon carbide   White alumina

FIGURE 5.8
Examples of SiC and fused alumina.

Not surprisingly, since electrofused technology has been available for 100 years, many variations of the process exist both in terms of starting compositions and processing routes. Some examples are illustrated in Figure 5.8.

5.4.4 Alloying Additives

Additives are employed to modify the properties of alumina as described below. Examples of additives include chromium oxide, titanium oxide, zirconium oxide, and vanadium oxide.

5.4.5 Pink Alumina

The addition of chromium oxide produces pink alumina. White alumina is alloyed with <0.5% chrome oxide to give the distinctive pink hue of pink alumina. The resulting grain is slightly harder than white alumina, while addition of a small amount of TiO_2 increases its toughness. The resultant product is a medium-sized grain available in elongated, or blocky but sharp, shapes.

5.4.6 Ruby Alumina

Ruby alumina has a higher chrome oxide content of 2%–3% and is more friable than pink alumina. The grains are blocky, sharp edged, and extremely cool cutting, making them popular for tool room and dry grind application on steels (e.g., ice skate sharpening).

Grinding power is inversely proportional to chrome content up to a level of about 5% after which excessive softening of the grain occurs.

Vanadium oxide has also been used as an additive giving a distinctive green hue.

5.4.7 Single-Crystal White Alumina

Grain growth is closely controlled in a sulfide matrix. The alumina is separated out by acid leaching without crushing. The grain shape is nodular which aids bond retention, while the elimination of crushing reduces mechanical defects from processing.

5.4.8 Postfusion Processing Methods

As mentioned above, the type of particle reduction method can greatly affect the resulting grain shape. Impact crushers like hammer mills will create a blocky shape, while roll crushers will cause more splintering. It is further possible using electrostatic forces, to separate sharp shapes from blocky grains to provide grades of the same composition but very different cutting action.

5.4.9 Postfusion Heat Treatment

The performance of an abrasive can also be altered by heat treatment, particularly for brown alumina. The grit is heated to 1100–1300°C, depending on grit size, in order to anneal cracks and flaws created by the crushing process. This can enhance toughness by 25%–40%.

5.4.10 Postfusion Coatings

Finally, several coating processes exist to improve bonding of the grains in the grinding wheel. Red iron oxide is applied at high temperature to increase surface area for better bonding in resin cut-off wheels. Silane is applied for some resin bond wheel applications to repel coolant infiltration between bond and abrasive grit and thus protect the resin bond.

5.4.11 Zirconia–Alumina

Zirconia is added to alumina to refine the grain structure and produce a tough abrasive. At least three different alumina–zirconia compositions have traditionally been used in grinding wheels:

- 75% Alox, 25% ZrO_2
- 60% Alox, 40% ZrO_2
- 65% Alox, 30% ZrO_2, 5% TiO_2

Manufacture includes rapid solidification by pouring the melt onto chilled steel plates. This results in an ultrafine microstructure of metastable crystallites of tetragonal phase of zirconia trapped between layers of alumina and monoclinic zirconia. The stable form

FIGURE 5.9
Fused alumina–zirconia grain types and microstructure.

of zirconia is monoclinic which is also about 6% less dense than tetragonal. If a crack is initiated in a grain, the tip will begin to run but hit a phase of tetragonal zirconia which transforms to the monoclinc, expanding and sealing the crack tip to provide a very effective form of grain toughening. A similar toughening process occurs in zirconia-toughened alumina ceramics through the use of additives such as yttria which are now being increasingly incorporated into new BZZ grades (Poon and Frischuk 1984).

The resulting BZZ abrasives are fine grained, extremely tough, and give excellent life in medium to heavy stock removal applications such as hot and cold billet grinding in foundries, cut off, and railway track grinding. The wheels are able to achieve quite extraordinary stock removal rates exceeding that of other machining processes. This ability at first sight seems very surprising as the grain has very poor thermal properties; similar materials are used as thermal barrier coatings in jet engines and the grinding is done dry. In fact, its thermal properties are an advantage. BZZ wheels have grain sizes up to 6# (about the size of a small pea) held in a resin bond (Figure 5.9). Consequently, the chip size can be very large, comparable to machining and the specific energy consequently approaching that of machining. The temperatures at the interface, however, can be high as illustrated in the thermal analysis for cut off of a steel billet by Snoeys et al. (1978). The temperature at the grain/workpiece interface here is 1100°C, yet the grain is held in a bond processed and therefore stable at no more than 300°C (Figure 5.10). The grain acts to thermally insulate the bulk of the wheel, while pushing all the heat into the chip and softening it. The chips

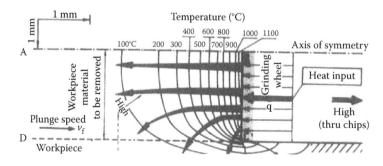

FIGURE 5.10
Representative isotherms from thermal modeling of cut-off process. $v_s = 100$ m/s, $v_f = 10$ mm/s, $b_w = 6$ mm, and $e_c = 5$ J/mm³. (From Snoeys, R. et al. 1978, *Annals of the CIRP*, 27(1), 571–581.)

FIGURE 5.11
Billet grinder. (Courtesy of MidWest Machines.)

will interact with the resin matrix to some extent and erode it maintaining a constant grain clearance. G ratios are typically in the range of 1–3, while Q' values can be >100 mm^2/s.

The thermal barrier effect also explains why certain workpiece materials such as titanium, with much lower thermal diffusivity than steels, comparable to BZZ, require special wheel grades and lower cutting speeds. The interface gets hotter, as witnessed by white hot chips as opposed to yellow chips, and the work material drives heat back into the wheel degrading the resin bond under comparable grind conditions. A billet grinder is shown in Figure 5.11.

5.5 Chemical Precipitation and/or Sintering of Alumina

5.5.1 Importance of Crystal Size

A limitation of the electrofusion route is that the resulting abrasive crystal structure is very large; an abrasive grain may consist of only one to three crystals. Consequently, when grain fracture occurs, the resulting particle loss may be a large proportion of the whole grain. This results in inefficient grit use. One way to avoid this is to dramatically reduce the crystallite size. Data published by Webster and Tricard (2004) show that hardness and toughness increase as crystallite size decreases to micron or submicron sizes (Figure 5.12).

5.5.2 Microcrystalline Grits

The earliest grades of microcrystalline grits were produced in 1963 (U.S. Patent 3,079,243) by compacting a fine-grain bauxite slurry, granulating to the desired grit size, and sintering at 1500°C. The grain shape and aspect ratio can even be controlled by extruding the slurry, as shown in Figure 5.12.

These grains could only be extruded at the coarsest sizes suitable for rough grinding, where they were used broadly prior to the introduction of BZZ. Interestingly, they are still

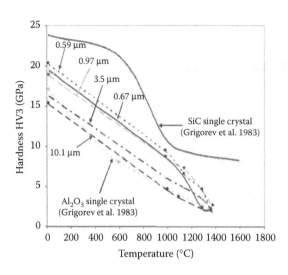

FIGURE 5.12
Sintered alumina rods for rough grinding. (Based on Webster, J. A., Tricard, M. 2004, *Innovations in Abrasive Products for Precision Grinding*. CIRP Innovations in Abrasive Products for Precision Grinding Keynote STC G.)

a grain of choice for grinding material such as titanium. It might be conjectured that the high aspect ratio allows the grain to be anchored deeper into the resin bond holding the wheel together when the wheel interface is subject to excessive heat (Table 5.2).

5.5.3 Seeded Gel Abrasive

The most significant development, however, probably since the invention of the Higgins furnace, was the release in 1986 of seeded gel (SG) abrasive by The Norton Company (U.S. Patent 4,312,827 1982; U.S. Patent 4,623,364 1986). This abrasive was a natural outcome of the wave of technology sweeping the ceramics industry at that time to develop high strength engineering ceramics using chemical precipitation methods. In fact, this class of abrasives is commonly termed "ceramic." SG is produced by a chemical process whereby MgO is first precipitated to create 50-nm-sized alumina–magnesia spinel seed crystals

TABLE 5.2

Mechanical Properties of Typical Fused and Sintered Alumina and SiC Abrasives

Abrasive	Hardness Knoop	Relative Toughness	Shape/Morphology	Applications
Green SiC	2840	1.60	Sharp/angular/glassy	Carbide/ceramics/precision
Black SiC	2680	1.75	Sharp/angular/glassy	Cast iron/ceramics/ductile nonferrous metals
Ruby alox	2260	1.55	Blocky/sharp-edged	High speed steel and high alloy steel
White alox	2120	1.75	Fractured facets/sharp	Precision ferrous
Brown alox	2040	2.80	Blocky/faceted	General purpose
Alox/10%ZrO	1960	9.15	Blocky/rounded	Heavy-duty grinding
Alox/40%ZrO	1460	12.65	Blocky/rounded	Heavy-duty/snagging
Sintered alox	1370	15.40	Blocky/rounded/smooth	Foundry billets/ingots

FIGURE 5.13
Multicomponent wheel structure containing ceramic, fused, and bubble materials.

in a precursor of boehmite. The resulting gel is dried, granulated to size, and sintered at 1200°C. The grains produced are composed of a single-phase α-alumina structure with a crystallite size of about 0.2 μm. Again, defects from crushing are avoided; the resulting abrasive is unusually tough but self-sharpening, because fracture now occurs at the micron level. The abrasive was so tough that it had to be blended with regular fused abrasive at levels as low as 5% to avoid excessive grinding forces. Typical blends are now:

- 5SG (50%)
- 3SG (30%)
- 1SG (10%)

These blended abrasive grades can increase wheel life by a factor up to 10 more than regular fused abrasives, although manufacturing costs are also higher. The patents on the basic material are now expired and various forms of SG material are available from a number of sources. For many wheel makers, the secondary abrasive blended with the SG is the primary focus for performance optimization and can be multicomponent containing various fused alox grain types and hollow bubble material (Figure 5.13).

The grain shape can also be controlled to surprising extremes by the crushing and other diminution processes adopted; from the very blocky to the very elongated as illustrated in Figure 5.14. The extreme elongated versions are more typical of grain for sanding and deburring belts, where the grain is vertically orientated using an electrostatic charge.

5.5.4 Sol–Gel Abrasives

In 1981, actually prior to the introduction of SG, 3M Co. introduced a sol–gel abrasive material, called Cubitron®, for use in coated abrasive fiber discs. This was again a submicron chemically precipitated and sintered material; but unlike SG was a multiphase composite structure that did not use seed grains to control crystallite size. (The value of the material for grinding wheel applications was not recognized until after the introduction of SG.) In the manufacture of Cubitron, alumina is coprecipitated with various modifiers such as

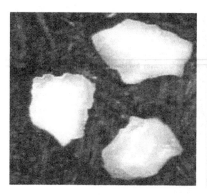

FIGURE 5.14
Grain shape controlled by crushing techniques.

magnesia, yttria, lanthana, and neodymia to control microstructural strength and surface morphology upon subsequent sintering. For example, one of the most popular materials, Cubitron 321, has a microstructure containing submicron platelet inclusions which act as reinforcements somewhat similar to a whisker-reinforced ceramic (Bange and Orf 1998). The resulting crystallite size is more typically in the 1–2 μm range resulting in a somewhat softer but more friable material than SG. Examples of ceramic grain microstructures are shown in Figure 5.15.

Direct comparison of the performance of SG and Cubitron is difficult because the grain is merely one component of the grinding wheel. SG, due to its crystallite size is harder (21 GPa) than Cubitron (19 GPa). Anecdotal evidence in the field suggests that wheels made from SG give longer life but Cubitron is freer cutting. This can make Cubitron the preferred grain in low stock removal rate applications. In response to this, Norton more recently launched an SG grain called NQ containing inclusions or microvoids (Garg 2002). This maintains the superior hardness of the submicron crystallite size but allows the grain to be micro trued to produce sharp fractured but durable grain edges with low cutting forces.

5.5.5 Agglomerated Grain Vortex™

The most recent introduction to the abrasive grain family is agglomerated grain as exampled by Vortex by the Norton Company (Figure 5.16; Bright and Wu 2004, 2008; Orlhac 2009). This fills a gap in crystallite size between the submicron SG to the fused single crystal but also offers some very unique properties. The grain is produced by sintering fine, angular, fused alumina feed in a high-temperature glass bond. The resulting agglomerates not only break down in a controlled manner leaving sharp cutting edges, but the agglomerate grains are *porous* and provide a high level of permeability to coolant. This makes the grain both a highly effective low-force primary grain and offers very attractive properties for use as a secondary filler in conjunction with a ceramic primary grain.

5.5.6 Extruded SG Abrasive

Control of grain shape of ceramic grain has been developed, in parallel with microstructural control, using extrusion or other molding techniques while still in the gel state. Norton has taken this concept to an extreme and in 1999 introduced TG and TG2 (extruded

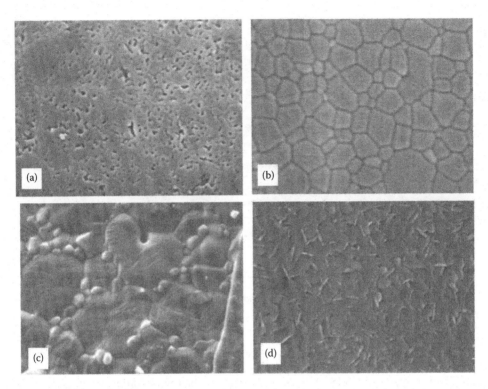

FIGURE 5.15
Examples of ceramic grain processing microstructures. (a) Unseeded pure alumina sintered gel with large uncontrolled grain growth. (b) Norton SG alumina with controlled microstructure. (c) Unseeded sintered alumina gel with magnesia additions. (d) 3M Cubitron 321 with magnesia and rare earth oxide additions.

SG) grains in products called Targa® and ALTOS® (Figure 5.17; Norton 1999a,b). TG grain had an aspect ratio of 4:1, while TG2 had an aspect ratio of 8:1. TG2 grains have the appearance of rods or "worms" due to these high aspect ratios. The needle shape should give the lowest value for an "r"-shape factor in the undeformed chip model and therefore the largest chip size and lowest specific cutting energy.

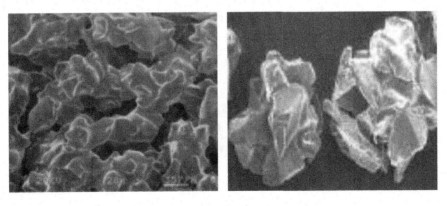

FIGURE 5.16
Norton Vortex™ agglomerated alox grain.

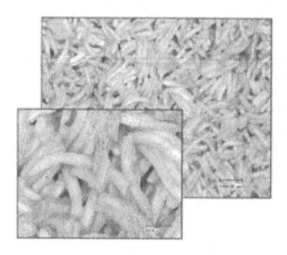

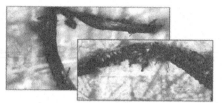

Swarf from grinding with TG2 grain

Swarf from grinding with standard alumina grain

FIGURE 5.17
TG2 extruded grain and associated large chip production. (From Hashimoto, F. 2012, Manufacturing technology for ultra large component. *2012 Saint Gobain Grinding Research Symposium*, Worcester, USA, November 8, 2012.)

More recently, 3M has provided a triangular shape grain (Cubitron) version of sol–gel grain as shown in Figure 5.18. Although its r-shape factor is somewhat higher, depending on orientation, it too achieves high stock removal rates and large chip sizes (Graf 2013). Shape and orientation are expected to play a large role in wheel development in the coming years.

The natural packing characteristics of high aspect ratio shapes are of major interest across many industries from catalysts to transporting carrots (Zhang 2006). The natural packing density for a low aspect ratio shape with a size distribution typical of abrasive grain is about 50%. However, the isotropic packing density for a needle shape with an

FIGURE 5.18
3M Cubitron 2 triangular-engineered grain.

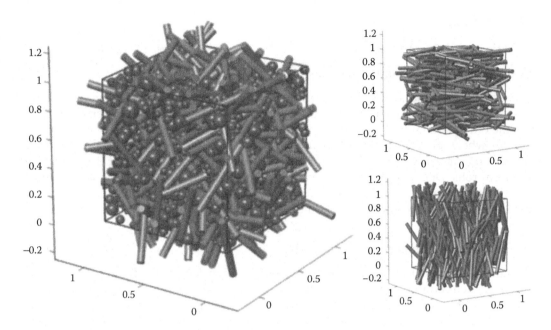

FIGURE 5.19
Packing of high aspect ratio grains and mix ratio blends. (From Zhang, W. 2006, Experimental and computational analysis of random cylinder packings with applications. PhD thesis. Louisiana State University.)

aspect of 10:1 is only 30%. For a grinding wheel, this results in a high strength, light-weight structure with enhanced permeability for coolant. The grains touch each other at only a few points where bond also concentrates like "spot welds" (Figure 5.19).

The product offers potential for both higher stock removal rates and higher wheel speeds due to the strength and density of the resulting wheel body (Klocke et al. 2000). The packing density depends on the grain orientations being isotropic. If layering of the grain occurs, then density increases. The TG2 grain, as with SG, is generally reduced to 30%–50% of the abrasive content, and a secondary filler, in particular the high porosity Vortex grain, of low aspect ratio is added to maintain an isotropic wheel structure, reduce the number of active cutting points, and conserve structural permeability.

5.5.7 Future Trends for Conventional Abrasives

Ceramic grain is now rapidly dominating the conventional abrasive market and has shown a much higher level of innovation in composition and shape in the last decade than that for CBN abrasive. In a number of cases, it has actually reversed the trend toward CBN usage in, for example, nickel alloy grinding for aerospace. Further developments can be expected both in shape, agglomerate compositions, crystallite size optimization, and control of packing and orientation. The introduction by 3M of the Cubitron 2 triangular grain is a clear indication that this technology will become increasingly competitive to the advantage of industry.

The production of electrofused product will continues to shift more and more from traditional manufacturing sites to those with good availability of electricity, such as around the Great Lakes of the United States of America and Norway, to lower cost, growing economies such as China and Brazil.

5.6 Diamond Abrasives

5.6.1 Natural and Synthetic Diamonds

Diamond holds a unique place in the grinding industry. Being the hardest material known, it is not only the abrasive choice for grinding the hardest, most difficult materials but also it is the only material that can true and dress all abrasive wheels effectively. Diamond is the only wheel abrasive that is still obtained from natural sources. Although synthetic diamond dominates in wheel manufacture, natural diamond is preferred for dressing tools and form rolls. Diamond materials are also used increasingly as wear surfaces for applications such as end stops and workrest blades on grinding machines. In these types of application, diamond can give 20–50 times the life of tungsten carbide.

5.6.2 Origin of Diamond

Diamond is created by the application of extreme high temperatures and pressures to graphite. Such conditions occur naturally at depths of 120 miles in the upper mantle or in heavy meteorite impacts. Diamond is mined from Kimberlite pipes that are the remnant of small volcanic fissures typically 2–50 m in diameter where magma has welled up in the past. Major producing areas of the world include South Africa, West Africa (Angola, Tanzania, Zaire, Sierra Leone), South America (Brazil, Venezuela), India, Russia (Ural Mountains), Western Australia, and most recently Canada. Each area and even each individual pipe will produce diamonds with distinct characteristics.

5.6.3 Production Costs

Production costs are high, 13 million tons of ore on average must be processed to produce 1 ton of diamonds. Much of this cost is supported by the demand for diamonds by the jewelry trade. Since World War II, the output of industrial grade diamond has been far outstripped by demand. This spurred the development of synthetic diamond programs initiated in the late 1940s and 1950s (Maillard 1980).

5.6.4 Three Forms of Carbon

The stable form of carbon at room temperature and pressure is graphite, where the carbon atoms are arranged in a layered structure (Figure 5.20). Within the layer atoms are positioned in a hexagonal lattice. Each carbon atom is bonded to three others in the same plane with the strong sp^3 covalent bonding required for a high hardness material. However, bonding between the layers is weak being generated from Van de Waals forces only, and resulting in easy slippage and low friction. In fact, pure graphite is highly abrasive because although there is low friction between the layers, the edges of individual sheets have dangling bonds that are highly reactive. It is only the presence of water vapor in the air of dopants added to the graphite that neutralizes these sites and makes graphite a low friction surface. Diamond, which is metastable at room temperature and pressure, has a cubic arrangement of atoms with pure sp^3 covalent bonding with each carbon atom bonded to four other carbon atoms. There is also an intermediate material called wurtzite or hexagonal diamond, where the hexagonal layer structure of graphite has been distorted above and below the layer planes but not quite to the full cubic structure. The material is

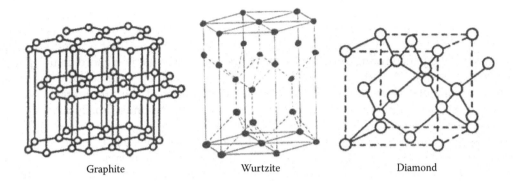

Graphite Wurtzite Diamond

FIGURE 5.20
Common structures of carbon.

nevertheless almost as hard as the cubic form. Within the last 10 years, nanodiamond has even been discovered during drilling for oil.

5.6.5 The Shape and Structure of Diamond

The principal crystallographic planes of diamond are the cubic (100), dodecahedron (011), and octahedron (111). The relative rates of growth on these planes are governed by the temperature and pressure conditions, together with the chemical environment both during growth and, in the case of natural diamond, during possible dissolution during its travel to the earth's surface. This in turn governs the diamond stone shape and morphology.

The phase diagram for diamond/graphite is shown in Figure 5.21.

5.6.6 Production of Synthetic Diamond

The direct conversion of graphite to diamond requires temperatures of 2500 K and pressures of >100 kbar. Creating these conditions was the first hurdle to producing man-made

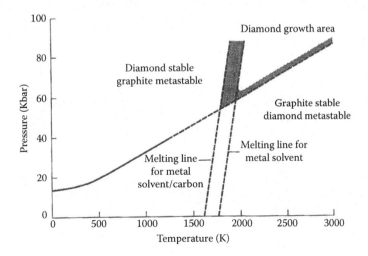

FIGURE 5.21
Phase diagram for carbon.

diamonds. The General Electric (GE) Co. achieved this through the invention of a high pressure/temperature gasket called the "belt" and announced the first synthesis of diamond in 1955. Somewhat to their surprise, it was then announced that a Swedish company ASEA had secretly made diamonds 2 years previously using a more complicated six-anvil press. ASEA had not announced the fact because they were seeking to make gems and did not consider the small brown stones they produced, the culmination of their program. De Beers announced their ability to synthesize diamonds shortly after GE in 1958 (De Beers Industrial Diamond Div. 1993, 1999).

The key to manufacture was the discovery that a metal solvent such as nickel or cobalt could reduce the temperature and pressure requirements to manageable levels. Graphite has a higher solubility in nickel than diamond has; therefore, at the high process temperatures and pressures, the graphite dissolves in the molten nickel and diamond then precipitates out. The higher the temperatures, the faster is the precipitation rate and the greater the number of nucleation sites. The earliest diamonds were grown fast at high temperatures and had weak, angular shapes with a mosaic structure. This material was released by GE under the trade name RVG® for "Resin Vitrified Grinding" wheels. Most of the early patents on diamond synthesis have now expired and competition from emerging economies have driven down the price of this type of material to as little as $400/lb, although quality and consistency from these sources is often still sometimes questionable.

5.6.7 Controlling Stone Morphology

By controlling the growth conditions, especially time, and nucleation density, it is possible to grow much higher quality stones with well-defined crystal forms; cubic at low temperature, cubo-octahedra at intermediate temperatures, and octahedra at the highest temperatures (Figure 5.22).

The characteristic shape of good-quality natural stones is octahedral, but the toughest stone shape is cubo-octahedral. Unlike in nature, this can be grown consistently by manipulation of the synthesis process. This has led to a range of synthetic diamond grades, typified by the MBG® series from GE and the PremaDia® series from de Beers 1999, which are the abrasives of choice for saws used in the stone and construction industry, and for glass grinding wheels.

5.6.8 Diamond Quality Measures

The quality and price of the diamond abrasive grain grade is governed both by the consistency of shape and the level of entrapped solvent in the stones. Since most of the blockiest abrasive is used in metal bonds processed at high temperatures, the differential thermal expansion of metal inclusions in the diamond can lead to reduced strength or even fracture. Other applications require weaker phenolic or polyimide resin bonds processed at

Octahedral   →   Cubo-octahedral   →   Cubic

FIGURE 5.22
Growth morphologies of diamond.

much lower temperatures and use more angular, less thermally stable diamonds. Grit manufacturers, therefore, characterize their full range of diamond grades by room temperature toughness (toughness index, TI), thermal toughness after heating at, for example, 1000°C (thermal toughness index, TTI), and shape (blocky, sharp, or mosaic). Included in the mid range, sharp grades are both crushed natural as well as synthetic materials.

5.6.9 Diamond Coatings

Diamond coatings are common. One range includes thick layers or claddings of electroplated nickel, electroless Ni–P, copper, or silver at up to 60 wt%. The coatings behave as heat sinks, while increasing bond strength and keeping abrasive fragments from escaping. Electroplated nickel, for example, produces a spiky surface that provides an excellent anchor for phenolic bonds when grinding wet. Copper and silver bonds are used more for dry grinding, especially with polyimide bonds, where the higher thermal conductivity outweighs the lower strength of the coating (Jakobuss 1999). Attention should be paid to wheel material data safety sheets to confirm chemical composition to ensure any coating used does not present a contaminant problem. For example, silver contamination may be a problem in grinding of titanium alloys.

Coatings can also be applied at the micron level either as a wetting agent or as a passive layer to reduce diamond reactivity with the particular bond. Titanium is coated on diamonds used in nickel-, cobalt-, or iron-based bonds to limit graphitization of the diamond while wetting the diamond surface. Chromium is coated on diamonds used in bronze- or WC-based bonds to enhance chemical bonding and reactivity of the diamond and bond constituents.

Finally, for electroplated bonds, the diamonds are acid etched to remove any surface nodules of metal solvent that would distort the plating electrical potential on the wheel surface leading to uneven nickel plating or even nodule formation. It also creates a slightly rougher surface to aid mechanical bonding.

5.6.10 Polycrystalline Diamond

Since 1960, several other methods of growing diamond have been developed. In 1970, DuPont launched a polycrystalline material produced by the sudden heat and pressure of an explosive shock. The material was wurtzitic in nature and produced mainly at micron particle sizes suitable more for lapping and polishing than grinding or as a precursor for polycrystalline diamond (PCD) monolithic material (Figures 5.23 through 5.25).

The year 1970 saw the introduction of PCD blanks that consisted of a fine-grain-sintered diamond structure bonded to a tungsten carbide substrate. The material was produced by the action of high temperatures and pressures on a diamond powder mixed with a metal solvent to promote intergrain growth. Since it contained a high level of metal binder, it could be readily fabricated in various shapes using electro discharge machining (EDM) technology. Although not used in grinding wheels, it is popular as reinforcement in form dress rolls and for wear surfaces on grinding machines. Its primary use though is in cutting tools.

5.6.11 Diamond Produced by Chemical Vapor Deposition

In 1976, reports began to come out of Russia of diamond crystals being produced at low pressures through chemical vapor deposition (CVD). This was treated with some skepticism in the West, even though Russia had a long history of solid research on diamond.

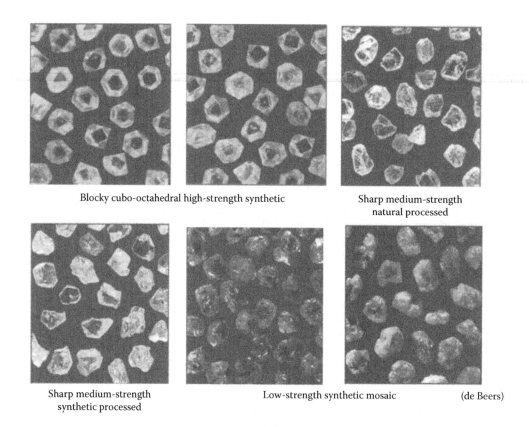

Blocky cubo-octahedral high-strength synthetic   Sharp medium-strength
                                                         natural processed

Sharp medium-strength   Low-strength synthetic mosaic   (de Beers)
synthetic processed

FIGURE 5.23
Typical diamond grit shapes, morphologies, and coatings.

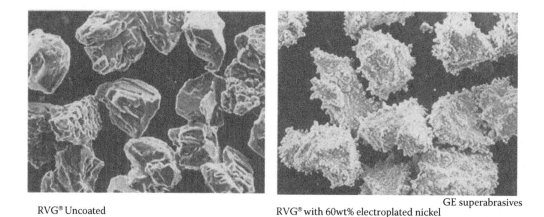

RVG® Uncoated   RVG® with 60wt% electroplated nickel   GE superabrasives

FIGURE 5.24
Effect of coating on surface morphology of diamond grain.

| Shock wave diamond grain | Ultra-detonated diamond (UDD) |
| | (courtesy ABC Superabrasives, Division Saint Gobain Ceramics) |

FIGURE 5.25
Examples of shock wave-produced diamond grains.

However, within 5 years, Japan was also reporting rapid growth of diamond by CVD at low pressures and the product finally became available in commercial quantities by about 1992. The process involves reacting a carbonaceous gas in the presence of hydrogen atoms in near vacuum to form the diamond phase on an appropriate substrate. Energy is provided in the form of hot filaments or plasmas at >800°C to dissociate the carbon and hydrogen into atoms. The hydrogen interacts with the carbon and prevents any possibility of graphite forming while promoting diamond growth on the substrate. The resulting layer can form to a thickness of >1 mm (Figure 5.26).

5.6.12 Structure of CVD Diamond

CVD diamond forms as a fine crystalline columnar structure. There is a certain amount of preferred crystallographic orientation exhibited; more so than, for example, PCD but far less than in single-crystal diamond. Wear characteristics are therefore much less sensitive to orientation in a tool. Again, the CVD diamond is not used as an abrasive but is proving very promising when fabricated in the form of needle-shaped rods for use in dressing tools and rolls. Fabrication with CVD is slightly more difficult, as it contains no metal solvents to aid EDM wire cutting and diamond wetting also appears more difficult and must be compensated by the use of an appropriate coating. More recently, the CVD process is now capable of producing single-crystal diamond, although the bulk of industrial product at this time remains multicrystalline.

5.6.13 Development of Large Synthetic Diamond Crystals

In the last 20 years, increasing effort has been placed on growing large synthetic diamond crystals at high temperatures and pressures. The big limitation has always been that press time and hence cost goes up exponentially with diamond size. The largest saw grade diamonds are typically 20/30#. The production of larger stones in high volume, suitable for tool and form-roll dressing applications, is not yet cost competitive with natural diamond. However, it has been recently proved to be an exception to this, namely the introduction,

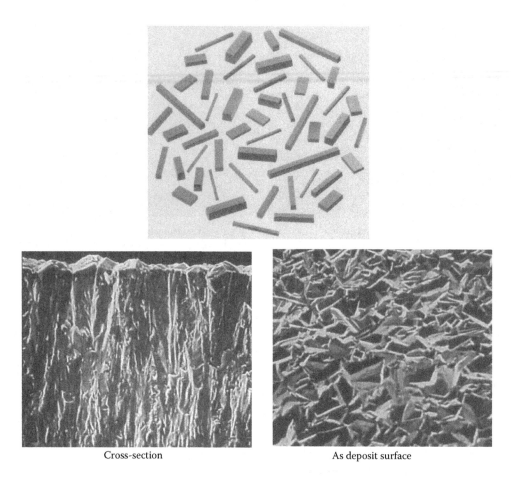

Cross-section   As deposit surface

FIGURE 5.26
CVD diamond samples and microstructure. (From Gigel, P. 1994, *Finer Points*, 6(3), 12–18. With permission.)

first by Sumitomo, of needle diamond rods produced by the slicing up of large synthetic diamonds. The rods are typically less than 1 mm in cross section by 2–5 mm long (similar in dimensions to the CVD diamond rods discussed above) but orientated along the principal crystallographic planes to allow optimized wear and fracture characteristics when orientated in a dressing tool.

Several companies now supply a similar product (Figures 5.27 and 5.28).

In the course of developing large synthetic diamond growth for industrial applications, it has proved possible to tightly control impurity levels in the source carbon material, in particular the minimization of nitrogen. Nitrogen occurs in all diamond. In synthetic diamond, it is present as dispersed interstitials giving the characteristic yellowish color, but in natural diamond, the prolonged time and temperature during growth allow the diamond to migrate and form platelet sheets. These reduce the resistance to crack propagation. The reduction in nitrogen has been found to raise the Knoop hardness of diamond in certain orientations by up to 30% (Sumitomo 2013). Since wear resistance is exponentially dependent on hardness, this is expected to have a significant impact on dresser wear resistance especially with increased usage of dressable superabrasive and ceramic grain wheels (Figure 5.29).

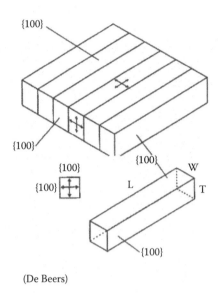

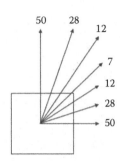

Relative wear rates as function of needle orientation for rotary diamond truers made with orientated diamond needles (Yokogawa and Furugawa, 1994)

(De Beers)

FIGURE 5.27
Monocrystal diamond needles cut with controlled crystallographic orientation for enhanced repeatability and life in dressing tools. (From De Beers Industrial Diamond Div. 1993, Monocrystal diamond product range. Commercial brochure. With permission.)

5.6.14 Novel Diamond Abrasive Grain

In line with conventional abrasive, there is increased interest in the use of agglomeration technology to engineer diamond abrasives with unique properties. For example, Foxmet (Sheridan 2013) published data on metal matrix diamond agglomerates that offer both controlled cutting actions and increased chip clearance, while also taking advantage of low-cost diamond feedstock (Figure 5.30).

Nanazyte™ is an example of an agglomerate for diamond (and other abrasive types including alox and SiC) produced for enhanced polishing properties. It consists of 20–80 μm hollow shells of micron diamond grains held together by a nanoparticle binder (Figure 5.31). In metallurgical polishing, for example, it can reduce a five-stage slurry lapping process down to a single-stage polishing in 25% of the processing time.

5.6.15 Demand for Natural Diamond

Even with the dramatic growth in synthetic diamond, the demand by industry for natural diamond has not declined. If anything, the real cost of natural diamond has actually increased especially for higher quality stones. The demand for diamonds for jewelry is such that premium stones used in the 1950s for single-point diamonds are now more likely to be used in engagement rings; while very small gem quality stones once considered too small for jewelry, and used in profiling dressing discs, are now being cut and lapped in countries such as India. With this type of economic pressure, it is not surprising that the diamonds used by industry are those rejected by the gem trade because of color, shape, size, crystal defects such as twins or naats, or excessive inclusion levels; or are the processed fragments from, for example, cleaving gems. Although significant quantities of processed material are still used in grinding wheel applications, it is the larger stones used in single-point and form-roll dressing tools that is of most significance. Here, the quality of

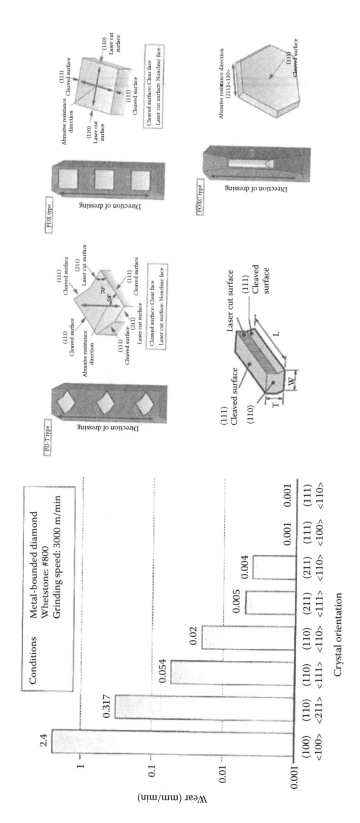

FIGURE 5.28
Commercial offerings of monocrystal diamond needles. (From Sumitomo. 2013, *Synthetic Single Crystal Diamond SUMICRYSTAL*. Catalog, Sumitomot Electric Hardmetal. With permission.)

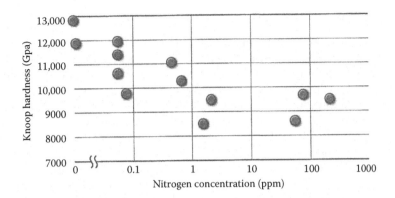

FIGURE 5.29
Effect of nitrogen content on SUMICRYSTAL diamond hardness. (From Sumitomo. 2013, *Synthetic Single Crystal Diamond SUMICRYSTAL*. Catalog, Sumitomo Electric Hardmetal. With permission.)

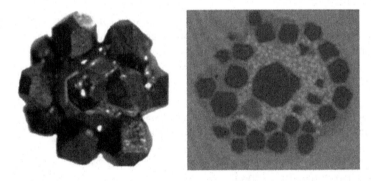

FIGURE 5.30
Diamond metal matrix agglomerated grain Foxmet S.A. (From Sheridan, C. 2013, Cutting versatility with diamond agglomerates. *Intertech 2013 Conference Proceedings*, IDA Paper 1337.)

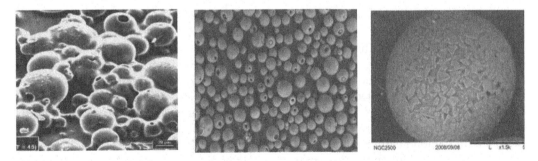

FIGURE 5.31
Hollow sphere Nanazyte™ agglomerated diamond.

the end product depends on the reliability of the diamond source and of the ability of the tool maker to sort diamonds according to requirements. The highest quality stones will be virgin as mined material. Lower quality stones may have been processed by crushing and/or ball milling, or even reclaimed from old form dressing rolls or drill bits, where they had previously been subjected to high temperatures or severe conditions.

5.6.16 Forms of Natural Diamond

Natural diamond grows predominantly as the octahedral form that provides several sharp points optimal for single-point diamond tools. It also occurs in a long stone form, created by the partial dissolution of the octahedral form as it ascended to the earth's surface. These are used in dressing tools such as the Fliesen® blade developed by Ernst Winter and Son. It should be noted though that long-stone shapes are also produced by crushing and ball milling of diamond fragments; these will have introduced flaws which significantly reduce strength and life. The old adage of "you get what you pay for" is very pertinent in the diamond tool business.

Twinned diamond stones called maacles also occur regularly in nature. These are typically triangular in shape (Figure 5.32). The twinned zone down the center of the triangle is the most wear-resistant surface known and maacles are used both in dressing chisels as well as reinforcements in the most demanding form-roll applications.

5.6.17 Hardness of Diamond

The hardness of diamond is a difficult property to define for two reasons. First, hardness is a measure of plastic deformation but diamond does not plastically deform at room temperature. Second, hardness is measured using a diamond indenter. Measuring hardness in this case is therefore akin to measuring the hardness of soft butter with an indenter made of hard butter. Fortunately, the hardness of diamond is quite sensitive to orientation and using a Knoop indenter; a distorted pyramid with a long diagonal seven times the short diagonal orientated in the hardest direction, gives somewhat repeatable results. The following hardness values have been obtained (Field 1983):

- (001) plane [110] direction 10,400 kg/mm^2
- (001) plane [100] direction 5700 kg/mm^2
- (111) plane [111] direction 9000 kg/mm^2

5.6.18 Wear Resistance of Diamond

More important than hardness is mechanical wear resistance. This is also a difficult property to pin down because it is so dependent on load, material, hardness, speed, etc. Wilks and Wilks (1972) showed that when abrading diamond with diamond abrasive, wear resistance increases with hardness but the differences between orientations are far more extreme. For example, on the cube plane, the wear resistance between the [100] and the [110] directions varies by a factor of 7.5 giving good correlation with wear data of needle diamonds reported in Figure 5.27. In other planes, the differences were as great as a factor 40 sometimes with only relatively small changes in angle. This has again been verified and exploited in diamond employed for dressers by, for example, Sumitomo (2013).

Not surprisingly, diamond gem lappers often talk of diamond having "grain"-like wood. Factors regarding the wear resistance of diamond on other materials in a machining process such as grinding, however, must include all possible attritious wear processes including thermal and chemical.

5.6.19 Strength of Diamond

Diamond being so hard is also very brittle. It can be readily cleaved along its four (111) planes. Its measured strength varies widely due in part to the nature of the tests but also

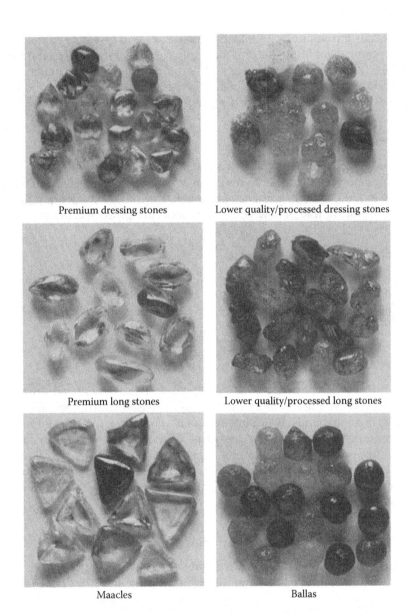

Premium dressing stones   Lower quality/processed dressing stones

Premium long stones   Lower quality/processed long stones

Maacles   Ballas

FIGURE 5.32
Natural industrial diamonds. (From Henri Polak Diamond Corp. 1979, Commercial brochure. With permission.)

because it is heavily dependent on the level of defects, inclusions, and impurities present. Not surprisingly, small diamonds (with smaller defects) give higher values for strength than larger diamonds. The compressive strength of top quality synthetic diamond (100#) grit has been measured at 1000 kg/mm^2.

5.6.20 Chemical Properties of Diamond

The diamond lattice is surprising pure, the only other elements known to be incorporated are nitrogen and boron. As discussed above, nitrogen is present in synthetic diamonds at up to 500 parts per million in single substitutional sites and gives the stones their characteristic yellow/green color but in natural diamond the nitrogen migrates and forms aggregates of platelets, and the diamond becomes the colorless stone found in nature. Synthetic diamond contains up to 10% included metal solvents, while natural diamond usually contains inclusions of the minerals in which it was grown (e.g., olivine, garnet, and spinels).

5.6.21 Thermal Stability of Diamond

Diamond is metastable at room temperatures and pressures; it will convert to graphite given a suitable catalyst or sufficient energy. In a vacuum or inert gas, diamond remains unchanged up to 1500°C; in the presence of oxygen, it will begin to degrade at 650°C. This factor plays a significant role in how wheels and tools are processed in manufacturing.

5.6.22 Chemical Affinity of Diamond

Diamond is readily susceptible to chemical degradation from carbide formers, such as tungsten, tantalum, titanium, and zirconium, and true solvents of carbon which includes iron, cobalt, manganese, nickel, chromium, and the group VIII platinum and palladium metals.

5.6.23 Effects of Chemical Affinity in Manufacture

This chemical affinity can be both a benefit and a curse. It is a benefit in the manufacture of wheels and tools, where the reactivity can lead to increased wetting and therefore higher bond strengths in metal bonds. For diamond tool manufacture, the reactant is often part of a more complex eutectic alloy (e.g., copper–silver, copper–silver–indium, or copper–tin), in order to minimize processing temperature, disperse, and control the active metal reactivity, and/or allow simplified processing in air. Alternatively, tools are vacuum brazed. For metal-bonded wheels, higher temperatures and more wear-resistant alloy bonds are used but fired in inert atmospheres.

5.6.24 Effects of Chemical Affinity in Grinding

The reactivity of diamond with transition metals, such as nickel and iron, is a major limitation to the use of diamond as an abrasive for machining and grinding these materials. Thornton and Wilks (1978, 1979) showed that certainly in single-point turning of mild steel with diamond, chemical wear was excessive and exceeded abrasive mechanical wear by a factor of 10^4. Hitchiner and Wilks (1984) showed that difference when turning nickel was $>10^5$. Turning pearlitic cast iron, however, the wear rate was only 10^2 greater. Furthermore, the wear on pearlitic cast iron was actually 20 times less than that measured using CBN tools. Much less effect was seen on ferritic cast iron which unlike the former material contained little free carbon; in this case, diamond wear increased by a factor of 10, when turning workpieces of comparable hardness.

5.6.25 Grinding Steels and Cast Irons with Diamond

It is generally considered, as the above results imply, that chemical-thermal degradation of the diamond prevents it being used as an abrasive for steels and nickel-based alloys but that under certain circumstances free graphite in some cast irons can reduce the reaction between diamond and iron to an acceptable level. For example, in honing of automotive cast iron cylinder bores, which is performed at very similar speeds (2 m/s) and cut rates to that used in the turning experiments mentioned above, diamond is still the abrasive of choice outperforming CBN by a factor of 10. However, at the higher speeds (80 m/s typical) and temperatures of cylindrical grinding of cast iron camshafts, the reverse is the case.

5.6.26 Thermal Properties

Diamond has the highest thermal conductivity of any material with a value of 600–2000 W/mK at room temperature falling to 70 W/mK at 700°C. These values are 40 times greater than the thermal conductivity of alumina. Much mention is made in the literature of the high thermal conductivity of both diamond and CBN, and the resulting benefits of lower grinding temperatures and reduced thermal stresses. Despite an extremely high thermal conductivity, if the heat capacity of the material is low, it will simply get hot quickly. Thermal models for moving heat sources, as shown by Jaeger (1942), employ a composite transient thermal property. The transient thermal property is $\beta = \sqrt{k\rho c}$, where k is the thermal conductivity, ρ is the density, and c is the thermal heat capacity.

The value of β for diamond is 6×10^4 W/mK compared to $0.3–1.5 \times 10^4$ W/mK for most ceramics, including alumina and SiC, and for steels. Copper has a value of 3.7×10^4 W/mK due in part to a much higher heat capacity than that of diamond. This may explain its benefit as a cladding material and wheel filler material.

Steady-state conditions are quickly established during the grain contact time in grinding. This is because the heat source does not move relative to the grain. The situation is similar to rubbing a finger across a carpet. It is the carpet that sees the moving heat source and stays cool, rather than the finger that sees a constant heat source and gets hot. In grinding, the abrasive grain is like the finger and the workpiece is like the carpet. In this case, it is the thermal conductivity of the grain that governs the heat conducted by the grain rather than the transient thermal property (Rowe et al. 1996). For nonsteady conduction, a time-constant correction is given by Rowe and Black (Marinescu et al. 2004, Chapter 6). The application of thermal properties to calculation of temperatures is discussed in more detail in Chapter 17 on external cylindrical grinding.

The coefficient of linear thermal expansion of diamond is 1.5×10^{-6}/K at 100°C increasing to 4.8×10^{-6}/K at 900°C. The values are significant for bonded wheel manufacturers who must try to match thermal expansion characteristics of bond and grit throughout the firing cycle.

For further details on the properties of diamond, see Field (1979, 1983).

5.7 Cubic Boron Nitride

5.7.1 Development of CBN

Cubic boron nitride or CBN is the final and most recent of the four major abrasive types, and the second hardest superabrasive after diamond. Trade names include Borazon®, from GE Superabrasives (now called Diamond Innovations) who first synthesized it commercially, Amborite® and Amber Boron Nitride® after de Beers, or in Russian literature as Elbor, Cubonite or β-BN.

Boron nitride at room temperatures and pressures is made using the reaction:

$$BCl_3 + NH_3 \rightarrow BN + 3HCl$$

The resulting product is a white slippery substance with a hexagonal-layered atomic structure called HBN (or α-BN) similar to graphite but with alternating nitrogen and boron atoms. Nitrogen and boron lie either side of carbon in the periodic table, and it was postulated that high temperatures and pressures could convert HBN to a cubic structure similar to diamond. This was first shown to be the case by a group of scientists under Wentdorf at GE in 1957. The first commercial product was released 12 years later in 1969 (Figure 5.33).

Both the cubic (CBN) and wurtzitic (WBN or γ-BN) forms are created at comparable pressures and temperatures to those for carbon. Again, the key to successful synthesis was the selection of a suitable solvent to reduce conditions to a more manageable level. The chemistry of BN was quite different to carbon; for example, bonding was not pure sp^3 but 25% ionic, and BN did not show the same affinity for transition metals. The successful solvent/catalyst turned out to be any one of a large number of metal nitrides, borides, or oxide compounds of which the earliest commercial one used (probably with some additional doping) was Li_3N. This allowed economic yields at 60 kbar, 1600°C, and <15 min cycle times.

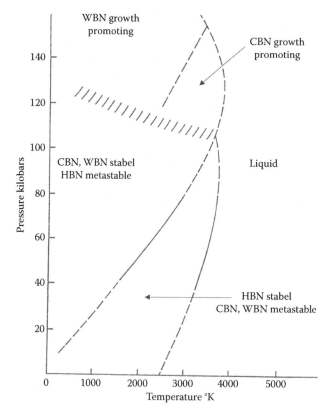

FIGURE 5.33
Phase diagram for boron nitride.

5.7.2 Shape and Structure of CBN

As with diamond crystal growth, CBN grain shape is governed by the relative growth rates on the octahedral (111) and cubic planes. However, the (111) planes dominate and furthermore, because of the presence of both B and N in the lattice, some (111) planes are positive terminated by B atoms and some are negative terminated by N atoms. In general B (111), plane growth dominates and the resulting crystal morphology is a truncated tetrahedron. Twinned plates and octahedra are also common (Figure 5.34). The morphology can be driven toward the octahedral or cubo-octahedral morphologies by further doping and/or careful control of the pressure–temperature conditions (Figure 5.35).

5.7.3 Types of CBN Grains

As with diamond, CBN grain grades are most commonly characterized by toughness and by shape. Toughness is measured both at room temperatures and at temperatures up to >1000°C comparable to those used in wheel manufacture, the values being expressed in terms of a TI and TTI. The details of the measurement methods are normally proprietary but in general, grains of a known screened size distribution are treated to a series of impacts and then rescreened. The fraction of grain remaining on the screen is a measure of the toughness. For TTI measurements, the grains may be heated in a vacuum or a controlled atmosphere, or even mixed with the wheel bond material which is subsequently leached out. TI and TTI are both strongly influenced by doping and impurity levels. Additional degradation of the grain within the wheel bond during manufacture can also occur due to the presence of surface flaws that may be opened up by penetration of bond.

The surface roughness of CBN is a more pronounced and critical factor than for diamond in terms of factors influencing grinding wheel performance. A rough angular morphology provides a better, mechanical anchor. Of the examples illustrated in Figure 5.34, GE type 1 abrasive is a relatively weak irregular crystal. The coated version GE type II abrasive used in resin bonds has a simple nickel-plated cladding. However, GE 400 abrasive is a tougher grain with a similar shape but with much smoother, flaw-free faces. The coated version GE 420 is therefore first coated with a thin layer of titanium to create a chemically bonded roughened surface to which the nickel cladding can be better anchored.

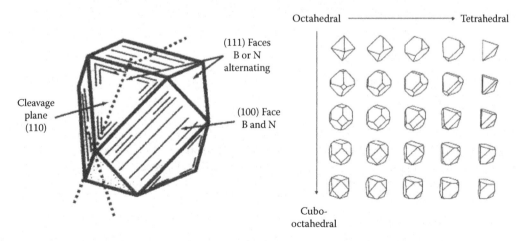

FIGURE 5.34
CBN crystal growth planes and morphology. (After Bailey, M. W., Juchem, H. O. 1993. *Industrial Diamond Review*, 3, 83–89; de Beers. 1998.)

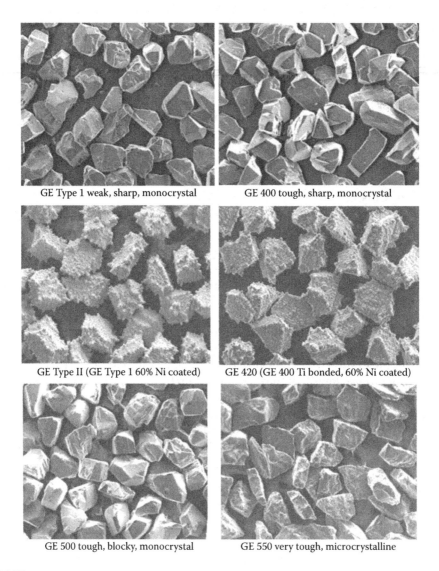

GE Type 1 weak, sharp, monocrystal   GE 400 tough, sharp, monocrystal

GE Type II (GE Type 1 60% Ni coated)   GE 420 (GE 400 Ti bonded, 60% Ni coated)

GE 500 tough, blocky, monocrystal   GE 550 very tough, microcrystalline

FIGURE 5.35
Examples of CBN grain types and morphologies. (From GE Superabrasives, 1998. *Understanding the Vitreous Bonded Borazon CBN System*. General Electric Borazon® CBN Product Selection Guide. Commercial brochure. With permission.)

Only a relatively few grades of CBN are tough and blocky with crystal morphologies shifted away from tetrahedral growth. The standard example is GE 500 used primarily in electroplated wheels. De Beers also has material, ABN 600, where the morphology has been driven toward the cubo-octahedral (De Beers Industrial Diamond Div. 1993, 1999).

5.7.4 Microcrystalline CBN

GE superabrasives developed a grit type GE 550 that is a microcrystalline product; this could be considered the "SG" of CBN grains. It is extremely tough and blocky and wears by microfracturing. However, just like SG grains, it also generates high grinding forces and is therefore limited to use in the strongest bonds, such as bronze metal, for high force/grit applications especially honing. It has also been used in limited quantities in plated applications. One problem with its microcrystalline nature is that the surface of GE 550 is much more chemically reactive with vitrified bonds. Similar materials, some described as "ultrafine-crystalline" are now available from a number of sources. Ichida et al. (2008) reports these can generate mirror-like finishes using relative coarse grain sizes. More recently, patents are appearing where the microfracture of single grains is being controlled by grain growth and orientation effects, for example, Ji (2014).

5.7.5 Sources and Costs of CBN

The manufacture of CBN has been dominated by diamond innovations (GE Superabrasives) in the United States of America, by de Beers from locations in Europe and South Africa, and by Showa Denko, Iljin, and Tomei from the far East. Russia and Romania have also been producing CBN for over 40 years, and China has rapidly become the most important player in terms of volume. Historically, consistency had been in question with materials from some of these latter sources but with intermediate companies in many cases controlling the QC aspects of the materials to the end user, they have become a very real low-cost alternative to traditional suppliers. Currently, CBN costs are of the order of $1500–5000/lb or at least three to four times that of the cheapest synthetic diamond.

5.7.6 Wurtzitic Boron Nitride

As with carbon, wurtzitic boron nitride (WBN) has also been produced by explosive shock methods. Reports of commercial quantities of the material began appearing in about 1970 (Nippon Oil and Fats 1981) but its use has again been focused more on cutting tool inserts with partial conversion of the WBN to CBN, and this does not appear to have impacted the abrasive market.

5.7.7 Hardness of CBN

The hardness of CBN at room temperature is approximately 4500 kg/mm^2. This is about half as hard as diamond and twice as hard as conventional abrasives.

5.7.8 Wear Resistance of CBN

The differences in abrasion resistance are much more extreme. A hardness factor of 2 can translate into a factor of 100 > 1000 in abrasion resistance depending on the abrading material. The author remembers as a research student under Wilks when the first CBN samples were supplied for abrasion resistance measurements using the same technique used for measuring the wear resistance of diamonds, the CBN was so soft in comparison to diamond that it was impossible to obtain a value on the same wear scale. As with diamond, the key is the total wear resistance to all attritious wear processes.

Like diamond, CBN is brittle; but it differs in having six (110) rather than four (111) cleavage planes. This gives a more controlled breakdown of the grit especially for the truncated tetrahedral shape of typical CBN grains. The grain toughness is generally much less than that of blocky cubo-octahedral diamonds. This combined with its lower hardness provides the very useful advantage that CBN wheels can be dressed successfully by diamond rotary tools.

5.7.9 Thermal and Chemical Stability of CBN

CBN is thermally stable in nitrogen or vacuum to at least 1500°C. In air or oxygen, CBN forms a passive layer of B_2O_3 on the surface which prevents further oxidation up to 1300°C. However, this layer is reactive with water, or more accurately high-temperature steam at 900°C, and will allow further oxidation of the CBN grains following the reactions (Carius 1989; Yang et al. 1993):

$$2BN + 3H_2O \rightarrow B_2O_3 + 2NH_3$$

$$BN + 3H_2O \rightarrow H_3BO_3 + N_2 > 900°C$$

$$4BN + 3O_2 \rightarrow 2B_2O_3 + 2N_2 > 980°C$$

$$B_2O_3 + 3H_2O \rightarrow 2H_2BO_3 > 950°C$$

5.7.10 Effect of Coolant on CBN

There is considerable debate as to whether this reacton is sufficient to explain the extraordinary difference in life of CBN, especially in plated single-layer wheel, between grinding in oil or water coolant that can be as great as 10 times or even 1000 times. Gift et al. (2006), for example, identifies wheel loading as being significantly higher in water-based coolant and may explain some of the more extreme differences. Thermal shock from the greater thermal conductivity has also been suggested as a factor.

Regardless of root cause, the effect is dramatic and comprehensive across all workpiece materials as illustrated in Table 5.3 which gives comparative life values for surface grinding with CBN wheels (Carius 2001).

CBN is also reactive toward alkali oxides; not surprisingly in light of their use as solvents and catalysts in CBN synthesis. The B_2O_3 layer is particularly prone to attack or dissolution by basic oxides such as Na_2O by the reaction (GE Superabrasives 1988):

$$B_2O_3 + Na_2O \rightarrow Na_2B_2O_4$$

Such oxides are common constituents of vitrified bonds and the reactivity can become extreme at temperatures above 900°C affecting processing temperatures for wheels (Yang et al. 1993).

TABLE 5.3

Effect of Coolant Type on CBN Wheel Performance

Workpiece	Synthetic Light Duty 2%	Soluble Heavy Duty 10%	Straight Oil
M2	X	1.7X	5X
M50	X	4X	16X
T15	X	1.7X	3X
D2	X	1.3X	11X
52–100	X	10X	14X
410 ss	X	25X	44X
IN 718	X	8X	50X

5.7.11 Effect of Reactivity with Workpiece Constituents

CBN does not show any significant reactivity or wetting by transition metals such as iron, nickel, cobalt, or molybdenum until temperatures in excess of 1300°C. This is reflected in a low rate of wear when grinding these materials with CBN abrasive in comparison with wear of diamond abrasive. CBN does show marked wetting by aluminum at only 1050°C and also with titanium. As demonstrated in wetting studies of low temperature silver–titanium eutectics, CBN reacts readily at 1000°C to form TiB_2 and TiN (Benko 1995). This provides an explanation of why in grinding aerospace titanium alloys such as Ti–6Al–4V, CBN wheels wear typically up to five times faster than diamond wheels (Kumar 1990). By comparison, the wear rate using the alternative of SiC abrasive is 40 times greater than CBN. This is a further example of the need to consider the combined effects of the mechanical, chemical, and thermal wear processes as much as abrasive cost.

Pure, stochiometrically balanced, CBN material is colorless, although commercial grades are either a black or an amber color depending on the level and type of dopants present. The black color is believed to be due to an excess (doping) of boron.

5.7.12 Thermal Properties of CBN

The thermal conductivity of CBN is almost as high as that of diamond. At room temperature, thermal conductivity is 200–1300 W/mK, and the transient thermal property $\beta = 2.0 \times 10^4$ to 4.8×10^4 J/m^2sK. The thermal expansion of CBN is about 20% higher than diamond.

5.8 Future Grain Developments

Research is accelerating both in existing alumina-based grain technology and in new ultrahard materials. In the group of ceramic processed alumina materials Saint-Gobain released SG in 1986 (U.S. Patent 4,623,364) followed by extruded SG in 1991 (U.S. Patent 5,009,676). More recently in 1993, Treibacher released an alumina material with hard filler additives (U.S. Patent 5,194,073). Electrofused technology has also advanced. Pechiney produced an Al–O–N grain, "Abral™" produced by the cofusion of alumina and AlON followed by slow solidification. It offered much higher thermal corrosion resistance relative to regular alumina while also having constant self-sharpening characteristics akin to ceramic processed materials but softer acting (Roquefeuil 2001).

New materials have also been announced with hardness approaching CBN and diamond. Iowa State University announced in 2000 an Al–Mg–B material with a hardness value comparable to CBN (U.S. Patent 6,099,605). Dow Chemical patented in 2000 an Al–C–N material with a hardness value close to diamond (U.S. Patent 6,042,627). In 1992, University of California patented some α-C_3N_4 and β-C_3N_4 materials that may actually be harder than diamond (U.S. Patent 5,110,679). Whether any of these materials eventually proves to have useful abrasive properties and can be produced in commercial quantities has yet to be seen. Nevertheless, there will undoubtedly be considerable advances in abrasive materials in the coming years.

References

Bailey, M. W., Juchem, H. O. 1993, The advantages of CBN grinding: Low cutting forces and improved workpiece integrity. *Industrial Diamond Review*, 3, 83–89.

Bange, D. W., Orf, N., 1998, *Tooling & Production*, March, 82–84.

Benko, E., 1995, *Ceramics International*, 21, 303–307.

Bright, E., Wu, M., 2004, 2008, Porous abrasive articles with agglomerated abrasives. U.S. Patent 6679758 January 20, 2004. U.S. Patent 7422513 September 9, 2008.

Carius, A. C., 1989, *Modern Grinding Technology*. SME Novi, MI, October 10, 1989.

Carius, A. C., 2001, CBN abrasives and the grindability of PM materials. *Precision Grinding & Finishing in the Global Economy—2001 Conference Proceedings, Gorham*, Oak Brook, IL, October 1, 2001.

De Beers Industrial Diamond Div., 1993, Monocrystal diamond product range. Commercial brochure.

De Beers Industrial Diamond Div., 1999, Premadia diamond abrasives. Commercial brochure.

De Pellegrin, D. V., Corbin, N. D., Baldoni, G., Torrance, A. A., 2002, The measurement and description of diamond particle shape in abrasion. *Wear*, 253, 1016–1025.

De Pellegrin, D. V., Corbin, N. D., Baldoni, G., Torrance, A. A., 2008, Diamond particle shape: Its measurement and influence in abrasive wear. *Tribology International*, 42/1, 160–168.

De Pellegrin, D. V., Stachowiak, G.W., 2004, Evaluating the role of particle distribution and shape in two-body abrasion by statistical simulation, *Tribology International*, 37, 255–270.

FACT. *Diamond*. Trade Publication.

Field, J. E., 1979, *The Properties of Diamond*. Academic Press, London.

Field, J. E., 1983, *Diamond—Properties and Definitions*. Cavendish Lab, Cambridge University, UK.

Garg, A. K. 2002, Abrasive grain. U.S. Patent 6391072, May 21, 2002.

GE Superabrasives, 1998, *Understanding the Vitreous Bonded Borazon CBN System*. General Electric Borazon® CBN Product Selection Guide. Commercial brochure.

Gift, F. C., Misiolek, W. Z., Force E., 2006, Fluid performance study for groove grinding a nickel-based superalloy suing electroplated CBN grinding wheels. *JMSE*, 126, 451–458.

Gigel, P., 1994, *Finer Points*, 6(3), 12–18.

Graf, W., 2013, *Precision Shaped Grains*. AGMA Gear USA Trade Publication.

Hashimoto, F., 2012, Manufacturing technology for ultra large component. *2012 Saint Gobain Grinding Research Symposium*, Worcester, USA, November 8, 2012.

Henri Polak Diamond Corp., 1979, Commercial brochure.

Heywood, J., 1938, *Grinding Wheels and Their Uses*. Penton Co.

Hitchiner, M. P., McSpadden, S., 2004, *Evaluation of Factors Controlling CBN Abrasive Selection for Vitrified Bonded Wheels*. Advances in Abrasive Technology VI, Trans Tech Publications Ltd., 267–272.

Hitchiner, M. P., Wilks, J., 1984, Factors affecting chemical wear during machining. *Wear*, 93, 63–80.

Ichida, Y., Sato, R., Ueno, H., NAgsao, S., 2008, High efficiency mirror grinding using ultra-crystalline CBN grains. *Proceedings of the Euspen International Conference*, Zurich.

Jaeger, J. C., 1942, Moving sources of heat and the temperature at sliding contacts. *Proc. R. Soc. New South Wales* 76, 203.

Jakobuss, M., 1999, Influence of diamond and coating selection on resin bond grinding wheel performance. *Precision Grinding Conference*, Chicago, IL, USA, June 15–17, 1999.

Ji, S., Long, C., Sowers, S., Zhang, K., 2014, Single crystal CBN featuring microfracturing during grinding. U.S. Patent Appl. 2014 0090307.

Kitajima, M., Hand, S. S., Kamiya, A., Suzuki, A., Sekiya, Y., Inagaki, T., 1994, CBN Grinding Wheel. U.S. Patent 5,364,422, November 15, 1994.

Klocke, F., Mueller, N., Englehorn, H., 2000, Abrasive Magazine, June/July 24–27.

Kumar, K. V., 1990, *SME 4th International Grinding Conference*, Dearborn, MI, MR90-505.

Maillard, R. (Editor), 1980, *Diamonds—Myth, Magic and Reality*, Crown Publishing, New York.

Marinescu, I., Rowe, W. B., Dimitrov, B., Inasaki, I., 2004. *Tribology of Abrasive Machining Processes.* William Andrew Publishing Norwich, NY.

Nippon Oil and Fats Co, Ltd., 1981, WURZIN (wBN) Tool. Commercial document, c1981.

Norton. 1999a, Project Optimos—Grind in the Fast Lane. Commercial brochure.

Norton, 1999b, Project Altos. Commercial brochure.

Orlhac, X., 2009, Abrasive articles with novel structures and methods for grinding, U.S. Patent7544114, June 9, 2009.

Poon, S., Frischuk, R. W., 1984, Aluimina zirconia abrasive, U.S. Patent 4457767, July 3, 1984.

Roquefeuil, F., 2001, ABRAL: A new electrofused Alon grain for precision grinding. *Precision Grinding & Finishing in the Global Economy—2001, Conference Proceedings*, Gorham, Oak Brook, IL, October 1, 2001.

Rowe, W. B., 2014, *Principles of Modern Grinding Technology*, 2nd edition. Elsevier Press, Oxford, UK and worldwide.

Rowe, W. B., Morgan, M. N., Black, S. C. E., Mills, B., 1996, A simplified approach to thermal damage in grinding, *Annals of the CIRP*, 45(1), 299–302.

Sheridan, C., 2013, Cutting versatility with diamond agglomerates. *Intertech 2013 Conference Proceedings*, IDA Paper 1337.

Snoeys, R., Leuven, K. U., Maris, M., Wo, N. F., Peters, J., 1978, Thermally induced damaged in grinding. *Annals of the CIRP*, 27(1), 571–581.

Sumitomo, 2013, *Synthetic Single Crystal Diamond SUMICRYSTAL*. Catalog, Sumitomot Electric Hardmetal.

Thornton, A. G., Wilks, J., 1978, Clean surface reactions between diamond and steel. *Nature*, 8/24/78, 792–793.

Thornton, A. G., Wilks, J., 1979, Tool wear and sold state reactions during machining. *Wear*, 53, 165.

Tymeson, M. M., 1953, *The Norton Story*. Norton Co., Worcester, MA.

Viernekes, N., 1987, CBN ceramic-bonded abrasive wheels for semi-automated grinding processes. *Wälzlagertechnik* 1987-1, 30–34.

Webster, J. A., Tricard, M., 2004, *Innovations in Abrasive Products for Precision Grinding*. CIRP Innovations in Abrasive Products for Precision Grinding Keynote STC G.

Wilks, E. M., Wilks, J., 1972, The resistance of diamond to abrasion. *Journal of Physics D: Applied Physics* 5, 1902–1919.

Yang, J., Kim, D., Kim, H., 1993, Effect of glass composition on the strength of vitreous bonded c-BN grinding wheels. *Ceramics International* 19, 87–92.

Zhang, W. 2006. Experimental and computational analysis of random cylinder packings with applications. PhD thesis. Louisiana State University, Louisiana, USA.

6

Specification of the Bond

6.1 Introduction

Wheel bond systems can be divided into two types: those holding a single layer of abrasive grain to a solid steel core, and those providing a consumable layer many grains thick with the abrasive held within the bond. The latter may be mounted on a resilient core or produced as a solid monolithic structure from the bore to the outer diameter. This chapter deals with the different types of bonding structures employed in grinding wheel design and the effects on wheel performance.

6.2 Single-Layer Wheels

Single-layer abrasive technologies provide a multi-billion dollar market dominated by sandpaper, polishing and superfinishing film, deburring, and fettling wheels. Applications encompass all the metalworking industries as well as paint finishing, marine boat sanding, and woodworking to name just a few others. It is a market several times larger than that of bonded grinding wheels, and still dominated by conventional abrasives but nonetheless technologically sophisticated in many aspects, such as grain structure, placement, and substrate construction. Sadly, a discussion of this topic is outside the scope of this handbook (although a comparable handbook on this topic is long overdue!).

Single-layer wheels for precision grinding applications are generally limited to superabrasives because of the economics of wheel life. They can be subdivided into plated wheels fabricated at temperatures of <100°C, and brazed wheels fabricated at temperatures as high as 1000°C. The following discussion applies in general both to plated cubic boron nitride (CBN) and to plated diamond wheels, although in practice CBN dominates the precision grinding market and is the central focus below.

6.3 Electroplated and Electroless-Plated Single-Layer Wheels

6.3.1 Structure of an Electroless-Plated Layer

Electroplated wheels consist of a single layer of superabrasive grains bonded to a precision-machined steel blank using nickel deposited by an electroplating, or

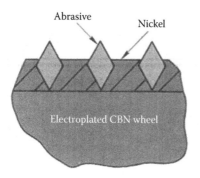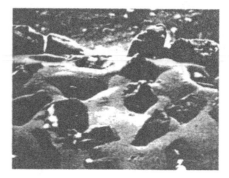

FIGURE 6.1
Schematic of an electroplated CBN wheel section and the appearance of the actual surface of such a wheel.

increasingly, by an electroless plating process (Figure 6.1). The plating depth is controlled to leave about 50% of the abrasive exposed. The biggest issues with electroplated wheels are depth variation, nodule production, and overplating due to different electrical potentials created by changes in wheel profile and metal impurities present in certain abrasives especially synthetic diamond (Sato et al. 1997). Most grain supplier will provide grades that have been etched to remove surface impurities. Nevertheless, plating times can be greatly extended for consistency. Electroless plating overcomes some of these issues. Electroplate compositions tend to be pure nickel but numerous other compositions have been reported for increased hardness and toughness depending on the application; these include, for example, Ni–Co–Mn (Li et al. 2002) and Ni–W (Nishihara et al. 2006). Electroless nickel compositions tend to be Ni–P or Ni–B based but can incorporate additional filler such as solid lubricants (Meyer 2008).

6.3.2 Product Accuracy

The accuracy and repeatability of the process is dependent on many factors. The blank must be machined to a high accuracy and balanced, the surface prepared appropriately (chemically activated); ideally blank profile tolerances are maintained to within 2 µm and wheel runout maintained to within 5 µm (McClew 1999).

The abrasive is generally resized to provide a tighter size distribution than that used in other bond systems. This is to avoid any high spots and better control grain aspect ratio. The abrasive is applied to the blank by various proprietary methods to produce an even and controlled density distribution. There is an initial "tacking" or attachment stage, where the grain is initially secured to the blank followed by a slower prolonged plate rate, where the nickel is allowed to build up evenly and stress free. Figure 6.2 shows the typical distribution of exposed grain heights on a plated surface for a standard D126 grain distribution. The target height is 63 µm but due to such issues as

1. Inconsistent plate depth from voltage potential variations
2. Grain shape issues, for example, high-aspect ratio grains lying flat against the blank (low exposure, grains buried)
3. Variations in tack plating thickness depth and time grain attached in tacking process

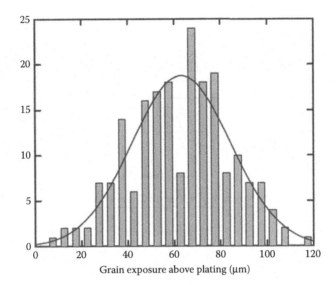

FIGURE 6.2
Typical grain exposure distribution for a standard D126 grit size. (From Shi, Z., Malkin, S., 2003, *Annals of the CIRP*, G04, 2, 267–271.)

The grain exposure can range from zero to the full grain size. The goal in tight tolerance applications is to reduce the width of the bell curve substantially by

1. Resizing the grain to subsieve sizes to eliminate any oversized grain
2. Shape sorting to narrow the range of aspect ratios of grains
3. Postconditioning (also termed "dressing," "truing," or "shaving"), where an amount equivalent to approximately 5%–7% of the nominal grit size is removed to produce a well-defined grit protrusion height above the plating

With good control of plate thickness, this helps to control and/or define a usable layer depth. Postconditioning also significantly reduces the initial break-in effects and stabilizes surface finish issues illustrated in Figure 6.3.

6.3.3 Wear Resistance of the Bond

The hardness, or more accurately, the wear resistance of the nickel is controlled by changes to the bath chemistry as mentioned above. Nitride coating, similar to coatings used on cutting tools, has also been reported to further improve the wear resistance of the nickel but data have been mixed indicating that performance parameters are not yet understood (Bush 1993; Julien 1994). Solid lubricant coatings on the wheel surface have also been reported to increase life.

6.3.4 Grit Size and Form Accuracy

The size of the grit must be allowed for when machining the required form in the blank. This will be different to the nominal grit size and dependent on the final aspect ratio of the particular grit feed stock. For example, Table 6.1 gives standard values for an abrasive with an aspect ratio of 1.4.

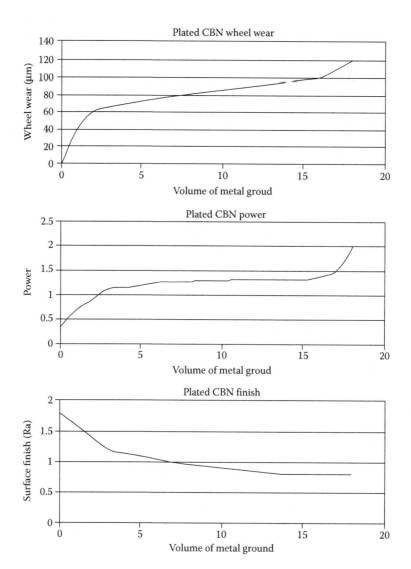

FIGURE 6.3
Typical performance characteristics of plated wheels (B126 grit size).

6.3.5 Wheel Wear Effects in Grinding

One major attraction of plated wheels is the fact that they do not require dressing, and therefore, eliminate the need for an expensive diamond form-roll and dressing system. However, plated wheels present challenges to the end user due to the effects of wheel wear. Figure 6.3 illustrates changes in grinding power, workpiece roughness, and wheel wear with time for a typical precision-plated wheel when CBN grinding aerospace alloys.

Initially, the surface roughness is high, as only the very tips of the grits are cutting. The power then rises rapidly together with an associated rapid rate of wheel wear and a drop in roughness. The process tends to stabilize, with wear flat formation being balanced by fracture, unless the grinding conditions are too aggressive. This leads to a much more

TABLE 6.1

Direct Plating Grit Size Allowances

FEPA	U.S. Mesh	Form Allowance (in.)	Form Allowance (μm)	Surface Concentration (ct/in.²)	Surface Concentration (ct/cm²)
B854	20/30#	0.0370	940		
B602	30/40#	0.0260	660	2.34	0.363
B427	40/50#	0.0180	455	1.8	0.279
B301	50/60#	0.0130	330	1.5	0.233
B252	60/80#	0.0110	280	1.4	0.217
B181	80/100#	0.0080	203	1.14	0.177
B151	100/120#	0.0066	168	1	0.155
B126	120/140#	0.0056	142	0.8	0.124
B107	140/170#	0.0046	117	0.67	0.104
B91	170/200#	0.0039	99	0.56	0.087
B76	200/230#	0.0034	86	0.47	0.073
B64	230/270#	0.0030	76	0.4	0.062
B54	270/325#	0.0026	66	0.33	0.051
B46	325/400#	0.0023	58	0.28	0.043

protracted period of time when the rates of change of all three variables are reduced by up to a factor 10. Failure occurs when power levels finally become so high that burn occurs, or the plating and grain are stripped from the core. This latter effect is particularly concerning because in most cases, it cannot yet be detected in advance, or predicted easily except by empirical data from production life values from several wheels. Certainly, attempts based on power or AE have not proved successful. One exception to this is a claim by Bremer et al. (2005) from General Motors that they could detect the onset of failure grinding camshafts and crankshafts by monitoring normal force though the infeed axis torque from a linear motor. In their example, the normal force jumped 40% from one part to the next prior to stripping (Figure 6.4).

6.3.6 Grit Size and Form-Holding Capability

Tables 6.2 and 6.3 give values for typical form-holding capabilities and roughness as a function of grit size for standard precision-plated and postplated conditioned wheels. Roughness values will vary somewhat dependent on workpiece hardness. The values indicated are those for grinding aerospace alloys in the hardness range 30–50 HrC using CBN abrasive.

6.3.7 Wheel Break-In Period

The phenomenon of a break-in period associated with a high rate of wear of a new wheel is particularly important when trying to hold tolerances of <0.001" (25 μm). Tables 6.2 and 6.3 give the break-in depth for both precision and conditioned wheels. As can be seen, with conditioning of the wheel surface to remove the tips of the grits, it is possible to virtually eliminate the break-in period. Although conditioning can double the price of the wheel, when trying to hold tolerances of <0.0005" (12 μm), it can be easily justified by increasing wheel life by an order of magnitude.

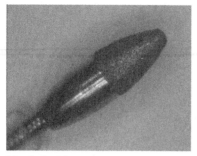

Plated wheel for removing plaque from the walls of an artery in the heart

Multi-wheel grinding of circlip grooves in transmission gear shaft

Deck grinding of engine blocks

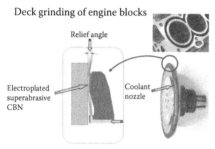

Grinding to replace milling of engine cylinder blocks

CBN wheel for rough grinding of crankshaft to replace turn broaching

FIGURE 6.4
Illustrate diverse applications of electroplated CBN wheels.

TABLE 6.2

Standard Precision-Plated-Wheel Form Capabilities

Grit Size FEPA	Grit Size U.S. Mesh	Minimum Radius (mm)	Precision Allowance Blank (μm)	Precision Tolerances "+/−" (μm)	Precision Roughness Ra	Precision Break-In Depth (μm)
B852	20/30	2	920	100		
B602	30/40	1.5	650	90		
B501	35/40	1.3	650	80		
B427	40/50	1	500	60	160	63
B301	50/60	0.8	400	40	125	
B252	60/80	0.7	290	30	85	40
B213	70/80	0.6	260	30	75	
B181	80/100	0.5	220	30	63	35
B151	100/120	0.4	190	30	38	
B126	120/140	0.3	165	25	35	
B107	140/170	0.26	140	25	32	
B91	170/200	0.23	130	25	32	20
B76	200/230	0.2	100	25	28	
B64	230/270	0.18	90	20	25	
B54	270/325	0.15	75	20	22	
B46	325/400	0.12	65	20	20	10

TABLE 6.3

Conditioned Plated-Wheel Form Capabilities

FEPA	Grit Size U.S. Mesh	Minimum Radius Capability (mm)	Condition Allowance Blank (μm)	Condition Tolerances "+/−"	Condition Roughness Ra	Condition Break-In Depth (μm)
B852	20/30	2	850	10		<5
B602	30/40	1.5	600	10		<5
B501	35/40	1.3	600	10		<5
B427	40/50	1	450	8	80	<5
B301	50/60	0.8	360	6	70	<5
B252	60/80	0.7	250–280	5	60	<5
B213	70/80	0.6	240–250	5	50	<5
B181	80/100	0.5	200	5	40	<5
B151	100/120	0.4	165	5	32	<5
B126	120/140	0.3	140–150	5	28	<5
B107	140/170	0.26	110	5	25	<5
B91	170/200	0.23	95	5	22	<5
B76	200/230	0.2	80	5	NA	<5
B64	230/270	0.18	70	4	NA	<5
B54	270/325	0.15	60	4	NA	<5
B46	325/400	0.12	50	3	NA	<5

NA—Not available.

6.3.8 Summary of Variables Affecting Wheel Performance

The number of variables for a given wheel specification that can make a significant impact on performance are quite limited. The plating thickness is held within a narrow band. The homogeneity of the plating is controlled by the plating rate and anode design. Care should be taken to avoid nodule formation especially around tight radii. Such areas are also the most prone to wear; this can be reduced by the use of electroless nickel–phosphorus for increased strength, evenness of plating, and maximum plate hardness. The biggest variable is the grit itself and how it wears under the prevailing grinding conditions. If the grit is too weak, then fracture and rapid wheel wear occur. If the grit is too tough, wear flats buildup and burn ensues.

6.3.9 Effect of Coolant on Plated Wheels

Another major factor is coolant. When grinding aerospace alloys with CBN in oil the high lubricity of the coolant ensures a slow but steady build up of wear flats. The lubricity of water-based coolant is much lower thus causing more rapid wear of the grain tips. However, water has much higher thermal conductivity than oil and induces thermal shock in the abrasive leading to weakening and fracture. When used wheels are examined, it is found that those used for grinding in oil may have lasted several times longer than those used in water yet they still have a high proportion of their layer depth remaining. Similar wheels used for grinding in water may have worn completely down to the nickel substrate. One conclusion is, therefore, to use a tougher grit when grinding in water but a slightly weaker grit in oil. An alternative would be to reduce the surface concentration of CBN when grinding in oil. Yet another factor raised in the literature is that there is far more tendency for the wheel to load in water-based coolant (Gift and Misiolek 2014). The question is, therefore, complicated but to date no water-based coolant has matched a good quality mineral oil. Where end users have tried to convert from oil- to water-based coolant for environmental issues, the drop in life has been on several occasions as extreme as 25× to 1000× grinding ferrous metals including nickel-based and steel alloy components, for example (Dasch et al. 2007).

Oil coolant is, therefore, required in most cases to obtain the necessary life to make a plated process competitive over alternate methods. Plated CBN in particular has proved extremely cost effective in aircraft engine subcontractors with low batch volumes of parts requiring profile tolerances in the 0.0004″–0.002″ as well as high-speed rough grinding of transmission shafts, camshafts and crankshafts, and gear profiles. But note the limitations on minimum finish and radius-holding capabilities shown in Tables 6.2 and 6.3.

6.3.10 Reuse of Plated Wheels

Used plated wheels are generally returned to the manufacturer for strip and replate. The saving is typically about 40%, the steel core can be reused at least five to six times if the wheels are not abused (e.g., pushed beyond point where plating strips in the grind), and with careful control of the stripping process. Blanks can also be remachined.

6.4 Brazed Single-Layer Wheels

Electroplating is a low-temperature process (<100°C) in which the plating holds the abrasive mechanically. Consequently, the plating depth required to anchor the abrasive needs to be at least 50% of the abrasive height. An alternative process is to chemically bond the CBN to the steel hub by brazing using a relatively high-temperature metal alloy system based on, for example, Ni/Cr with trade names such as MSL™ (metal single layer) from Saint-Gobain Abrasives (Peterman n.d.; Chattopadhyay et al. 1990; Lowder and Evans 1994). Use of a chemical bonding method allows a much greater exposure of the abrasive and hence an increased usable layer depth. It also gives greater chip clearance and lower grinding forces. However, brazing occurs at temperatures up to a 1000°C and that can degrade the grit toughness and distort steel blanks. Braze also wicks up around the grit placing it under tensile stress upon cooling and thus further weakening it. Consequently, the biggest markets for brazed wheels tend to be for high-stock removal roughing operations of materials such as fiber glass, brake rotors, and exhaust manifolds, fettling of iron castings or applications with form tolerances >0.002″ (50 μm). For a schematic of a brazed CBN wheel section and the appearance of the actual surface, see Figure 6.5.

6.4.1 Controlled Grain Spacing

One other market that spurred brazed diamond technology was the semiconductor industry, in particular, as an option to eliminate "pop out" or diamond loss in electroplated chemomechanical polishing (CMP) conditioning pads for wafer grinding (Ohi 2004). Polishing of silicon wafers is a highly demanding and expensive process. CMP conditioning pads dress the fine diamond cup wheels that polish the actual wafer generating both the initial wheel flatness and regenerating an appropriate cutting action. Brazing ensured a strong bonding with the diamond under high normal forces. Time and cost could also be

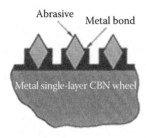

FIGURE 6.5
Schematic of brazed CBN wheel section and appearance of the actual surface of such a wheel.

invested in the manufacture to control the spacing and even orientation of the diamonds, where researchers studied highly regular, fully randomized, and clustered structures. Additionally, resin- and diamond-like coatings were developed to protect the dresser surface from the chemicals in the polishing slurry (Sung and Sung 2008). Highly randomized grain distributions such as Norton's "self-avoiding random distribution" gave highly uniform *texture*, while very regular arrays of grains could give *surface finishes* approaching regular-bonded diamond wheels.

Considerable interest has since been raised in extending this technology to grinding as a tool to replace milling due to the ability to handle high-stock removal rates with low surface roughness.

Controlled grain spacing offers considerable promise for difficult to grind materials such as soft rubber composites, aluminum alloys, and polycarbonates to tough advanced materials such as carbon fiber reinforced composite (CFRP) (see Figures 6.6 and 6.7). Various methods are mentioned in the literature to generate regular grain arrays including simple grid and template screens to microreplication and glue dot printing (Burkhard 2002; Yuan and Gao 2008).

Efforts have even been made to orientate the grain points vertically using an electrostatic charge (Ohashi 2012); a technique common to sandpaper manufacture but much more difficult to apply for high-temperature brazing with a conducting medium.

The resulting tools have been called "engineered grinding tools" (Pinto et al. 2008) characterized by a predetermined and controlled arrangement of abrasive grains. In general, it has been found that the roughness of the ground surface is controlled by the protrusion height distribution and is independent of the maximum protrusion height. The surface

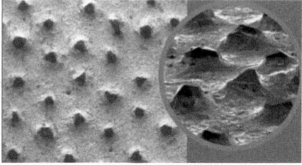

FIGURE 6.6
Example of regular grit spacing achieved for a single-layer wheel.

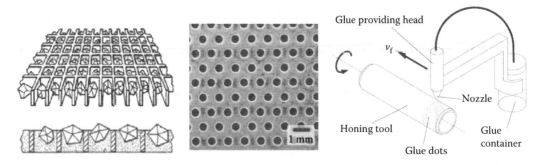

FIGURE 6.7
Examples of techniques for achieving controlled grain spacing.

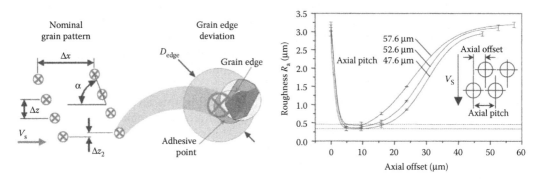

FIGURE 6.8
Effect of grain positions on roughness.

quality can therefore be optimized by tailoring the distribution to be uniform. In a wheel with an arrayed arrangement of grains, the finish is a strong function of the axial offset between adjacent rows of grains (see Figure 6.8). Grain shape/orientation is also critical (Koshy et al. 2003).

Coolant application is also improved, reducing grinding temperatures and can be considered an extension of older processes of slotting plated wheels by reducing slot widths to essentially a grain row.

Homogeneity of grain distribution has recently carried over to influence developments in bonded wheels especially in vitrified bonds.

6.5 Vitrified Bond Wheels for Conventional Wheels

6.5.1 Application of Vitrified Bonds

Vitrified bond alumina wheels represent nearly half of all conventional wheels and are employed for the great majority of precision high-production grinding applications. Vitrified superabrasive technology, especially for CBN, is the fastest growing sector of the precision grinding market but is still less than 20% of the market total. Figure 6.6 shows roughing and finishing of exhaust manifolds using metal single-layer and vitrified CBN wheels (Figure 6.9).

FIGURE 6.9
Roughing and finishing of exhaust manifolds using metal single-layer and vitrified CBN wheels. (Photo courtesy of Campbell Grinders.)

6.5.2 Fabrication of Vitrified Bonds

Vitrified bonds are essentially glasses made from high-temperature sintering of powdered glass frits, clays, and chemical fluxes such as feldspar and borax. The attractions of vitrified bonds are their high-temperature stability, brittleness, rigidity, and their ability to support high levels of porosity in the wheel structure. The mixture of frits, clays, and fluxes are blended with abrasive and a binder, such as dextrin and water. The mixture is pressed in a mold usually at room temperatures. The binder imparts sufficient green strength for the molded body to be mechanically handled to a kiln, where it is fired under a well-controlled temperature/time cycle in the range of 600–1300°C depending on the abrasive and glass formulation. The frit provides the actual glass for vitrification, the clays are incorporated to provide green strength up to the sintering temperature, while the fluxes control/modify the surface tension at the abrasive grain–bond interface. Clays and flux additions therefore control the amount of shrinkage which, except for the very hardest of wheel grades, is kept to a minimum. It should also be noted that the pressing stage in wheel manufacture, which is done either to a fixed pressure or fixed volume, provides a controlled volume of porosity after firing.

Some typical conventional vitrified wheel specifications are given in Table 6.4 that comply with standard coding practice.

6.5.3 Structure and Grade of Conventional Vitrified Wheels

In addition to the grit type and size discussed above, it can be seen that two other factors are key to the wheel specification: *grade* or hardness designated by a letter, and *structure* designated by a number (see Table 6.4).

6.5.4 Mixture Proportions

To understand how these factors relate to the physical properties of the wheel, consider first how loose abrasive grains pack together under pressure. If grains with a standard size

TABLE 6.4

Commercial Examples of Wheel Designations

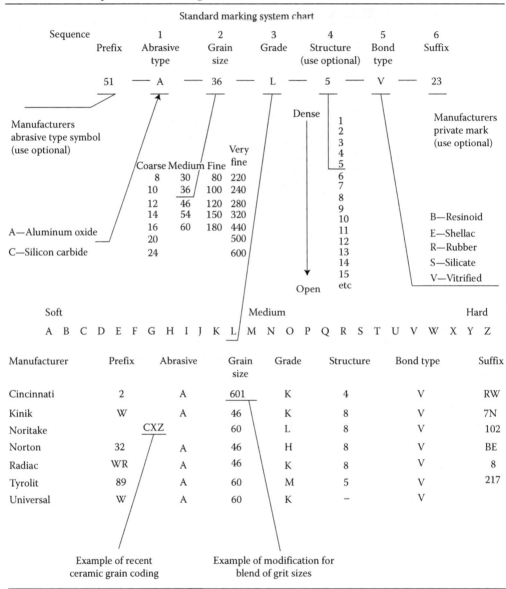

Manufacturer	Prefix	Abrasive	Grain size	Grade	Structure	Bond type	Suffix
Cincinnati	2	A	601	K	4	V	RW
Kinik	W	A	46	K	8	V	7N
Noritake		CXZ	60	L	8	V	102
Norton	32	A	46	H	8	V	BE
Radiac	WR	A	46	K	8	V	8
Tyrolit	89	A	60	M	5	V	217
Universal	W	A	60	K	–	V	

Example of recent ceramic grain coding

Example of modification for blend of grit sizes

distribution are poured into a container and tamped down, they will occupy about 50% by volume. It will also be noticed that each grain is in contact with its neighbors resulting in an extremely strong and rigid configuration. Now consider the effect of adding the vitrified bond to this configuration. The bond is initially a fine powder and fills the interstices between the grains. Upon sintering, the bond becomes like a viscous liquid that wets and coats the grains. There is usually actual diffusion of oxides across the grain boundary resulting in chemical and as well as physical bonding. If, for example, 10% by volume of a vitrified bond had been added, then a porosity of 40% would remain. The size and shape of individual pores are governed by the size and shape of the grains. The percentage of abrasive that can be packed into a given volume can be increased to greater than 60% by broadening the grain size distribution. The volume of abrasive can also be reduced to as low as 30%, while maintaining mutual grain contact by changing the shape of the abrasive. For example, long, needle-shaped (high-aspect ratio) abrasive grains have a much lower packing density than standard grain (DiCorletto 2001).

6.5.5 Structure Number

However, consider the situation where the grit volume of a standard grit distribution is now reduced from 50%. The most obvious effect is that immediately some of the grains stop being in contact. The integrity and strength of the whole can now only be maintained in the presence of the bond that fills the gaps created between the grains and provides the strength and support. These points are called bond posts and become critical to the overall strength and performance of the wheel. As the grit volume is further reduced the bond posts become longer and the structure becomes weaker. Not surprisingly therefore the abrasive volume percent is a critical factor and is designated by the wheel manufacturers as *structure number*. For example, Coes (1971) gives the following association between grit structure number and abrasive percent for Norton brand wheels.

Structure number	0	1	2	3	4	5	6	7	8
Abrasive vol.%	68	64	60	58	56	54	52	50	48

Kinik (n.d.) reported for their brand of wheels that structure number is related to abrasive volume by

Structure number	0	1	2	3	4	5	6	7	8	9	10	11	12	13	14
Abrasive vol.%	62	60	58	56	54	52	50	48	46	44	42	40	38	36	34

Each supplier uses slightly different notation and most are not generally reported for competitive reasons.

6.5.6 Grade of Conventional Vitrified Wheels

With the abrasive volume defined, the remaining volume is shared between the bond and porosity. The bond bridges can obviously be strengthened by increasing the amount of bond to make them thicker. The greater the amount of bond present, the lower the porosity and the harder the wheel will act. The actual definition of *grade* will again vary from supplier to supplier. For some, it is simply a direct correlation to porosity; for others, it is a more complicated combination of porosity percent P and structure number S. Malkin (1989) gives one supplier's system where the grade letter is correlated according to

$$\text{Grade} \propto 43.75 - 0.75P + 0.5S$$

This definition is designed to make grinding performance characteristics relate to grade (e.g., burn, dressing forces, and power) so that grinding performance changes more predictably from one grade letter to the next.

The interplay can also be seen in Figure 6.10, which gives the porosity level for various Norton grade letter/structure number combinations.

More formally, vitrified bond systems are described by ternary-phase diagrams that map the allowed bond/grain/porosity combinations as shown in the example in Figure 6.11 (DiCorletto 2001).

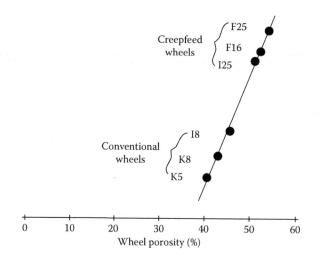

FIGURE 6.10
Porosity for typical Norton grade/structure combinations. (Based on Engineer, F. et al., 1992, *Journal of Engineering for Industry* 114, 61–66.)

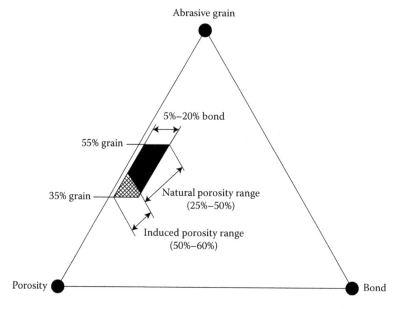

FIGURE 6.11
Ternary phase diagram showing operating range for alox vitrified bond systems. (DiCorletto, Saint-Gobain Abrasives.)

6.5.7 Fracture Wear Mode of Vitrified Wheels

In addition to the size of the bond bridge, the fracture mode is also critical. The bond must be strong enough to hold the grains under normal grinding conditions, but under higher stress, it must allow the grain to fracture in a controlled way. The bond should not be so strong relative to the grit strength that the abrasive glazes and leads to burn (Figure 6.12).

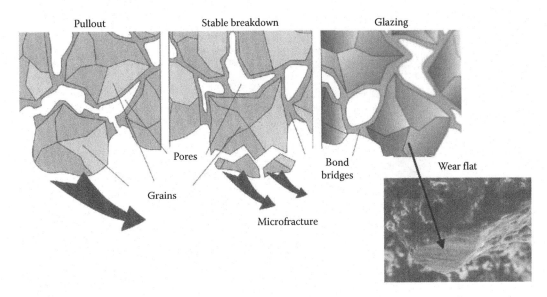

FIGURE 6.12
Pullout, stable breakdown, and glazing regimes in grinding. (Based on a drawing by Rappold, 2002, *Cylindrical Grinding*. Rappold—Winterthur, Trade Brochure, 02/2002 #136551.00.)

One method to regulate this is by adding fine quartz or other particles to the bond to control crack propagation. Another is to recrystallize the glass creating nucleation centers that act in a similar fashion.

6.5.8 High-Porosity Vitrified Wheels

The primary attraction for producing wheels with high structure numbers is to allow the highest levels of porosity to be produced, while still maintaining structural integrity. This provides for very good coolant access and chip clearance in the grinding process. However, it is very difficult to maintain green strength and the integrity of the pores during manufacture of the wheel without additives to act as structural supports or "pore formers." These are typically either hollow particles such as bubble alumina, glass beads, or mullite which remain an integral part of the wheel structure but break open at the grinding surface, porous grains such as Vortex™ or fugitive materials such as napthalene, sawdust, or crushed walnut shells that burn out in the firing process. Hollow particles maintain a strong and coherent wheel structure, while fugitive fillers leave a structure with a high permeability that allows coolant to be carried deep into the wheel. Fugitive pore formers also allow a great flexibility in the shape and size of the pore as shown in Figure 6.13. In particular, pore formers allow pore sizes much greater than the grit size to be readily induced.

6.5.9 Multiple Pore Size Distributions

Wheel manufacturers such as Universal Grinding Wheel (Saint-Gobain Abrasives) have taken this concept further, and produced wheels with multiple pore former size distributions to create both macroporosity for high permeability, and microporosity for controlled fracture of the bond. This type of wheel, with trade names such as Poros 2™, has proved very effective for creep feed grinding where coolant delivery into the grinding contact zone is critical for avoidance of burn (see Figure 6.14).

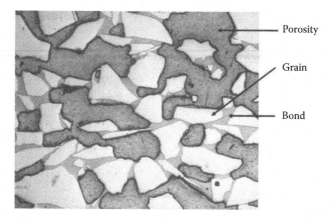

FIGURE 6.13
Polished surface of an induced porosity vitrified wheel structure. (Resin has been used to infiltrate the pores for sample preparation.)

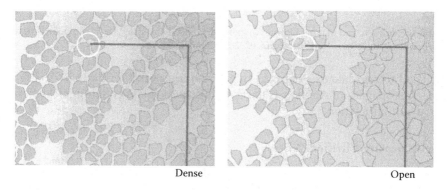

Dense   Open

FIGURE 6.14
Comparison of regular induced porosity wheels and Poros 2™ dual-structure wheels.

6.5.10 Ultra-high-Porosity Vitrified Wheels

The introduction of extruded seeded gel needle-shaped grains has provided another opportunity for creating extremely porous and permeable structures. The natural packing density of grains with an aspect ratio of 8:1 is about 30% by volume. Norton (Saint-Gobain Abrasives) has recently developed a product called Altos™ with a totally interlinked porosity as high as 65%–70%. The structure contains only a few percent of bond, but is nevertheless very strong because the bond migrates and sinters at the contact points between grains acting analogous to "spot welds." The high structural permeability allows prodigious amounts of coolant to be carried into the grind zone. This type of wheel gives probably the highest stock removal rates of any vitrified wheel, higher even than those possible with vitrified CBN, together with excellent G ratios for a conventional abrasive. Impressive numbers have also been reported with the cubitron 2 grain from 3M (2012). Major opportunities are being found in a number of industries including grinding difficult burn-sensitive materials such as nickel-based alloys for the aerospace and land-based power generation industries, as well as on soft steels for rough grinding of gears to replace hobbing.

6.5.11 Combining Grade and Structure

In very broad terms wheel grades E through I are considered soft and are usually used with high structure numbers (11–20 with induced porosity) for creep feed and burn sensitive applications. Grades J through M are considered medium grade usually used with lower structure numbers for steels and regular cylindrical and internal grinding. Very hard wheels are produced for applications such as ball bearing grinding. These wheels are X or Z grade and can contain as little as 2% porosity. Specifications of this hardness are produced by either hot pressing or by over-sintering such that the bond fills all the pores. In this type of application, structural numbers can vary from 8 up to >24. Their use is limited to relatively few specialist applications such as the grinding of ball bearings.

6.5.12 Lubricated Vitrified Wheels

The pores may also be filled with lubricants such as sulfur, wax, or resin by impregnating regular wheel structures after firing. Sulfur, a good high-temperature extreme pressure lubricant, is common in the bearing industry for internal wheels, although becoming less popular due to environmental issues.

6.6 Vitrified Bonds for Diamond Wheels

6.6.1 Introduction

Vitrified bond selection for diamond must take a number of considerations into account that places different demands relative to conventional wheels.

These are primarily the effects of

1. Hard work materials
2. Low chemical bonding
3. High grinding forces
4. Reactivity with air at high temperatures

These considerations are discussed as follows.

6.6.2 Hard Work Materials

Materials ground with diamond tend to be hard, nonmetallic, and brittle materials. Therefore, there are limited issues with wheels loading up with grinding debris and wheel porosity can be relatively low. On the other hand, hard workpiece debris is likely to cause much greater bond erosion than other work materials. Therefore, either the bond erosion resistance must be higher or, more practically, a lot more bond must be used.

6.6.3 Low Chemical Bonding

Diamond does not show significant chemical bonding with components in a vitrified bond. The bond must therefore rely primarily on mechanical bonding sometimes enhanced with various diamond grain coatings either to improve wetting or mechanical anchorage.

6.6.4 High Grinding Forces

Grinding forces with diamond can be very high and efforts are made to limit the forces by reducing the number of cutting points by significantly lowering the volume of diamond from 50%. This introduces the term *concentration* which is a measure of the volume of superabrasive per unit volume of wheel; 200 concentration is equivalent to 8.8 ct/cm³ by weight or 50% by volume. Most diamond wheels are typically 12–100 concentration.

6.6.5 Diamond Reactivity with Air at High Temperatures

Diamond reacts with air at temperatures above 650°C. Therefore, the wheels must either be fired at low temperatures, or in an inert or reducing atmosphere. Very low-temperature bonds, however, were traditionally very prone to dissolution in water. That limited shelf life and made air firing unattractive. A simple method was, therefore, developed to manufacture wheels at higher temperatures by hot pressing using graphite molds. The graphite generated a reducing carbon-rich atmosphere locally. Since the mold strength was low the bond had to be heated above regular sintering temperatures to limit pressing pressures. Consequently, the bonds fully densified with <2% open porosity. Pockets could be generated in the wheel by adding soft lubricants materials such as graphite or HBN that wears rapidly on exposure at the grinding surface; fugitive fillers were also added to burn out during firing. Nevertheless, the wheels had a major limitation—their bond content was so high they could not be automatically dressed using diamond tools. As such they fell into the same category as metal and resin bonds (see below) that had to be trued and then subsequently conditioned using dressing blocks. On many grinders, the process is semi- or even fully automated with skilled operators.

This type of bond structure had been standard for many years, and used extensively in applications such as double disk and polycrystalline diamond (PCD) grinding (see Figure 6.15), although it is now superseded in most cases by newer porous bond technologies (see Figure 6.16).

FIGURE 6.15
Hot pressed fully densified vitrified diamond bond structure.

FIGURE 6.16
A porous vitrified fine-grain diamond structure employed for grinding of polycrystalline diamond inserts.

6.6.6 Porous Vitrified Diamond Bonds

Dressable porous cold-pressed vitrified diamond bond technology has been driven by the increased use of PCD and carbide for cutting tools, the growth of engineering ceramics, and the introduction of carbide HVOF coatings to replace chrome. Vitrified diamond bonds, much used in conjunction with micron sizes of diamond grit, are employed for edge grinding of PCD and polycrystalline cubic boron nitride (PCBN) cutting tools, thread grinding of carbide taps and drills, and fine grinding and centerless grinding of ceramics, for example, seals and some diesel engine applications. A number of these bonds are now being dressed automatically with rotary diamond dressers on the grinder without subsequent conditioning.

6.7 Vitrified Bonds for CBN

6.7.1 Introduction

When CBN was introduced into the market in 1969, its cost naturally lent itself to being processed by wheel makers that knew how to handle expensive abrasive—namely diamond wheel makers, using the dense hot-pressed vitrified systems described above. Unfortunately, these had none of the properties, such as chip clearance and dressability, required for high-production grinding of steels where CBN would prove to be most suited.

Furthermore, vitrified bonds used by conventional wheel makers were so reactive that they literally dissolved away all the CBN into the bond by converting it into boric oxide. Grit suppliers tried to counter this by producing CBN grains with thin titanium coatings on them. Unsurprisingly, it took 10 years and numerous false starts before porous vitrified bonds with the capability of being dressed automatically were finally presented to the market. While some manufacturers still pursued hot-pressed bonds with high fugitive or other filler content (Li 1995), the majority developed controlled reactivity cold-pressed bonds using methods common to processing of conventional vitrified wheels. Just as with conventional abrasives, it was possible to modify the bond formulations to obtain just sufficient reactivity and diffusion to create strong wetting and bonding.

6.7.2 Requirements for Vitrified CBN Bonds

The demands of vitrified bonds for CBN differ again from those for either conventional or diamond bonds. Typical wheel supplier specifications, in compliance with standard coding practice, are shown in Table 6.5. The wheel specification format is dictated by the standard practices of the diamond wheel industry. As such, the hardness is expressed as a grade letter but wheel structure is often not given, or described in only the vaguest of terms. As with vitrified diamond, concentration plays a key role in controlling the number of cutting points on the wheel face. Concentrations for CBN wheels, however, tend to be higher than in diamond wheels at up to 200 concentration (50% by volume) especially for internal and many cylindrical grinding applications. This limits the structural number to a relatively narrow range.

6.7.3 CBN Wheel Structures

Typical porous vitrified CBN wheel structures are shown in Figure 6.17.

The sequence of photographs in many ways represent the development of CBN bonds over the last 30 years. The initial wheels were very dense structures with porosity levels of the order of 20%. With the high cost of CBN, performance was focused on achieving maximum possible wheel life. With the development of cylindrical grinding applications for burn-sensitive hardened steels in the 1980s, the porosity levels rose to 30%. More recently, with the rapid expansion of CBN into aerospace and creep feed applications, porosity levels have risen to the order of 40%. Further development in this area appears key to several wheel makers, for example (Noichl n.d.).

6.7.4 Grades of CBN Wheels

A comparison with porosity levels in Figure 6.13 shows clearly that the grade of CBN wheel for a given application, even with the development of higher porosity structures, is much denser than for conventional wheels. This is hardly surprising, as CBN is so expensive, it must be held for a much greater period of time, even if higher levels of wear flats are created. This is possible because of the high thermal diffusivity of CBN relative to both alox and most workpiece materials. A number of wheel manufacturers do try to mark up CBN wheels grades to be close to those of conventional wheels for the purpose of helping end users more familiar with conventional wheels specify a given wheel for an application. However, it must be understood that usually dressing forces will be much higher both because of the grain hardness and of the additional bond, while hydrodynamic forces from the coolant will also be higher because of the lower porosity. This will place additional challenges on the system stiffness and create a need for new strategies for achieving part tolerances. Some relief is becoming available in terms of higher strength bonds that allow porosity levels to shift back toward those of alox wheels again but this is often counteracted by higher working wheel speeds.

TABLE 6.5

Commercial Examples of Vitrified CBN Wheel Specifications

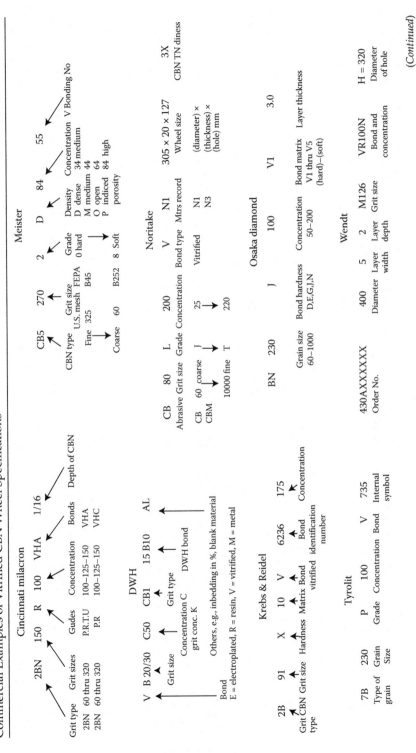

(Continued)

Q7

TABLE 6.5 (*Continued*)

Commercial Examples of Vitrified CBN Wheel Specifications

Efesis (FAG)

Example 853 B64 P 8 V7153 C192
　　　　　(1) (2) (3) (4) (5) (6)

(1) Abrasive type (B = CBN)
(2) Grit size (μm)
(3) Grade
(4) Structure
(5) Vitrified bond
(6) Concentration

TVMK (Toyoda Van Moppes)

Example B 200 N 150 V BA 3.0
　　　　　(1) (2) (3) (4) (5) (6) (7)

(1) Grit type
(2) Grit size (U.S. Mesh)
(3) Hardness (grade)
(4) Concentration
(5) Bond type
(6) Bond feature
(7) Layer depth (mm)

Unicorn (Universal)

Example 1B 126 M 150 V SS
　　　　　(1) (2) (3) (4) (5) (6)

(1) Grit type
(2) Grit size (μm)
(3) Grade
(4) Concentration
(5) Vitrified bond
(6) Bond system

Norton

Example 1B 220/1 J 175 VX322C
　　　　　(1) (2) (3) (4) (5)

(1) Grit type
(2) Grit size
(3) Grade
(4) Concentration
(5) Vitrified bond system

Unicorn (Indimant)

Example 49 B126 V36 W2J6V G 1M
　　　　　(1) (2) (3) (4) (5) (6)

(1) Grit type
(2) Grit size (μm)
(3) Concentration (V36 = 150 conc.)
(4) Vitrified bond system
(5) Grade
(6) Internal coding

Winter

Example B 64 VSS 34 26 G A18C V360
　　　　　(1) (2) (3) (4) (5) (6) (7) (8)

(1) Grit type
(2) Grit size (μm)
(3) Vitrified bond system
(4) Structure
(5) Bond code
(6) Grade
(7) Mfg codes
(8) Concentration (V360 = 150 conc.)

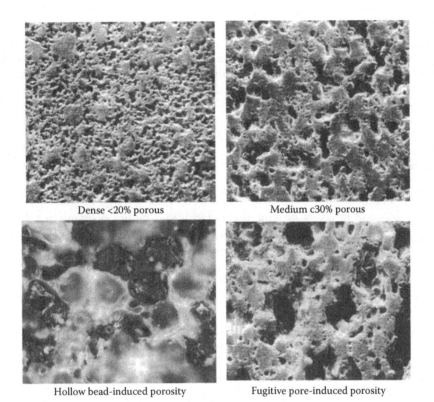

FIGURE 6.17
Vitrified CBN wheel structures (polished surfaces).

6.7.5 Firing Temperature

The actual glass bonds and manufacturing techniques used for vitrified CBN wheels are highly proprietary and there is rapid development still in progress. General Electric in 1988, for example, recommended that bonds with CBN should not be fired at temperatures over 700°C (GE Superabrasives 1988). Yet, Yang in 1998 for a nominally identical bond composition subsequently found the optimum firing temperature to be 950°C (Yang 1998).

6.7.6 Thermal Stress

An important factor is to match thermal expansion characteristics of the glass with the abrasive (e.g., Balson 1988), or to optimize the relative stress developed between bond and grain in the sintering process. Yang et al. (1993) reported this could be readily optimized by adjustments to minor alkali additives primarily sodium oxide.

6.7.7 Bond Mix for Quality

As with bonds for conventional abrasives, bond strength can be improved by the introduction of microinclusions for crack deflection either in the raw materials or by recrystallization of the glass, for example, (Valenti 1992). With the far greater demand for life placed on CBN vitrified bonds and the narrower working range of grades available, quality control of composition and particle size of the incoming raw materials, and the firing cycles used to sinter the bonds are critical. It has very often been process resilience, as demonstrated by batch-to-batch consistency in the finished wheels, which has separated a good wheel specification from a poor one.

6.8 Resin Bond Wheels

Resin covers a broad range of organic bonds fabricated by hot pressing at relatively low temperatures, and characterized by the soft nature of cutting action, low-temperature resistance and structural compliance. The softest bonds may not even be pressed but merely mixed in liquid form with abrasive and allowed to cure. Concepts of grade and structure are very different to vitrified bonds. There is no interlocked structure with bond bridges (because there is minimal porosity), but rather an analogy would be to compare the grains to currants in a currant bun. Retention is dependent on the localized strength and resilience of the bond surrounding the grain and very sensitive to localized temperatures created in the grind zone and the chemical environment. For example, the bond is susceptible to attack by alkali components in coolants.

Resin bonds can be divided into three classes based on strength/temperature resistance (Figure 6.18). These are namely:

- Plastic
- Phenolic resin
- Polyimide resin

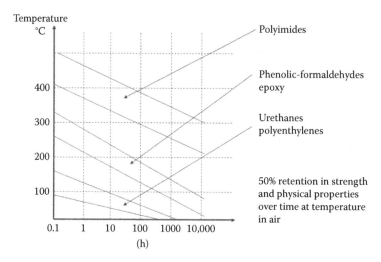

FIGURE 6.18
Temperature/time properties of resins.

6.9 Plastic Bonds

These provide the softest wheels made using epoxy or urethane type bonds. Used with conventional abrasives, plastic wheels are popular for double disk and cylindrical grinding. At one time, prior to the introduction of vitrified CBN, these were the primary wheels for grinding hardened steel camshafts where they gave both a very soft grinding action and a compliance that helped inhibit the generation of chatter. They are still popular in the knife industry and in job shops for grinding burn-sensitive steels. Manufacturing costs and cycle times are low so pricing is attractive and delivery times can be very fast.

For superabrasive wheels, plastic bonds appear limited to ultrafine grinding applications using micron diamond grain for the glass and ceramics industries. Again its compliance offers an advantage of finer surface finish capabilities but wheel life is limited.

6.10 Phenolic Resin Bonds

6.10.1 Introduction

Phenolic bonds represent the largest market segment for conventional grinding wheels after vitrified bonds, and dominate the rough grinding sector of the industry for snagging and cut-off applications. The bonds consist of thermosetting resins and plasticizers that are cured at around 150–200°C. The bond type was originally known as "Bakelite" and for this reason still retains the letter "B" in most wheel specifications. Grade or hardness is controlled to some extent by the plasticizer and use of fillers.

6.10.2 Controlled-Force Systems

Unlike vitrified wheels, most resin wheels are used under controlled pressure, that is, controlled force rather than fixed infeed systems and very often at high speed.

6.10.3 Abrasive Size

Abrasive size is usually used to control recommended grade. Finer grit wheels remove material faster for a given pressure but wear faster and are used to avoid the excessive porosity that would be required in a coarse wheel to get cutting action. Porosity reduces burst speed and allows grits to be easily torn out. With higher available pressures, coarser grit sizes can be used (Table 6.6). Glass fibers are also added to reinforce cut-off wheels for higher burst strength.

6.10.4 Benefits of Resilience

Resin bonds are also used for precision applications where its resilience provides benefits of withstanding interrupted cuts and better corner retention. One such area is flute grinding of steel drills where the wheel must maintain a sharp corner and resist significant side forces. There have been enormous improvements in life and removal rates over the last 10 years with the introduction of SG and most recently TG abrasives. Some are capable of

TABLE 6.6

Standard Pressure Ranges for Conventional Resin Bond Cut-Off Wheels

Operation	Pressure Range	Abrasive Size
Portable grinder	10–25 lbf	16–36#
Floorstand	25–100 lbf	12–20#
Swing frame	100–200 lbf	10–16#
Remote control machine	200–2000 lbf	8–14#

Source: Pressure controlled grinding—based on Coes, L. Jr, 1971, *Applied Mineralogy 1—Abrasives*, Springer-Verlag, New York.

grinding at $Q' > 100$ mm^3/mm s, while still producing several drills between dresses. This is one example where advances in conventional engineered abrasives are competing very successfully with emerging CBN technologies.

6.10.5 Phenolic Resin Bonds for Superabrasive Wheels

For superabrasive wheels, phenolic resin bonds represent the earliest, and most popular, bond type particularly for diamond wheels and especially for toolroom applications. The bonds were originally developed for diamond with the introduction of carbide tooling in the 1940s. Their resilience made them optimal for maintaining tight radii, while withstanding the impact of interrupted cuts typical of drill, hob, and broach grinding. To prevent localized temperature rise, the abrasive is typically metal coated to act as a heat sink to dissipate the heat. In addition, high volumes of copper or other metal fillers may be used to increase thermal conductivity and heat dissipation.

Not surprisingly, phenolic resin bonds were quickly adopted with the introduction of CBN in 1969, and phenolic resin bonds predominate for the steel tool industry (Craig 1991).

6.10.6 Wheel Marking Systems for Resin Bonds

Because the basic technology is so mature, the number of wheel makers is too numerous to list. However, the marking system for wheels is covered by standards such as ANSI B74-13 shown below for the United States or JIS B 4131 for Japan (Table 6.7; Koepfer 1994).

Many wheel makers are located close to specific markets to provide quick turnaround. Alternatively, many are sourced from low-cost manufacturing countries. The key to gaining a commercial advantage in this type of competitive environment is application knowledge either by the end user developing a strong database and constant training, or using the knowledge of the larger wheel makers with strong engineering support.

6.11 Polyimide Resin Bonds

6.11.1 Introduction

Polyimide resin was developed by DuPont in the 1960s originally as a high-temperature lacquer for electrical insulation. By the mid-1970s, it had been developed as a

TABLE 6.7

ANSI Standard Marking Systems for Superabrasive Wheels

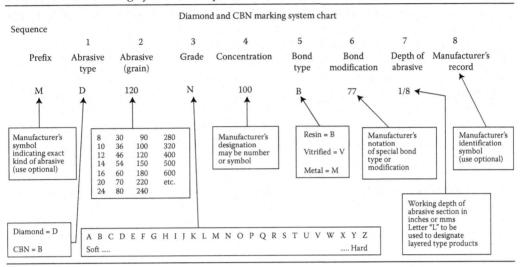

Diamond and CBN marking system chart

cross-linked resin for grinding wheels giving far higher strength, thermal resistance and lower elongation than conventional phenolic bonds. The product was licensed to Universal Diamond Products (Saint-Gobain Abrasives) and sold under the trade name of Univel™, where it came to dominate the high-production carbide grinding business especially for flute grinding. Polyimide has 5–10 times the toughness of phenolic bonds and can withstand temperatures of 300°C for 20 times longer. Its resilience also allows it to maintain a corner radius at higher removal rates or for longer times than phenolic resin (Figure 6.19).

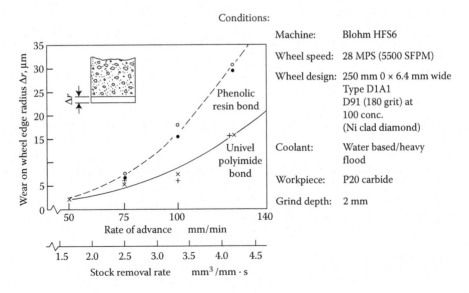

Conditions:	
Machine:	Blohm HFS6
Wheel speed:	28 MPS (5500 SFPM)
Wheel design:	250 mm 0 × 6.4 mm wide Type D1A1 D91 (180 grit) at 100 conc. (Ni clad diamond)
Coolant:	Water based/heavy flood
Workpiece:	P20 carbide
Grind depth:	2 mm

FIGURE 6.19

Comparison of the performance of phenolic and polyimide bonds.

6.11.2 Cost Developments and Implications

Polyimide bonds, for reason of cost, are limited to superabrasives and are most effective with diamond abrasives on carbide. Because the wheels are so tough, they will highlight any weakness in the machine such as spindle play or backlash in the infeed system. They also require a spindle power of at least 7.5 kW/cm linear contact width (25 hp/in.).

In the past few years, with the expiry of various patents, alternate sources for polyimide resin have become available. They are significantly less expensive than the Du Pont-based process but to date have not quite matched the performance. However, the price/performance ratio is still very attractive making polyimide resin bonds cost-competitive relative to phenolic resin bonds in a broader range of applications.

6.11.3 Induced Porosity Polyimide

In some applications, the Univel product has proved so tough in comparison to regular phenolic bonds that induced porosity techniques from vitrified bond technology have been used to improve the cutting action.

6.12 Metal Bonds

6.12.1 Introduction

Metal represents the toughest and most wear resilient of bond materials, and is almost exclusively used with superabrasives. Much of this is for stone and construction, glass grinding, and honing. As such, it is the largest user of synthetic industrial diamond, but falls outside the scope of this book, being very much roughing operations.

6.12.2 Bronze Alloy Bonds

Metal bonds for production grinding has traditionally been based on bronze in the copper–tin alloy range of 85:15–60:40 with various fillers and other small alloy components like chromium. Metal bonds are the most resilient and wear resistant of any of the bonds discussed but also create the highest grinding forces, and the most problems in dressing. Their use has been limited to thin wheels for dicing and cut off, profile grinding, fine grinding at low speeds, and high-speed contour or peel grinding. This latter process is dependent on maintaining a well-defined point on the wheel and therefore the maximum wheel life. However, in many cases involving CBN, metal has been replaced by vitrified bond, even at the sacrifice of some wheel life, in order to improve the ease of dressing. Introductions of novel dressing techniques such as the STUDER-WireDress™ EDM process, ECM or even pulsed laser dressing may start to reverse this trend (Weingartner et al. 2012; Walter et al. 2012).

6.12.3 Porous Metal Bonds

Porous metal bonds in one form or another have been under development since the 1960s (Chalkley and Thomas 1972). Metal bronze bonds become more brittle and weaker as the tin content is increased. Raising the porosity level to the point of having interconnected porosity while still maintaining a reasonable strength was achieved by creating a sintered metal skeleton and vacuum impregnating the pores with resin to create a hybrid bond. This was sold under trade names such as Resimet™ from Van Moppes (Saint-Gobain Abrasives). This type of bond was extremely successful for dry grinding applications on tool steels and carbide. It was freer cutting and gave longer life than resin, requires no conditioning, while the metal bond component offered an excellent heat sink.

In the 1980s, brittle metal bond systems began to emerge for effectively holding the diamond grains while having just sufficient porosity that profiles could be formed in the wheel automatically by crush dressing using steel or carbide form rolls.

6.12.4 Crush Dressing

Bonds suitable for crush dressing, sold under trade names such as Crushform™ developed by Van Moppes-IDP (Saint-Gobain Abrasives) (Daniel 1983; Barnard 1985, 1989a,b), were of particular interest to the carbide tool insert market. However, there were some problems with this type of wheel. The dress process did not leave the wheel in a free-cutting state and therefore the surface had to be subsequently conditioned using dressing sticks or brushes. This was readily resolved as shown in Figure 6.20. Note the horizontal brush infeed in the top left picture.

The bigger problem, however, was the extreme forces generated in dressing. Where the use of crushable metal bond wheels has been successful, such as at OSG in Japan (Yoshimi and Ohshita 1986), special high-stiffness grinders have had to be built specifically for their use.

FIGURE 6.20
"Crush-form" dressing of porous metal bond diamond wheel for form grinding of carbide tools. (Courtesy of Saint-Gobain Abrasives.)

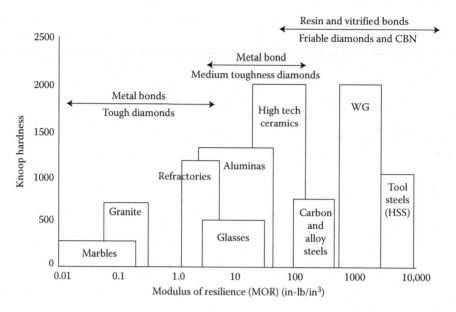

FIGURE 6.21
Jakobuss (1999) illustrates superabrasive bond and grain selection as a function of workpiece resilience.

As such the use of crushable metal bonds has been limited awaiting advances in standard production machine tool stiffness. During the period 1990–2010, vitrified technology was therefore substituted in most cases (Pung 2001) but researchers still recognized and explored the potential for metal bonds in terms of strength and thermal conductivity if higher levels of porosity could be introduced and CTE issues could be overcome. The result has been a renaissance in new freer cutting, dressable, metal bond technologies, especially for diamond, such as Scepter™ (McSpadden et al. 1999) and Paradigm™ from Norton that is currently revolutionizing applications such as carbide tool flute grinding, peel grinding, PCD centerless grinding, and carbide thread grinding. The market in this product is expected to grow dramatically and is already replacing a significant portion of the older polyimide resin technologies due to its higher stock removal capabilities.

Figure 6.21 based on Jakobuss 1999, illustrates superabrasive bond and grain selection as a function of workpiece resilience.

6.13 Other Bond Systems

There are several older traditional bond systems used with conventional abrasives. These include the following.

6.13.1 Rubber

Rubber bonds introduced in the 1860s are still used extensively for regulating wheels for centerless grinding and some reinforced grades for wet cut-off grinding. Manufacturing is becoming an increasing problem for environmental reasons, and alternatives such as epoxy are being substituted where possible.

6.13.2 Shellac

Shellac or "elastic" p-bonded wheels were first made in 1880, and due to a combination of elasticity and resilience, probably represent the best wheel for producing fine, chatter-free, finishes for grinding of steel rolls for the cold strip steel mills and paper industries. Shellac comes from fluid exuded by insects onto themselves as they swarm cassum or lac trees in India. As such it is highly variable both in availability and properties depending on the weather conditions and species. On occasion, a single wheel maker can consume 10% of the entire world's production. Not surprisingly, many wheel makers have sought alternative solutions to grinding applications.

6.13.3 Silicate

Silicate bonds were first produced around 1870 by mixing wet soda of silicate with abrasive, tamping in a mold, drying and baking. It is still popular in certain parts of the world by reason of its simplicity and low cost of manufacture. The wheels are generally used for large face wheels.

References

3M, 2012, 3M Cubitron 2 gear grinding wheels, Commercial brochure.

Balson, P. C., 1988, Vitreous bonded cubic boron nitride abrasive article. US Patent 3,986,847, October 19, 1976.

Barnard, J. M., 1985, Creep feed grinding using crushform and dressable superabrasive wheels, *Superabrasives '85 SME Conference*, Chicago, IL, MR85-292.

Barnard, J. M., 1989a, Part 1 Crushable CBN and diamond wheels. *IDR*, 1/89, 31–34.

Barnard, J. M., 1989b, Part 2 Crushable wheels—case histories. IDR, 4/89, 176–178.

Bremer, S., Hucker, S., Burgess, P., 2005, Plated grinding wheel life maximization method. US Patent 6,932,675, August 23, 2005.

Burkhard, G. Machining with a defined arrangement of hard metal grains *3rd International Studer Grinding Symposium*, January 31, 2002, Thun, Switzerland.

Bush, J., 1993, Advanced plated CBN grinding technology. *IDA Diamond and CBN Ultrahard Materials Conference*. Windsor, Canada. September 29–30.

Chalkley, J., Thomas, D. Improvements in abrasive wheels and other abrasive tools. UK Patent GB1279413A June 28, 1972.

Chattopadhyay, A. K. et al., 1990, On performance of chemically bonded single-layer CBN grinding wheel. *Annals of the CIRP* 39(1), 309–312.

Coes, L. Jr, 1971, *Applied Mineralogy 1—Abrasives*, Springer-Verlag, New York.

Craig, P., 1991, The age of resin isn't history, *Cutting Tool Engineering*, June, 94–97.

Daniel, P., 1983, Crushform wheels can be formed in your plant, *IDR* 6/83.

Dasch, J., D'Arcy, J., Hanna, H., Yin, Y., Kopple, R., Salmon, S., 2007. High-speed grinding with electroplated CBN wheels using oil vs. waterbased fluids. *ISAAT 2007/SME International Grinding Conference*, 1–7.

DiCorletto, J., 2001, Innovations in abrasive products for precision grinding. *Precision Grinding and Finishing in the Global Economy—2001 Conference Proceedings*, Gorham, October 1, 2001, Oak Brook, IL.

GE Superabrasives, 1988, Understanding the Vitreous Bonded Borazon CBN System, General Electric Borazon® CBN Product Selection Guide. Commercial brochure.

Gift, F., Misiolek, W., 2014, Fluid performance study for groove grinding a nickel based superalloy using electroplated CBN grinding wheels, *Journal of Manufacturing Science and Technology* 126, 451–458.

Jakobuss, M., 1999, Influence of diamond and coating selection on resin bond grinding wheel performance. *Precision Grinding Conference*, Chicago, IL, USA, June 15–17.

Julien, 1994, Titanium nitride and titanium carbide coated grinding tools and method thereof. US Patent 5,308,367, May 3, 1994.

Kinik, n.d., Grinding wheels Catalog #100E. Trade catalog.

Koepfer, C., 1994, Grit, glue—technology tool, Modern Machine Shop, December, 53.

Koshy, P., Iwasaksi, I., Elbestawi, M., 2003, Surface generation with engineered diamond grinding wheels; insights from simulations *Annals of the CIRP*, 2, 353–356.

Li, R., 1995, Improved vitrified abrasive bodies, WO Patent WO 95/19871, July 27, 1995.

Li, Y., Li, G., Jing, H., He, Y. New type of matrix material for the manufacture of electroplated diamond tools, *IDR* 4/2002, 59–62.

Lowder and Evans, 1994, Process for making monolayer superabrasive tools, US Patent 5,511,718, November 4, 1994 (see also US 3,894,673 and Chichester, UK 4,018,576).

Malkin, S., 1989, *Grinding Technology*, Ellis Horwood, Chichester, UK.

McClew, D., 1999, Technical and economic considerations of grinding aerospace alloys with electroplated CBN superabrasive wheels, *Precision Grinding '99. Gorham International*, Chicago, June.

McSpadden, S.B. et al., 1999, Performance study of Scepter™ metal bond diamond grinding wheel, *Precision Grinding Conference*, Chicago, IL, USA, June 15–17.

Meyer, J., 2008, Eigenschaften und Anwendungen von Chemisch Nickel-Dispersionsschichten (Properties and applications of electroless nickel composite coatings), Materialwissenschaft und Werkstofftechnik 39(12), 958–962.

Nishihara, K., Onikura, H., Ohnishi, O., 2006, Development of production device for Ni-W electroplated microgrinding tools and machining characteristics with the fabricated tools, *8th International Conference on Progress of Machining Technology*, 357–360.

Noichl, H., n.d., What is required to make a grinding wheel specification work? IGT Grinding Forum, University of Bristol.

Ohashi, K., Kawasuji, Y., Shinji, Y., Samejima, Y., Ogawa, S., Tsukamoto, S., 2012. Fundamental study on setting of diamond abrasive grains, *Advanced Materials Research* 565, 40–45.

Ohi, T., 2004. Trends and future developments for diamond CMP pad conditioners, *IDR* 1, 14–20.

Peterman, L., n.d., ATI Techview PBS® vs. Electroplating. Trade paper.

Pinto, F.W., Vargas, G.E., Wegener, K., 2008, Simulation for optimizing grain pattern on engineered grinding tools, *CIRP Annals-Manufacturing Technology* 57, 353–356.

Pung, R., 2001, Enhancing quality and productivity with vitrified superabrasive products, *Precision Grinding and Finishing in the Global Economy 2001. Gorham International*, October 1, 2001, Oak Brook, IL, USA.

Rappold, 2002, Cylindrical grinding, Trade brochure, Rappold-Winterthur 02/2002 #136551.00.

Sato, K., Suzuki, K., Tokyama, T., Kagawa, H., 1997, Some problems in the manufacture of an electrodeposited diamond wheel, *Journal of Materials Processing Technology* 63, 829–832.

Shi, Z., Malkin, S., 2003, An investigation of grinding with electroplated wheel, *Annals of the CIRP* G04, 2, 267–271.

Sung, C., Sung, M., 2008, The brazing of diamond, *Intertech 2008 Conference Presentation*, May 5–7, 2008.

Valenti et al., 1992, Glass–ceramic bonding in alumina/CBN abrasive systems, *Journal of Material Science* 27, 4145–4150.

Walter, C., Rabiey, M., Warhanek, M., Jochun, N., Wegener, K., 2012, Dressing and truing of hybrid bonded CBN grinding tools using a short-pulsed fiber laser, *CIRP Annals-Manufacturing Technology* 61, 279–282.

Weingartner, E., Roth, R., Kuster, F., Boccadoro, M., Fiebelkorn, F., 2012, Electrical discharge dressing and its influence on metal bonded diamond wheels, *CIRP Annals-Manufacturing Technology* 61, 183–186.

Yang, J. et al., 1993, Effect of glass composition on the strength of vitreous bonded c-BN grinding wheels, *Ceramics International* 19, 87–92.

Yang, J., 1998, The change in porosity during the fabrication of vitreous bonded CBN tools, *Korean Ceramic Society* 35/9, 988–994.

Yoshimi, R., Ohshita, H., 1986, Crush-formable CBN wheels ease form grinding of end mills, *Machine and Tool Blue Book*, April, 69–72.

Yuan, H., Gao, H. 2008, On the control and optimization of abrasive distribution pattern on grinding tool surfaces, *International Journal of Materials and Product Technology* 31(1), 72–80.

7

Dressing

7.1 Introduction

Understanding the procedures and mechanisms of dressing grinding wheels is critical to obtaining optimum performance in grinding. The available dressing methods are numerous and confusing—even the basic terminology varies from one manual or paper to another. For the purposes of this discussion, the following terms will be used:

1. *Truing*—Creating a round wheel concentric to the axis of wheel rotation and generating, if necessary, a particular profile on the wheel face. It is also to clean out any metal embedded or "loaded" in the wheel face. A further function is to obtain a new set of sharp cutting edges on the grains at the cutting surfaces.
2. *Conditioning*—Preferential removal of bond from around the abrasive grits.
3. *Dressing*—Truing the wheel and conditioning the surface sufficient for the wheel to cut at the required performance level.

Many people use the term dressing to mean conditioning but with most of the high production grinding occurring today truing and conditioning are simultaneous processes and are referred to in combination as "dressing." In Europe, conditioning can mean dressing or truing, sharpening can mean conditioning, and profiling can be used for truing (Pricken 1999).

All wheels require dressing, with the exception of electroplated wheels although, even here, the surface may be initially trued 10–20 μm or occasionally conditioned lightly with a dressing stick to remove the loaded metal. The focus of this chapter, however, is on bonded wheels especially vitrified. These bonds are popular because their porous, crushable bonds allow dressing in a single automatic operation.

Dressing processes for conventional wheels can be divided into two distinct classes:

1. Dressing with stationary diamond tools
2. Dressing with rotary diamond truers that offer much longer tool life

The simplest to begin with are stationary tool processes.

7.2 Traverse Dressing of Conventional Vitrified Wheels with Stationary Tools

7.2.1 Nomenclature

The terminology for the various parameters involved in dressing is as follows:

a_d = dressing depth of cut (or dress infeed amount) per pass

b_d = effective contact width of the dressing tool

n_s = grinding wheel revolution per minute (rpm)

f_d = axial tool traverse across grinding wheel surface, feed/rev

U_d = dressing overlap ratio

v_{fd} = axial tool traverse feed velocity

7.2.2 Single-Point Diamonds

The simplest tool is the single-point diamond. Typical designs of the diamond are shown in Tables 7.1 through 7.3. The majority of tools are shape A with a corner of rough unlapped diamonds. In general, these corners are well defined for repeatable dress action for flat wheel forms. The diamond is buried in a metal matrix with about one-third of the diamond exposed. High-quality diamonds tend to have up to four usable points and the tools can be returned to the toolmaker for resetting. Although the initial cost is higher, this is usually the most cost-effective choice unless tools are being abused. In this case, much lower-quality throwaway tools are recommended. The diamond weight can vary from a standard 1/4 up to 2 carat (ct) for aggressive or heavy dressing.

7.2.3 Diamond Size

General recommendations for diamond size are based on the wheel size as presented in Table 7.4.

TABLE 7.1

Standard Single-Point Diamond Shapes

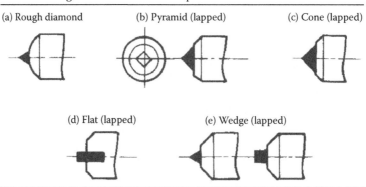

TABLE 7.2

Standard Single-Point Diamond Top-End Shapes of Tool Shanks

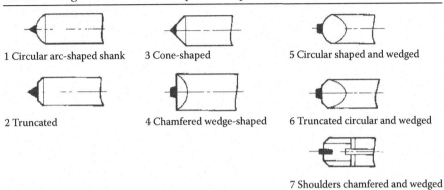

1 Circular arc-shaped shank   3 Cone-shaped   5 Circular shaped and wedged

2 Truncated   4 Chamfered wedge-shaped   6 Truncated circular and wedged

7 Shoulders chamfered and wedged

TABLE 7.3

Standard Single-Point Tool Shank Shapes

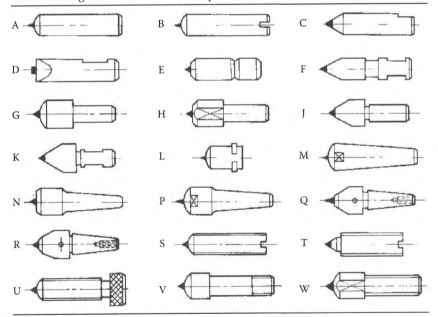

TABLE 7.4

**Diamond-Size Recommendations for
Single-Point Tools Based on Wheel Diameter**

Up to 3″	1/5 Carat
3″ to 7″	1/4 Carat
8″ to 10″	1/3 Carat
11″ to 14″	1/2 Carat
15″ to 20″	3/4 Carat
Over 20″	1 Carat

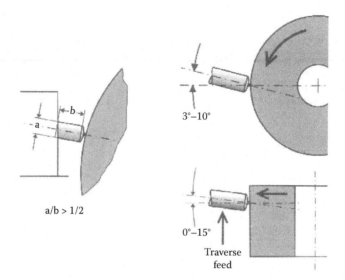

FIGURE 7.1
Dressing configuration using single-point diamonds. (From Rappold, 2002, *Cylindrical Grinding*. Rappold–Winterthur, Trade Brochure, 02/2002 #136551.00.)

7.2.4 Scaif Angle

The diamond is mounted on a holder and held at a scaif angle against the wheel rotation and with the traverse motion as illustrated in Figure 7.1.

The single point cuts a thread across the face of the wheel fracturing or dislodging grains and bond and leaving a fresh topography on the wheel surface (Figure 7.2).

7.2.5 Cooling

Copious coolant should be applied as the diamond is heat sensitive and the toolholder should have its own coolant nozzle. The coolant supply must be turned on before commencing a dressing pass. If the coolant is turned on during a pass, the diamond may be damaged by severe thermal shock.

7.2.6 Dressed Topography

The resulting roughness is governed, in simplistic terms, by the height of profile δ resulting from the overlap of the tool radius from one rotation of the wheel to the next. This height

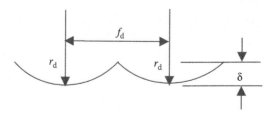

FIGURE 7.2
Dress profile generated by single-point tool.

should always be less than the dress depth in order to avoid non cleanup of the wheel surface at each pass and a resulting poor appearance in the ground part. The value for δ is controlled by the traverse rate, depth of cut, and tool radius. Significant research has been carried out to predict surface topography from tool profiles and wheel/tool interaction kinetics (e.g., Torrance and Badger 2000).

7.2.7 Dressing Feed and Overlap Ratio

Grinding engineers employ empirical approximations and guidelines to determine dressing feed rate. Depending on the dressing depth, the tool is assigned an effective cutting width b_d, which is assumed swept out on the wheel at each revolution. An overlap ratio U_d is then assigned for each type of operation and allows the axial dressing feed rate v_{fd} to be determined.

$$U_d = \frac{b_d}{f_d} = \frac{b_d \cdot n_s}{v_{fd}}$$

For a typical single-point dressing, dressing width b_d is about 0.5–1.0 mm and the following values for overlap ratio U_d may be used for general applications. These values are applicable to most traverse dressing operations.

Rough grinding $U_d = 2$–3
Medium grinding $U_d = 3$–4
Finish grinding $U_d = 6$–8

In practice, many problems are caused by setting the dressing feed rate too slow. This is equivalent to assigning a value of overlap ratio that is far too high. The result is rapid wear of the diamond and damage or glazing of the abrasive grains in the wheel. The grinding forces will be too high and the wheel will burn load and/or suffer grain pullout.

An alternative method, since the grinding severity is based on grit size in the wheel, is to set the effective contact width/rev to half the average abrasive grit size (Table 7.5).

TABLE 7.5

Grit Size Values for Calculating Dress Traverse Rates

FEPA Designation Standard	ANSI Grit Size	U.S. Grit Number	Average Size (mm)	Average Size (in.)
301	50/60	50	0.3	0.012
251	60/70	60	0.25	0.01
213	70/80	80	0.225	0.009
181	80/100	100	0.175	0.007
151	100/120	120	0.15	0.006
126	120/140	150	0.125	0.005
107	140/170	180	0.1	0.004
91	170/200	220	0.0875	0.0035
76	200/230	240	0.075	0.003
64	230/270	280	0.0625	0.0025
54	270/325	320	0.045	0.0018

7.2.8 Dressing Depth

The depth of cut will also control roughness. However, the maximum dress depth should be kept under 30 µm for regular alumina wheels, after which there is little change only increased tool wear. For ceramic-based abrasives, the maximum dress depth should be under 20 µm. The reason for this is to microfracture the grain, a concept that takes on even greater importance in a subsequent section below dressing cubic boron nitride (CBN). Fused alumina with its large crystallite size will macrofracture under dresser impact so each dress is essentially creating a whole new freshly fractured surface of cleaved grains. Ceramic grains with micron-sized crystallite microstructure cleave at the micron level so it is possible to resharpen each individual grain with a series of microridges. Since ceramic grains are harder than fused grains, these edges will resist wear significantly longer so it is possible to get the same life or longer but dressing at a depth of 5–10 µm instead of 25 µm. The minimum dress amount will depend on machine accuracy and stability, wheel wear, finish, etc., but may be as low as 2 µm.

7.2.9 Dressing Forces

For a typical K grade, conventional wheel dressing forces with single-point diamonds are typically in the range of 30–80 N normal to the wheel with a cutting force coefficient of about 0.25. Although these forces are lower than in other tools to be discussed below, the tool should nevertheless be well clamped in the holder and not overextended. Note the requirement $a/b > 1/2$ in Figure 7.1.

7.2.10 Dressing Tool Wear

Single-point tools wear relatively quickly compared with multipoint dressing tools. A tool is typically worn out when the wear flat at the tip exceeds about 0.6 mm. One advantage of having the tool tilted to the axis of the wheel is that the tool can be rotated in the holder to keep the tip sharp. Caution should be applied, however, in that standard commercial tools often do not have the tool accurately centered in the holder. This can cause the operator to continually chase size after each rotation. The result is that the tool does not get rotated but rather is thrown away.

   As the dressing tool wears, it loses its sharpness. This can affect the grinding process in a number of ways. Grinding forces and power may be reduced due to the dislodgment of abrasive grains and severe grain fracture using a blunt diamond, as shown in Figures 7.3 and 7.4. However, this is not necessarily good news. Workpiece roughness is greatly increased and the grinding process becomes more variable as a dressing tool wears. This is the opposite of the requirement for close control of tolerances (Chen 1995; Marinescu et al. 2013; Rowe 2014). With dressing tool wear, there is also increased risk of dressing chatter. This is due to vibration of the diamond while dressing and leads to very poor wheel topography and chatter marks.

7.2.11 Rotationally Adjustable Tools

Tools are available called "rotoheads" or Norton's "U-dex-it" that are specifically designed so that the head can be rotated without loosening the tool in the holder. The diamond is centrally positioned in the holder to within 25 µm.

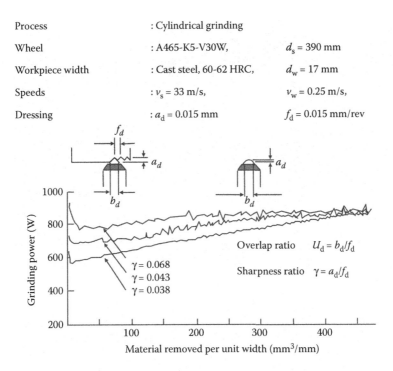

Process : Cylindrical grinding

Wheel : A465-K5-V30W, d_s = 390 mm

Workpiece width : Cast steel, 60-62 HRC, d_w = 17 mm

Speeds : v_s = 33 m/s, v_w = 0.25 m/s,

Dressing : a_d = 0.015 mm f_d = 0.015 mm/rev

Overlap ratio $U_d = b_d/f_d$

Sharpness ratio $\gamma = a_d/f_d$

$\gamma = 0.068$
$\gamma = 0.043$
$\gamma = 0.038$

FIGURE 7.3
Effect of dressing tool wear on grinding power and grinding wheel wear. (After Chen, X., 1995, Strategy for selection of grinding wheel dressing conditions. PhD thesis, Liverpool John Moores University, UK.)

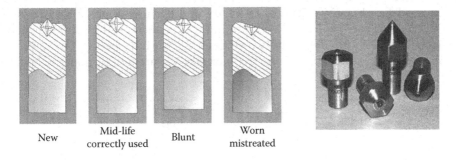

New | Mid-life correctly used | Blunt | Worn mistreated

FIGURE 7.4
Problems in tool wear due to poor tool monitoring. (Courtesy of Saint-Gobain Abrasives.)

7.2.12 Profile Dressing Tools

For profiling applications chisel-shaped tools with well-defined radii are used. These are used on profile dressing units such as Diaform® which were traditionally pantograph-based although now more often use two-axis CNC motion (Figure 7.5).

7.2.13 Synthetic Needle Diamonds

The wear of single-point diamonds in profile dressing leads to problems with changing dress conditions because of the increasing cross section of the flat generated. The

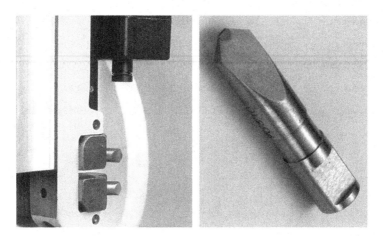

FIGURE 7.5
Diamond blades for Diaform® wheel traverse profiling.

introduction of synthetic needle diamonds provide a solution in the form of a constant cross section. This type of tool has seen increasing usage as a replacement for higher-quality single points. They are used both as a single stone and more commonly as a blade tool with up to four stones in a row. When specifying the diamond stone size, b_d is assumed to be the width of the diamond. This will vary somewhat based on the orientation of the stones which will also affect finish (Figure 7.6).

Few guidelines have been published for the use of stationary tools for generating profiles on the wheel face. The Winterthur Company (Winterthur 1998) recommends 0.6 mm stones are used for abrasive grit sizes <80/100#, and 0.8 mm stones for coarser wheels with an overlap ratio U_d value of 4. The number of stones used is dependent on the wheel diameter and width. Typically two stones are used for wheels <4 in., three stones for <20 in., and four stones for >20 in. The Noritake Company (Noritake n.d.) gave the

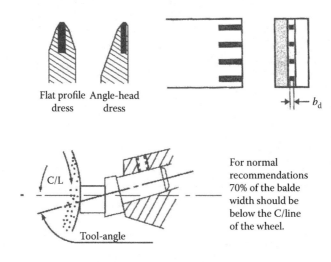

Flat profile   Angle-head
dress   dress

b_d

C/L

Tool-angle

For normal recommendations 70% of the balde width should be below the C/line of the wheel.

FIGURE 7.6
Needle diamond dressing blades and application. (From Unicorn International, n.d., D25 *Tool Selection Guide for Anglehead Grinding Machine*. With permission.)

following recommendations based on the wheel surface area and wheel grit size (Table 7.6 and Figure 7.7).

By offsetting the position of the stones, it is possible to use this type of blade dresser for dressing simple profiles such as angle-head wheels. The tool should be prelapped to the appropriate angle to limit break-in times.

7.2.14 Natural Long Diamond Blade Tools

Prior to the introduction of synthetic needles, natural long stones were used. These have trade names such as the Fliesen® tool from Winter (Saint-Gobain Abrasives). Some tools of this style have multiple layers of diamonds to maximize tool life, but care needs to be taken in their design to avoid changes in dress behavior when transitioning from one layer to another (Figure 7.8).

Blade tools have double the truing forces of single-point diamonds (50–150 N depending on wheel grade and grit size) but can handle depths of cut up to 50 μm for roughing of regular abrasive. For ceramic type abrasives, the maximum dress depth is 25–20 μm. b_d is

TABLE 7.6

Recommendations for Stationary Dressing Tools Using Needle Diamonds

Wheel Diameter Number × Wheel Width (in.²)		Number of Stones	Abrasive Grit Size	Needle Diamond Face Width (in. × 10⁻³)
	<5	1	36#–60#	36–44
>5	<20	2	60#–120#	24–36
>20	<60	3	120#–150#	20–24
>60	>120	4	150#+	16–20
>120		5		

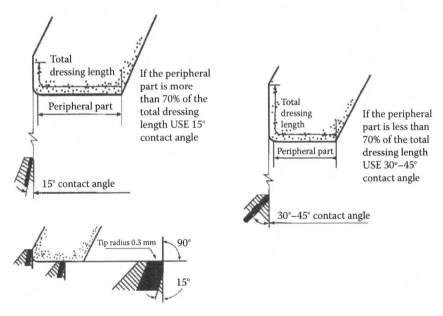

FIGURE 7.7
Application of needle blade tools for angle-head wheel dressing. (Courtesy of Saint-Gobain Abrasives.)

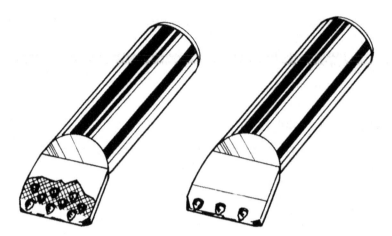

FIGURE 7.8
Examples of blade tools with natural long stones.

typically 0.75–1.0 mm. As with single points, the tools should be biased by up to 30° in the direction of traverse.

7.2.15 Grit and Cluster Tools

Finally, for the roughest dressing of large cylindrical or centerless wheels, there are grit tools and cluster impregnated tools (Figure 7.9).

Grit tools represent the most cost-effective dressing tool for the commonest applications using straight wheels. Clusters consist of a single layer of five to seven large natural diamonds semi-exposed on a round flat surface held in a sintered metal matrix. As with other tools, they are inclined at up to 15° with the tool center line intersecting the wheel center line. Their large head diameter results in fast traverse rates for reduced dress cycle times relative to other tools.

Grit tools consist of a consumable layer of diamond grains held in a highly wear-resistant sintered metal matrix. The tools wear progressively over time, exposing new grains. Diamond size selection is governed by the abrasive grain size in the wheel, while the tool width b_d (and length L) is dictated by the diameter and width of the wheel (Table 7.7). Note that the tool now consists of a random collection of diamond cutting points, whose action

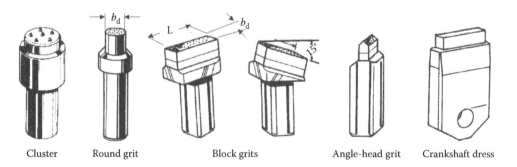

Cluster   Round grit   Block grits   Angle-head grit   Crankshaft dress

FIGURE 7.9
Examples of standard cluster and grit tool configurations.

TABLE 7.7

Grit Tool Recommendations for Conventional Wheels

Diamond Grit Size	Abrasive Grit Size	Type	b_d	L	Application
18/20	46#	Round	1/4″	N/A	small toolroom
20/25	54#	Round	3/8″	N/A	medium toolroom
20/30	60#	Block	1/4″	1/2″	<20″φ × <10″ wide
20/30	80#	Block	1/4″	3/4″	>20″φ × <10″ wide
30/40	100#	Block	3/8″	1/2″	<20″φ × >10″ wide
40/50	120#	Block	3/8″	3/4″	>20″φ × >10″ wide

 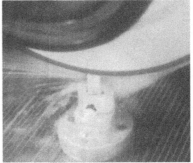

(Pricken)

FIGURE 7.10
Single-point and blade tool dressing—note copious use of coolant.

will depend on their exposure during any point in the life of the tool. Also the tool will wear a radius to the shape of the wheel. Overall the process is not as consistent as a single point, but in most cases acceptable and offset by the fact the tools are cheap, easy to make, and long lasting. Dressing forces with grit tools, however, must be respected; forces are typically five to eight times greater than those for single-point diamonds. The tool must, therefore, be clamped extremely rigidly with little or no overhang. Minimum dress depth is 10 μm because of the relatively dull dress action. They can handle dress depths of up to 125 μm dressing conventional alumina wheels and 50–25 μm with ceramic-type abrasives (Figure 7.10).

7.2.16 Form Blocks

In addition to the stationary tools for traverse dressing, full forms can be dressed simultaneously using form blocks (Figure 7.11). These are blocks that have a layer of diamond either sintered or directly plated and molded to the form required in the wheel. They are used especially in surface grinding where the block is set on the table at the same height as the finished ground height. The reciprocating stroke length is adjusted so that it dresses the wheel before finish grind. The blocks are either molded to the full form required or supplied as standard shapes for flexibility in tool room applications. Dimensional form accuracies can be held to about ±5 μm, minimum radius capability is 75 μm.

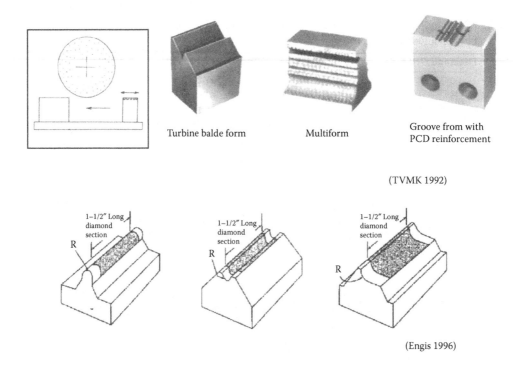

Turbine balde form   Multiform   Groove from with PCD reinforcement

(TVMK 1992)

1–1/2″ Long diamond section

R

1–1/2″ Long diamond section

R

1–1/2″ Long diamond section

R

(Engis 1996)

FIGURE 7.11
Examples of block dressers for profile dressing alox wheels.

7.3 Traverse Dressing of Superabrasive Wheels with Stationary Tools

7.3.1 Introduction

Perhaps, the most widely sought after—but as yet unavailable—stationary tool is the one that can dress high-performance vitrified CBN wheels. The problem is that single-point and needle diamonds wear much too quickly for most superabrasive wheels at the speeds the wheels operate at. Grit tools leave the wheels too dull and create too much pressure. There are a couple of exceptions, however; dressing small and/or low-concentration wheels.

7.3.2 Jig Grinding

Jig grinding, such as on Moore jig grinders, use a range of CBN and diamond wheels. The process has a relatively low-stock removal requirement because most applications are still performed dry and the wheels must be mounted on long quills to get deep into, for example, mold cavities. The wheel spindle is usually pneumatically driven and can be slowed to a few hundred rpm for dressing making the wheel act extremely soft. Using small diameter, high porosity wheels, the author has even had acceptable life with single-point diamond tools dressing *dry* (although this would not be the optimum process, it can do a job).

7.3.3 Toolroom Grinding

The other area is again in the toolroom targeted at grinding steels such as D2, A2, and M2 on low power reciprocating surface grinders. Several wheel makers have produced

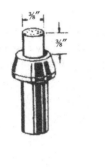
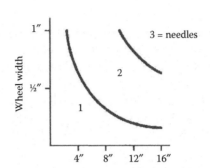

Grit tool dressing		Needle diamond Blade dressing	
CBN grit	Diamond grit	CBN grit	Diamond needle
Size	Size	Size	Size
80#−120#	40/50#	50#−80#	0.8 mm
120#−180#	50/60#	80#−140#	0.6 mm
180#−325#	60/80#	140#−325#	0.5 mm
	(Universal)		(Noritake)

FIGURE 7.12
Grit tool and needle blade tool recommendations for dressing low-concentration vitrified CBN wheels.

products such as Memox® from Noritake, CBLite®, and Vitrazon TR® from Norton (Saint-Gobain Abrasives) that are low-concentration vitrified CBN wheels with relatively high porosity. At low-stock removal rates typical of a surface grind operation ($Q'_w \leq 1$ mm³/ mm/s), they can still achieve a G ratio of 500–1000. The attraction of this type of wheel is that it can remove material at a rate as fast or faster than that of a conventional wheel, since in most cases the process is limited more by spindle power and machine stiffness, while an unskilled operator can set a grinder up to remove a given depth of stock without having to compensate repeatedly for wheel wear. This type of wheel has been dressed with single points, but more often grit tools and needle diamond blade tools. The following recommendations are found in the trade literature (Figure 7.12).

Depth of cut per pass should be kept in the range of 2–5 μm. Dressing forces may be as high as 100 N so rigid tool support is again critical.

Resin CBN and diamond wheels can be trued with similar small grit tools (or "nibs") to those used for vitrified CBN. Carius reports that diamond form blocks are also used to dress resin CBN wheels (Carius 1984). CVD diamond needle diamond blades are also used. However, these wheels need to be subsequently conditioned which is discussed in a separate section below. Vitrified bond and porous metal bond diamond wheels, if containing a high porosity, may also be trued with diamond nibs. Dense hot-pressed wheels, however, must be trued and conditioned with conventional abrasive wheels and blocks.

7.4 Uniaxial Traverse Dressing of Conventional Wheels with Rotary Diamond Tools

7.4.1 Introduction

Rotary diamond tools are the industry's answer to life issues with stationary tools and are in many ways, the rotary equivalents to single points, blades, grit tools, and form blocks.

A rotary diamond tool (also called "truer," "dresser," or "roll") consists of a disk with diamond in some form held on the periphery driven on a powered spindle. Life is significantly enhanced because of the 100-fold increase in diamond now available. However, the rotary motion also provides additional benefits in terms of dressing action. In particular, the relative speed of the dressing roll to the wheel, known as the dressing speed ratio or sometimes as the crush ratio, has a major impact on the conditioning action occurring during dressing. The simplest method we will consider is uniaxial dressing, where the axis of the wheel and the dresser spindle are parallel.

7.4.2 Crush or Dressing Speed Ratio

As indicated in Figure 7.13, all the parameters used for stationary tools are still important. In addition, there is the crush ratio defined as the surface speed ratio of the roll to the wheel or $q_d = v_d/v_s$.

In 1968, Schmitt (1968) produced the seminal study on the effect of crush ratio on conditioning of conventional vitrified wheels. The work was focused on plunge dressing with formed diamond rolls that will be discussed below. However, the research clearly illustrated the effect of crush ratio on finish and dressing forces, as shown in the trend graphs in Figure 7.14.

More recently, Takagi and Liu have studied the effect of crush ratio by analyzing the velocity vector of the impact between the diamond in the roll and the abrasive grains and assigning a "truer penetration angle θ" (Takagi and Liu 1996). When θ is small or negative, the force is essentially shear and grain wear is attritious, but as the crush ratio (q_d) approaches +1, the force becomes increasingly compressive and leads to large scale crushing of grain and bond.

Crush ratio must be considered a key parameter in dressing. For a unidirectional (+ve) crush ratio, the finish and forces change significantly over the range of +0.4 to +1. However, the dressing forces also increase dramatically leading to higher roll wear, stiffer machine

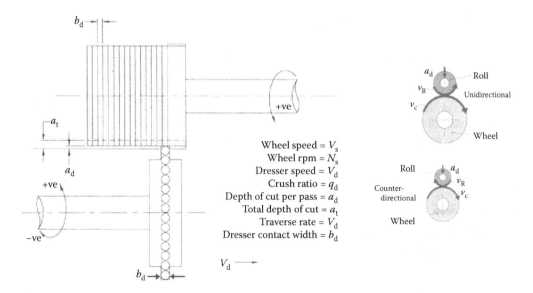

Wheel speed = V_s
Wheel rpm = N_s
Dresser speed = V_d
Crush ratio = q_d
Depth of cut per pass = a_d
Total depth of cut = a_t
Traverse rate = V_d
Dresser contact width = b_d

FIGURE 7.13
Dress parameters for uniaxial traverse dressing.

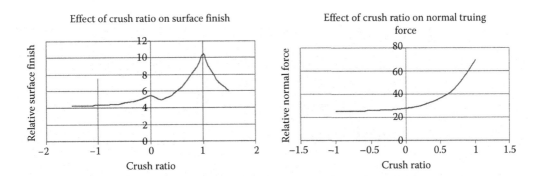

FIGURE 7.14
Effect of dressing speed ratio on surface roughness and dressing force.

requirements, and higher torque dresser motors. Diamond truer wear climbs so dramatically that it is usually recommended not to exceed +0.9. For most conventional wheel applications with traverse dressing, the wheel and machine characteristics, especially dresser designs, are such that most applications run counter-directional (–ve) operating in the range of –0.4 to –0.8. Also, the depths of cut taken can usually generate by contact geometry alone the required finish in spite of a lack of crush action. The situation, however, is rather different for CBN or form roll dressing as will be described in the subsequent sections.

Rotary dressing introduces both increased flexibility and increased potential for problems. The diamond disk is now rotating introducing balance issues and the potential for chatter and a resulting "orange-peel" appearance to the ground surface. Fractional multiples of the roll/wheel rpm can induce chatter and should be avoided. These are not just simple ratios such as 1:2 or 1:3 but can be as subtle as, for example, 7:13 or 5:11. Small adjustments in wheel or roll rpm can have a major impact on the quality of the ground surface.

7.4.3 Single-Ring Diamond and Matrix Diamond Disks

The rotary diamond differs from a stationary tool in that it is not cutting a continuous thread in the wheel but, consisting as it does of a ring of exposed diamond points, it is cutting a series of "divets" out of the grinding wheel. For a truing disk with a single ring of diamonds, the overlap factor U_d is dependent on the diamond spacing in order to ensure complete cleanup of the wheel face (Figure 7.15).

Also, for a well-defined spacing of diamonds, if a stone is missing or misplaced, it can set up repetitive patterns on the face of the wheel which transfers to the ground surface. The truer designs to be discussed below therefore fall into two categories: those with a series of accurately spaced diamonds akin to a rotary blade dresser, and truers with a totally random distribution of diamonds in a metal matrix akin to rotary grit tools.

Disk dressers are the rotary equivalent of the blade tool. They contain a ring of diamond held in a sintered or brazed matrix, and lapped to a precise form. Traditionally, the diamonds were high-quality long natural stones but are now being replaced in many cases by PCD and more recently CVD materials. Companies such as Dr. Kaiser and Norton (Saint-Gobain Abrasives) have specialized in their manufacture. Typical roll tolerances and a range of forms as given by Dr. Kaiser are shown in Figures 7.16 and 7.17.

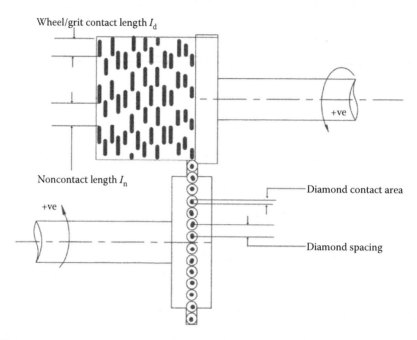

FIGURE 7.15
Wheel surface appearance generated by a single diamond ring rotary truer traversing with an overlap factor of one.

| Brazed PCD | Sintered CVD | Sintered radius | Sintered flat |

FIGURE 7.16
Examples of various traverse diamond truer designs.

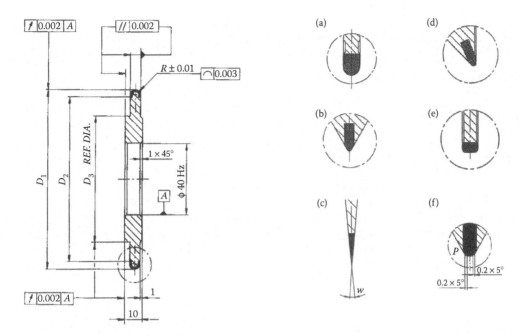

FIGURE 7.17
Diamond traversing rotary disks with defined contact geometry. (a,b) General profiling, (c) fine pitch profiling, (d) o.d. and single face, (e) o.d. and slot face, and (f) fine straight dress. (From Kaiser, n.d., *High-Precision Diamond Profile Rolls*, Trade Brochure. With permission.)

The use of this type of roll is reserved for the highest precision operations with tight finish requirements <0.4 Ra. The rolls are expensive but can hold radii as small as 200 μm for >10° included angle and 100 μm for >30° included angle. Larger radii are held to ±10 μm allowing a precise value to be entered into a CNC control to generate an accurate wheel profile.

7.4.4 Dressing Conditions for Disk Dressers

Traverse rates should be calculated from actual geometry of the disk. For disks of a given tip radius r and depth of dress a_d, simple geometry give the effective contact width as $b_d = 2[(2r - a_d)a_d]^{1/2}$. The same rules for overlap factor U_d as a function of b_d apply as for stationary tools. Dress infeed amounts should be limited to the range of 5–20 μm. Truing forces are low and, for the small radii disks, comparable to single-point dressing. Consequently, the dresser spindle motors require relatively modest power (<0.2 kW) and stiffness requirements resulting in compact units that can be readily fitted or retrofitted to the grinder.

7.4.5 Synthetic Diamond Disks

Synthetic diamond disks are expensive but the initial outlay can be compensated for by the fact that, if wear is properly monitored before becoming catastrophic, the disks can be relapped up to 40 times (Kaiser n.d.).

7.4.6 Sintered and Impregnated Rolls

Less expensive are sintered, impregnated, or "infiltrated" rolls which consist of a molded layer of diamond abrasive grains. These will contain a random distribution of diamonds.

The rolls tend to be of a relatively large radius >2 mm which may be relapped two to three times, or a flat profile with a consumable layer of 2–5 mm.

7.4.7 Direct-Plated Diamond Rolls

Another low-cost, throwaway alternative for some applications is direct-plated diamond with similar profiles to sintered rolls.

7.4.8 Cup-Shaped Tools

Cup shapes are used as well as disks as illustrated by the example in Figure 7.16. A cup-shaped tool is used tilted to the wheel face at an angle usually defined more by space availability for the motor than by the optimum dress geometry. Cups are either used where space is confined such as in internal grinding, or where an outer diameter and face must be dressed. The other situation for their use is with low torque dresser motors (see below). In the case of sintered cups, the tilt angle is prelapped in the face of the cup to avoid break-in issues.

7.5 Uniaxial Traverse Dressing of Vitrified CBN Wheels with Rotary Diamond Tools

7.5.1 Introduction

The rules for dressing vitrified CBN wheels are similar in many ways to those for described for conventional wheels. The same concepts of crush ratio, traverse rates, effective contact width, and depth of cut apply. The changes that must be made to the dressing conditions relate to the greater hardness, toughness and cost of the abrasive, and the greater hardness of the bond.

7.5.2 Dressing Depth

First and foremost, the depth of cut per pass with CBN is greatly reduced. This is in large part an economic requirement and the effect of this is to reduce the maximum surface roughness due to geometric effect from the truer. That, combined with the harder wheel grade, makes a higher crush ratio necessary and a more aggressive truer design to compensate. Whereas most conventional wheel applications run with a negative crush ratio, CBN is generally dressed with a crush ratio of +0.4 to +0.9. The only exceptions to these are where dressing is expressly required to lower finish to a minimal value (e.g., fuel injector seat grind), or because of a lack of dresser spindle motor torque.

7.5.3 Crush Ratio

Crush ratio can have a profound effect of the dressing action. Ishikawa and Kumar (1991) reported a study on dressing of vitrified-bonded wheels containing coarse grade 80# GE 1 abrasive. They distinguished between three forms of grit fracture "micro," "medium," and "macro" as illustrated in the micrographs in Figure 7.18. It was determined that at a modest

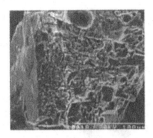  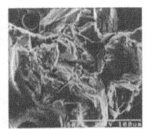

Microfracture   Medium fracture   Macrofracture

FIGURE 7.18
Fracture modes of 80# GE Type 1 Abrasive. (From Ishikawa, T., Kumar, K., 1991, Conditioning of vitrified CBN superabrasive wheels. *Suerpabrasives '91 Conference Proceedings SME.* June 11–13, 1991, Chicago, USA, 7.91–7.106.)

q_d = +0.2, there was a definite shift from predominantly a microfracture regime at a_d = 1 μm to a macrofracture regime at a_d = 3 μm; changes to a_d as small as 0.5 μm had a significant impact on grind power and finish. Microfracture led to a high-surface abrasive concentration and therefore a higher wheel life but also relatively high grinding forces; macrofracture with its lower surface concentration of sharper abrasives led to lower wheel life but lower grinding energy. As the crush ratio was increased from +0.2 to +0.8 the level of macrofracture increased dramatically to dominate the process accompanied by increased bond loss. This result is important because coarse grade GE 1 abrasive was the workhorse of cylindrical grinding and it therefore defined the required dressing infeed accuracy and crush ratio requirements for the earliest grinders designed specifically for vitrified CBN.

These results are specific to a particular grade and size of CBN that is relatively easy to fracture. It is, therefore, to be expected that a tougher grade of CBN, or blockier shape or finer grit size would require either a higher crush ratio or deeper depth of cut to achieve the same degree of grit fracture. Evidence for this is suggested in the work by Takagi and Lui (1996) who found that when dressing much tougher 80# GE 500 abrasive at 5 μm depth of cut, microfracture still dominated at q_d = 0.5; only at q_d = 0.9 was this replaced by macrofracture.

With the generally tougher but sharper, grades of CBN now in use dress depths/pass have increase somewhat from perhaps 2 to –3 μm, although this is more based on anecdotal field results than published research studies.

7.5.4 The Dressing Affected Layer

The other effect of very fine dress infeed depths is related to the fact that the dress depth becomes comparable to or significantly less than the depth to which the surface of the wheel has been affected by previous grinding. The surface of any grinding wheel is significantly modified compared to its bulk structure. The dressing process fractures and removes abrasive particles and bond to reduce the surface concentration of both.

Yokogawa was the first to describe this affected layer which he termed "Tsukidashiryo," also known as "Active Surface Roughness"; this can vary in depth from a few microns to over thirty (Yokogawa 1983). For most medium- to high-stock removal applications, once grinding begins the abrasive metal chips will wear the bond preferentially and further increase the affected depth (Figure 7.19). This effect is accompanied by a drop in grinding forces and a rise in surface roughness and is most striking for the first few parts after dress. Figure 7.20 illustrates the expected trend for various crush ratio parameters based on Jakobuss and Webster (1996).

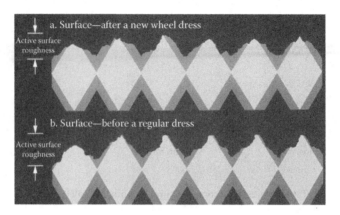

FIGURE 7.19
Concept of "active surface roughness."

One study by Fujimoto et al. (2006) is quite insightful on the matter. The study follows the forces, finish, and wear of B181 grits grinding typical tool steel in water-based coolant. Their results combined with field observations suggest the following.

1. An initial wear period of 2–3 µm with a rapid decrease in grinding force and high wear associated with instability in some freshly generated cutting edges creating new cutting edge, and bond loss. There is also a rapid increase in finish from a typical 0.3 Ra for B181 grit.

2. A "steady-state region" where grain microfracture dominates but where attritious wear resulting in wear flat formation begins to develop. The extra force/grain begins to induce some macrofracture.

3. By the time the wheel wear depth has reached 15–20 µm macrofracture now dominates and the surface finish exceeds 0.7 Ra and wheel wear may accelerate again limiting the wheel wear allowed between dresses.

These observations are in line with field results in water, where typically most precision cylindrical grinding is carried out with B91–B181 grit size and in most cases, the wheel wear is limited to 10–20 µm between dresses based on grit size. Oil coolant can have a

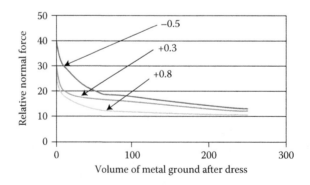

FIGURE 7.20
Effect of crush ratio on normal grinding forces.

significant effect in slowing the rise in finish and on occasion forces may actually climb with time until dressing is required, driven by, for example, burn, and reducing the wear permitted between dresses.

In summary, the recommend dress parameters for vitrified CBN for precision cylindrical grinding are

1. *Dress depth/pass* of 1–4 μm with most applications at 2–3 μm. Make adjustments in 0.5 μm increments if machine control allows

2. *Total dress depth* 10–15 μm (rule of thumb keep value <10% of grit size)

3. *Crush ratio* +0.3 to +0.9 with most application at +0.4 to +0.6 (dresser wear increases rapidly for crush ratio of >+ 0.8 and can become unacceptable on wider wheels)

4. *Traverse rate*: Use an overlap factor of 2–5. (Defining the true contact width of the dresser can sometimes be difficult as the diamond width and dresser widths can be different. Since most are designed to be very narrow it is often found that, rule of thumb, use a traverse rate in the range of (wheel rpm × (0.5–0.8) CBN grit size).)

5. *Dresser war compensation*: The rotary diamonds used to dress vitrified CBN wheels are designed to wear in order to self-sharpen and maintain a good dress action. For a typical wheel 350 mm diameter by 20 mm wide, this wear can be 5%–10% of the total dress infeed (that is, 0.5–1 μm). If this is not compensated in the controller then over time it will become significant cost factor. On wider wheels, the author has seen wheels removed from the grinder as worn out with up to a third of the CBN layer still left

6. *Wheel speed*: Always dress at the same wheel speed used for grinding for speeds over 50 m/s. Avoid work/wheel rpm integer ratios

For internal grinding and other applications using grit sizes in the B46—B91 range, many of the above described effects still occur but on a much smaller scale due to the reduced grit sizes.

7.5.5 Touch Dressing

When dressing a conventional wheel a similar break-in period occurs but is generally too rapid to be observed, and the dress infeed amount is such that most or all of the layer affected by grinding is removed and a fresh surface layer is created each dress. Not so for CBN where the dress depth/pass a_d is only 4 μm or less and the total dress amount is perhaps 10 μm. A brand new wheel straight after the first dress will have its shallowest affected depth, which will increase with grinding. A CBN wheel is most likely to cause issues such as burn grinding the very first part after a new wheel dress. For the second dress if too little material is removed, the parts/dress achieved will be reduced whereas if too much is removed the surface returns to that of a new wheel. In general, a balance has to be struck dependent on the particular grinding process in question. What is clear is that not only is the dress depth of cut per pass important governed by the fracture characteristics of the abrasive, and requiring machine control accuracy of <0.5 μm, but the total depth of cut is also critical governed by the active surface roughness requiring dress to dress repeatability of 2–3 μm.

Achieving this level of accuracy obviously requires machine tools that are accurate enough to infeed repeatably at this small increment. CBN-capable grinders have mechanical slide systems with AC servo/ballscrews or linear motors with an infeed resolution of 0.1 μm. Even with this level of accuracy, however, there is still the problem of thermal stability which can cause positional errors of the diamond relative to the wheel of up to 100 μm. In the 1980s, therefore techniques were developed to detect the contact of the dresser with the wheel *for flat profiles*. By far the most sensitive and reliable to date are those based on acoustics. Sensors have been developed by several companies capable of detecting a dress depth of cut of <0.25 μm. The systems detect and filter sound in the frequency range of 50–400 kHz adjustable for different grit sizes and wheel speeds. The actual output is an integrated power signal at a defined frequency based on grit size and noise. A major limitation on the signal/noise ratio (S/N ratio) has been bearing noise from the spindle and a number of strategies have been developed to minimize this problem. These include mounting the sensor adjacent to the dresser and using hydraulic fluid as an acoustic coupling Figure 7.21, the use of coolant as an acoustic coupler, or by mounting the sensor on the wheel head. Coolant coupling does give a very good S/N signal when working properly but any air bubbles entrapped in the fluid will instantly kill the signal and has been a major issue. Wheel head mounted sensors are becoming popular and are often combined with wheel balancing. They can work well but the sensor is a long way from the source of the signal and can be greatly attenuated compounded by several interfaces and even damping from the wheel hub material itself, for example, carbon fiber or bakelite. Dresser head sensors are close to the signal source. They were originally bolted directly next to the diamond away from the bearing with an external pick up. These are still in common use, although the problem is every time the roll is changed the gap between the sensor and pickup has to be reset and is very sensitive to position. The preferred method is now to mount the sensor and pickup internally in the dresser spindle directly under the diamond, where the gap is set just once by the spindle manufacturer.

A dress procedure using an acoustic sensor is detailed in Figure 7.22.

1. The dresser is initially set back a safe distance from the wheel, typically 50–100 μm and multiple passes at dress depth of 5–10 μm/pass are made in a search mode. The infeed amount is generally governed by cycle time issues, that is, making the fewest number of passes.

2. On first contact with the worn wheel surface, the infeed amount drops to 2–5 μm/pass.

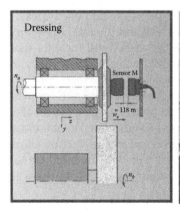

FIGURE 7.21
External dresser-mounted acoustic touch sensor.

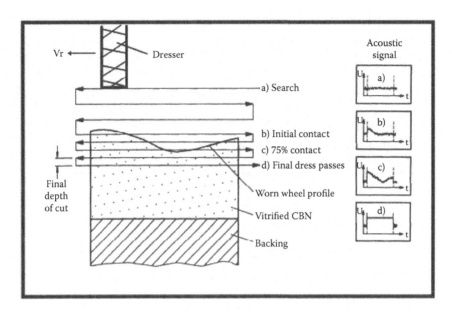

FIGURE 7.22
Acoustic sensor dressing strategy. (After Dittel, Walter, 1996, *Acoustic Control Systems,* Trade Brochure, Walter Dittel GmbH.)

3. At 75% full face contact (i.e., signal above a pre set trip level), the wheel drops into the final dress depth/pass.

4. A pre-set number of passes are made at final infeed a_d. The signal from the sensor should be a solid line.

The accuracy of the process is actually governed more by the infeed amount per pass during the initial search stage and is usually a compromise with cycle time issues. As mentioned above, a value must also be entered into the machine control unit to compensate for truer wear. This value must be determined empirically by monitoring wheel diameters readous of the CNC controller before and after dress for several dress cycles, or by physically measuring the wheel with a pie tape every 50 dresses.

Suzuki (1984) was one of the first to report the use of acoustic sensors for dressing CBN wheels for actual production applications, where he used it to true resin CBN wheels to grind camshafts before vitrified CBN was widely available. The technology was then transferred to the first vitrified CBN-capable camlobe grinders, the Model GCH 32 from Toyoda Machine Works (TMW). Suzuki overcame the problem of dresser spindle bearing noise by isolating the (Marposs) acoustic sensor a short distance away from the dresser and having a separate touch feeler (steel pin on a spring) attached to it to touch the wheel. The pin was plunge fed into the wheel at 2 μm increments. After the diamond trued the wheel the feeler retouched the wheel to calibrate the relative position of feeler and diamond (Figure 7.23).

Since the feeler and the dresser were close together, it was assumed that thermal movements would not create significant errors. A value was required in the machine control to compensate for diamond wear *and* pin wear in order to keep track of wheel diameter. The primary variance in the dress amount was the pin wear that was governed by the infeed increment of 2 μm. The sensor proved extremely repeatable, although prone initially to false signals from the presence of coolant. This was resolved by carrying out the actual touch sensing portion of the cycle dry. The system has been very successful for camshaft

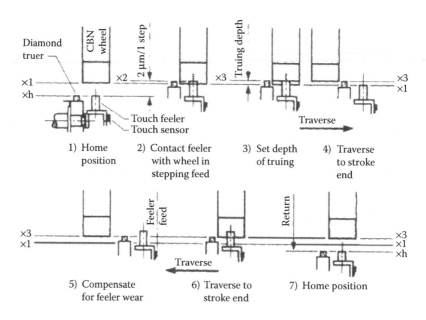

FIGURE 7.23
TMW strategy for AE enhanced rotary dressing. (From Suzuki, I., 1984, Development of camshafts and crank-shafts grinding technology using vitrified CBN wheels. *SME*, MR84-526.)

and crankshaft grinding and is functioning on several hundred machines worldwide (Hitchiner 1997).

Since all high-speed vitrified CBN wheels are segmented, they are especially prone to slight changes in shape with wheel speed. Dressing should therefore *always* be carried out at the same wheel speed as for grinding for operating speeds >50 m/s to avoid chatter.

7.5.6 Truer Design for Touch Dressing

Attention must also be paid to the diamond truer design. Since CBN is so much harder than conventional abrasive, the truer will wear much more per dress. This makes profiling truers with precisely lapped geometries uneconomic for most applications. Certainly for all flat form dressing applications, a truer with a consumable diamond layer is required.

7.5.7 Impregnated Truers

An obvious solution, based on conventional wheel experience, would be to use an impregnated truer. The difficulty is that these truers are designed such that the matrix wears just enough to keep the diamond exposed. Conventional wheels create a lot of abrasive swarf which erodes the matrix but little diamond wear. In contrast, CBN wheels create little swarf and hence matrix wear but cause a much higher degree of diamond attritious wear. Consequently, the matrix must be somewhat less wear resistant for CBN applications but still retain the diamond. An alternatives is to use porous metal bond or even porous vitrified diamond bonds such as the hDD technology from Meister

The width of a regular impregnated truer is also a problem in light of the discussion regards grit fracture above. An impregnated truer will have an effective contact width b_d many times wider than the diamond size and so will have diamond grains dispersed

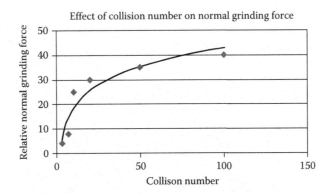

FIGURE 7.24
Effect of collision number in dressing on grinding force.

randomly throughout. Consequently, during dressing, if we picture the situation illustrated in Figure 7.15 but now with a much wider width, some CBN grains may be hit several times while others remain untouched. The first hit will be at the full truing depth and fracture the CBN grit but the following grains will be at a much shallower depth and merely glaze it again. The total number of hits is termed the "collision number." The effect is very apparent in Figure 7.24 which plots the effect of collision number on normal grinding force under a range of dressing and grinding conditions (Brinksmeier and Çinar 1995). A single ring of diamond grains is clearly the best option.

It is possible to estimate the equivalent of this single layer and multiples thereof for various truer contact widths and diamond grit sizes of impregnated truers as a function of diamond concentration (Table 7.8).

Although these guidelines have worked well for general applications, there was a need for a sharper dress for burn sensitive materials and aggressive removal rates. Several solutions have been developed, some proprietary, several covered by patents.

Most of these consist of a compact or single layer of diamond. For example, Hitchiner (1997) reported an impregnated layer of diamond sandwiched between two steel side plates for support. Although the overall width is 0.080," the actual diamond layer is only about a grain thick. Figure 7.25 shows an x-ray photograph through a truer showing the individual diamonds. The design is suitable primarily for dressing of flat profiles. Winter (Saint-Gobain Abraives) patented a truer design based on holding the diamond to the side of a steel support by direct plating (Winter EP 116668). This method allows

TABLE 7.8

Recommendations for Impregnated Diamond Truer Compositions

| Diamond Size | Concentration for 1 mm Contact Width | | | To Dress |
	Sharp	Medium	Dull	
D501	100	150	–	B181—B151
D301	80	125	160	B126—B91
D181	60	90	120	B91—B64
D126	47	75	95	B76—B54
D91	35	43	72	B46

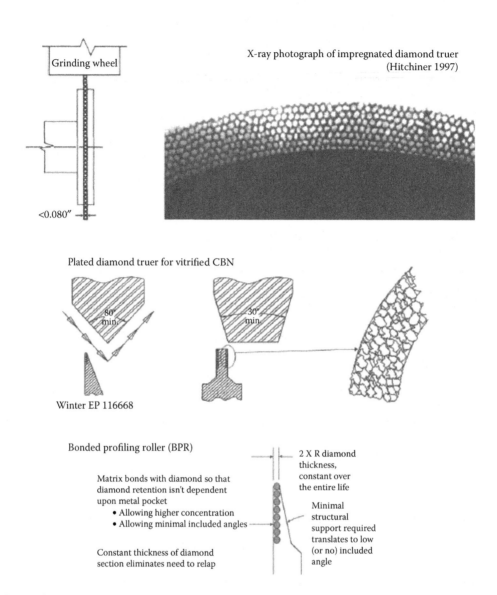

FIGURE 7.25
Strategies for aggressively dressing vitrified superabrasive wheels. (Courtesy of Saint-Gobain Abrasives.)

simple contouring in addition to dressing flat profiles. Finally Norton (Saint-Gobain Abrasives) developed a method called BPR™ or bonded profile roller, using the brazed plated process to produce a truer which is only a diamond grain wide (Norton 1993). This maintains a constant tip radius as it wears and can be used for quite complex profiling.

7.5.8 CVD Needle Diamond Disks

Finally, there are truers based on the use of CVD needle diamonds. These would be expected to give the most consistent, effective dress of any truer design for dressing of flat

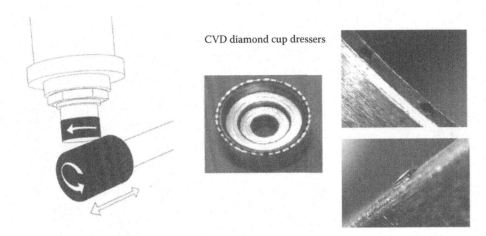

FIGURE 7.26
Needle ("prismatic") diamond dressing cup design and application.

profiles. For these to function correctly, the diamond must be above the level of the matrix in which it is held. This limits their usable depth to about 0.5 mm without significant fracturing occurring. However, they can be reexposed two to three times. Unfortunately, even with the diamond flush with the matrix, the needles are often prone to fracture and, based on their high cost, it seems more economic to use a simple impregnated or BPR style roll in most cases. The exception is in small wheels with very tight microfinish requirements (Figure 7.26).

Total truing force with these styles of dresser is of the order of 2–10 N for typical cylindrical grinding conditions to <2 N for internal grinding.

Mounting rotary dressing disks on the work head instead of a dedicated dresser spindle dressing is becoming more common as stiff, high speed, work heads are applied to applications such as peel grinding (Figure 7.27).

The use of acoustic sensors become more of a challenge for *profile dressing*, that is, where the wheel position must be determined in two planes. The basic problem is two planes but

FIGURE 7.27
Work head mounted dresser on a Weldon peel grinder.

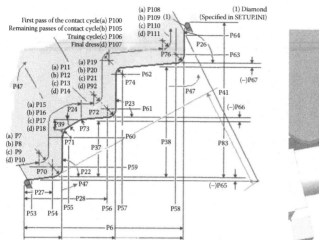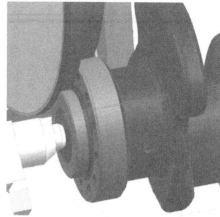

FIGURE 7.28
Example of program complexity dressing in two planes.

one signal. It is therefore necessary to make two independent touch contacts on orthogonal surfaces. In addition, dress rates will change based on the contact radius, that is, will need to slow around radii. All this combined with adding dresser wear compensation can make the process quite complex as illustrated in Figure 7.28 which is from a Landis control screen dressing a wheel for crank flange end.

The grinding process is highly effective achieving up to 1000 parts/dress such that dressing only has to occur once every shift or day. Extra time can therefore be taken to get the process under tight control without adding significantly to the average part cycle time. Orthogonal points for initial contact are first established on surface off the active grinding surface top right and bottom left. The traverse rate for each surface is controlled by a separate parameter for each pass of the wheel with care on radii to avoid excessive speed and spiral chatter.

7.6 Cross-Axis Traverse Dressing with Diamond Disks

7.6.1 Introduction

Cross-axis dressing has often been considered a poorer dressing method. It has historically been applied to situations such as the retrofitting of older internal grinders from single-point diamond to rotary dressing where space does not allow a large enough dresser spindle motor for the required torque to operate in a uniaxial orientation, or it is simply impossible to orientate the dresser spindle otherwise as typified by Figure 7.29. The axis of the dresser is orientated at 90° to the wheel axis. The method has also been attempted to profile dress conventional wheels for grinding crankpins with blend radii and sidewall grinding but the slightest error in height position relative to the wheel center results in poor surface quality on the part.

FIGURE 7.29
Cross-axis dressing of an internal vitrified CBN wheel.

The process, however, has been revisited for dressing vitrified CBN wheels using the styles of dressers indicated above.

7.6.2 Traverse Rate

The dress action produces only shear so is never as effective as uniaxial dressing with a high +ve crush ratio. The traverse rate is dependent on dressing disk diameter ϕ_d and depth of cut a_d. Simple geometry gives an optimum traverse rate of:

$$v_{\text{fad}} \approx 1.5 N_s \cdot (\phi_d \cdot a_d) \frac{1}{2}$$

where N_s is the wheel rpm.

Hence the process is, to a first approximation, independent of CBN or diamond grit size, or dresser rpm. However, as the technique has become more common, there are indications that slowing the dresser rpm will increase surface roughness. Dressers developed for uniaxial dressing also work well for cross-axis dressing, particularly the thin impregnated diamond disk design described above. The most successful application of cross-axis dressing to date has come from CNC profile grinding for applications such as punch grinding and high-speed contour grinding. An example of this is shown in Figure 7.30 which is a photograph of a Weldon (Weldon Solutions, York, Pennsylvania) high-speed grinder tooled for cylindrical profile grinding at 130 m/s. The acoustic sensor is mounted in the wheel head and monitors both the dress process and the grind process, as well as adding crash protection. The dresser touches on the o.d. and face at the start of the dress sequence to determine wheel position in x and z planes and compensate for thermal movement.

Cross-axis dressing is the most cost-effective method of profile dressing where the contour allows its use. One additional benefit is it gives clearance to dress profiles of over 180°. This allows, for example, back angle relief to be dressed on the sides of a 1A1R shape wheel when high-speed contour grinding shaft diameters with shoulders.

FIGURE 7.30
Cross-axis dressing on a Weldon 1632 high-speed grinder equipped with AC servo electric dresser (Saint-Gobain Abrasives) and wheel spindle-mounted AE sensor (Dittel).

7.7 Diamond Form Roll Dressing

7.7.1 Manufacture and Design

Traverse dressing of profiles, especially with the required frequency when using conventional wheels, has a major limitation—cycle time. Modern high production, precision grinding relies on rapid dress times achieved by plunging a truer coated with diamond conforming to the required profile. Rotary form truers or rolls can be classified into two common categories:

1. RPC rolls, or reverse-plated construction, produced by a precision electro forming process
2. Infiltrated rolls produced by high temperature furnacing

It is important that the end user understands the manufacture and properties of each type of roll in order to select the best product for the application. Usage in the market is split about 50:50 between the two. There are many general usage recommendations but little regards the specifics of manufacturing for proprietary reasons.

One source has published a series of photographs to illustrate their process that can give insight (Figure 7.31; TVMK n.d.)

After design of the required form, and any modifications required for final shape correction in the mold, the profile is cut on the inside of a mold. In the example shown below the form is first generating an EDM-wire cut form tool which is then plunged into a graphite-based material making up the mold. Other proprietary mold materials and CNC machine processing methods are used depending on the particular manufacturer. Diamonds are then tacked onto the cut surface of the mold. "Hand-set" rolls have diamonds placed in very specific patterns to control finish. Traditionally these have been set laboriously by hand Figures 7.31b and 7.32, although automated robotic techniques are now reported as shown in Figure 7.31c.

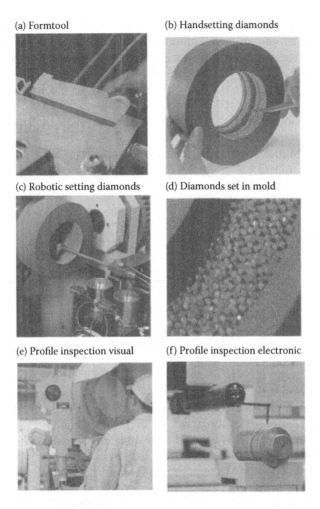

(a) Formtool   (b) Handsetting diamonds

(c) Robotic setting diamonds   (d) Diamonds set in mold

(e) Profile inspection visual   (f) Profile inspection electronic

FIGURE 7.31
Aspects of diamond roll manufacturing. (Courtesy of TVMK.)

FIGURE 7.32
Examples of reference handset diamond roll pattern blocks.

Alternatively a high density of diamond is packed on to the face either by pouring or, for higher densities, by centrifuging. Handset patterns are used for low force applications or rough finish requirements. High density or "random set" diamond rolls are used on stiff dressing systems for maximum roll life. Dresser spindle stiffness issues tend to limit roll form widths for this latter style to about 6" on most grinders. In the case of sintered rolls very often, the diamond is premixed with the metal (tungsten–iron) powder prior to packing to give a diamond section thickness of approximately 1.5 mm (Decker 1993).

In addition to the regular stones, which run in size from 18/20# to 40/50#, additional evenly spaced stones (maacles, long stones, PCD, etc.) are often added to reinforce profile areas of weakness such as tight radii.

Processing can now take one of two routes.

7.7.2 Reverse Plating

For RPC rolls, the coated mold is placed in a nickel-plating tank and a shell of nickel is allowed to build up around the diamond. The plating process can take up to a month in order to avoid internal stresses or gassing, and to allow the contour to be faithfully followed. After this time, a steel core is fitted using a low temperature alloy to attach it and the mold broken open. The whole process occurs at or relatively close to room temperatures which minimizes distortion. However, the shell is thin and does not take a lot of abuse (Figure 7.33).

For infiltrated rolls, the core is fitted prior to processing and a tungsten–iron-based powder is packed between the core and the diamonds. The whole is then furnaced at several hundred degree Celsius. The process is much quicker than plating but the higher temperatures cause greater distortion and form error.

After mold break out, the bore is ground concentric to the o.d. with ±2 μm. Depending on the required tolerances, most rolls are then lapped where necessary to correct profile errors and reinforce key areas. Modern processing methods are such that RPC rolls do not necessarily require lapping in many instances to produce the required form tolerances. Certainly the lower processing temperatures result in low distortion. Therefore, the amount of and variation in lap, and hence consistency of performance, from one roll to another is less. But lapping, accompanied by some mechanical or chemical exposure of the diamond, is often critical to control the diamond surface density throughout the profile.

After lapping, the roll is balanced, and a coupon cut to confirm profile. Standard tolerances on profiles are typically 40%–70% of that allowed on the component. The process is capable of 2 μm on geometrical form tolerances, ±2 μm on lengths. Angular tolerances are held to ±2 min. Tighter tolerances are achievable, down to 0.75 μm on radial profiles for the bearing industry. Tolerancing has become so tight in recent years that it is often necessary for the roll maker to buy the same model of profile inspection equipment as used by the customer in order to get correlation in inspection.

7.7.3 Infiltrated Rolls

Infiltrated rolls are used for operations requiring fast roll deliveries, and for abusive applications, especially where operator skill levels are a concern. The tough tungsten iron construction can take more impact abuse than the thin nickel shell of the reverse-plated roll design. Infiltrated rolls are also used for roughing applications using sparse handset

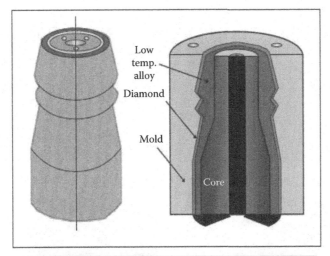

FIGURE 7.33
Reverse-plated diamond roll manufacture. (Courtesy of Norton.)

diamond patterns. The tough matrix of the infiltrated roll can better withstand erosion between the diamonds by the loosened abrasive grains. For this reason, silicon carbide wheels are most often dressed with infiltrated rolls.

7.7.4 Reverse-Plated Rolls

Reverse-plated rolls are used for finishing operations, for maximum roll life under good process control, and for applications with good system stiffness. Reverse-plated rolls will generally have high-density random diamond coverage to protect the matrix from erosion (Figure 7.34).

The diamond concentration can vary over the roll depending on the profile. For example, for profiles with a shallow angle to the axis of plunge where, heat from rubbing is an issue, the concentration can be cut significantly in the case of handset rolls, or by adding diamond free areas (Figure 7.35). Similarly for groove grinding, multiple roll assemblies may be used to provide diamond free areas to eliminate burn (Figure 7.36).

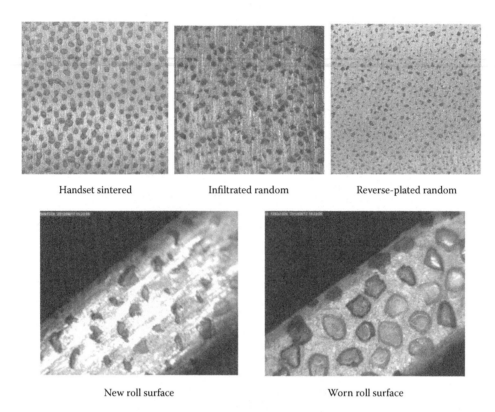

Handset sintered   Infiltrated random   Reverse-plated random

New roll surface   Worn roll surface

FIGURE 7.34
Surface appearance of typical diamond roll constructions.

FIGURE 7.35
Diamond free areas on shoulder of reverse-plated roll.

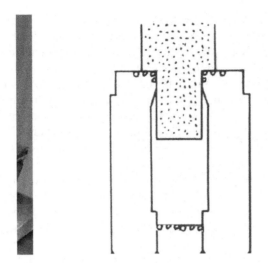

FIGURE 7.36
Three-piece diamond roll set for groove grinding. The wheel continues to plunge to lower diamond surface.

7.7.5 Dress Parameters for Form Rolls

This discussion will consider first the dressing of conventional abrasive wheels with form diamond rolls. This can then be extrapolated to include vitrified CBN. The dressing process entails plunging the roll into the wheel at a fixed infeed rate in mm/min or mm/rev of wheel at a fixed crush ratio followed by a fixed dwell time. The infeed rate per rev is analogous to dress depth per pass in traverse dressing. The crush ratio has a direct analogy with traverse dressing principles, while the dwell time can perhaps be related to overlap factor, that is, number of turns of the wheel that the roll is in contact at the end of the infeed cycle.

The dress infeed can be one of three configurations as shown in Figure 7.37. For the first, the axes for the roll, wheel and workpiece are all parallel. This is the easiest for checking the form accuracy and designing the roll. However, the grind is likely to be prone to burn and corner break down when grinding surfaces that are perpendicular to the component axis. This can be relieved to some extent as described above by reducing diamond

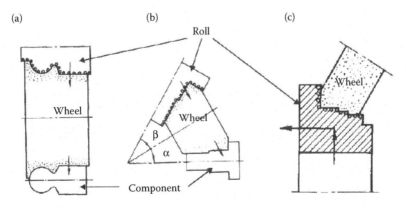

FIGURE 7.37
Dress methods for diamond form rolls. (a) Parallel plunge, (b) angle approach, and (c) plunge and wipe.

concentration but is still far from ideal. The second is an angle approach, the roll axis and wheel axis are no longer parallel in order to optimize the angle of approach of the roll to minimize burn. Finally, the third approach is a combination of angle approach followed by a traverse movement or "wipe." This is usually done to minimize dressing resistance especially where the dresser spindle might otherwise be laboring. It also improves surface finish and gives longer roll life. This discussion will be focused on parallel plunge approach.

Plunge dressing may be discontinuous occurring after a given number of parts, or performed continuously throughout the grinding cycle. This is very common in surface form grinding with heavy cuts, where it is known as "continuous dress creep feed" or CDCF.

1. *Dressing depth for form rolls.* The depth of cut in plunge dressing with alox wheels is typically >5 µm/rev which is 25%–50% less than in most traverse dressing operations. The surface finish and cutting action is therefore more dependent on the design of the roll. Rezeal et al. presented the effect of diamond spacing on surface finish under continuous dress conditions as shown in Figure 7.38 (Rezeal et al. n.d.). The results illustrate that diamond spacing and dress depth both have a significant influence.

2. *Diamond spacing for form rolls.* Diamond spacing is rarely if ever defined as such by roll makers. Instead the roll print will have a diamond size and ct/cms to define surface coverage (Table 7.9). Diamond spacing can be estimated from these values; it can be readily shown that diamond spacings are in fact all significantly less than

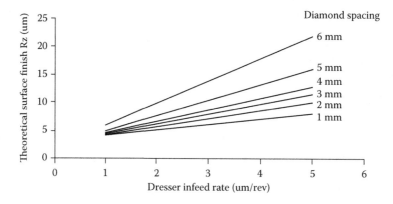

FIGURE 7.38
Theoretical surface roughness as a function of diamond spacing.

TABLE 7.9

Typical Diamond Coverage Values for Diamond form Rolls

Diamond Size	Diamonds/Carat	Diamond Coverage (ct/cm²)		
		Dense	Medium	Sparse
18/20#	110	2.3	2.0	1.6
20/25#	140	2.1	1.8	1.5
25/30#	250	1.7	1.5	1.3
30/35#	360	1.5	1.3	1.1
35/40#	615	1.2	1.0	0.8
40/45#	1225	0.9	0.7	0.5

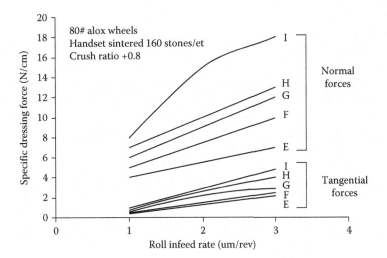

FIGURE 7.39
Forces in plunge roll dressing of alox wheels.

1 mm, suggesting the dress infeed per rev plays a more critical role. The dress infeed per rev is limited by the system stiffness, wheel grade, and dresser spindle power (Figure 7.39).

3. *Dressing forces for form rolls.* Rezeal et al. (n.d.) compared the dressing forces for handset coarse sintered rolls with reverse-plated fine mesh synthetic diamond rolls and found the latter dressed with up to twice the force (Figure 7.40). The dressing force coefficient was about 0.2. These values can be used to calculate dresser motor capacity. However, caution should be used when designing equipment for unidirectional dressing as considerably more motor power is required to resist the force from a roll speeding up driven by the wheel (up to a factor 3!), than

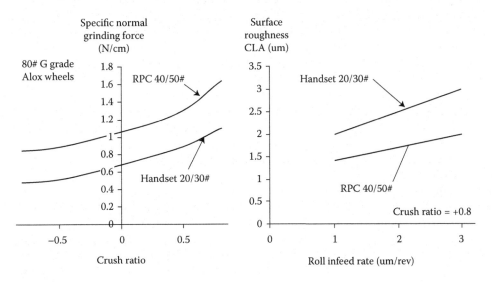

FIGURE 7.40
Effect of roll manufacturing method and diamond size on grinding forces and surface roughness.

that required for opposing a given force in counter-directional dressing. Also, the values given above are for flat forms; at least an additional factor 2 must be assumed for deep profiles.

4. *Infeed rate for form rolls.* For a noncontinuous dress on a typical alox wheel the infeed rate will be in the range of 0.2–2 μm/rev depending on the machine stiffness. After the infeed of the machine axes has been completed, there will be a programmed dwell period while the system relaxes. For a standard production grinder, this should occur within 0.5 s. Pahlitzsch and Schmidt (1969) reported that surface roughness reached a minimum value after 80–150 revolutions of the wheel at dwell on the roll. This provides the end user with a working range of dwell times. Excessive dwell times should be avoided to prevent premature roll wear and the development of chatter. Also, when dressing the newer ceramic-based wheel specifications, dresser motor load can become a problem for wide wheels; while conventional abrasive wheels require about 0.5 Hp/in. (15 W/mm) wheel width, ceramic wheels require double this in the dresser motor.

Available information on the plunge roll dressing of vitrified CBN wheels is sparse, although growing rapidly as a production technique. The primary applications for this process are in the aerospace industry using relatively porous wheels <6 in. wide, and in the fuel injection and bearing industries using narrow (<1 in.) wide wheels. In both cases the limitations are in the machine and spindle stiffnesses.

7.7.6 Dress Parameters for Form CBN Wheels

Hitchiner (1998, 1999) gives typical parameters for dressing of CBN wheels for grinding aircraft blades and vanes. Crush ratio values of +0.6 to +0.8 are used as for conventional wheels but infeed rates are limited to 0.1–0.25 mm/min or 0.03–0.20 μm/rev. which are 10 times less than for conventional wheels. On stiff purpose-designed machines, the dwell time is kept to a minimum, that is, zero dwell time is programmed in the CNC control, while the actual value is limited by the machine control response and inertia; perhaps 0.1 s. For a modern stiff grinder, the relaxation time is as little as 0.35 s while for older, weaker grinders a dwell time of up to 2 s has been necessary to generate a round wheel. More comment is made on this in the discussion on dressing spindles (Chapter 15).

Efforts have been made to employ acoustic sensors for touch dressing. Care has to be taken to allow for the fact that contact has to be resolved in two planes. The dressing arrangement in Figure 7.41 was presented by Landis (Waynesboro, Pennsylvania) at the IMTS (Tool Show) in Chicago in 1996 for dressing vitrified CBN angle-approach wheels. The acoustic sensor is mounted to the left of the diamond roll with diamond free areas to relieve dressing pressure. The roll shape provided two orthogonal planes on which to touch the wheel. An alternative approach on more complicated forms is to have a wheel edge and side from which to touch.

Issues with system relaxation times and programmed dwell times, be it with conventional or CBN abrasive, inevitably lead to situations where the wheel has less than ideal sharpness. To overcome this, several methods have been applied to give a very brief period of contact. These rely upon sweeping the roll past the wheel in a sliding motion analogous to using a stationary block dresser. The method, with the dresser mounted on a linear slide, has been used on, for example, Matrix grinders for ballnut grinding for the last 30 years. An alternative method, having the dresser spindle mounted on a swing arm. The technique had advantages of offering the minimum contact time but field reports suggest

FIGURE 7.41
Landis 2SE with form roll dressing arrangement for vitrified CBN. (Courtesy of Landis, Waynesboro, Pennsylvania.)

stability problems rough dressing conventional wheels. A problem with this approach is that numerous spark out passes must be made to ensure the entire wheel face has been contacted and eliminate chatter. This then becomes analogous to traverse dressing with a wide impregnated roll. Certain parts of the wheel are hit numerous times while other areas could go almost untouched.

7.7.7 Handling Diamond Rolls

Diamond rolls are high-precision tools and must be treated as such. The following are recommended procedures for assembling a roll on a spindle shaft.

1. *Standard roll/spindle assembly tolerances.*

    Standard toleranced assemblies refer to rolls with bore tolerances of 2.5–7 µm over nominal mounting on shafts with nominal to 2.5 µm undersize giving an average "loose" fit of 5 µm. The shaft itself should have a runout condition of ≤1.25 µm TIR, and after assembly in the spindle bearings of ≤2.5 µm TIR.

    a. Diamond rolls should be mounted in a clean environment removed from the production area, by a person properly trained to deal with the tolerance level involved.

    b. A fine grit oilstone should be rubbed lightly on the end faces of the diamond roll and any spacers to ensure that there are no burrs or raised metal due to previous handling.

    c. A film of gauge oil should be sprayed in the bores and then wiped clean with a lint free cloth to ensure the absence of dust or grit before assembling.

    d. The shaft should be treated as step c.

    e. Carefully align the diamond roll bore to the shaft. It is critical that the roll be started in a straight fashion otherwise the roll can jam on the shaft before engaging the full bore. A slight lead of 5 µm over 40 mm is extremely helpful.

        The diamond roll should slide down the shaft with minimal pressure until it bottoms out.

f. An aid to ease of assembly is to place the roll in warm water (40–60°C) to minimally expand the bore and ease assembly. *Do not* use a hot plate or hot air gun as this will distort the roll or, in the case of reverse-plated rolls, melt the alloy holding the diamond layer to the core.

g. Never force the diamond roll onto the shaft if unable to assemble. Inspect the bore and shaft for actual size or inspect for "raised" metal contact.

h. Complete mounting by assembling clamping nut or mounting screws, using care not to over tighten, to approximately 130 N m (100 ft lbs) on nut or 4 N m (3 ft lbs) on typical 10–32 screw.

i. Inspect assembly for runout by rotating in bearings before mounting in machine (if possible). Runout condition should not exceed 5 μm TIR on diamond rolls using diamond free indicating bands at ends of roll.

For disassembly, the roll should again be removed from the shaft in a clean environment, away from the production area. The shaft end and diamond roll end faces should be wiped clean to prevent dirt becoming trapped during removal and scoring the shaft or roll bore. If proper assembly procedure was followed removal should be accomplished by simply sliding diamond rolls off shaft. Never force the roll especially by the use of hammers of any kind.

2. *Line-fit roll/spindle assembly tolerances.*

Line-fit toleranced assemblies refer to rolls with bore tolerances of nominal to –2.5 μm under nominal mounting on shafts with nominal to 2.5 μm undersize giving an average fit of zero or line fit. The shaft itself should have a runout condition of ≤1.25 μm TIR, and after assembly in the spindle bearings of ≤2 μm TIR.

a. It is generally recommended that the roll manufacturer assembly and disassemble line-fit rolls. It is mandatory that the assembly be carried out in a clean environment, preferably temperature controlled, by trained individuals.

b. A fine grit oilstone should be rubbed lightly on the end faces of the diamond roll and any spacers to ensure there are no burrs or raised metal due to previous handling.

c. A film of gauge oil should be sprayed in the bores and then wiped clean with a lint free cloth to ensure the absence of grit before assembling.

d. Chill the shaft for approximately 30 min using cold tap water or ice packs to contract the mating diameter. Warm the diamond roll in hot water (65°C max) for 2–5 min to minimally expand the bore.

e. Quickly, but carefully, align the diamond roll bore in the shaft and slide the roll down until bottomed out.

f. Turn the roll slowly on the shaft, maintaining a downward pressure, until temperature is equalized and the diamond roll becomes immovable.

g. Complete mounting by assembling clamping nut (using care not to overtighten) to approximately 130 N m (100 ft lbs). Locking screws are not recommended because of the lack of time to accurately align the bolt pattern.

h. Inspect assembly for runout by rotating in the bearing assembly before mounting in the machine (if possible). Runout should not exceed 2.5 μm TIR on the diamond rolls using the diamond free indicating diameters at the ends of roll.

For disassembly, the roll should again be removed from the shaft in a clean environment, away from the production area. The shaft end and diamond roll end faces should be wiped clean to prevent dirt becoming trapped during removal and scoring the shaft or roll bore. Cool assembly in cold tap water or with ice packs for 30 min, then run hot water over the diamond roll only before quickly removing the diamond roll assembly. Never force the roll especially by the use of hammers of any kind.

7.8 Truing and Conditioning of Superabrasive Wheels

Nonporous superabrasive cannot in general be dressed with diamond tooling. Truing can be performed for some softer CBN bonds, such as resin, using diamond nibs or rotary diamond traversing disks but in general most wheels are trued and conditioned using conventional abrasive blocks or wheels. In the case of diamond, this can sometimes be completed in one operation to be effectively a dress process, otherwise two separate grades of conventional abrasive are chosen; the first is with a comparable or larger grit size to true the wheel, the second with a grit size half that of the superabrasive to condition the wheel by eroding the bond only. The processes can be carried out wet or dry and in general coolant is used only if used in grinding.

In its simplest form the dressing arrangement is simply to infeed dressing sticks into the wheel (cylindrical or cutter grind applications) or pass the wheel over the block (surface grind applications). Alternatively a mechanical or electrical brake truer device is used (Figure 7.42).

On the simpler mechanical version the dressing wheel is driven by the grinding wheel. The brake truer contains a set of weights which move out centrifugally as the rotational speed increases until they brake by making contact with inner wall of the unit. This allows a speed differential to be maintained between dressing and grinding wheel Figure 7.43 shows an example. The electric version merely has an AC motor instead to regulate speed and is used for small, thin or fine mesh wheels that provide insufficient torque to drive the mechanical version. When using the brake truer, infeed rates can start

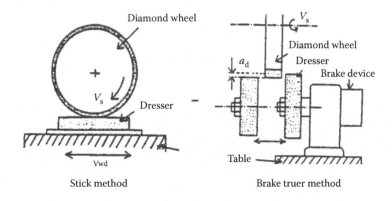

Stick method   Brake truer method

FIGURE 7.42
Conditioning processes with abrasive stick and wheel. (From Inasaki, I., 1989, *Annals of the CIRP*, 38/1, 315–318.)

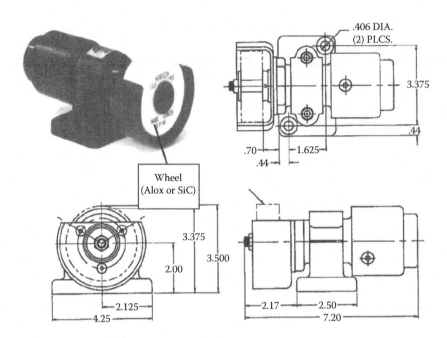

FIGURE 7.43
Brake truer device. (From Norton, 1993, Superabrasive truing and dressing devices. *Catalog*, 118.)

at up to 50–75 μm/pass at 2 m/min for roughing depending on the wheel grade before bringing the infeed amounts down for final flatness.

Traditional dressing, truing and conditioning grades of stones for both diamond and CBN resin wheels are given in Table 7.10. Also, recent developments in engineered ceramic grains indicate much higher removal rates and hence shorter dress times for both truing and conditioning may be achievable using these grains in conditioning wheels and blocks but this has not yet been well documented.

In many large shops resin wheels are dressed off-line using a purpose built devices such as the example in Figure 7.44 usually with optic display to monitor progress.

Dressing of diamond wheels over 300 mm in diameter can be especially time consuming. One method to reduce dress time for cylindrical grinding applications, is to use the work drive as a dresser motor and mount the dresser wheel in a fixture or arbor driven between centers. The diamond wheel speed should be about 25 m/s while the dresser wheel should be at 20%–40% this speed running unidirectional. Traverse rates should be about 0.1 m/min with infeed depths of 15–25 μm. After truing with a SiC wheel, the diamond surface should be conditioned with white alox sticks (Figure 7.45).

There was a considerable interest and research into stick infeed methods for conditioning cylindrical CBN wheels in the 1980s and early 1990s prior to the optimization of vitrified CBN rotary truing methods and establishment of functional grinding parameters for high production grinding. For example, Juchem (1993) reported data conditioning resin bonded CBN with white alumina sticks. At a constant infeed rate it was found that the initial grinding force after dress increased initially with dress time then fell to a steady-state value. Optimal dress conditions with the minimum of wheel wear occurred when the forces just reach this steady-state condition. At this point the bond was optimally eroded without over-exposing the abrasive grains (Figure 7.46).

TABLE 7.10

Recommended Dressing Wheel Grades for Truing with Brake Truer Devices

Abrasive	Band	Grit Size	Operation	Wet	Dry
Diamond	Resin	80#–120#	DRESS	WA60L	GC60L
		150#–320#	DRESS	WA120L	GC60L
		400#–800#	DRESS	WA325L	GC60L
	Metal	80#–120#	DRESS	WA46N-R	GC46N-R
		150#–320#	DRESS	WS80N-R	GC80N-R
		400#–800#	DRESS	WA230N-R	GC230N-R
	Vitrified	80#–120#	DRESS	WA80N	GC80N
		150#–320#	DRESS	WA150N	GC150N
		400#–800#	DRESS	WA400N	GC400N
CBN	Resin	80#–120#	TRUE	GC60J-N	GC60J-N
		150#–320#	TRUE	GC120J-N	GC120J-N
		400#–800#	TRUE	GC325J-N	GC325J-N
		80#–120#	CONDITION	WA220G	WA220G
		150#–320#	CONDITION	WA400G	WA400G
		400#–800#	CONDITION	WA800G	WA800G
	Metal	80#–120#	TRUE	WA46J-N	GC46J-N
		150#–320#	TRUE	WA80J-N	GC80J-N
		400#–800#	TRUE	WA230J-N	GC230J-N
		80#–120#	CONDITION	WA220G	WA220G
		150#–320#	CONDITION	WA400G	WA400G
		400#–800#	CONDITION	WA800G	WA800G
	Vitrified	80#–120#	TRUE	WA80N	GC80N
		150#–320#	TRUE	WA150N	GC150N
		400#–800#	TRUE	WA400N	GC400N
		80#–120#	CONDITION	WA220G	WA220G
		150#–320#	CONDITION	WA400G	WA400G
		400#–800#	CONDITION	WA800G	WA800G

Source: Compiled from several sources especially Diamant Boart America, 1991, 5214 Universal truing and dressing unit—Instruction and operation manual.

Materials for conditioning wheels are not limited to conventional abrasives. Soft mild steel <100HrB and molybdenum are both used in thrufeed centerless grinding for conditioning resin diamond wheels. The material is fed as bars through the grinder generating long stringy chips that erodes the resin matrix.

An alternative is to treat the surface of the wheel with slurry containing loose abrasive grain. Several systems exist where the surface is blasted directly with a high pressure slurry jet (Kataoka et al. 1992) or fed between a steel roll and a wheel (Hanard 1985).

The first CBN cam grinders used resin CBN wheels (Figure 7.47), which were trued with a rotary diamond then conditioned with either an alumina stick (Fortuna 1991) or with slurry fed between the wheel and a crush roll (TMW). The process was extremely cost effective from the aspect of abrasive cost/part. The only problem was the conditioning process because it was hard to control relative to the simpler dressing process required with the vitrified CBN wheel technology that superseded it (Renaud and Hitchiner 1991).

FIGURE 7.44
Offline dresser for resin, hybrid, and dense superabrasive wheels.

(a)

(b)

FIGURE 7.45
Online dressing systems for resin wheels: (a) Spindle-mounted SiC wheel. (b) Auxiliary spindle-mounted SiC wheel.

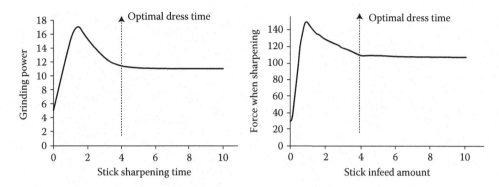

FIGURE 7.46
Optimization of stick dressing by monitoring stick pressure and initial grinding forces.

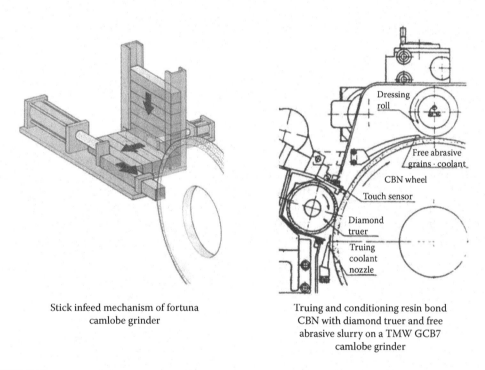

Stick infeed mechanism of fortuna camlobe grinder

Truing and conditioning resin bond CBN with diamond truer and free abrasive slurry on a TMW GCB7 camlobe grinder

FIGURE 7.47
Examples of conditioning processes on early CBN-capable camlobe grinders.

Conditioning with the correct block grade produces a very well exposed abrasive wheel surface, and even for vitrified CBN this can be greater than by simply rotary diamond dressing. This area is critical for low-stock removal applications such as finish double disk grinding (Hitchiner et al. 2001). Chen describes how to open up a vitrified CBN wheel surface by grinding and "touch dressing" and how to avoid the problem of closing up the wheel surface by dressing too deep (Chen et al. 2002). Problems experienced when dressing techniques for conventional vitrified abrasive are employed for vitrified CBN include: high grinding forces, rapid consumption of the abrasive layer, poor grinding results, and shortened redress life.

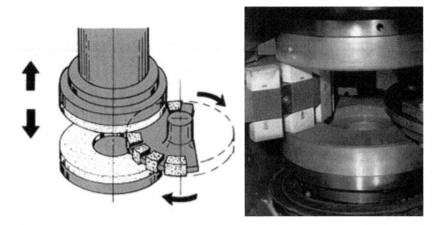

FIGURE 7.48
Koyo block truing/conditioning process for superabrasive double disk wheels. (Courtesy of Koyo Machine USA. Novi, Michigan.)

Koyo Machine offer a range of vertical spindle double disk grinders designed specifically around the use of superabrasive wheels for finish grinding of tight toler-ance components, for example, the fuel injection, hydraulic pump, and gear and ceram-ics industries (Figure 7.48). Double disk grinding is characterized by low chip loads and high normal forces due to the high contact area of the wheels. Consequently with CBN there is little grit pull out and fracturing or bond erosion and the abrasive grains glaze giving progressively lower finishes and higher forces until either flatness is lost or burn occurs. The more exposed the abrasive the longer the time between dresses. The machines therefore use a rotary block dressing method where white alumina blocks are passed between the wheels. This also makes the machines flexible to use resin even metal as well as vitrified-bond wheels.

References

Brinksmeier, Çinar, 1995, Characterization of dressing processes by determination of the collision number of the abrasive grits. *Annals of the CIRP*, 44(1), 299–304.

Carius, A. C., 1984, Preliminaries to success—Preparation of grinding wheels containing CBN. SME Technical Paper MR84-547.

Chen, X., 1995, Strategy for selection of grinding wheel dressing conditions. PhD thesis, Liverpool John Moores University, UK.

Chen, X., Rowe W. B., Cai, R., 2002, Precision grinding using CBN wheels. *International Journal of Machine Tools and Manufacture.*, 42, 585–593.

Decker, D. B., 1993, Truing and dressing grinding wheels with rotary dressers. *Finer Points*, 5/(4), 6–10.

Diamant Boart America, 1991, 5214 Universal truing and dressing unit—Instruction and operation manual.

Dittel, Walter, 1996, *Acoustic Control Systems*. Walter Dittel GmbH, Trade brochure.

Fortuna, 1991, *Automated Camshaft Grinding*. FORTUNA Werke, Trade brochure.

Fujimoto, M., Ichida, Y., Sato, R., Morimoto, Y., 2006, Characterization of wheel surface topography. *JSME Internationa Journal*, Series C, 49(1), 106–113.

Hanard, M. R., 1985, Production grinding of Cam Lobes with CBN. *SME Conf Proc. "Superabrasives '85"* p4-1–4-11, USA.

Hitchiner, M. P., 1997, Camshaft lobe grinding and the development of vitrified CBN technology. *Abrasives Magazine*, Aug/Sept, 12–18.

Hitchiner, M. P., 1998, Dressing of vitrified CBN wheels for production grinding. *Ultrahard Materials Technology Conference*, May 28, 1998, Windsor, Ontario, 139–174.

Hitchiner, M. P., 1999, Grinding of aerospace alloys with vitrified CBN. *Abrasives Mag.*, December/January, 25–35.

Hitchiner, M. P., Willey, B., Ardelt, A., 2001, Developments in flat grinding with superabrasives. Precision Grinding & Finishing in the Global Economy. *Conference Proceedings Gorham*, November 11, 2001, Oak Brook, Illinois.

Inasaki, I., 1989, Dressing of resinoid bonded diamond grinding wheels. *Annals of the CIRP*, 38/1, 315–318.

Ishikawa, T., Kumar, K., 1991, Conditioning of vitrified CBN superabrasive wheels. *Suerpabrasives '91 Conference Proceedings SME.* June 11–13, 1991, Chicago, Illinois, 7.91–7.106.

Jakobuss, M., Webster, J., 1996, Optimizing the truing and dressing of vitrified-bond CBN grinding wheels. *Abrasives Magazine*, August/September, 23.

Juchem, H. O., 1993, Conditioning of ultrahard abrasive grinding tools. *Finer Points*, 5(Pt4), 21–27.

Kaiser, n.d., *High-Precision Diamond Profile Rolls*. Trade brochure.

Kataoka, S. et al., 1992, Dressing method and apparatus for super abrasive grinding wheel. US Patent, 5,168,671, December 8

Marinescu, I., Rowe, W. B., Dimitrov, B., Ohmori, H., 2013, *Tribology of Abrasive Machining Processes*, 2nd edition. Elsevier, William Andrew Imprint, Oxford, UK.

Noritake, n.d., LL-Dresser. Trade brochure.

Norton, 1993, Superabrasive truing and dressing devices. *Catalog*, 118.

Pahlitzsch, G., Schmidt, R., 1969, Wirkung von Korngrosse und –konzentration beim Abrichten von Schleifscheiben mit diamantbestucken Rollen, wt-Zeitschrift fur industrielle Fertigung, 59, Jahrgang, Heft 4 Seite 158–161.

Pricken, W., 1999, Dressing of vitrified bond wheels with CVDRESS and MONODRESS. *IDR*, 3/99, 225–231.

Rappold, 2002, *Cylindrical Grinding*. Rappold–Winterthur. Trade brochure 02/2002 #136551.00.

Renaud, W., Hitchiner, M. P., 1991, The development of Camshaft Lobe grinding with vitrified CBN. *SME Conference Proceedings*, Superabrasives '91, MR91-163.

Rezeal, S. M., Pearce, T. R. A., Howes, T. D., n.d., *Comparison of Hand-Set and Reverse Plated Diamond Tollers Under Continuous Dressing Conditions*, IGT, University of Bristol, UK.

Rowe, W. B., 2014, *Principles of Modern Grinding Technology*, 2nd edition. Elsevier, UK and USA.

Schmitt, R., 1968, Truing of grinding wheels with diamond studded rollers. Dr-Ing Dissertation TU Braunschweig 1968.

Suzuki, I., 1984, Development of camshafts and crankshafts grinding technology using vitrified CBN wheels. *SME*, MR84-526.

Takagi, J., Liu, M., 1996, Fracture characteristics of grain cutting edges of CBN wheel in truing operation. *Journal of Materials Processing Technology*, 62, 396–402.

Torrance, A. A., Badger, J. A., 2000, The relation between the traverse dressing of vitrified grinding wheels and their performance. *International Journal of Machine Tools and Manufacture*, 40, 1787–1811.

TVMK, 1992, *Products Guide*, Toyoda Trade Brochure Van Moppes Ltd.

TVMK, n.d., Proposal & Creation Toyoda Van Moppoes Ltd Company Guidance. n.d. Trade brochure.

Unicorn International, n.d., D25 *Tool selection Guide for Anglehead Grinding Machine*. Trade brochure.

Universal Superabrasives, 1994, Diamond Dressing Tools. Trade brochure 1994.

Winter, Patent EP 116668.

Winterthur, 1998, Corporation Precision grinding wheels. Trade brochure.

Yokogawa, Yonekura, 1983, Effects of Tsukidashiryo of Resinoid bonded borazon CBN wheels on grinding performance. *Bulletin of JSPE*, 17(2), 113–118.

8

Grinding Dynamics

8.1 Introduction

8.1.1 Loss of Accuracy and Productivity

Vibrations can cause serious problems in grinding processes leading to loss of machining accuracy and loss of productivity. Of the various types of vibration, chatter vibration is one of the most crucial ones because it reduces form accuracy as well as increasing surface roughness of the ground parts. Form accuracy and low surface roughness are two of the main targets to be attained by grinding. In addition, productivity is lost because material removal rate has to be reduced as a way of suppressing chatter. A great deal of research has been conducted aimed at achieving a clear understanding of the mechanism of chatter vibration and consequently at developing practical suppression methods.

8.1.2 A Need for Chatter Suppression

The ultimate aim of research on grinding chatter is to develop practical methods for suppressing chatter vibrations while maintaining high productivity. While there have been some proposals from research laboratories to meet this requirement, few have been successfully applied in industry. Thanks to significant developments in sensing and control technologies available today, it seems that the necessary tools for developing methods of suppressing chatter vibrations are being provided.

8.2 Forced and Regenerative Vibrations

8.2.1 Introduction

There are basically two types of vibration in grinding processes: forced vibrations and self-excited vibrations (Figure 8.1).

8.2.2 Forced Vibration

Out of balance and eccentricity of the grinding wheel are the main causes of forced vibrations arising internally within the grinding machine (Inasaki and Yonetsu 1969; Gawlak 1984). The wheel as a source of vibration can be relatively easily identified through frequency measurement. The main concern with wheel-induced vibration is how to eliminate out of

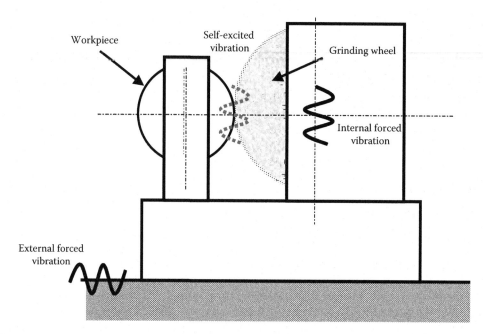

FIGURE 8.1
Chatter vibrations in grinding.

balance and wheel runout. There are a number of other sources of forced vibration such as vibration from belts and hydraulic devices integrated into a grinding machine (Rowe 2014). Forced vibrations are usually thought of as harmonic in nature. However, forced vibrations can also be impulsive. An impulsive vibration occurs, for example, when a feed drive is brought to a halt with a jerk (Rowe 2014). This can lead to size and roundness errors.

External forced vibrations may sometimes be transmitted through the floor from other machines such as nearby drop forges and underground trains. External vibrations tend to be low frequency and are usually suppressed by attention to grinding machine mountings (Tobias 1965).

8.2.3 Regenerative Vibration

A great deal of effort has been made to understand the mechanisms of self-excited chatter vibration in grinding. There are various conceivable reasons for process instability, for example, gyroscopically induced vibration of the grinding wheels. However, the most common causes of grinding chatter are simpler in nature (Hahn 1963; Rowe 2014).

Chatter in centerless grinding is another example of a more complex case due to the nature of the work support geometry (Rowe and Koenigsberger 1965; Rowe 2014). Centerless grinding is analyzed in Chapter 19.

The regenerative effect is considered to be a major cause of self-excited vibration in grinding. Regenerative vibration is similar to the self-excited vibration experienced in cutting processes (Inasaki et al. 1974). Due to the rotational motion of the workpiece during the material removal process, the waves generated on the workpiece surface caused by the relative vibration between the grinding wheel and the workpiece results in a change of depth of cut after one revolution of the workpiece. The phase shift between the surface waves (outer modulation) and the current relative vibration (inner modulation) makes the process unstable when a certain condition is satisfied. A characteristic feature of grinding chatter is that such regenerative effect possibly exists on both the workpiece and the grinding wheel surfaces (Gurney 1965; Inasaki 1975). This complicates self-excited grinding chatter. We can distinguish the two types of regenerative vibration as follows.

- *Work-regenerative chatter.* The waves generated on the workpiece surface through the regenerative effect grow quite rapidly; therefore, this type of chatter vibration is regarded as one of the constraints when the setup parameters are determined.
- *Wheel-regenerative chatter.* On the other hand, waves generated on the grinding wheel surface grow rather slowly due to higher wear resistance of the grinding wheels; therefore, this type of chatter is a determinant for wheel life. When the vibration amplitude builds up to a certain critical limit, it is considered that the grinding wheel has reached the end of its redress life and those waves should be removed through trueing and dressing.

8.3 The Effect of Workpiece Speed

The development of regenerative chatter amplitude in cylindrical grinding is schematically illustrated in Figure 8.2. When the workpiece velocity is extremely high, of the order of some 10 m/min, or the chatter frequency is low, vibration with large amplitude can be observed at the beginning of grinding even if a newly dressed grinding wheel is used. In addition, significant chatter marks can be observed with the naked eye on the workpiece surface suggesting that the work-regenerative effect is the main reason for this vibration.

The occurrence of this chatter is significantly influenced by the combination of setup parameters as shown in Figure 8.3 (Sugihara et al. 1980a). When the workpiece speed is low, or chatter frequency is high, vibration cannot be detected at the beginning of grinding; however, the amplitude increases gradually as the grinding time advances. In this case, the chatter marks are not easily seen with the naked eye on the ground surface.

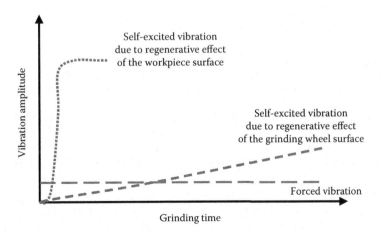

FIGURE 8.2
Vibration phenomena in grinding.

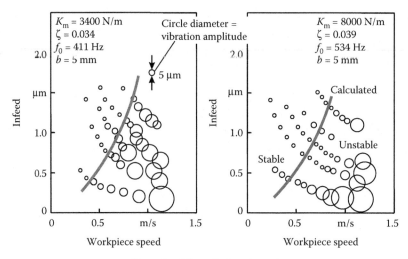

K_m: static stiffness, ζ: damping ratio,
f_0: natural frequency, and b: grinding width

FIGURE 8.3
Stability limit of cylindrical plunge grinding.

However, the surface roughness perpendicular to the grinding direction deteriorates a lot. The rate of amplitude increase is affected by the combination of setup parameters and the type of grinding wheel used. The chatter frequency, which is closely related to the natural frequency of the mechanical system, and the workpiece velocity have a dominant effect on the occurrence of the two different types of regenerative chatter vibration. Furthermore, it can be generally said that grinding processes are in most cases, unstable in terms of the grinding wheel regenerative chatter (Inasaki et al. 1974). In other words, the speed of the vibration development is a matter of concern with respect to this type of chatter.

Other important factors to be taken into account in terms of grinding chatter are the elastic deformation of the grinding wheel (Brown et al. 1971; Inasaki 1975a) and the geometrical interference between the grinding wheel and the workpiece. The influence of the former factor on the stability will be discussed in Section 8.4.

8.4 Geometrical Interference between Grinding Wheel and Workpiece

A large grinding wheel smoothes out rapid changes in workpiece profile preventing high-frequency vibrations from building up to large amplitudes. The smoothing effect is termed grinding wheel interference (Rowe and Barash 1964). The waves generated on the workpiece as well as on the grinding wheel surfaces are the envelope of the relative vibration between them. To start with, the waves generated on the workpiece surface will be considered. In this case, the waves are the envelope of the periphery of the grinding wheel. As far as the following conditions are satisfied, the amplitudes of the relative vibration and the waves generated on the workpiece surface are identical: low vibration frequency, small relative amplitude, and high workpiece velocity. However, once the critical limit determined with the above-mentioned parameters is exceeded, the amplitude of waves generated on the workpiece surface becomes smaller than that of the relative vibration. In other words, the envelope curve is attenuated.

Assuming that the amplitude of the relative vibration and the waves are y and a_w, respectively, the following relationship can be derived (Inasaki 1975):

$$G_{e0} = \frac{a_w}{y}$$

$$= \frac{1}{2}\left(1 - \cos\sqrt{\frac{y_{cr}}{y}}\,\pi\right) \quad \text{for } \frac{y_{cr}}{y} < 1$$

$$= 1 \quad \text{for } \frac{y_{cr}}{y} \geq 1 \tag{8.1}$$

where

$$y_{cr} = \frac{v_w^2}{\omega^2}\frac{2(d_w \pm d_s)}{d_w d_s} \tag{8.2}$$

and y_{cr} is a critical amplitude, v_w is the workpiece speed, ω is the angular chatter frequency, d_w is the workpiece diameter, and d_s is the grinding wheel diameter. The plus sign in Equation 8.2 is for cylindrical external grinding, the minus sign for internal grinding, and $d_w = \infty$ for surface grinding. When $y_{cr} < y$, the amplitude of waves becomes smaller than that of the relative vibration. Otherwise, both amplitudes are identical. As for the waves generated on the grinding wheel, the critical amplitude can be obtained by replacing the workpiece speed v_w with the grinding wheel speed v_s. Therefore, the critical amplitude is much larger for the waves generated on the grinding wheel, because the wheel speed is much higher than the workpiece speed.

The calculated results from Equation 8.1 are given in Figure 8.4 (Inasaki 1975). The geometrical interference is a strong nonlinear term in the grinding dynamics (Inasaki et al. 1974).

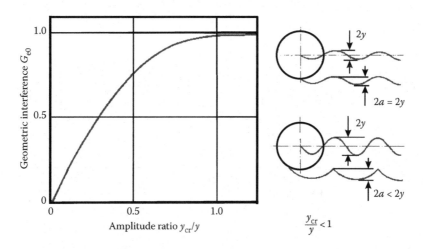

FIGURE 8.4
Geometrical interference.

It is clear from Equations 8.1 and 8.2, waves with large amplitude and higher frequency cannot be generated on the workpiece surface because the critical amplitude becomes small.

8.5 Vibration Behavior of Various Grinding Operations

The vibration behavior in cylindrical, internal, and surface grinding processes differ in significant ways (Inasaki 1977a). In the case of internal and surface grinding, the chatter frequency is, in most cases, related to the natural frequency of the grinding wheel spindle system because the dynamic stiffness of internal grinding spindles is more often lower than that of the workpiece system. This is not the case for cylindrical grinding, where the dynamic stiffness of the workpiece system is usually lower than that of the grinding wheel spindle system. In addition, chatter vibration caused by the regenerative effect on the workpiece surface seems less likely to develop in surface grinding. This may be partially due to the fact that the phase shift between the inner and the outer modulation is not necessarily constant because of the uncertainty in the workpiece reciprocating motions. In all three cases, however, the probability of chatter will be greater for plunge processes than for traverse or cross-feed processes. In plunge grinding, 100% of the existing surface waviness acts to stimulate waviness in the next pass of the grinding wheel. In a traverse or cross-feed grinding process, the overlap factor is less than 100% and might only be 20% which greatly reduces the risk of chatter (Tobias 1965; Rowe 2014).

There are two types of grinding operations: plunge grinding and traverse grinding. The stability analysis becomes slightly more complex for the traverse grinding process because the different contact condition between the grinding wheel and the workpiece should be taken into consideration along with the wheel width (Shimizu et al. 1978). Figure 8.5 shows an example of the stability limit in cylindrical traverse grinding (Sugihara et al. 1980a,b).

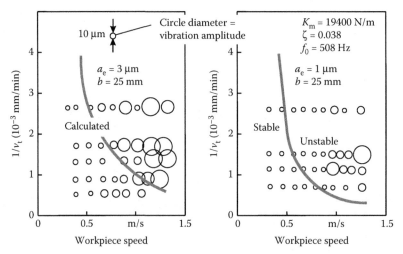

K_m: static stiffness, ζ: damping ratio, f_0: natural frequency,
b: grinding width, a_e: depth of cut, and v_t: traverse speed

FIGURE 8.5
Stability limit in cylindrical traverse grinding.

8.6 Regenerative Self-Excited Vibrations

8.6.1 Modeling of Dynamic Grinding Processes

A mathematical model of dynamic grinding process can be established taking the factors shown in Figure 8.6 into account. Additional parameters in grinding dynamics, not needed in cutting dynamics, are the contact stiffness of the grinding wheel and the grinding damping. A comprehensive block diagram for representing the dynamic grinding process is complex. Therefore, the process is divided into two extreme cases; the dynamic grinding process model for work-regenerative chatter and the model for the wheel-regenerative chatter. This simplification can be made possible by taking the geometrical interference into account. When the condition $y_{cr}/y \geq 1$ in Equation 8.1 is satisfied, only the work-regenerative effect need to be considered, and hence the grinding wheel regenerative effect can be ignored. This is due to the following two reasons:

1. The work-regenerative effect has a large effect on the process stability.
2. The development of grinding wheel regeneration is much slower than that of workpiece regeneration.

On the other hand, for the case of $y_{cr}/y \ll 1$, the work-regenerative effect can be ignored, because the amplitude of waves generated on the workpiece surface is much smaller than the amplitude of the relative vibration. However, the regenerative effect on the grinding wheel surface must be considered in the stability analysis.

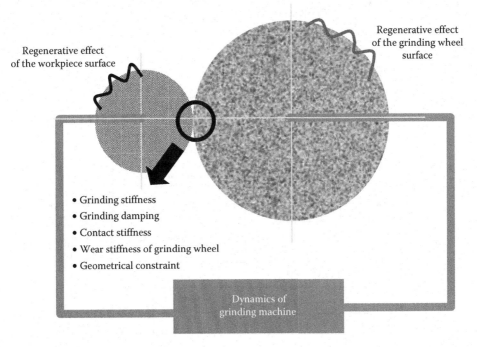

Regenerative effect of the workpiece surface

Regenerative effect of the grinding wheel surface

- Grinding stiffness
- Grinding damping
- Contact stiffness
- Wear stiffness of grinding wheel
- Geometrical constraint

Dynamics of grinding machine

FIGURE 8.6
Factors affecting grinding dynamics.

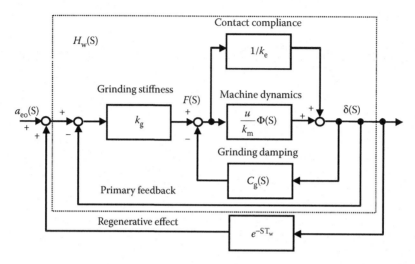

FIGURE 8.7
Block diagram for workpiece regenerative chatter.

Based on the above simplification, the block diagrams for the dynamic grinding process are depicted in Figures 8.7 and 8.8 for work-regenerative chatter and for grinding wheel-regenerative chatter, respectively (Inasaki 1977b).

8.6.2 Grinding Stiffness and Grinding Damping

A simplified linear dynamic grinding system can be represented by masses, springs, and damping elements. Based on the following simple grinding force model, which says that the normal grinding force F_n is proportional to the material removal rate:

$$F_n = \lambda b \left(\frac{v_w}{v_s} a \right)^\varepsilon \tag{8.3}$$

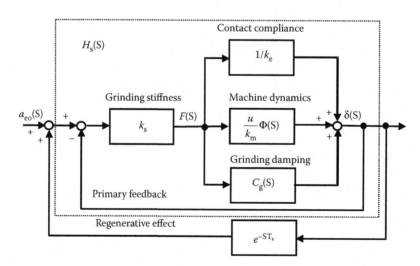

FIGURE 8.8
Block diagram for grinding wheel regenerative chatter.

The dynamic grinding force model can be derived as (Inasaki 1977a):

$$F_n(t) = k_g a(t) + c_g \dot{x}(t) \tag{8.4}$$

where by differentiating Equation 8.3:

$$k_g = \lambda \varepsilon b \frac{v_w}{v_s} \left(\frac{v_w}{v_s} a \right)^{\varepsilon-1} \tag{8.5}$$

$$c_g = \lambda \varepsilon b \frac{1}{v_s} \left(\frac{v_w}{v_s} a \right)^{\varepsilon-1} \sqrt{\frac{a}{((1/d_s) \pm (1/d_w))}} \tag{8.6}$$

where λ is the constant, b is the grinding width, v_w is the workpiece speed, v_s is the grinding speed, a is the wheel depth of cut, ε is the exponent, and $\dot{x}$ is the mutual approach speed between the grinding wheel and the workpiece. The plus sign in the denominator is for external cylindrical grinding, the minus sign for internal grinding, and $d_w = \infty$ for surface grinding.

The constant k_g given by Equation 8.5 is the grinding stiffness, which is the coefficient between the grinding force and the depth of cut. The grinding damping c_g is given by Equation 8.6 and is the coefficient between the grinding force and the mutual approach speed between the grinding wheel and the workpiece. The grinding damping increases in proportion to the length of the contact between the wheel and the workpiece. Generally speaking, the grinding system becomes less stable as the grinding stiffness increases, while increase of grinding damping makes the system more stable.

In order to analyze chatter vibration caused by the grinding wheel regenerative effect, the grinding stiffness given by Equation 8.5 should be replaced with the wear stiffness of the grinding wheel. As a first order approximation, the wear stiffness is obtained as:

$$k_s = k_g G \frac{v_s}{v_w} \tag{8.7}$$

where G is the grinding ratio. Taking practical values of the grinding ratio and the speed ratio into account, it is confirmed that the wear stiffness of the grinding wheel is much higher than the grinding stiffness. Therefore, as far as the analysis of the chatter vibration caused by the workpiece regenerative effect is concerned, the wear stiffness of the grinding wheel can be assumed to be infinite.

8.6.3 Contact Stiffness

A characteristic feature of grinding dynamics is that the elastic deformation of vitrified and resin-bond grinding wheels is too large to be neglected and furthermore, it has a significant influence on the process stability. The contact stiffness of the grinding wheel is defined as the relationship between the normal compressive force F_n and the elastic deformation of the grinding wheel δ induced at the contact zone. Some models have been proposed for theoretically calculating the elastic deformation of grinding wheels by applying the Hertzian elastic contact theorem (Figures 8.9 and 8.10; Brown et al. 1971; Inasaki 1975).

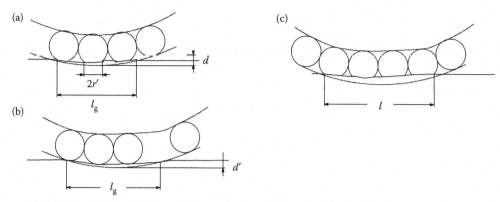

(a) Deflection of grit-workpiece contact assuming the grinding wheel remains circular

(b) Deflection of the wheel–workpiece contact assuming the grains are undistorted

(c) Combined effect of (a and b)

FIGURE 8.9
A deformation model for the grinding wheel. (From Brown, R. H., Saito, K., Shaw, M. C., 1971, *Annals of the CIRP*, 19, 105–113.)

Furthermore, it has been found both from theory and from experiments that grinding wheel deflection is significantly increased due to wheel roughness (Rowe et al. 1993).

According to those analyses, it is suggested that the deformation δ is given by the following equation:

$$\delta = F_n^{\rho} \tag{8.8}$$

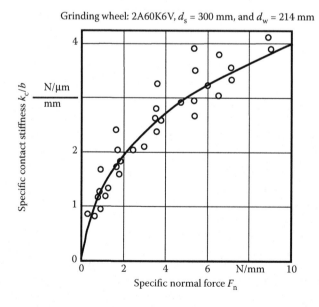

FIGURE 8.10
Specific contact stiffness of the grinding wheel.

where $0 < \rho < 1$.

Therefore, the contact stiffness obtained through differentiating Equation 8.8

$$k_c = \frac{dF_n}{d\delta} \tag{8.9}$$

has a nonlinear characteristic of the hard-spring type. By combining Equations 8.3 and 8.9, the contact stiffness can be expressed as a function of the grinding setup parameters. For example, the increase of the speed ratio v_w/v_s and the depth of cut results in an increase of the contact stiffness.

The contact stiffness has also been experimentally investigated (Younis 1972; Inasaki 1977a). Of course, an increase of grinding wheel hardness corresponds to an increase of the contact stiffness.

8.6.4 Dynamic Compliance of the Mechanical System

The dynamic compliance of the mechanical system, which consists of the workpiece, the grinding wheel, and the grinding machine, is represented by

$$\frac{u}{k_m} \Phi(j\omega) \tag{8.10}$$

and

$$u = \cos\alpha \cos(\alpha - \beta) \tag{8.11}$$

where α is the angle between the direction of the depth of cut and the direction of the natural frequency mode of the mechanical system, β is the angle between the direction of the depth of cut and the direction of the resultant grinding force, k_m is the static stiffness of the mechanical system, and $\Phi(j\omega)$ is the nondimensional dynamic compliance.

The practical grinding machine has many degrees of freedom; however, in most cases, the model is simplified as having a single degree of freedom in the theoretical investigation. It is important to notice that the orientation factor u has a significant influence on the resultant dynamic compliance.

8.6.5 Stability Analysis

The stability limit of the self-excited chatter vibration in plunge cylindrical grinding can be obtained by substituting $s = j\omega$ into the characteristic equations based on Figures 8.7 and 8.8. Influence of the depth of cut, the workpiece speed, and the grinding width on the stability limit is illustrated in Figure 8.11 (Inasaki 1977a) for typical chatter vibrations of both types: the regenerative effect on the workpiece surface and the regenerative effect on the grinding wheel surface.

Grinding stiffness k_g is very important for the onset of work-regenerative chatter. The grinding stiffness depends on the work material and the nature of the grinding wheel. Hard action grinding wheels produce higher grinding forces than soft grinding wheels and mean that a smaller infeed rate will lead to chatter. This is demonstrated by the stability limit for work-regenerative chatter in plunge cylindrical grinding (Rowe 2014).

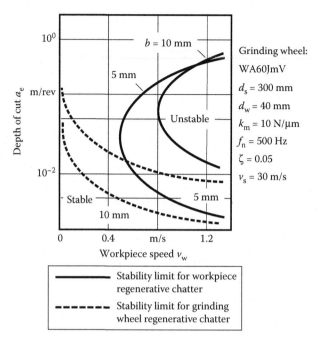

FIGURE 8.11
Typical stability limits for grinding chatter.

$$1 + (1 - e^{-j\omega T_w}) \cdot \frac{k_g b}{k_m(j\omega)} = 0 \tag{8.12}$$

where ω is the vibration frequency, $\omega T_w = 2\pi\omega/\Omega_w = 2\pi n$, Ω_w is the work rotational frequency in radians per second, and the function $k_m(j\omega)$ is the frequency-dependent system stiffness. Solutions for Equation 8.12 can be found for any number of waves, $n = 1, 2, 3$, and so on. The solution for each number of waves is given by a chatter frequency above the natural frequency.

At low work speeds, the unstable regions overlap for large numbers of waves and instability occurs when the grinding stiffness/machine stiffness ratio $k_g/k_{mo} > 0.105$. Grinding stiffness k_g reduces greatly at low work speeds, high wheel speeds, and with narrow grinding contact width as can be seen from Equation 8.3. The factor λ reflects the hardness of the grinding action of the grinding wheel and work material combination. Equation 8.3 explains improved stability at low work speeds. The stabilizing effect is enhanced by improved grinding process damping at low work speeds as reflected in Equation 8.4. Grinding damping reduces the dynamic magnifier Q. With $Q = 3$, the minimum stiffness ratio for onset of chatter is increased to $k_g/k_{mo} = 0.48$ compared with $k_g/k_{mo} = 0.105$ for $Q = 10$ (Figure 8.12).

With respect to work-regenerative chatter vibration, absolute stability can be attained when the workpiece speed is sufficiently low. On the other hand, wheel-regenerative chatter vibration has a large area of instability, that is, most of practical grinding conditions exist in the unstable region. Therefore, as far as the chatter vibration caused by the regenerative effect on the grinding wheel surface is concerned, it is necessary to know the rate of increase of the vibration amplitude.

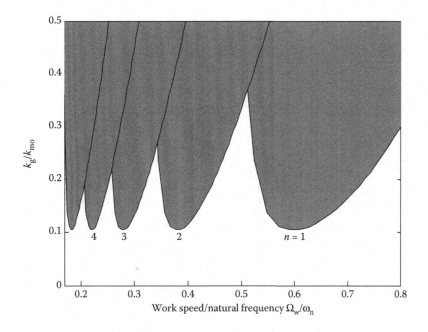

FIGURE 8.12
Stability limits with light machine damping $Q=10$. (From Rowe, W. B., 2014, *Principles of Modern Grinding Technology*, 2nd edition. Elsevier, Oxford and New York.)

The positive real part of the roots of the characteristic equation is the index for the rate of increase of the vibration amplitude, while the imaginary part indicates the chatter frequency. Some calculated examples of the roots distribution are shown in Figures 8.13 and 8.14 (Inasaki 1977a). The important results deduced from those figures are

- The increase rate of vibration amplitude for the wheel-regenerative chatter is much slower than that of the workpiece regeneration type.
- The roots exist with the constant interval of $1/T_s$ or $1/T_w$ in the imaginary axis, where T_s and T_w are the rotational period of the grinding wheel and the workpiece, respectively. This result explains the fact that the chatter vibration observed is accompanied with an amplitude modulation.
- The chatter frequency is always higher than the natural frequency of the mechanical system.

Figure 8.15 shows the calculated results of the positive real parts for the grinding wheel-regenerative chatter. It is assumed here that the instability occurs at the frequency that gives the maximum positive real part. From this result, the following conclusions are deduced:

- The development of chatter vibration becomes faster with larger depth of cut, larger grinding width, lower workpiece speed, and higher grinding wheel speed. With respect to the effect of depth of cut, however, it is necessary from the practical point of view, to consider the increase of vibration amplitude against the amount of cumulative material removed. The calculated result shows that a larger

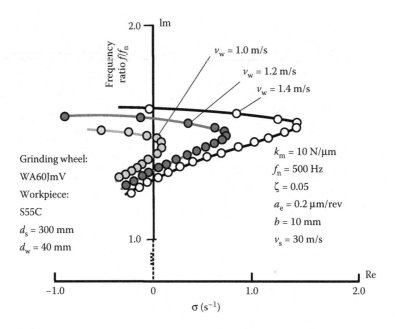

FIGURE 8.13
Roots of workpiece regenerative chatter.

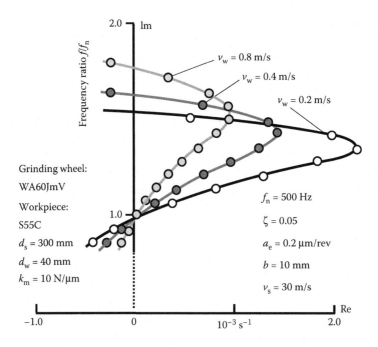

FIGURE 8.14
Roots of grinding wheel regenerative chatter.

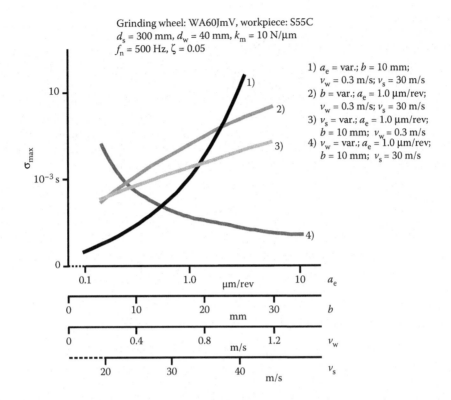

FIGURE 8.15
Parameter-related rate of increase of the chatter amplitude.

amount of material can be removed before the grinding wheel comes to the end of its redress life with larger depth of cut.

- The rate of increase in vibration amplitude decreases with a decrease in the contact stiffness and increase in the wear stiffness of the grinding wheel. This result means that the influence of the grinding wheel hardness is complex. An increase of stiffness and damping in the mechanical system reduces the rate of chatter development.

A similar analysis can be conducted for internal as well as surface grinding (Inasaki 1977a).

The stability analysis of traverse grinding is slightly more complex than that of plunge grinding. However, some theoretical and experimental investigations have been conducted for the workpiece regenerative chatter in cylindrical grinding (Shimizu et al. 1978). Important conclusions were

- The process tends to be more unstable under the condition of lower traverse speed, higher workpiece speed, larger grinding wheel width, and smaller depth of cut.
- Chatter frequency increases with increases of traverse speed, grinding wheel width, depth of cut, and workpiece speed.

8.7 Suppression of Grinding Vibrations

In order to suppress vibrations in grinding, it is necessary to identify whether it is forced vibration or self-excited vibration. Figure 8.2 provides a possibility for identifying the type of vibration in grinding. If the vibration is detected while the machine idles, it is forced vibration. Vibrations with higher frequency than the grinding wheel rotational frequency are, in most cases, regenerative chatter. Vibration observed at the beginning of grinding, just after dressing, is more likely to be workpiece regenerative chatter. Grinding wheel-regenerative chatter appears after a considerable time of grinding.

8.7.1 Suppression of Forced Vibrations

The most significant source of forced vibration in grinding is unbalance of the grinding wheel. In order to suppress the adverse effect of forced vibration on the grinding process, unbalance of the grinding wheel should be detected using a vibration sensor, followed by balancing the grinding wheel (Kaliszer 1963; Trmal and Kaliszer 1976; Gawlak 1984). Figure 8.16 shows an example of a balancing method based on liquid injection into a wheel flange pocket (Horiuchi and Kojima 1986). Elimination of unbalance of the grinding wheel is essential to meet the requirement of higher grinding accuracy.

Eccentricity of the grinding wheel is another significant source of forced vibration. This can be eliminated through trueing the grinding wheel; however, trueing and balancing should be repeated alternating several times in order to completely eliminate the vibration source because trueing possibly generates an additional unbalance in the grinding wheel.

Sources of forced vibration can usually be located through frequency analysis of the vibrations. For example, forced vibrations caused by unbalance and eccentricity of the grinding wheel have a frequency component that corresponds to the wheel rotational frequency.

8.7.2 Suppression of Self-Excited Chatter Vibrations

The ultimate goal of research is to develop practical means for suppressing vibrations. Based on understanding of the chatter principle, a number of practical methods have

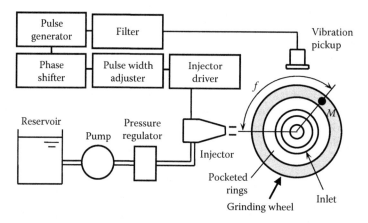

FIGURE 8.16
Automatic balancer for grinding wheels.

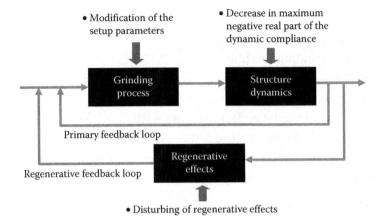

FIGURE 8.17
Principles of suppression of regenerative chatter.

been proposed. The methods can be categorized into one of three strategies shown in Figure 8.17:

- Modification of the grinding conditions
- Increase of the dynamic stiffness of the mechanical system
- Disturbing the regenerative effect

Figure 8.17 represents a block diagram of the simplified dynamic grinding system that consists of the grinding stiffness, the mechanical system, and the regenerative feedback loop. The block diagram is valid for workpiece regenerative chatter.

A stability analysis based on the dynamic model depicted in Figure 8.17 can be achieved as shown in Figure 8.18. The dynamic compliances of the mechanical system are represented by vector loci on the complex planes, while straight lines parallel to the imaginary axes represent the material removal process. Instability occurs when both lines have

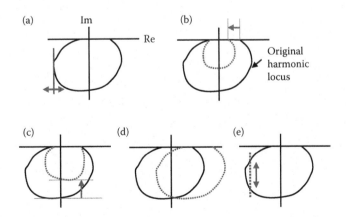

FIGURE 8.18
Strategies for suppressing regenerative chatter. (a) Modification of grinding conditions. (b) Increase of static stiffness decrease of orientation factor. (c) Increase of damping. (d) Shift of harmonic locus. (e) Disturbing regenerative effects.

intersections. Based on those representations, methods for suppressing the regenerative chatter vibration can be further categorized as follows:

1. Modification of grinding conditions, Figure 8.18a.
2. Increase of the dynamic stiffness of the mechanical system.
   2.1 Increase of the static stiffness, Figure 8.18b.
   2.2 Decrease of the orientation factor, Figure 8.18b.
   2.3 Increase of the damping, Figure 8.18c.
3. Shifting the vector locus of the dynamic compliance to positive real part, Figure 8.18d.
4. Disturbing the regenerative effect, Figure 8.18e.

With respect to Method 1, decrease of the grinding stiffness is the most straightforward because it results in shifting the line parallel to the imaginary axis to the left, and consequently, the intersections of both lines can be avoided. Decrease of the grinding width and the workpiece speed meets this requirement (Figure 8.11).

Increase of the static stiffness or decrease of the orientation factor is effective for shrinking the vector loci and consequently for improving the dynamic performance of the grinding machines, Methods 2.1 and 2.2. Figure 8.19 shows an influence of the orientation factor on the stability (Inasaki et al. 1974). In this series of grinding tests, the cross section of the workpiece center was modified from circular to rectangular and its orientation angle was changed. Interestingly, the critical limit in terms of grinding width changes depending on the orientation angle of the workpiece center. Influence of the static stiffness and the orientation factor on the stability is significant; therefore, it is worthwhile to take this effect into account at the design stage of a grinding machine.

Another strategy for improving the dynamic performance of the mechanical system is to increase the damping by adding some kind of dampers. Methods are divided into two kinds: application of passive (Tönshoff and Grosebruch 1988) and active dampers. Figure 8.20 shows an example of passive damper application (Hong et al. 1990). In this case, the damper is attached to the wheel head of the cylindrical grinding machine. Passive dampers are effective only when they are optimally tuned to the main mechanical system,

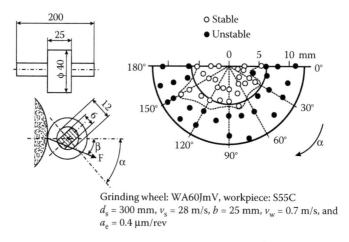

Grinding wheel: WA60JmV, workpiece: S55C
$d_s = 300$ mm, $v_s = 28$ m/s, $b = 25$ mm, $v_w = 0.7$ m/s, and $a_e = 0.4$ μm/rev

FIGURE 8.19
Influence of the orientation factor on chatter stability.

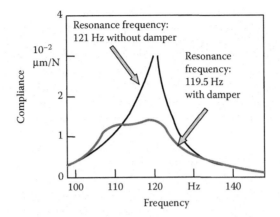

FIGURE 8.20
Effect of a passive damper for reducing structure compliance.

and the vibration characteristic of the mechanical system does not change significantly during the operation. However, application of active dampers is more flexible and can cope with a change in the vibration characteristics of the mechanical system. Figure 8.21 shows an example of the active damper application (Weck and Brecher 2001).

Shifting the vector locus to the right on the complex plane, Method 3 can be achieved by adding a spring element between the workpiece and the grinding wheel. It is essential here to add only the spring element without any additional mass to the system. The resultant vector locus after the attachment of the spring element k_a is:

$$\Omega(j\omega) = \frac{1}{k_a} + \frac{u}{k_m}\Phi(j\omega) \tag{8.13}$$

This idea for suppressing the chatter vibration can be to some extent realized by decreasing the contact stiffness of the grinding wheel (Sexton and Stone 1981). Figure 8.22 shows an example of the grinding wheel modified for this purpose.

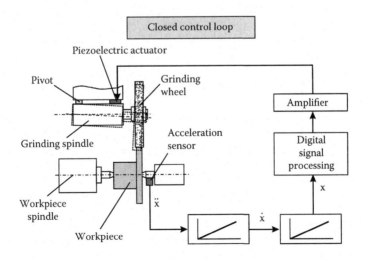

FIGURE 8.21
Application of active damping in plunge external grinding.

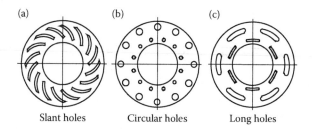

FIGURE 8.22
Flexible grinding wheel for suppressing chatter vibration.

A practical method for suppressing self-excited vibration is to intentionally disturb the phase shift between the inner and the outer modulation, Method 4. This idea can be put into practice by periodically varying the rotational speed of either the workpiece or the grinding wheel. The former method is effective for suppressing workpiece regenerative chatter (Inasaki 1977b), while the latter method is effective for suppressing grinding wheel regenerative chatter (Hoshi et al. 1986). Figures 8.23 and 8.24 show the effect of varying the workpiece rotational speed and the grinding wheel rotational speed, respectively. The methods introduced here are effective and practical; however, the applications are restricted to rough grinding because varying the rotational speed may have some adverse effect on the surface quality of the ground parts.

8.8 Conclusions

The analysis provided in this chapter explains the characteristics of forced and self-excited vibrations in grinding. It has been shown that reducing work speed and increasing wheel

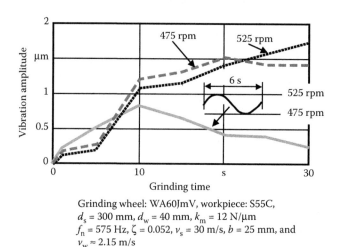

Grinding wheel: WA60JmV, workpiece: S55C,
$d_s = 300$ mm, $d_w = 40$ mm, $k_m = 12$ N/μm
$f_n = 575$ Hz, $\zeta = 0.052$, $v_s = 30$ m/s, $b = 25$ mm, and
$v_w \approx 2.15$ m/s

FIGURE 8.23
Chatter suppression by varying workpiece rotational speed.

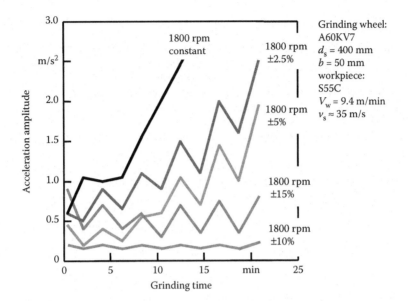

FIGURE 8.24
Chatter suppression by varying grinding wheel speed.

speed reduce the tendency to chatter. With respect to grinding machine design, it is essential to increase stiffness and damping in order to achieve high dynamic stability. The benefits of a softer action grinding wheel are the reduction in grinding forces. There are also benefits in introducing compliance at the grinding wheel surface. Other possible methods of reducing vibrations in grinding including the application of passive and active damping.

From the theoretical point of view, a full description of chatter is complex. This is mainly due to the difficulty of identifying the required dynamic characteristics of the grinding process, of the grinding machine, and of the grinding wheel. In addition, those characteristics are likely to change during operation of the grinding process. Taking those difficulties into account, it is considered that a sophisticated chatter suppression system is required that consists of monitoring and control of chatter to meet the requirement for precision grinding processes. Advanced sensors and actuators available today are generally expected to make the achievement of chatter control possible. For the roughing process, however, periodical change of the grinding wheel rotational speed as well as workpiece rotational speed appears to be a practical solution.

References

Brown, R. H., Saito, K., Shaw, M. C., 1971, Local elastic deflections in grinding. *Annals of the CIRP*, 19, 105–113.

Gawlak, G., 1984, Some problems connected with balancing of grinding wheels. *Journal of Engineering for Industry*, 106, 233–236.

Gurney, J. P., 1965, An analysis of surface wave instability in grinding. *Journal of Mechanical Engineering Science*, 7/2, 198–209.

Hahn, R. S., 1963, Grinding chatter—Causes and cures. *The Tool and Manufacturing Engineer*, September, 74–78.

Hong, S. K., Nakano, Y., Kato, H., 1990, Improvement of dynamic characteristics of cylindrical grinding machines by means of dynamic dampers. *Proceedings of the 1st International Conference on New Manufacturing Technology*, Chiba (Japan), 413–418.

Horiuchi, O., Kojima, H., 1986, A new liquid-injection type automatic balancer for the grinding wheel (in Japanese). *Japan Society of Political Economy*, 52(2), 713–718.

Hoshi, T., Matsumoto, S., Mitsui, S., Horiuchi, O., Koumoto, Y., 1986, Suppression of wheel-regenerative grinding vibration by alternating wheel speed (in Japanese). *Japan Society of Political Economy*, 52(10), 1802–1807.

Inasaki, I., 1975, Ratterschwingungen beim Außen-rund-Einstechschleifen (Chatter vibration in external cylindrical plunge grinding). *Werkstatt und Betrieb* 108/6, 341–346.

Inasaki, I., 1977a, Regenerative chatter in grinding. *Proceedings of the 18th MTDR Conference*, 423–429, UK.

Inasaki, I., 1977b, Selbsterregte Ratterschwingungen beim Schleifen, Methoden zu ihrer Unterdrückung (Self-excited chatter vibration in grinding). *Werkstatt und Betrieb* 110/8, 521–524.

Inasaki, I., Yonetsu, S., 1969, Forced vibrations during surface grinding. *Bulletin of JSME* 12/50, 385–391.

Inasaki, I., Yonetsu, S., Shimizu, T., 1974, Selbst-erregte Schwingungen beim Aussenrundeinstechschleifen (Self-excited vibration in external cylindrical plunge grinding). *Annals of the CIRP* 23/1, 117–118.

Kaliszer, H., 1963, Accuracy of balancing grinding wheels by using gravitational and centrifugal methods. *Proceedings of the 4th International MTDR Conference, Advances in MTDR* 395–415, Birmingham University, UK.

Rowe, W. B., 2014, *Principles of Modern Grinding Technology*, 2nd edition. Elsevier, Oxford and New York.

Rowe, W. B., Barash, M. M., 1964, Computer method for investigating the inherent accuracy of centerless grinding. *International Journal of Machine Tool Design and Research*, 4, 91–116.

Rowe, W. B., Koenigsberger, F., 1965, The work-regenerative effect in centreless grinding. *International Journal of Machine Tool Design and Research*, 4, 175–187.

Rowe, W. B., Morgan, M. N., Qi, H. S., 1993, The effect of deformation on contact length in grinding. *Annals of the CIRP*, 42(1), 409–412.

Sexton, J. S., Stone, B. J., 1981, The development of an ultrahard abrasive grinding wheel which suppresses chatter. *Annals of the CIRP*, 30/1, 215–218.

Shimizu, T., Inasaki, I., Yonetsu, S., 1978, Regenerative chatter during cylindrical traverse grinding. *Bulletin of JSME*, 21/152, 317–323.

Sugihara, K., Inasaki, I., Yonetsu, S., 1980a, Stability limit of regenerative chatter in cylindrical plunge grinding—A proposal of the practical stability limit equation (in Japanese). *Japan Society of Political Economy*, 46(2), 201–206.

Sugihara, K., Inasaki, I., Yonetsu, S., 1980b, Stability limit of regressive chatter in cylindrical traverse grinding (in Japanese). *Japan Society of Political Economy*, 46(3), 305–310.

Tobias, S. A., 1965, *Machine Tool Vibration*. Blackie, London.

Tönshoff, H. K., Gosebruch, H., 1988, Verstellbarer passiver Dämpfer für Schwingungen in Außenrund-schleifmaschinen. *VDI-Z*, 130/5, 57–60.

Trmal, G., Kaliszer, H., 1976, Adaptively controlled fully automatic balancing system. *Proceedings of the 17th International MTDR Conference* 85–92, Birmingham, UK.

Weck, M., Brecher, C., 2001, The essential difference of the chatter phenomena between processes with defined and undefined cutting edges, *Technical Presentation in CIRP-STC "G."* Paris, January.

Younis, M. A., 1972, Theoretische und praktische Untersuchung des Ratterverhaltens beim Außenrundschleifen (Theoretical and Practical investigation of chatter in external cylindrical grinding). *Industrie-Anzeiger*, 94/59, 1461–1465.

9

Grinding Wheel Wear

9.1 Types of Wheel Wear

9.1.1 Introduction

As a result of process forces and temperatures during grinding, a grinding wheel is subject to modification by a process of wheel wear. Wear leads to changed process conditions and quality deviations in the component. Figure 9.1 shows three different types of grinding wheel wear: profile deviation, roundness deviation, and changes in grinding wheel sharpness.

In plunge grinding, where the wheel profile is reproduced in the ground component, profile deviations lead to workpiece shape defects. In the case of longitudinal grinding, profile deviations lead to screw thread undercuts. Roundness deviations make the machine system vibrate by dynamic alternating forces, which cause chatter marks to be machined on the component.

A loss of grinding wheel sharpness leads to higher grinding forces and temperatures, which may entail dynamic and thermal deflections between the grinding wheel and workpiece as well as uncontrolled grinding processes lead to chatter marks on the component. Finally, there will be shape and position errors as well as dimensional deviations on the component. These modifications of the grinding wheel in the course of the grinding process are due to wear and result from microscopic changes in the abrasive grains and alterations in the chip space.

9.2 Wheel Wear Mechanisms

Wheel wear results from material loss at the wheel surface, which can be traced back to mechanical contact between the wheel moving relative to the workpiece or any other body such as the dressing tool. Wear effects can be ascribed to the following main mechanisms: abrasion, adhesion, tribochemical reactions, surface disruption, and diffusion (DIN 50320 1979; DIN 50323 1988).

9.2.1 Abrasive Wheel Wear

As a prerequisite of abrasive wear, the surface of one of the two interacting partners of the abrasive process must be penetrated and a tangential movement must take place between them. The result is plastic and elastic deformations with groove and chip formation in

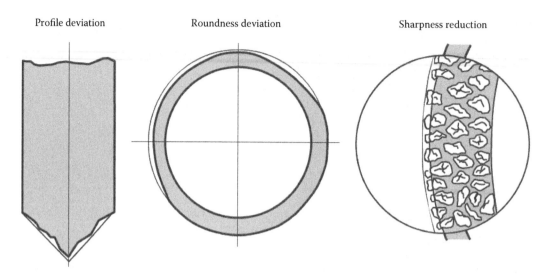

FIGURE 9.1
Types of grinding wheel wear.

the microrange. Grooving wear dominates, when hard workpiece material particles or loose particles of grain in the contact zone lead to surface changes in the grinding wheel (Engelhorn 2002).

9.2.2 Adhesive Wheel Wear

Adhesion wear is based on an atomic bond at a microcontact surface between the active partners of the wear process through microwelding. This bond is very strong, which means that shearing through the relative movement of the active parts takes place at a different place than that of the original microcontact surface, usually inside the partner with the lower strength. Chemical adhesion is based on atomic interaction through thermally induced diffusion processes. In contrast, in mechanical adhesion, the surfaces of the active parts are engaged in the microrange, while high temperatures lead to surface deformation (Telle 1993).

9.2.3 Tribochemical Wheel Wear

In the case of tribochemical wear, chemical reactions take place either between the active partners of the wear process or with the surrounding environmental medium. These chemical reactions cause changes in the boundary layer properties, which lead to adhesion of reaction products on the abrasive grain and to grain damage. Factors of a tribochemical reaction are chemical affinity between the active partners and ambient conditions such as temperature, pressure, and concentration.

9.2.4 Surface Disruptions

Surface disruptions can be traced back to mechanical thermal alternating pressures opening up grain boundaries and cleavage planes. This leads to structure changes, fatigue, cracks, and separation of single particles causing breakage and cutting material failure (Zum Gahr 1987; Telle 1993).

9.2.5 Diffusion

Prerequisites of diffusion processes in the working zone are the adequate activation energy and a sufficient chemical potential of the active partners. The diffusion accounts for a thermal activation of single atoms, which, as a result, change places. This causes material loss, and impurity atoms are inserted into the grain surface, which might lead to a loss of hardness. Surface diffusion processes can be divided into intercrystalline diffusion along the grain boundaries and transcrystalline diffusion into the grain volume.

9.3 Wear of the Abrasive Grains

9.3.1 Types of Grain Wear

Overlap between the above-mentioned wear mechanisms leads to changes in the abrasive grain. These wear types are depicted in Figure 9.2. They basically can be divided into:

- Flattening
- Microcrystalline grain splintering
- Partial grain break off
- Total grain break off (Peklenik 1958)

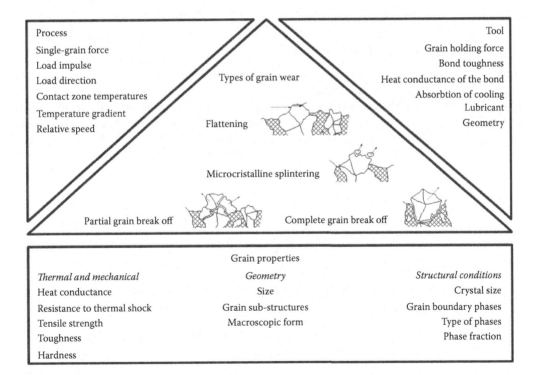

FIGURE 9.2
Influences on grain wear. (From Anon., 2003a. *Diamant- und CBN-Schleifscheiben*. Firmeninformation der Fa. Winter & Sohn GmbH, Norderstedt, www.winter-diamantwerkz-saint-gobain.de. With permission.)

9.3.2 A Combined Wear Process

The strength of the particular wear process depends on the process parameters and on the grain and bond properties. The character of the wear process is governed by contributions from thermal and mechanical wear, which are determined by machining parameters, cooling and lubrication conditions, and process kinematics. Grinding wheel properties are determined by the stability and thermal diffusivity of the grinding wheel bond. Additionally, the basic porosity of the bond has a crucial influence on lubricant absorption and thus on the thermal conditions within the working zone. The abrasive grains of the abrasive medium differ in terms of hardness, tensile strength, and ductility. The fracture and splintering behavior can thereby be controlled by screening procedures during abrasive manufacture and the synthesis process (Juchem and Martin 1989; Jackson and Hayden 1993; Uhlmann and Stark 1997).

9.3.3 Grain Hardness and Temperature

The hardness of the abrasive grains also depends on the process conditions. Figure 9.3 shows the hardness of polycrystalline sintered corundum grains with changing process temperature. With increasing temperature, abrasive grain hardness declines. At 800°C, it is approximately 25% compared to room temperature. As well as sintered corundum, diamond grains show a significant hardness decline above 700°C. The wear resistance of the grains depends not only on the hardness at ambient temperature, but more importantly on the hardness at the operating contact temperatures. In contrast to the named abrasive grains, cubic boron nitride (CBN) shows only very low hardness decline with increasing temperature, which makes it preferable for high-speed grinding processes.

9.3.4 Magnitude of the Stress Impulses

The strength of the grain wear process is related to the magnitude of the stress impulses of the abrasive interactions. As an example, Figure 9.4 shows the abrasive grains of a D126 St50 grinding wheel with a concentration of C90 during grinding silicon carbide. In the left part of the image, the wear is typified by a wear flat developed on the grain. By increasing the magnitude of the stress impulses by superposing an ultrasonic oscillation, the wear

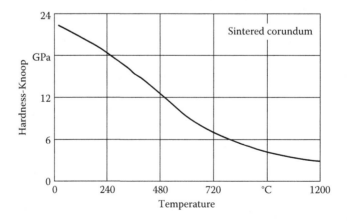

FIGURE 9.3
Hardness against temperature in polycrystalline sintered corundum abrasive grains. (From Anon., 2003b. *Firmeninformation der Fa.* Hermes Schleifmittel GmbH, Hamburg. With permission.)

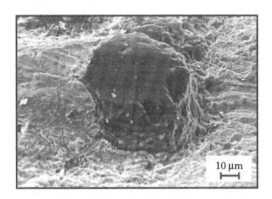 

FIGURE 9.4
Grain flattening and splintering. (From Uhlmann, E., Daus, N. 2000. *Proceedings of the 7th International Symposium.* Cer. Mat. Com. Eng. Goslar, 19–21/06. With permission.)

process can be changed leading to splintering under otherwise similar process conditions, thus generating new sharp grain cutting edges.

9.3.5 Growth of Grain Flats

Reasons for the flattening of single grains are the above-mentioned mechanisms of abrasion, adhesion, tribochemical reaction, and diffusion. Flats only develop at the grain tip area at low single-grain forces and high process temperatures. Hence, the grinding wheel specification must be adjusted for the process conditions. A grinding wheel specified for reciprocating grinding cannot be sensibly used in other fields of application. Therefore, if the wheel is used in creep feed grinding or in internal cylindrical grinding, grain flats may be increased due to decreased single-grain forces and increased process temperatures (Uhlmann and Stark 1997).

Figure 9.5 shows a model of a D126 C100 grinding wheel surface in a topography section of 1×1 mm^2. The topography is depicted first for flattened grains and second for process conditions with normal splintering behavior. All other characteristics of the grinding wheel remain unchanged.

A clear change of parameters can be seen in the tip area. Flattening can only be verified by these parameters. The high material rate in the tip area leads to a high reduced peak height Rpk and to a high peak area A_1 on the flattened grain. The parameters of the groove area are not affected by flattening.

9.3.6 Grain Splintering

A further grain wear type is microcrystalline grain splintering. This wear type is caused by microcracks resulting from mechanical and thermal tensions. These microcracks lead to a microfracture, or even to partial grain break off.

9.3.7 Grain Breakout

In the case of total grain break off, whole abrasive grains are detached from the bond. The reason is a mechanical overstress of the bond due to excessive grain protrusions, as

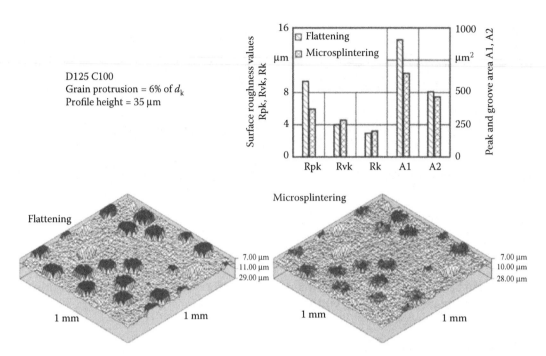

FIGURE 9.5
Influence of grain flattening and microcrystalline splintering on the grinding wheel topography.

well as by excessive process temperatures. In these cases, the grain retention forces are smaller than the process forces. If a grinding wheel bond is subject to thermal overstress, especially in the case of a resin bond, the bond might soften. It is also possible for the decomposition temperature of the binding material to be reached. The grain break off force decreases with a growing grain protrusion (Yegenoglu 1986).

9.3.8 Bond Softening

Figure 9.6 shows two grinding wheel topographies, the right one showing a cumulative grain break off due to increasing bond softening. At constant bond wear, this leads to a deeper embedding of the abrasive grain. The result in the present case is a decreasing number of cutting edges and increased stress on the surrounding grains. If bond wear increases due to bond softening, this leads to increased radial wear of the wheel.

Figure 9.6 also shows the parameters of the grinding wheel topography. It can be clearly seen how the groove parameters, valley area A_2 and reduced valley depth Rvk grow with increasing grain break off.

9.3.9 Effect of Single-Grain Forces

Wear occurs on the basis of single-grain forces and process temperatures (Marinescu et al. 2004). Continuously, sharp grain cutting edges are favorable for the grinding process and for a high-grinding ratio G. This requires microcrystalline grain splintering.

Wear types depend on the thermal mechanical grain stress. Figure 9.7 presents the qualitative depiction of grain flat growth, microcrystalline splintering, partial grain break off, and total grain break off against the thermal and mechanical grain stress.

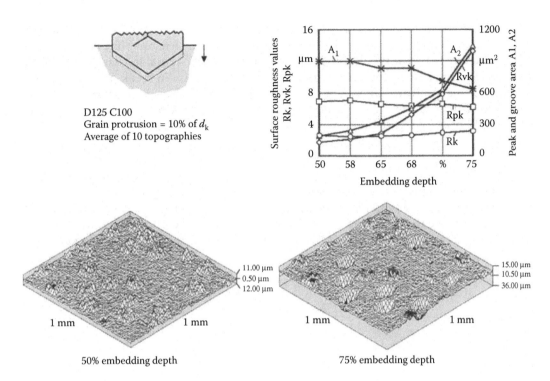

FIGURE 9.6
Topography behavior with increasing grain break off.

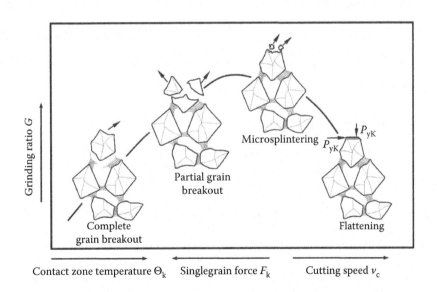

FIGURE 9.7
Strength of wear type against the process conditions. (From Uhlmann, E., Stark, C. 1997. *Potentiale von Schleifwerkzeugen mit mikrokristalliner Aluminiumoxidkörnung*. Beitrag 58. Jahrbuch Schleifen, Honen, Läppen und Polieren. With permission.)

To achieve the desired microcrystalline grain splintering, an initial force is necessary that depends on the grain properties as well as on the grain structure. If this initial force is not achieved, the wear type shifts to grain flattening. If the initial force is exceeded, the wear mechanism shifts over partial grain break off to total grain break off. Through the selection of the grain specification and by changing the grinding material concentration and bond structure, the grinding wheel can be adjusted to the desired single-grain forces achieving the necessary initial force for microcrystalline splintering.

9.3.10 Wear by Deposition

Besides the above-mentioned wear types, wear by deposition may also occur. Workpiece material residues are deposited under high pressure in the chip space, where they are held by undercut. Since these depositions are built up over several cutting edges, no cutting is possible any more with these grains (Lauer-Schmaltz 1979). This process is also known as wheel loading.

9.4 Bond Wear

9.4.1 Introduction

Not only the abrasive grain, but also the grinding wheel bond is increasingly subject to wear. The reason is abrasion by ground material particles, which have an abrasive effect on the binding material. With increasing wear, the bond is set back. In the case of long-chipping materials, this bond damage may occur at the grain cutting edges through the flowing chip. Whereas in the case of short-chipping materials, wear occurs through a lapping process in the chip space.

For efficient grinding, new multilayer grinding wheels usually have a grain protrusion of 20%–30% of the nominal grain diameter. This grain protrusion is necessary for the cutting process to evacuate the removed material volume and to let the cooling lubricant reach the active area.

9.4.2 Balancing Grain and Bond Wear

Constant process conditions require constant grain protrusion above the bond level. This implies a uniform grain and bond wear. Besides the specification of grain and bond, the application criteria are decisive for the wear balance. A balance occurs in the so-called self-sharpening range, when blunt grains constantly detach from the bond giving way to succeeding sharp grains as new cutting edges in the grinding process. Thus, the grinding wheel is constantly ready to work (Figure 9.8; Warnecke et al. 1994; Anon. 2003a).

If the grinding wheel is badly adjusted, the grain and bond wear are unbalanced. If the bond wear is too low compared to the grain wear, the grinding layer becomes blunt with insufficient grain protrusion. The process behavior is characterized by high process forces with thermal and mechanical overstress. This results in thermal damage and chatter marks on the component.

If bond wear is excessive in a super-sharp grinding wheel, the embedded depth of the grains decreases and the grain-holding forces with it. The result is excessive radial wear

Insufficient bond wear   Self-sharpening range

FIGURE 9.8
Grinding layer of a cubic boron nitride grinding wheel with resin bond with too low bond wear and in the self-sharpening range. (From Anon., 2003a. *Diamant- und CBN-Schleifscheiben. Firmeninformation der Fa.* Winter & Sohn GmbH, Norderstedt. Available at www.winter-diamantwerkz-saint-gobain.de. With permission.)

of the grinding wheel, which makes the process uneconomic. Hence, the ideal bond is not that with the lowest wear, but the one with the best adjustment to grain wear.

Grinding with continuous in-process sharpening has been developed as a reaction to topographic changes in the grinding wheel during the process leading to nonstationary process behavior. This technology allows machining processes, which, under conventional process conditions, lead to system overstress. Moreover, a specific control of the parameter level is possible for nearly all grinding tasks (Spur 1989; Tio 1990; Cartsburg 1993; Liebe 1996).

In recent years, vitrified bonded diamond and CBN grinding wheels became more and more important, due to their comparatively easy dressability and their outstanding porosity in comparison to electroplated or resin-bonded wheels, which allows higher material removal rates through higher chipping volume and higher contact coolant volume flow rates. Thus, the adjustment bond wear at vitrified bonded wheels is less important, compared to single-grain forces leading to microcrystalline splintering or rather self-sharpening effect and thus achieving maximum grinding ratios.

9.5 Assessment of Wheel Wear

9.5.1 Microtopography

The current microtopography of the grinding tool can be judged directly by measurement or reproduction, or indirectly by analyzing the process effects on the work result. The best-known methods for a direct judgment of the grinding wheel topography are the measurement of the tool surface by gauging, for example, by laser triangulation, profile method, or making imprints for a judgment under the microscope. The direct methods known today have the disadvantage that they can only be realized by intervening into the process disturbing the thermal balance (Brinksmeier and Werner 1992; Tönshoff et al. 1998; Warnecke 2000; Marinescu et al. 2004).

9.5.2 Profile Wear

Through its varying size and strength at the profile edges of the grinding wheel during the process, increasing microwear leads to an increase in macrowear. These profile deviations entail quality deviations on the component. It is especially at the exposed profile tips of the grinding wheel where the process stress is the highest and edge rounding occurs. Figure 9.9 shows the profile wear of a D126 C50 diamond-grinding wheel with resin bond grinding an SSiC-ceramic. The grinding wheel profile angle is $\alpha_s = 45°$.

The right-hand part of the image shows the increasing edge rounding in the course of the process. In this range, the wear volume of the grinding wheel profile also increases. The loss of grinding wheel volume, observable as the difference between the grinding wheel volume in the newly profiled state and that after the subsequent profiling, is characteristic for the wheel life up to the total consumption of the grinding layer. It is composed of the volume worn in the grinding process and that removed during dressing. Since the grinding wheel reobtains the required shape in the profile dressing process, the maximum radial wear can be determined for the radial loss. The loss volume can be calculated by multiplying this sum by the geometry parameters of the grinding wheel. Finally, the dressing volume can be calculated by knowing the other two volumes. It gives information on the required regeneration effort of the profile in the profile dressing process, representing a decisive cost factor especially when using superabrasive grinding wheels (Malkin 1989; Liebe 1996; Klocke et al. 1997).

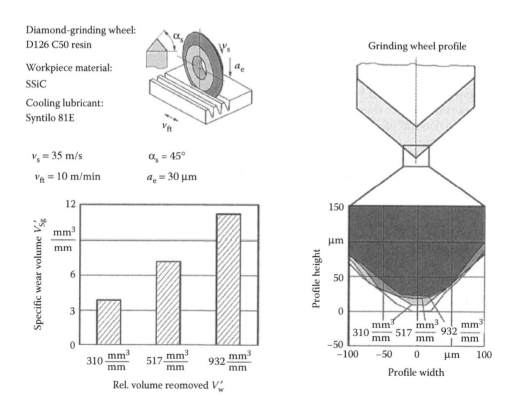

FIGURE 9.9
Profile wear of a diamond-grinding wheel with resin bond machining silicon carbide.

References

Anon., 2003a. *Diamant- und CBN-Schleifscheiben*. Firmeninformation der Fa. Winter & Sohn GmbH, Norderstedt. Available at www.winter-diamantwerkz-saint-gobain.de.

Anon., 2003b. *Firmeninformation der Fa*. Hermes Schleifmittel GmbH, Hamburg.

Brinksmeier, E., Werner, F., 1992. Monitoring of grinding wheel wear. *42nd General Assembly of CIRP*, Aix-en-Provence, F, August 23–29, Ann. CIRP 41.

Cartsburg, H., 1993. Hartbearbeitung keramischer Verbundwerkstoffe. PhD thesis, TU-Berlin, Hanser, München.

DIN 50320, 1979. *Verschleiß, Begriffe, Systemanalyse von Verschleißvorgängen, Gliederung des Verschleißgebietes*. Beuth Verlag, Berlin.

DIN 50323, 1988 and 1993. *Tribologie; Verschleiß, Begriffe. Teil 1 Nov. 1988, Teil 2 (Entwurf) Nov. 1993, Deutscher Normenausschuss*. Beuth Verlag, Berlin.

Engelhorn, R., 2002. Verschleißmerkmale und Schleifeinsatzverhalten zweiphasig verstärkter Sol-Gel-Korunde. PhD thesis, RWTH Aachen.

Jackson, W. E., Hayden, S. C., 1993. Quantifiable diamand characterization techniques: Shape and compressive fracture strength. *Proceedings of the Diamond and CBN Ultrahard Materials Symposium*, Windsor, Canada.

Juchem, H. O., Martin, J. S., 1989. Verbrauch an Diamant und CBN-Körnungen steigt stetig. *IDR* 23, 2.

Klocke, F., Hegener, G., Muckli, J., 1997. *Innovative Schleifwerk Zeuge sichern Wettbewerbsvorteile. VDI-2*. vol. 139, 7–8, Springer VDI Verlag, Düsseldorf, Germany.

Lauer-Schmaltz, H., 1979. Zusetzung von Schleifscheiben. PhD thesis, RWTH Aachen.

Liebe, I., 1996. Auswahl und Konditionierung von Werkzeugen für das Außenrund-Profilschleifen technischer Keramiken. PhD thesis, TU-Berlin.

Malkin, S., 1989. *Grinding Technology*. Ellis Horwood, New York.

Marinescu, I. D., Rowe, W. B., Dimitrov, B., Inasaki, I., 2004. *Tribology of Abrasive Machining Processes*. William Andrew Publishing, Norwich, New York.

Peklenik, J., 1958. Untersuchungen über das Verschleißkriterium beim Schleifen. *Industrie-Anzeiger* 80(27), S. 397–402.

Spur, G., 1989. *Keramikbearbeitung—Schleifen, Honen, Läppen, Abtragen*. Hanser, München.

Telle, R., 1993. Werkstoffentwicklung und Materialverhalten moderner Schneidkeramiken. in *Werkzeuge für die moderne Fertigung*; Hrsg.: W. Bartz, Technische Akademie Esslingen, vol. 370, Expert Verlag, Esslingen, Germany.

Tio, T. H., 1990. Pendelplanschleifen nichtoxidischer Keramiken. PhD thesis, Verlag Hanser, TU-Berlin.

Tönshoff, H., Karpuschewski, B., Andrae, P., Türich, A., 1998. Grinding performance of superhard abrasive wheels—Final report concerning CIRP-Co-Operative work in STG G. *Annals of the CIRP* 47, 2.

Uhlmann, E. 1996., Entwicklungsstand von Hochleistungsschleifwerkzeugen mit mikrokristalliner Aluminiumoxidschleifkörnung. *Proc. 8. Int. Braunschweiger Feinbearbeitungskolloquium*. 24–26/04, Braunshweig University.

Uhlmann, E., Daus, N., 2000. Ultrasonic assisted face grinding and cross-peripheral grinding of ceramics. *Proceedings of the 7th International Symposium*. Cer. Mat. Com. Eng. Goslar, 19–21/06.

Uhlmann, E., Stark, C., 1997. *Potentiale von Schleifwerkzeugen mit mikrokristalliner Aluminiumoxidkörnung*. Beitrag 58. Jahrbuch Schleifen, Honen, Läppen und Polieren.

Warnecke, G., 2000. *Zuverlässige Hochleistungskeramik. Abschlussbericht zum BMBF-Verbundprojekt Prozesssicherheit und Reproduzierbarkeit in der Prozesskette keramischer Bauteile*. Kaiserlautern.

Warnecke, G., Hollstein, T., König, W., Spur, G., Tönshoff, H. K., 1994., *Schleifen von Hochleistungskeramik—Werkstoff, Anwendung, Bearbeitung, Qualität*. zugl. *Abschlussbericht BMFT Verbundprojekt Schleifen von Hochleistungskeramik*. Verlag TÜV Rheinland, Köln.

Yegenoglu, K., 1986. Berechnung von Topographiekenngrößen zur Auslegung von CBN-Schleifprozessen. PhD thesis, RWTH Aachen.

Zum Gahr, K. H., 1987. *Microstructure and Wear of Materials*. Elsevier Science, Amsterdam.

10

Coolants, the Coolant System, and Cooling

10.1 Introduction

Coolant is a term generally used to describe grinding fluids for cooling and lubrication in grinding. The main purpose of grinding fluids is to minimize mechanical, thermal, and chemical impact between the abrasive material and the workpiece material during the abrasive process. The lubricating effect of a grinding fluid reduces friction between the abrasive grains and the workpiece, as well as between the bond and the workpiece. A second effect of a grinding fluid is the direct cooling of the grinding contact zone, where temperatures are highest through the absorption and transportation of the heat generated in the grinding process. Other effects of a grinding fluid are the evacuation of chips from the contact zone, bulk cooling of the workpiece outside the grinding contact zone, bulk cooling of the grinding machine, and corrosion protection (König and Klocke 1996; Marinescu et al. 2004, 2013; Rowe 2014). This chapter introduces basic factors and coolant properties relating to the application of coolants in grinding. Basic elements of coolant systems for coolant delivery are also introduced. Particular aspects of coolant requirements and thermal aspects of the grinding process are also described in other chapters such as Chapter 7 (Dressing), Chapter 16 (Surface Grinding), and Chapter 17 (External Cylindrical Grinding). Basic factors relating to effectiveness and evaluation of process cooling by application of coolants is described in Section 10.11.

10.2 Basic Properties of Grinding Fluids

10.2.1 Basic Properties

The selection of a grinding fluid is of crucial significance for the achievement of favorable cooling and lubricating conditions. Type, base oil, additives, and concentration of the fluid are important for the efficiency of cooling and lubrication. Cooling and lubrication requirements are met in different ways by every particular grinding fluid. Depending on the contact conditions in the process, the cooling and lubricating properties of the applied grinding fluid have a substantial impact on the process and on the work result.

10.2.2 Primary Requirements

The primary requirements of a grinding fluid are good lubrication, good cooling and flushing performance, and high corrosion protection. Lubrication is not simply a mechanical effect: Lubrication involves complex tribochemical interactions. It therefore involves the materials of the tribochemical pair as well as the nature of the grinding fluid and the grinding environment (Marinescu et al. 2013; Rowe 2014).

10.2.3 Secondary Requirements

The secondary requirements are economic and efficient operation, operational stability (long life), and environmental protection.

It is imperative for grinding fluids to be compatible with environmental and human health, as well as being reliable in operation. Additional requirements of the fluid are

- Easily filtered and recycled
- Easy removal of the residual film from the workpiece, grinding wheel, and machine
- Possibility for a solid particle transport of the swarf removal
- Inhibits the foaming and mist formation
- Exhibits the low flammability
- Exhibits the good compatibility with the materials of the machine tool system

In case of water-composite fluids, mixing behavior and emulsifiability must be considered (Table 10.1).

The functional properties and the operational behavior of cooling lubricants are significantly influenced by physical–chemical properties. Thermal capacity and conductivity, evaporation heat, and viscosity are affected by the quantitative ratios of the base materials used. Additionally, the performance of a cooling lubricant can be adjusted by the addition of active substances and additives.

10.3 Types of Grinding Fluids

Grinding fluids are commercially available with different property profiles to meet the requirements of specific machining tasks. DIN 51 385 divides grinding fluids into

TABLE 10.1

Important Properties of Cooling Lubricants

Usage Properties	
Functional Properties	**Operational Behavior**
Lubrication effect (pressure absorption capacity)	Human and environmental compatibility (toxicity, odor, and skin compatibility)
Cooling effect	Resistance to aging and bacteria (stability)
Flushing effect (cleaning, chip transport)	Filterability, recyclability, mixing behavior, and emulsifiability
Corrosion protection	Washability, residual behavior, and solid particle transport capability
	Foam, fog behavior, and inflammability
	Compatibility with different materials

Source: From Brücher, T., 1996, Kühlschmierung beam Schleifen keramischer Werkstoffe. PhD thesis, Technische Universität Berlin. With permission.

- Water immiscible
- Water miscible
- Water-composite fluids

The more general fields of application of cooling lubricants are cutting and partial forming processes (DIN 51 385).

Water-immiscible cooling lubricants are generally not mixed with water for any application (DIN 51 385).

Water-miscible cooling lubricants are emulsifiable, emulsifying, or water-soluble concentrates to which water is added before use.

Water-composite cooling lubricants are ready-for-use composites of water-miscible cooling lubricants with water. Within the group of water-miscible cooling lubricants, DIN 51 385 subdivides into

- Oil-in-water emulsions
- Water-in-oil emulsions
- Cooling lubricant solutions

For cutting, mainly oil-in-water emulsions and solutions are used, whereas water-in-oil emulsions are less common (Eckhardt 1983).

There are differences within the group of water-immiscible cooling lubricants according to the fraction and the type of the active substances contained (Bartz 1978; VDI-Richtlinie 3396 1983). Classification within the group of water-miscible cooling lubricants is carried out according to the content of active substances or to droplet size in rough disperse and fine disperse emulsions, as well as in fine colloidal, micellar, and molecular disperse solutions (Bartz 1978; VDI-Richtlinie 3396 1983).

The group of water-immiscible cooling lubricants also comprises natural and synthetic hydrocarbons such as mineral oils or poly-alpha-olefins, synthetic and vegetable ester, as well as water- and oil-soluble polymers such as polyglycols or composites of these substances (König et al. 1993). To improve their lubricating properties and their pressure absorption capacity, chemically active extreme pressure (EP), substances, or polar agents binding the lubricating film can be added to base oils. Furthermore, water-immiscible and water-composite cooling lubricants may contain corrosion, foam, and oxidation inhibitors or antifog additives (VDI-Richtlinie 3396 1983; Korff 1991; König et al. 1993).

The oil-in-water emulsions are mainly stable disperse composites of water and mineral oil or esters, which contain finely dispersed oil droplets in a water phase (König et al. 1993). The appropriate concentration of cooling lubricant emulsions must be determined for every single case of application depending on the corrosion protection capacity of the emulsions and on the cutting conditions. Conventional emulsion concentrations for grinding are in the range of 2%–6% (Eckhardt 1983), or in special cases up to 20% (Klocke 1982). Oil and water can be amalgamated with the help of bipolar surface-active substances. These substances favorably dissolve at the interface of the oil and water phase of an emulsion reducing the surface tension. Emulsifiers reduce the natural striving of the disperse phase to minimize the surface area (Mang 1983; Spur 1983; Möller and Boor 1986; König et al. 1993; Kassack 1994). Through the variation of the emulsifier content or of the emulsifier type, different grades of dispersion can be set. Water-miscible solutions are made of completely water-soluble inorganic (e.g., water-soluble salts) or organic (e.g., polyglycols and boron acid amide) active agents for the improvement of corrosion protection and wetting capability, and are free of mineral oil (Möller and Boor 1986; Bartelt and Studt 1992; König et al. 1993). Usually, the cooling lubricant solutions are used as emulsions in low concentration (Eckhardt 1983). Water-composite cooling lubricants contain EP additives, polar active agents, stabilizers, solution agents, preservatives, and corrosion and foam inhibitors in order to improve their functional properties (Bartz 1978; Spur 1983; VDI-Richlinie 3396 1983; Möller and Boor 1986; König et al. 1993).

10.4 Basic Properties

10.4.1 Composition of Oil-Based Fluids

In many cases, mineral oils are used as the base material for cooling lubricants. Mineral oil bases consist of hydrocarbons with mostly paraffinic, naphthenic, or aromatic structures. Depending on the fraction of the paraffins, naphthenes, and condiments, a distinction can be made between paraffin-based, naphthene-based, and mixed base oils (Mang 1983; Möller and Boor 1986). The main advantages of paraffinic base oils in cooling lubricants are good viscosity to temperature behavior, high ageing resistance, low affinity to evaporation as well as high flashpoint, and low toxicity. In contrast to paraffin-based mineral oils, the advantages of naphthene-based oils are their good resistance against low temperatures, better thermal stability, higher moistening capacity, and active agent solubility (Möller and Boor 1986; Pfeiffer et al. 1993). Extracted and deparaffined solvent raffinates or hydrocrack oils are gaining even more importance due to a lower condiment ratio, less mist generation, and high resistance to ageing (König et al. 1993; Kassack 1994). Also, synthetically produced poly-alpha-olefins, polyglycols, and esters are increasingly used, which, in contrast to mineral oils, have higher viscosity indices, a lower affinity to evaporation, longer service life as a result of high thermal stability, and/or resistance to oxidation and high human compatibility. Additionally, esters have especially good lubricating characteristics (Mang 1983; König et al. 1993; Pfeiffer et al. 1993; Kassack 1994). General values of the kinematic viscosity of cutting and grinding oils are in the range of $v = 2.0$–45 mm^2/s at $40°$C (Mang 1983).

10.4.2 Physical Properties of Water-Based and Oil-Based Fluids

In addition to the concentrates, water-composite cooling lubricants contain a high percentage of water. The properties of this cooling lubricant group are crucially influenced by the quality of the mixing water used, upon which special requirements must be regarded concerning the nitrate, chloride, sulfate, and phosphate content, total hardness, pH value, and the microbial resilience (Figure 10.1; Möller and Boor 1986; Leiseder 1991; König et al. 1993; Pfeiffer et al. 1993).

The most important base materials of water-immiscible and water-composite cooling lubricants (water and mineral oil) have fundamentally different thermal physical properties (Table 10.2). Hence, the capacity of a cooling lubricant to carry away thermal energy from the grinding process through the heat absorption strongly depends on its water or mineral oil content. Basically, the cooling effect of cooling lubricants is defined by their heat conductivity, evaporation heat, specific heat, and wetting capacity. Due to their high

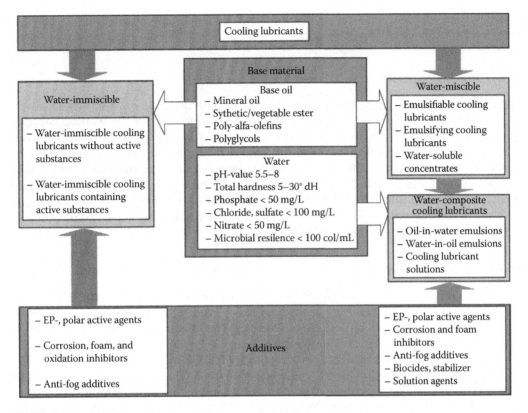

FIGURE 10.1
Classification and composition of cooling lubricants. (From Brücher, T., 1996, Kühlschmierung beam Schleifen keramischer Werkstoffe. PhD thesis, Technische Universität Berlin. With permission.)

water fraction, cooling lubricant solutions are characterized by an efficient cooling effect. Compared to water-immiscible cooling lubricants, oil-in-water emulsions have also a good cooling effect, which decreases in favor of a higher lubricating effect in case of increasing the oil fraction (Zwingmann 1979; König 1980; Eckhardt 1983; VDI-Richtlinie 3396 1983). The cooling effect of water-immiscible cooling lubricants is also strongly influenced by

TABLE 10.2

Physical Properties of Water and Mineral Oil

	Water	**Mineral Oil**
Density ρ at 20°C in kg/m³	998.2	ca. 870
Specific heat c_p at 20°C in J/gK	4.2	1.9
Heat conductivity λ at 20°C in W/mK	0.58	0.14
Evaporation heat Δh_v at 40°C in J/g	2260	210
Kinematical viscosity v at 40°C in mm²/s	0.6	approx. 2.0–45
Surface tension σ_o against air in mN/m	73	30

Sources: From Mang, T., 1983, *Die Schmierung in der Metallbearbeitung. Vogel Verlag, Würzburg;* VDI-Richtlinie 3396, 1983, *Kühlschmierstoffe für spanende Fertigungsverfahren.* Düsseldorf, VDI-Verlag; Möller, U. J., Boor, U., 1986, *Schmierstoffe im Betrieb.* VDI-Verlag, Düsseldorf. With permission.

viscosity. Low-viscosity cooling lubricants penetrate tight gaps much faster, hence this lubricant type is better at dissipating heat.

10.4.3 Rinsing Capacity

The rinsing or washing capability of cooling lubricants depends on viscosity and wetting capacity. The surface tension against the air is an indicator of the wetting capacity of liquids. At a surface tension against the air of approximately $\sigma_o = 30$ mN/m, the wetting capacity of mineral oil is superior to water. Through the addition of detergents, the surface tension of water can be approximated to be $\sigma_o = 72$ mN/m and can be reduced to $\sigma_o = 30$ mN/m (Zwingmann 1960). Generally, with decreasing viscosity, water-immiscible cooling lubricants exhibit better washing capability (Mang 1983; Spur 1983). Due to the low viscosity, water-miscible cooling lubricants show a superior rinsing capacity compared to water-immiscible products (Spur 1983).

10.4.4 Lubricating Capability

The lubricating capacity of a cooling lubricant depends basically on the additives it contains. Furthermore, the lubricating capacity of water-immiscible and water-miscible cooling lubricants is influenced through their viscosity. The kinematic viscosity of mineral oil is 15 times higher than that of water, and grinding oils generally have a better lubricating capacity than water-composite cooling lubricants (Kohblanck 1956; Zwingmann 1979). The lubricating effect of oil-in-water emulsions depends on the oil fraction it contains and increases with a larger percentage of oil. Since the cooling lubricant solutions are free of mineral oil, they have a lower lubricating capacity than emulsions (Eckhardt 1983). Due to the high temperatures and pressures in the contact zone, the pressure absorption capacity of water and mineral oil is not sufficient for the formation of a stable lubricating film. For this reason, additives are added to state-of-the-art cooling lubricants to improve the lubricating properties.

10.5 Additives

An important group of cooling lubricant additives is polar additives, whose molecules contain polar groups (Figure 10.2). Polar additives are mostly unsaturated hydrocarbon compounds such as fatty acids, fatty alcohol, and fatty acid esters. Due to their polarity, these agents firmly deposit on the workpiece surface and form an adhering lubricating film. Additionally, there are cases where chemical reactions occur between the material and the cooling lubricant additive developing metallic soaps acting as highly viscous, plastic lubricating films. Due to the low melting point of these metallic soaps of approximately 150°C, the impact of polar additives decreases with the increasing of temperatures (Zwingmann 1979; König 1980; Spur 1983; VDI-Richtlinie 3396 1983; Korff 1991; König et al. 1993; Kassack 1994). A further group of active agents are the so-called EP additives, which consist of phosphor and sulfur compounds. Additives previously used containing chlorine are barely applied nowadays due to ecological and physiological reasons (König et al. 1993; Kassack 1994). The EP additives generate metal phosphates or sulfides in the contact zone through chemical reactions with the workpiece surface. They act as solid lubricating

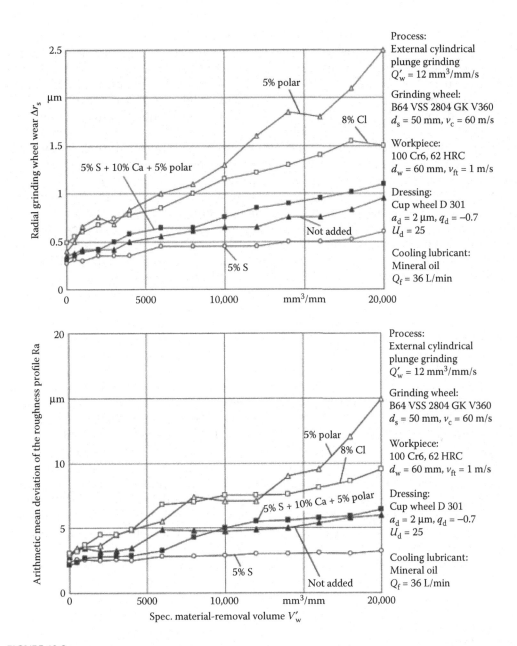

FIGURE 10.2
Influence of different additives on the grinding wheel wear and surface quality during cylindrical grinding. (From Heuer, W., 1992, *Außenrundschleifen mit kleinen keramisch gebundenen CBN-Schleifscheiben. Fortschrittberichte.* VDI, Reihe 2, Nr. 270. With permission.)

layers with high pressure resistance and low shear strength. The minimum temperature necessary for the occurrence of this reaction depends on the used agent. Phosphor additives have a temperature range of approximately 50–850°C, while sulfur-containing additives are active between approximately 500°C and 1000°C (Keyser 1974; Zwingmann 1979; König 1980; Zimmermann 1982; Spur 1983; Kassack 1994). To cover a wide temperature range, EP additives that act at high temperatures are combined with polar additives. This therefore acts at lower temperatures in the cutting process of metals. As an example, sulfur substrates on a fatty oil basis are often used in metal working (Korff 1991; Kassack 1994).

Other cooling lubricant additives are added to avoid corrosion through adsorption on the workpiece surface or chemical reactions with the faces of the workpiece. Alcanolamines and carbon or boron acids are often used as corrosion inhibitors. Due to the nitrosamines problem, sodium nitrides are no longer used as corrosion protection additives. Boron compounds in water-composite cooling lubricants are widely used as corrosion protection additives because of an additional protection against bacterial infection. The preservatives used in water-composite products for the control of microorganisms, bacteria, and fungus growth are toxic and cannot be considered harmless in physiological terms. Alongside boron acids, the most important groups of these agents are formaldehyde separators, phenols, and N/S heterocycles. In order to avoid foaming, polysilicons and acrylates are added, which have low surface tension and make foam quickly collapse. Antifog additives are mainly polymethacrylates and olefin copolymers, which lead to a recombination of the aerosol to fluid droplets causing the fog to condense near the point of its origin (Möller and Boor 1986; Korff 1991; König et al. 1993; Pfeiffer et al. 1993; Kassack 1994).

10.6 Application Results

Tangential grinding force and grinding power can be minimized and generated heat reduced by reducing friction using a cooling lubricant with a strong lubricating effect (Howes 1990; Brinksmeier 1991a). Successful cooling leads to quick heat dissipation keeping the active partners below a critical temperature. However, the shear resistance of the workpiece material is increased through the cooling of the active zone, again causing an increase in the process forces (Brinksmeier 1991a). Depending on its cooling and lubricating performance, the cooling lubricant has a significant influence on the achievable material-removal rate, grinding forces, grinding temperature, and on grinding wheel wear. Beyond the influence on process parameters, achievable surface quality and subsurface characteristics crucially depend on the type of cooling lubricant. It has also been reported that clogging of the grinding wheel depends on the type of cooling lubricant (Khudobin 1969; Tawakoli 1990). Against this background, a specific selection and adaptation of the cooling lubricant is necessary for particular machining tasks.

10.7 Environmental Aspects

The continuous increasing of the public awareness toward environmental protection, new ecological legal conditions, as well as increasing disposal costs have led to new approaches in manufacturing of grinding fluids. It has been recognized that inappropriate disposal and landfilling of cooling lubricants represent a serious hazard since deposition has a significant impact on the air, soil, and ground water. Moreover, dioxins are emitted to the environment. The "Act of Closed Substance Cycle Waste Management and Ensuring Environmentally Compatible Waste Disposal," which became effective in 1996 in Germany, focuses on the protection of natural resources through the avoidance and recycling of waste. The manufacturing processes are required by law to take place with a minimum input and particularly with a minimum consumption of resources emitting a minimum of harmful substances (Brinksmeier and Schneider, 1993). Hence, the entire cooling lubricant system represents a key starting point for an ecological design of the grinding process. The grinding fluid is fed in large amounts to the grinding machine and large quantities of abrasive slurry are generated in the fluid supply system.

Ecological and health aspects are resulting in a more frequent application of dry machining or of so-called "minimum quantity lubrication" systems. Increasing demands for improved product quality and economic grinding of parts, and a minimum amount of grinding fluid, are contradictory requirements (Klocke and Gerschwiler 1996; Heinzel 1999; Weinert 1999).

10.8 Supply System

10.8.1 Introduction

The design of a supply system and the selection of the feed parameters must meet the specific technological demands of the grinding process. Since the cooling of the grinding process primarily depends on the cooling lubricant supply to the contact zone, secondary cooling effects play only a minor role. This means the percentage of the cooling lubricant amount acting in the contact zone has to be set to a technologically required minimum.

10.8.2 Fluid Supply System Requirements

The cooling lubricant supply system is required to accomplish several different tasks during the machining process as well as during auxiliary process time or even during the off state of the grinding machine. Firstly, it has to provide an uninterrupted flow of cooling lubricant to the active zone. Moreover, it is required to store and transport the cooling lubricant maintaining a constant quality and temperature and with a sufficient quantity to execute the job of cooling, lubricating, flushing, and chip transport. In addition to economic requirements related to investment costs or maintenance cost, a number of further requirements must be met including operating safety especially when using oil as cooling lubricant (König et al. 1993; Brücher 1996; König and Klocke 1996).

In an industrial environment, it is a common approach to install a central or group circulation system that supplies a number of machine tools using the same cooling lubricant. These systems require the specification of a single cooling lubricant for all processes supplied but reduce the complexity for cleaning, cooling, controlling, and supply of the fluid, and, in addition, reduce the circulating volume of the cooling lubricant (Brücher 1996).

Centralized systems are composed of components transporting the fluid to the process (pumps, pipes, nozzles, measurement and control devices, and mixing devices), a return system (channels, pipes, and pumps), maintenance devices (filters, reservoirs, and monitoring devices), and equipment for slurry treatment (conveyor, chip crusher, centrifuges, and cleaning nozzles). Design of a cooling lubricant supply system strongly depends on the required flow and pressure of the fluid at the contact zone. By applying a particular nozzle form, its positioning and the required fluid pressure determine the total volume of cooling lubricant to be supplied. Additionally, the volume stored in the feed and return pipes, the volume contained in filter and tempering devices, a minimum reserve volume, and, if necessary, an additional volume for foam discharging have to be taken into account (VDI-Richtlinie 3035 1997).

Although cooling lubricant supply systems are often designed for either water-miscible or water-immiscible cooling lubricants, the application for both types and many different specifications of cooling lubricants has become an increasing industrial demand. This needs to be incorporated into the material choice for a cooling lubricant supply system, where it is generally recommended to avoid zinc-plated steel pipes or nonferrous fittings. In order to prevent the degradation of cooling lubricant or corrosion of machine components, the compatibility of all materials has to be ensured especially for such fluid types that change its physical and chemical properties over the course of time. Tanks for fresh and used fluid are required to store the entire cooling lubricant volume of the supply system in case of a machine stoppage that leads the fluid flowing back into those tanks due to a gravitation controlled piping system. Because of long distances between machine tools and central tanks, it is often necessary to install a separate backflow tank at each machine. Furthermore, the design of each tank should prevent the deposition of any solid residuals or backflow of fluid and offer a possibility to empty the tank completely (VDI-Richtlinie 3035 1997).

Depending on volume flow, fluid pressure, and contamination of the fluid, a number of different pumps can be used. A crucial feature of all types is the sealing of the pump against the contaminated cooling lubricant, which is presently done by axial face seals made of tungsten carbide including special CVD or PVD coatings. To minimize pressure drop, the supply pump should be placed as close as possible to the nozzles. The cross section of the connected piping system has to be adjusted to the particular flow conditions in order to prevent cavitation, which normally limits the flow velocity in intake pipes to 1.5 m/s and in pressure pipes to 2.5 m/s (VDI-Richtlinie 3035 1997).

Grinding swarf contaminates the grinding fluid and degrades the grinding operation itself and the lifetime of the fluid. Cleaning and conditioning of the fluid is accomplished by a number of different methods and principles. Chemical, biological, and, above all, mechanical contamination could be eliminated by sedimentation, filtering, centrifugation, and magnetic tape separators depending on the required degree of purity. Sedimentation is a reasonable method for coarse cleaning of the fluid; however, for finish grinding a multistage cleaning process using several different filter principles is recommended. The most common filters are tape filters and candle filters, both requiring a change and disposal of the filter within a fixed period of time. In addition to a very clean cooling lubricant, it is crucial for high-speed grinding applications and finish grinding operations to bring the supplied fluid at a precisely controlled temperature. The grinding heat needs to be dissipated while passing through the fluid supply system. The fluid needs to linger long enough in the supply system to allow heat dissipation or the heat has to be removed by an extra cooling system (Tawakoli 1990; Brücher 1996; König and Klocke 1996; VDI-Richtlinie 3035 1997).

It is essential in all grinding applications to control the fluid pressure or volume flow. In consideration to safety of the grinding process, an automatic machine stop is essential to cope with an unplanned pressure drop or significant flow reduction. Furthermore, it is desirable to check and control the quality of the supplied fluid by measuring the pH value or electrical conductivity. In grinding with water-immiscible cooling lubricants, the mist generated in combination with high process temperatures and glowing chips flying away from the contact zone could lead to an explosion. Therefore, safety devices, such as blowback flaps and fire extinguishers, have to be installed. For grinding with minimum lubrication systems, separate technical devices are required (Tawakoli 1990; VDI-Richtlinie 3035 1997).

10.8.3 Alternative Cooling Systems

The amount of the required cooling lubricant can be reduced by the use of alternative cooling lubricants. It has been shown that good grinding results can be achieved by using liquid nitrogen linked to minimum lubrication. Compared to grinding with oil, surface qualities are slightly higher; however, the wear of the grinding wheel is much lower. The lubrication of the grinding process with solid graphite is a further possibility. In this case, the machining forces are similar to those in grinding with a grinding fluid. If solid graphite lubrication is linked to dry machining, considerably lower process forces were found out, particularly in case of increased feed rates.

Besides the design of the supply system, the use of alternative cooling lubricants, and the reduction of the cooling lubricant quantity, there is a further potential for optimization of the entire cooling system, including the whole life cycle of the cooling lubricant. Since cooling lubricants are designed for a particular machining task, the complex processes of downstream cleaning and processing is not considered. Only cleaning of the filter system is included in the design. In this case, slurry in the filter system can be rehabilitated and a large part of the substances it contains can be recycled (Brinksmeier 1991a).

It is possible that the entire process will be integrated within the grinding machine in the future. After start-up, the grinding machine will be able to process a large part of the used cooling lubricant and refeed it to the process. Furthermore, residual material is discharged and abrasive residuals can be refeed to the production process.

10.9 Grinding Fluid Nozzles

10.9.1 Basic Types of Nozzle System

The performance and characteristics of the cooling lubricant nozzle for the supply liquid lubricant have a major influence on the grinding result. A number of different nozzles have been developed in order to meet the requirements of various applications. Some of these nozzles are described in detail in Chapter 16 on surface grinding and Chapter 17 on cylindrical grinding. Generally, there are three ways to distinguish types of nozzle systems (Heinzel 1999):

- By function (flooding or not flooding)
- By focusing (free jet nozzle, point nozzle, swell nozzle, and spray nozzle)
- By nozzle geometry (squeezed pipe, needle nozzle, and shoe nozzle)

The primary task of all nozzles is the delivery of lubricants to the active zone. The nozzle carries out this task through focusing and directing the lubricant jet as well as accelerating the liquid. Investigations show a positive effect on the cooling performance by focusing the lubricant jet, associated with minimizing the turbulence of the flow by a sharp-edged exit of the nozzle or an extended parallel outlet of the nozzle. Additionally, a minimum flow velocity must be generated, that is, by reducing the cross-sectional area at the nozzle outlet for wetting the wheel surface with cooling lubricant. The main obstacle to overcome for wetting the entire grinding wheel surface prior to its entry into the contact zone is a rotating air cushion around the grinding wheel (Marinescu et al. 2004). Due to the friction between the rough wheel surface and its surrounding atmosphere, a rotating air cushion is generated. The rotating air stream causes a permanent air flow away from the grinding wheel creating an air barrier. Especially at high cutting speeds, due to this effect, the grinding fluid is not able to reach the contact zone (Tawakoli 1990; Treffert 1995; Brücher 1996; Heinzel 1999; Beck 2001).

In order to rupture the rotating air cushion by the cooling lubricant itself, a significant amount of kinetic energy has to be spent. The primary cooling lubricant nozzle can be used applying a higher flow rate and a higher flow velocity, which needs a higher volume of cooling lubricant and significantly reduces the overall economic efficiency of the whole supply system. Further analysis of the jet velocity and flowrate requirements is given by Rowe (2014). Another solution is the usage of a second nozzle in a radial direction to the wheel in front of the inlet of the primary cooling lubricant nozzle. Further improvement can be achieved by employing one nozzle for the peripheral surface and two additional nozzles for the side faces of the grinding wheel.

Air guide plates closely aligned to the wheel surface can also be used to deflect the rotating air cushion away from the grinding wheel. Therefore, precise alignment and readjustment to a changing wheel profile or diameter are required (Tawakoli 1990; Brücher 1996; Heinzel 1999).

10.9.2 Jet Nozzle

At present, the most common type of cooling lubricant nozzle is the free jet nozzle aimed at flooding the entire contact zone. Being rather simple in design, this nozzle type is oriented in the tangential direction to the grinding wheel. In addition, the nozzle outlet should be positioned very close to contact zone. For maximizing the volume flow through the contact zone, the flow velocity should correlate with the peripheral velocity of the grinding wheel. This can be achieved by varying the volume flow, the flow pressure, and the outlet cross-sectional area. However, the maximum flow through the contact zone is essentially limited by the pore space of the grinding wheel or the grinding layer. The geometry of the free jet nozzle is independent from the wheel profile or its dimensions, which makes this nozzle type relatively flexible in application. A tangential nozzle with a small width is referred to as a point nozzle, whereas the combination of several point nozzles is called a needle or multipoint nozzle (Heinzel 1999).

10.9.3 Shoe Nozzle

An alternative nozzle design, combining elements for deflecting the rotating air cushion with a highly effective distribution of the cooling lubricant to the contact zone, is the so-called shoe nozzle. This nozzle type fits exactly to the wheel profile and encloses the grinding wheel on three sides. The rotating air cushion is deflected at the nozzle inlet allowing the complete wetting of the wheel surface with lubricant at the inner chamber of the shoe nozzle. The rotation of the grinding wheel itself compels the fluid to circumferential velocity. The total amount of the cooling lubricant supplied can be limited to the volume necessary to fill the whole pore space of the wheel surface (Figure 10.3). The nozzle geometry is determined by the grinding wheel profile and an adjustment to a changing wheel diameter; hence, there is only a limited flexibility in the application of those nozzles. Several

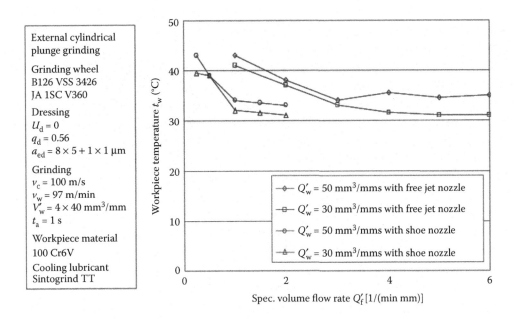

External cylindrical plunge grinding

Grinding wheel
B126 VSS 3426
JA 1SC V360

Dressing
$U_d = 0$
$q_d = 0.56$
$a_{ed} = 8 \times 5 + 1 \times 1\ \mu m$

Grinding
$v_c = 100$ m/s
$v_w = 97$ m/min
$V'_w = 4 \times 40$ mm³/mm
$t_a = 1$ s

Workpiece material
100 Cr6V

Cooling lubricant
Sintogrind TT

Workpiece temperature t_w (°C)

Spec. volume flow rate Q'_f [1/(min mm)]

- $Q'_w = 50$ mm³/mms with free jet nozzle
- $Q'_w = 30$ mm³/mms with free jet nozzle
- $Q'_w = 50$ mm³/mms with shoe nozzle
- $Q'_w = 30$ mm³/mms with shoe nozzle

FIGURE 10.3
Workpiece temperature versus specific volume flow rate of the cooling lubricant for a shoe nozzle. (From Beck, T., 2001, Kühlschmierstoffeinsatz beim Schleifen mit CBN. PhD thesis, RWTH Aachen. With permission.)

investigations prove the capability of this nozzle form to reduce the wheel wear as well as the thermal degradation of the boundary layer with a less required flow rate of the lubricant (Tawakoli 1990; Heinzel 1999; Beck 2001).

10.9.4 Through-the-Wheel Supply

A slightly different concept of lubricant distribution is the supply of cooling lubricant from the interior of the grinding wheel or the grinding layer. The fluid is fed into a chamber of the wheel body allowing the centrifugal force to distribute it through radial channels to the grinding layer. Through pores or gaps in the grinding layer, the cooling lubricant is directly provided to the contact zone. The technical complexity of this solution has so far prevented a broad application. Alternatively, a porous grinding wheel can be infiltrated by fluid supplied from a conventional external cooling nozzle, which will leave the wheel inside the contact zone due to the centrifugal force (Tawakoli 1990; Heinzel 1999).

10.9.5 Minimum Quantity Lubrication Nozzles

Minimum quantity lubrication (MQL) is aimed at reducing the amount of lubricant used for a grinding application. MQL nozzles have been the topic of several research projects. With the assistance of pressurized air, a mist of cooling lubricant is sprayed onto the surface of the grinding wheel. Ideally, only a thin film of fluid covers the wheel surface prior to its entry into the contact zone. Investigations concerning MQL report an increase of grinding forces, wheel wear, workpiece surface roughness, and onset of grinding burn at lower material-removal rates in contrast to conventional fluid supply systems. Furthermore, secondary functions of the fluid such as chip transport or cooling of the grinding machine have to be carried out through additional devices. However, certain grinding operations

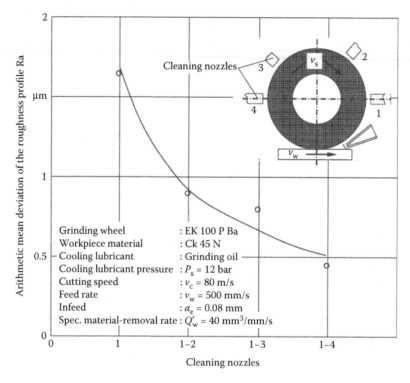

FIGURE 10.4
Influence of arrangement and number of cleaning nozzle on surface quality during surface grinding. (From König, W., Klocke, F., 1996, *Fertigungsverfahren Band 2—Schleifen, Honen,* VDI-Verlag, Düsseldorf. With permission.)

using MQL show some potentials for applications in an industrial environment such as rough grinding with plated metal bond cubic boron nitride (CBN) wheels (Heinzel 1999; Weinert 1999).

10.9.6 Auxiliary Nozzles

In addition to the nozzle supplying cooling lubricant to the contact zone, it is recommended to use auxiliary nozzles in radial direction to the grinding wheel. Their task is to remove chips and other loading from the wheel surface as well as to extinguish glowing chip particles. The effectiveness of these nozzles depends more on the fluid pressure than on the volume flow. A significant reduction of the surface roughness of ground workpieces could be achieved by applying two more of those cleaning nozzles (Figure 10.4; Vits 1985; König and Klocke 1996).

10.10 Influence of the Grinding Fluid in Grinding

10.10.1 Conventional Grinding

In comparison to dry grinding and grinding with emulsions, lower temperatures were measured in grinding with oil (Dederichs 1972). The cooling effect can be influenced by the occurrence of film boiling of the fluid in the contact zone, leading to an abrupt overheating of the workpiece and to thermal damage. Face grinding of steel with a water-based fluid showed that with a rising subsurface temperature, caused by an increase in depth of cut, the cooling lubricant increasingly evaporates above a depth of cut of $a_e = 35$ µm and temperatures higher than 100°C, leading to similar grinding temperatures as in dry grinding. There is a similar effect during grinding with oil, however, at temperatures higher than 300°C (Yatsui and Tsukada 1983; Howes et al. 1987; Howes 1990; Brinksmeier 1991a; Marinescu et al. 2004). Thus, in many cases, it appears to be more effective to reduce the grinding heat generated by using a grinding fluid with a good lubricating effect than to absorb an increased amount of heat with the help of a grinding fluid of a high specific heat capacity (Klocke 1982).

In addition to a superior material-removal rate and surface quality, wear of the grinding wheels and tangential grinding forces are lower in case of using the grinding oils than water-composite fluids. Furthermore, higher G ratio and better surface quality are achieved with grinding oil (Tönshoff and Jürgenharke, 1979; Althaus 1982). No uniform tendencies of the normal grinding force could be observed comparing the use of grinding oils and water-composite fluids (Figure 10.5; Zwingmann 1960; Gühring 1967; Keyser 1970; Sperling 1971;

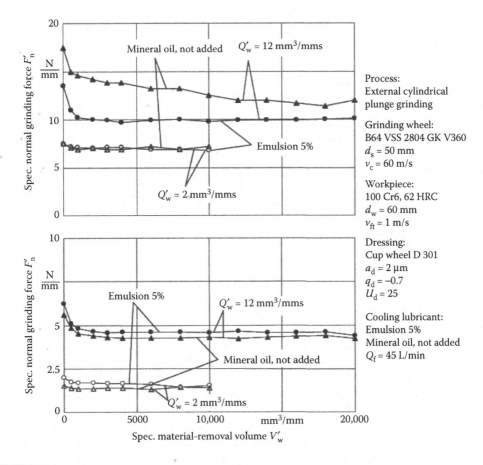

FIGURE 10.5
Influence of different cooling lubricants on grinding forces during cylindrical grinding. (From Heuer, W., 1992, *Außenrundschleifen mit kleinen keramisch gebundenen CBN-Schleifscheiben. Fortschrittberichte.* VDI, Reihe 2, Nr. 270. With permission.)

Peters and Aerens 1976; Oates et al. 1977; Polyanskov and Khudobin 1979; Tönshoff and Jürgenharke 1979; König 1980; Althaus 1982; Zimmermann 1982; Ott 1985; Vits 1985; Holz and Sauren, 1988; Kerschl 1988; Carius 1989; Brinksmeier 1991a,b; Heuer 1991, 1992; Ott 1991; Böschke 1993; Treffert 1995; Webster 1995; Heinzel 1999; Marinescu et al. 2004).

In external grinding of roller bearing steel 100Cr6 (62 HRC) with CBN grinding wheels, pure grinding oil generated a lower wear than 5% emulsion, with improved surface quality and lower tangential forces. Higher wear of CBN grinding wheels with water-composite fluids is often traced back to hydrolytic wear. Model tests, where CBN grains were heated up to 1000°C, showed grain-edge rounding, etching of the grain surface, and a loss of weight as a result of a chemical reaction between boron nitride and water leading to the development of boron acid and ammonia. Such wear behavior, however, could not be observed under normal grinding conditions. Generally, it is found that chemical wear in grinding with CBN grinding wheels using solutions or emulsions is of secondary significance (Triemel 1976; Heuer 1991).

Similar results were found in internal cylindrical grinding of hardened 100Cr6 steel (Figure 10.5). In this case, the normal cutting forces as well as the tangential force components were eventually higher using oil than with water-composite cooling lubricants, whereas in external cylindrical grinding a decrease in tangential cutting force was observable. Furthermore, higher G ratio and better surface quality are achieved with grinding oil (Tönshoff and Jürgenharke, 1979; Althaus 1982)

10.10.2 Influence of the Fluid in Grinding Brittle–Hard Materials

Despite the fundamentally different material-removal mechanisms, conclusions relating to fluids for brittle–hard materials are similar to those for ductile materials. In reciprocating face grinding of ceramic materials with diamond grinding wheels, there are advantages concerning surface quality and process behavior in case of using grinding oil compared to water composites. While the use of water-composite fluids is characterized by an increase of the normal force during the grinding of Al_2O_3 and HPSN, lubrication with grinding oil showed a process behavior with low and nearly constant normal grinding forces up to a specific material-removal volume of $V'_w = 780\,mm^3/mm$. Furthermore, there is lower radial wear of the grinding wheel during the grinding of these ceramics with grinding oil (Tio and Brücher 1988; Brücher 1993; Spur 1993). The obvious differences in the topography of Al_2O_3 surfaces generated by grinding with oil and emulsions suggest a considerable influence of the grinding fluid on the chip-formation mechanisms. In contrast to the surfaces generated by grinding oil, there are hardly any directional grinding marks on surfaces generated by water-composite fluids (Tio and Brücher 1988; Roth and Wobker 1991; Wobker 1992; Brücher 1993).

These findings are also confirmed for the face grinding of an aluminum oxide reinforced with 10% ZrO_2. Different surface structures are generated depending on the grinding fluid and nearly constant normal and tangential grinding forces occur. Moreover, the grinding wheel wear is lower if grinding oil is used (Brinksmeier 1991a; Heuer 1991; Roth and Wobker 1991; Wobker 1992). In contrast, if petroleum or petroleum fog was used for cooling and lubrication in the face grinding of different oxide ceramic materials in further research projects, lower normal forces occurred than with emulsion, emulsion fog, or compressed air. The lowest tangential force and the lowest wear, however, were measured in grinding with emulsion and emulsion fog. The achieved surface quality was almost independent from the grinding fluid in these investigations (Sawluk 1964). Grinding of HPSN and Al_2O_3/TiC gave different results. In these cases, higher normal forces were measured with grinding oil than with water-composite fluids. This fact was explained by elevated thermal stress of the grinding wheel. The wear of the grinding wheel was also lower with these materials when water-immiscible fluids were used (Brinksmeier 1991a; Heuer 1991; Roth and Wobker 1991; Wobker 1992).

There are hardly any differences between water-composite grinding fluids of different compositions in terms of grinding forces during face grinding with axial feed of $Al_2O_3+10 \%$ ZrO_2, HPSN, and Al_2O_3/TiC. However, wear of the grinding wheel was lower with emulsion than with a solution irrespective of the material (Wobker 1992). Friction and wear tests using a four-ball tester with hexadecanen (C16H18) and different additives showed a significant reduction of the friction value and the wear coefficient in case of the friction pairs Al_2O_3/Al_2O_3 and ZrO_2/ZrO_2. The most obvious influence in case of these pairs was shown after the addition of 0.1 mol% zinc dialkyl dithiophosphate as a cooling fluid additive. In case of the friction pairs SiC/SiC and Si_3N_4/Si_3N_4, the influence of additives was smaller. While marginally smaller friction and wear coefficients were observed in case of the silicon nitride tool under addition of zinc dialkyl dithiophosphate, too, no change of the friction and wear parameters was detected in case of silicon carbide (Bartelt and Studt 1992). The effects of an addition of fatty acids of different chain lengths were analyzed in sliding wear tests with hexadecan lubrication within the scope of other investigations. Fatty acids with six or more carbon atoms proved to be good lubrication additives. This effect results from thicker adsorption layers (Studt 1987).

In case of grinding spectacle lenses, superior cutting performance was achieved by using grinding oil, although the surface roughness was higher (Pfau 1987). The use of light grinding oil is better in the cutoff grinding of glass. Although the use of petroleum would lead to an extension of servicing periods up to 10%–15%, this fluid is not used due to fire hazard. In contrast, when face grinding glass blocks with diamond cup wheels, a 3%–5% synthetic water-composite fluid is used instead of oil (Seifarth 1987). In case of cutoff grinding of hard stone, studies show that grinding oil leads to a reduction of grinding forces and tool wear. Furthermore, a higher surface quality can be expected if grinding oil is used. Wear tests on grains of diamond grinding wheels indicated an increasing grain wear through grain flattening and splintering, while the highest number of damaged grains was observed in cutoff grinding with grinding oil (Tönshoff and Schulze 1980). Investigations of grinding of concentric circular grooves in marble showed only a minor influence of the composition of water-composite fluid (Gerhäuser and Laika 1977).

Investigations carried out a minimum volume flow required in the contact zone for grinding different ceramic materials (Figure 10.6). Supplying this specific contact zone volume flow, grinding wheel wear reaches a minimum. A further increase of the volume flow leads to negligible reduction of wear with clearly increased resistance against the supply of the cooling lubricant. Hence, this minimum contact zone volume flow is a significant criterion for the design of the coolant supply system (Brücher 1996).

10.10.3 High-Speed and High-Performance Grinding

In case of high-speed grinding of hardened 100Cr6 steel with electroplated CBN grinding wheels, the use of grinding oil instead of water-composite grinding fluids lead to a reduction of wear. At a cutting speed of $v_c = 140$ m/s and material-removal rates of $Q'_w = 2$–$80\,mm^3/mms$ normal as well as tangential force components are reduced when grinding with oil. The roughness of the grounded surface was slightly higher with oil in contrast to different emulsions. The reason for the positive result in terms of surface quality in grinding with emulsions is an accelerated wear of the grains, which leads to an increase of the active cutting edge number and chip thickness (Treffert 1995). In high-speed

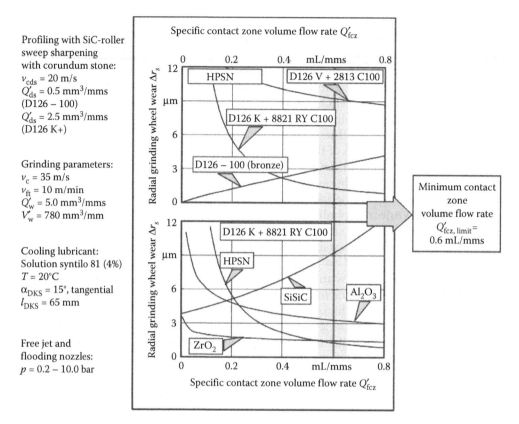

FIGURE 10.6
Determination of a minimum contact zone flow rate for different grinding conditions. (From Brücher, T., 1996, Kühlschmierung beam Schleifen keramischer Werkstoffe. PhD thesis, Technische Universität Berlin. With permission.)

grinding of materials of different hardness, with CBN wheels up to 50% higher grinding ratios can be achieved with grinding oil compared to a synthetic solution and an emulsion. However, the achievable advantages with grinding oil increased with a decreasing material hardness (Kerschl 1988).

In case of high-performance grinding of Ck45N steel with conventional corundum grinding wheels, the use of grinding oil at a cutting speed of $v_c = 60$ m/s led to an increase of the achievable cutting performance by approximately 200% in contrast to an emulsion, where the performance was limited by the occurrence of burn marks. At the same time, smaller normal forces and surface roughness were observed at an equal material-removal rate (Gühring 1967).

During the external cylindrical plunge grinding of Ck45N and 100Cr6 with corundum grinding wheels, lower tangential grinding forces and a lower grinding wheel wear were determined when grinding oils were used, independently of the material-removal rate. The normal component of the grinding force exhibits lower values during the grinding of the mentioned materials at elevated material-removal rates above $Q'_w = 3.0\,\mathrm{mm}^3/\mathrm{mms}$ with oil instead of water-composite fluids. In case of grinding with lower material-removal rates, the normal force level for both materials is higher with oils than with different emulsions. The reason is a higher grain-related normal force and a higher number of kinematic cutting edges due to increased plastic deformation. This effect is much stronger in case of ductile material Ck45N than in case of 100Cr6. Due to a wear-related leveling of the grinding wheel surface, grinding with poor lubricants entails an increase in the number of kinematic cutting edges at low material-removal rates. As a result, normal forces arising from water- or oil-based fluids become equal at an increasing chip volume and at a specific material-removal rate below $Q'_w = 3.0 \, \text{mm}^3/\text{mms}$.

Better surface quality was achieved in grinding with oil at specific material-removal rates of $Q'_w = 3.0 \, \text{mm}^3/\text{mms}$ in case of both materials (Figure 10.7; Vits 1985). A large uncut chip thickness causes excessive cutting-edge stresses leading to intensive self-sharpening of the wheel surface through grain breakage. In case of water-composite fluids, the sharpening of the grinding wheel is higher due to increased cutting-edge stress. As a consequence, the cutting edge number is lower, which leads to a higher surface roughness. In case of low material-removal rates, an inverse tendency can be observed at the beginning of the grinding process. In this range, which is characterized by small uncut chip thicknesses and increased grain flattening, a higher friction during grinding with solutions and emulsions initially leads to increased blunting of the grains and thus to higher cutting edge numbers, which, in turn, results in better surface quality. After the grinding-in phase, which leads to a balance of grain flattening, splintering, and breakout, a lower roughness of the surface can be observed while grinding with oil.

In case of using the water-composite, the process behavior strongly depends on the composition and the concentration, which also influences the cooling and lubricating properties (Bock 1993). In grinding of tempered steel, the specific grinding energy increases with a rising water fraction and with the use of solutions instead of emulsions (Dederichs 1972; Vits 1985). When emulsions were used for internal grinding of ball-bearing steel 100Cr6 with vitrified bond CBN grinding wheels, work results were improved at concentrations

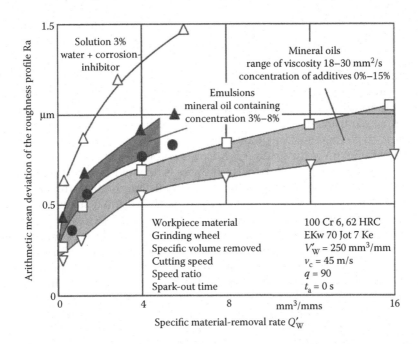

FIGURE 10.7
Influence of different types of cooling lubricants on surface quality during surface grinding. (From Vits, R., 1985, Technologische Aspekte der Kühlschmierung beim Schleifen. PhD thesis, RWTH Aachen. With permission.)

between 2% and 10% with increasing concentrate or oil content (Althaus 1982). In case of thread grinding with plated and resinoid bond grinding wheels, which reduce the grinding wheel wear, lower grinding forces, and superior surface quality were found with increasing the emulsion concentration (Klocke 1982). At using conventional grinding wheels to grind steel, less influence of the concentration of water-composite fluids was found. In fact, smaller normal and tangential forces, superior surface qualities, and less wear were found with water-composite fluids containing mineral oil. However, there was no clear influence of concentration on process or work-result parameters within the group of emulsions (Vits 1985).

The performance of water-immiscible fluids strongly depends on the viscosity. Decreased oil viscosity leads to a lower lubrication effect and increased viscosity causes a lower cooling effect. The best results can be expected from grinding oils with medium viscosity. This conclusion was reached comparing grinding oils for internal cylindrical grinding with viscosities of $v = 12, 48, 220,$ and 432 mm mm^2/s at 20°C. The oil of a viscosity of $v = 48$ mm^2/s had the highest grinding ratio. A viscosity increase of $v = 18$–30 mm^2/s, linked to a cutting speed of $v_c = 45$ m/s and grinding wheels of corundum (Vits 1985), as well as of $v = 7$–25 mm^2/s at a cutting speed of $v_c = 90$ m/s using CBN tools (Treffert 1995), had only a marginal effect on grinding forces and on surface quality during external cylindrical grinding of ball-bearing steel.

Beyond the mentioned parameters, the efficiency of water composite as well as of water-immiscible fluids can be affected by additives forming a lubricating film. Often so-called EP additives are used to realize chemical reactions at defined temperatures with the material of the workpiece. The stability of the reaction products and subsequently the work results mainly depend on the lubricant additives, the machined material and the process parameters (Ott 1991). Therefore, no general statements on the efficiency of single additives can be derived from the present investigations. In many cases, an additive adapted to the machining task improves the grinding process and the work result (Nee 1979; Klocke 1982; Vits 1985; Carius 1989; Brinksmeier 1991b; Heuer 1991, 1992; Spur et al. 1995). During the grinding of nickel-based alloys with CBN grinding wheels, for instance, grinding forces could be reduced and burning and chatter marks avoided through the addition of sulfur additives into an ester-base product (Spur et al. 1995). In case of external cylindrical grinding of 100Cr6, the addition of 5% sulfured fatty acid ester and the addition of polar components or chlorine paraffins in contrast to the unalloyed base oil led to the lowest grinding wheel wear and to the best surface quality (Heuer 1991, 1992).

Beyond the parameters of the grinding process and of the surface quality, the condition of the ground subsurface plays a key role in the assessment of the effectiveness of different grinding fluids. Althaus (1982) reported higher residual compressive stress at process start after grinding with oil than with emulsion. He used the residual stress as a criterion for the assessment of the subsurface state in internal grinding of 100Cr6 at a cutting speed of $v_c = 30$ m/s and a specific material-removal rate of $Q'_w = 1.0$ mm^3/mms with conventional corundum wheels as well as with vitrified CBN tools. With an increasing material-removal rate, the residual compressive stress increases even more with water-composite fluids than while grinding with oil (Althaus 1982). Those results are contrary to findings of novel research projects, which observed residual compressive stress, when emulsions were used for external cylindrical grinding of the same material with a cutting speed of $v_c = 60$ m/s and a specific material-removal rate of $Q'_w = 12$ mm^3/mms independently of the volume removed. This relation is interpreted by Brinksmeier (1991b) and Heuer (1992) against the background of the grind-in behavior of vitrified CBN grinding wheels. According to this, the grind-in process decelerated through reduced friction due to grinding with oil and leads to smaller chip spaces and to a higher thermal stress of the subsurface, reflected in a stress level shifted to residual tensile stress. If there are identical grinding wheel topographies, equal residual compressive stress is measured for water-composite and water-immiscible fluids, although a higher specific grinding energy is absorbed from the process if an emulsion is used due to the higher friction. This effect is compensated by the better cooling effect of this fluid. The differences from the results of Althaus can be explained by the much lower material-removal rate. Thus, due to sufficient chip space in the grinding wheel surface, the grind-in of the grinding wheel is insignificant (Brinksmeier 1991b). Other investigations report an unfavorable effect on the microhardness gradient normal to the surface during grinding with emulsion, characterized by the formation of a soft membrane. In this case, the oil could avoid a negative influence on the subsurface (Flaischlen 1977). Vits (1985) also reported a smaller depth of the affected zone after grinding 100Cr6 with grinding oils than with water-composite fluids.

The results listed in this chapter illustrate the need to be aware of the complexity of the tribochemical and tribomechanical influences in grinding. For further reading on the tribology of abrasive machining processes, the reader is referred to the book by Marinescu et al. (2013).

10.11 Reduction of Temperature by Convective Cooling

The previous section demonstrates that grinding power, grinding wear, and grinding temperatures depend strongly on complex tribological factors. However, given knowledge of measured grinding power, and removal rate in a particular grinding process, it is possible to evaluate convection factors for different coolants. The following section summarizes basic factors involved in the heat transfer process and explains differences between different grinding processes.

10.11.1 Convective Cooling

Convective cooling is a process where heat is directly absorbed into the coolant as the fluid passes through the grinding contact zone. Convective cooling is strongest at low and medium removal rates with low contact zone temperatures below the coolant boiling temperature and large contact length (Zhang Lei et al. 2013). Coolant is not equally effective in removing heat from all grinding processes. For example, in high efficiency deep grinding, there are examples where most of the grinding heat is directly removed by the grinding chips (Rowe and Jin 2001), whereas in low-temperature creep-feed grinding with deep cuts and long contact length, the largest share of the heat is usually removed by the coolant, as described in Section 16.5 on creep-feed grinding. Experimental evaluation of convective cooling requires consideration of heat partition, maximum contact temperature, thermal properties and contact length as illustrated in the following examples. While there are many thermal models available in the literature, the following model is probably the simplest that takes reasonable account of all heat flows. Further details of the analysis are available in Section 17.2.

10.11.2 Maximum Contact Temperature in Grinding

In the grinding contact area, heat is dissipated primarily through four heat sinks; the workpiece, the wheel grains, the chips, and the grinding fluid as represented in Figure 10.8.

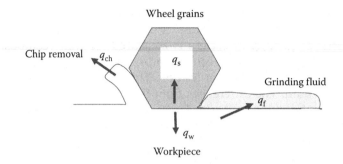

FIGURE 10.8
Heat flows to the workpiece, the wheel grains, the grinding chips, and the grinding fluid.

The grinding energy is converted into heat which is dissipated through these four heat flows within the grinding contact area and therefore the total process heat flow per unit area is $q = q_w + q_s + q_{ch} + q_f$. Analysis of the four heat flows produces the Rowe maximum workpiece temperature rise (Rowe 2014).

$$\Delta T = \frac{R_{ws}(q - q_{ch})}{h_w + h_f} \tag{10.1}$$

The cooling due to the grinding fluid is taken into account through the fluid convection factor h_f. Other terms are the wheel-work partition ratio $R_{ws} \approx 1/\left(1 + k_g/\beta_w \cdot \sqrt{r_0 \cdot v_s}\right)$, the grinding heat flow per unit area $q = P/b \cdot l_c$, the heat flow to the chips $q_{ch} = e_{ch} \cdot a_e \cdot v_w/l_c$, and where for ferrous materials, the specific energy absorbed by the chips $e_{ch} \approx 6 \text{ J/mm}^3$. The workpiece conduction factor $h_w = \beta_w \sqrt{v_w}/C\sqrt{l_c}$. The fluid convection factor h_f is described in Section 10.11.3.

10.11.3 Fluid Convection Factor h_f

The fluid convection factor in grinding is defined in terms of maximum temperature rise in the grinding contact zone $h_f = q_f/\Delta T$ and where q_f is the heat flow to the fluid per unit contact area. The heat convected to the grinding fluid increases with maximum workpiece contact temperature rise and fluid convection factor until the fluid boils and then completely burns out. It is found from grinding experiments that complete burnout occurs roughly at a temperature 50% greater than the fluid boiling temperature for water-based emulsions. In dry grinding, the fluid convection factor approaches zero.

There are two simple models that provide estimates of fluid convection factor. These are the *fluid wheel model* (FWM) and the *laminar flow model* (LFM) described in more detail by Rowe (2014). The simplest of these is the FWM of convection which assumes that the fluid acts as a solid wheel travelling at wheel speed and has the thermal properties of the grinding fluid. Applying sliding heat source theory,

$$h_f = \frac{\beta_f}{C} \cdot \sqrt{\frac{v_s}{l_c}} \quad \text{(Fluid wheel model)} \tag{10.2}$$

where $\beta_f = \sqrt{k_f \cdot \rho_f \cdot c_f}$ is the thermal heat property of the fluid for sliding contact. The factor $C \approx 1$ for high sliding speeds.

The LFM of convection is based on the boundary layer theory of laminar flow past a flat surface as described by Rogers and Mayhew (1967). The Rogers and Mayhew approach can be simplified by assuming the fluid speed within the contact area is equal to the grinding wheel speed. Another modification is to assume mean temperature in the contact area is two-thirds maximum workpiece temperature leading to

$$h_f = G_f \cdot \sqrt{\frac{v_s}{l_c}} \quad \text{(laminar flow model)} \tag{10.3}$$

where $G_f = 4 / 9 \cdot k_f^{2/3} \cdot \rho_f^{1/2} \cdot c_f^{1/3} \cdot \eta_f^{-1/6}$.

Both models predict that convection factor increases with wheel speed and reduces with contact length. However, the temperature rise for constant grinding power is reduced with increased contact length because of the greater reduction in heat flow per unit area.

The LFM gives lower values than the FWM. A criticism of the LFM is that recirculation must take place within the grinding wheel pores which tends to invalidate the assumption of a steady laminar stream past the surface. The boundary layer thickness is consequentially reduced and the convection factor increased.

Rowe (2014) calculated the following convection factors for an alumina grinding wheel speed of 30 m/s and typical coolant properties:

	Fluid Wheel Model	**Laminar Flow Model**
Water-based emulsion	87,700 W/m²K	27,960 W/m²K
Straight oil	28,170 W/m²K	4,804 W/m²K

Experimental values of convection factors were estimated from temperature measurements by Lin et al. (2009) at 36 m/s wheel speed, and by Barczak et al. (2010) at 25 and 45 m/s for various steel and cast iron workpieces. Further estimates were made by Zhang Lei et al. (2013). Results are shown in Figure 10.9 for a water-based emulsion. Experimental

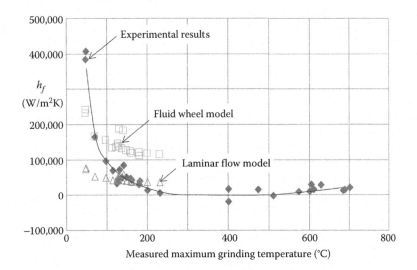

FIGURE 10.9
Fluid convection factors for various ferrous materials and a water-based emulsion. Based on Rowe (2014) thermal model. Complete fluid burnout occurs above 180°C.

values of convection factors rapidly reduce as boiling temperature is approached and go to zero above the burnout temperature as clearly shown. At even higher grinding temperatures, the measured convection factor starts to increase again due to radiation cooling. Obviously, predicted convection factors only have relevance below boiling temperatures. Above boiling temperature, convective cooling still occurs but rapidly decreases toward zero at approximately 180°C for a water-based emulsion.

References

Althaus, P. G., 1982, Leistungssteigerung beim Innenschleifen durch kubisches Bornitrid (CBN) und neue Maschinenkonzeptionen. PhD thesis, Universität Hannover.

Barczak, L. M., Batako, A. D., Morgan, M. N., 2010, A study of plane surface grinding under minimum quantity lubrication condition. *International Journal of Machine Tools and Manufacture*, 50, 977–985.

Bartelt, G., Studt, P., 1992, The effect of selected oil additives on sliding friction and wear of ceramic/ceramic couples lubricated with hexadecane. *Proceedings of the 5th Nordic Symposium on Tribologie*. Helsinki, Finland.

Bartz, W. J., 1978, Wirtschaftliches Zerspanen durch Kühlschmierstoffe. *Teil I und II, wt-Z. industrielle Fertigung*, 8, 471.

Beck, T., 2001, Kühlschmierstoffeinsatz beim Schleifen mit CBN. PhD thesis, RWTH Aachen.

Bock, R., 1993, Umweltfreundliche Kühlschmierstoffe. *Jahrbuch Schleifen, Honen, Läppen und Polieren. 57, 63, Vulkan.*

Böschke, K., 1993, *Der Kühlschmierstoff als Werkstoff.* wt-Werkstattstechnick, vol. 3, Springer Verlag, Düsseldorf, Germany.

Brinksmeier, E., 1991a, Aufgaben der Kühlschmierstoffe bei spanender Bearbeitung. Proc. "Kühlschmierstoffe in der spanenden Fertigung." des dt. Industrieforums f. Tech. (DIF), Frankfurt.

Brinksmeier, E., 1991b, Prozess- und Werkstückqualität in der Feinbearbeitung. *Fortschrittberichte.* VDI, Reihe 2, Nr. 234, VDI-Verlag, Düsseldorf.

Brinksmeier, E., Schneider, C., 1993, Bausteine für umweltverträgliche Feinbearbeitungsprozesse. Proc. 7. Braunschweiger Feinbearbeitungskolloquium, Hohe Prozesssicherheit, hohe Leistung, hohe Präzision.

Brücher, T., 1993, Kühlschmierung—ein wesentlicher Faktor für wirtschaftliche Schleifbearbeitung. Proc. "Wirtschaftliche Schleifverfahren," des dt. Industrieforums f. Technologie (DIF), Ratingen.

Brücher, T., 1996, Kühlschmierung beam Schleifen keramischer Werkstoffe. PhD thesis, Technische Universität Berlin.

Carius, A. C., 1989, *Effects of Grinding Fluid Type and Delivery on CBN Wheel Performance.* SME, Modern Grinding Technologie, Detroit, MI.

Dederichs, M., 1972, Untersuchung der Wärmebeeinflussung des Werkstücks beim Flachschleifen. PhD thesis, RWTH Aachen.

DIN 51 385, 2013, *Schmierstoffe Bearbeitungsmedien für die Umformung und Zerspanung von Werkstoffen—Begriffe.* Berlin, Beuth.

Eckhardt, F., 1983, Kühlschmierstoffe für die spanende Metallbearbeitung. Teil 1–11, TZ für Metallbearbeitung.

Flaischlen, E., 1977, Maßnahmen zur Vermeidung von thermischer Oberflächenschäden beim Schleifen—Beispiel aus der Praxis. *Jahrbuch Schleifen, Honen, Läppen und Polieren*, 48, 151, Vulkan.

Gerhäuser, W., Laika, K., 1977, Einflussgrößen auf den Verschleiß von Diamantwerkzeugen bei der Gesteinsbearbeitung. *Jahrbuch Schleifen, Honen, Läppen und Polieren*, 48, 341, Vulkan.

Gühring, K., 1967, Hochleistungsschleifen. PhD thesis, RWTH Aachen.

Heinzel, C., 1999, Methoden zur Untersuchung und Optimierung der Kühlschmierung beim Schleifen. PhD thesis, Universität Bremen.

Heuer, W., 1991, Potentiale der Kühlschmierung beim Schleifen mit hochharten Schleifstoffen. Proc. "Kühlschmierstoffe in der spanenden Fertigung." des deutschen Industrieforums f. Techn. (DIF), Frankfurt.

Heuer, W., 1992, *Außenrundschleifen mit kleinen keramisch gebundenen CBN-Schleifscheiben. Fortschrittberichte.* VDI, Reihe 2, Nr. 270.

Holz, R., Sauren, J., 1988, *Schleifen mit Diamant und CBN.* Ernst Winter & Sohn GmbH & Co., Hamburg.

Howes, T. D., 1990, Assessment of the cooling lubricative properties of grinding fluids. *Annals of the CIRP,* 39(1), 313.

Howes, T. D., Neailey, K., Harrison, J., 1987, Fluid film bioling in shallow cut grinding. *Annals of the CIRP,* 36(1), 223.

Kassack, J. F., 1994, Einfluss von Kühlschmierstoff-Additiven auf Werkzeugverschleiß, Zerspankraft und Bauteilqualität. PhD thesis, RWTH Aachen.

Kerschl, H.-W., 1988, Einfluss des Kühlschmierstoffes beim Hochgeschwindigkeitsschleifen mit CBN. *Werkstatt und Betrieb,* 12, 979.

Keyser, W., 1970, Kühlschmierung beim Schleifen. *IDR,* 3, 158.

Keyser, W., 1974, Kühlschmiermittel für die Feinstbearbeitung von Oberflächen. *Jahrbuch Schleifen, Honen, Läppen und Polieren.* 46, Ausgabe.

Khudobin, L. V., 1969, Cutting fluids and its effect on grinding wheel clogging. *Machines and Tooling,* 4, 54.

Klocke, F., 1982, *Gewindeschleifen mit Bornitridschleifscheiben.* Produktionstech. Berlin, Forschungsber. Für die Praxis, Bd. 30, Hanser Verlag.

Klocke, F., Gerschwiler, K., 1996, *Trockenbearbeitung-Grundlagen. Grenzen, Perspektiven. VDI-Berichte Nr. 1240.*

Kohblanck, G., 1956, *Kühlen und Schmieren in der Zerspantechnik. Teil 1 und 2 Fertigungstechnik 5 und,* 4(152), 205.

König, W., 1980, *Fertigungsverfahren. Bd. 2,* VDI-Verlag, Düsseldorf.

König, W., Klocke, F., 1996, *Fertigungsverfahren Band 2—Schleifen, Honen,* VDI-Verlag, Düsseldorf.

König, W. et al., 1993, *Kühlschmierstoff—Eine ökologische Herausforderung an die Fertigungstechnik. Wettbewerbsfaktor Produktionstechnik. Sonderausgabe für AWK,* VDI-Verlag, Düsseldorf.

Korff, J., 1991, Additive für Kühlschmierstoffe. Proc. "Kühlschmierstoffe in der spanenden Fertigung." des deutschen Industrieforums f. Techn. (DIF), Frankfurt, 21–22, Oktober.

Leiseder, M. L., 1991, *Metalworking Fluids,* Verlag moderne Industrie, Landsberg.

Lin, B., Morgan, M. N., Chen, X., Wang, Y. K., 2009, Study on the convective heat tranfer coefficient of coolant and the maximum temperature in the grinding process. *International Journal of Advanced Manufacturing Technology,* 42, 1175–1186.

Mang, T., 1983, *Die Schmierung in der Metallbearbeitung.* Vogel Verlag, Würzburg.

Marinescu, I. D., Rowe, W. B., Dimitrov, B., and Inasaki, I. 2004. *Tribology of Abrasive Machining Processes.* William Andrew Publishing, Norwich, NY.

Marinescu, I., Rowe, W. B., Dimitrov, B., Ohmori, I., 2013, *Tribology of Abrasive Machining Processes. 2nd edition,* Elsevier, Oxford, UK.

Marinescu, I. D., Rowe, W. B., Dimitrov, B., Inasaki, I. 2004. *Tribology of Abrasive Machining Processes.* William Andrew Publishing, Norwich, New York.

Möller, U. J., Boor, U., 1986, *Schmierstoffe im Betrieb,* VDI-Verlag, Düsseldorf.

Nee, A. Y. C., 1979, The effect of grinding fluid additives on diamond abrasive wheel efficiency. *International Journal of Machine Tool Design and Research,* 19, 21.

Oates, P. D., Bezer, H. J., Balfour, A. M., 1977, Bewertung von Kühlschmierstoffen für die Verwendung mit AMBER BORON NITRIDE—Schleifmitteln. *IDR,* 4, 221.

Ott, H. W., 1985, Kühlschmieren—Voraussetzung für kostengünstiges Schleifen und Abrichten. Proc. "Schleifen als qualitätsbestimmende Endbearbeitung." des VDI Bildungswerkes, Düsseldorf.

Ott, H. W., 1991, Kühlschmierstoffzusammensetzung und Prozessgrößen beim Schleifen. Proc. "Kühlschmierstoffe in der spanenden Fertigung." des deutschen Industrieforums f. Techn. (DIF), Frankfurt.

Peters, J., Aerens, R., 1976, An objective method for evaluating grinding coolants. *Annals of the CIRP*, 25(1), 247–251.

Pfau, A., 1987, Stand der Technik in der Bearbeitung von Brillengläsern. *Diamant Information M4, De Beers Industrial Diamoond Division, April*.

Pfeiffer, W. et al., 1993, *Kühlschmierstoffe—Umgang, Messung, Beurteilung*. Schutzmaßnahmen, BIA-Report 3, 91.

Polyanskov, Y. V., Khudobin, L. V., 1979, The effect of coolant on the surface finish of a ground surface in sparking-out. *Russian Engineering Journal*, 5, 46.

Rogers, G. F. C., Mayhew, Y. R., 1967, *Engineering Thermodynamics, Work and Heat Transfer*. 2nd edition, Longman, London and New York.

Roth, P., Wobker, H.-G., 1991, Schleifbearbeitung keramischer Werkstoffe. *Sprechsaal* 4, 254.

Rowe, W. B., Jin, T. 2001. Temperatures in high efficiency deep grinding. *Annals of the CIRP*, 50(1), 205–208.

Rowe, W. B., 2014, *Principles of Modern Grinding Technology*. 2nd edition, Elsevier, Oxford, UK.

Sawluk, W., 1964, Flachschleifen von oxidkeramischen Werkstoffen mit Topfscheiben. PhD thesis, TH Braunschweig.

Seifarth, M., 1987, Bearbeitung von optischem Glas mit Diamantwerkzeugen. *Diamant Information M4, De Beers Industrial Diamond Division, April*.

Sperling, F., 1971, Optimales Kühlschmieren beim Schleifen. *Industrial Anzeiger 87, 2150*.

Spur, G., 1983, Kühlschmierstoffe für die Metallzerspanung. *Lehrblätter/Fertigungstechnik, ZwF 12*, 585–586.

Spur, G., 1993, Werkstoffspezifische Schleiftechnologie—Schlüssel für erhöhte Prozessfähigkeit in der Keramikbearbeitung. *Jahrbuch Schleifen, Honen, Läppen und Polieren. 57. Ausgabe, 335*.

Spur, G., Niewelt, W., Meier, A., 1995, Schleifen von Superlegierungen für Gasturbinen—Einfluss des Kühlschmierstoffs auf das Arbeitsergebnis. *ZwF, 6*, 311.

Studt, P., 1987, Influence of lubrication oil additives on friction of ceramics under conditions of boundary lubrication. *Wear, 115*, 185.

Tawakoli, T., 1990, *Hochleistungs-Flachschleifen, Technologie. Verfahrensplanung und wirtschaftlicher Einsatz*. VDI-Verlag.

Tio, T. H., Brücher, T., 1988, Kühlschmierung bei der Schleifbearbeitung keramischer Werkstoffe. Proc. Arbeitskreises "Keramikbearbeitung," Produktionstechnisches Zentrum Berlin.

Tönshoff, H. K., Jürgenharke, B., 1979, *Innenschleifen kleiner Bohrungen. Jahrbuch Schleifen, Honen, Läppen und Polieren. 49. Ausgabe*.

Tönshoff, H. K., Schulze, R., 1980, Einfluss des Kühlmittels bei der Bearbeitung von Hartgestein. *IDR, 1*, 19.

Treffert, C., 1995, *Hochgeschwindigkeitsschleifen mit galvanisch gebundenen CBN-Schleifscheiben. Berichte aus der Produktionstechnik, Bd. 4/49*.

Triemel, J., 1976, Schleifen mit Bornitrid. *Fertigungstechnische Berichte. Bd. 6*.

VDI-Richtlinie 3035, 1997, *Anforderungen an Werkzeugmaschinen, Fertigungsanlagen und periphere Einrichtungen beim Einsatz von Kühlschmierstoffen*, VDI-Verlag, Düsseldorf.

VDI-Richtlinie 3396, 1983, *Kühlschmierstoffe für spanende Fertigungsverfahren*, VDI-Verlag, Düsseldorf.

Vits, R., 1985, Technologische Aspekte der Kühlschmierung beim Schleifen. PhD thesis, RWTH Aachen.

Webster, J., 1995, Selection of coolant type and application technique in grinding. Proceedings of Supergrind '95 "Developments in Grinding. Storrs, CT.

Weinert, K., 1999, *Trockenbearbeitung und Minimalmengenschmierung*, Springer Verlag, New York.

Wobker, H. -G., 1992, *Schleifen keramischer Schneidstoffe. Fortschrittberichte*. VDI, Reihe 2, Nr. 237.

Yatsui, H., Tsukada, S., 1983, Influence of fluid type on wet grinding temperature. *Bulletin of the Japan Society of Precision Engineering* 2, 133.

Zhang Lei, Rowe, W. B., Morgan, M. N., 2013, An improved fluid convection solution in conventional grinding. Proceedings of the Institution of Mechanical Engineers, Part B Journal of Engineering Manufacture, vol. 227, no. 6, pp. 332–338. Sage Publishing, UK.

Zimmermann, D., 1982, Kühlschmierstoffe für die Feinbearbeitung. *tz für Metallbearbeitung* 4, 16.

Zwingmann, G., 1960, *Schmier- und Kühlflüssigkeiten bei der Feinbearbeitung*, Schriftenreihe Feinbearbeitung, DEVA, Stuttgart.

Zwingmann, G., 1979, Kühlschmierstoffe für die spanende Metallbearbeitung. Teil 1 und 2. *Werkstatt und Betrieb* 6, 409, 483.

11

Monitoring of Grinding Processes

11.1 Need for Process Monitoring

11.1.1 Introduction

The behavior of any grinding process is complex. There are a large number of input variables and the whole process is transient, that is, the mechanisms change with time. The general need for a monitoring system is expressed by Figure 11.1.

11.1.2 Need for Sensors

An essential feature of any monitoring system is that there are sensors that can detect whether the process is running normally or abnormally. A monitoring system in an automatic grinding machine is incorporated into a control system. The system has to be able to correct the machine operational settings so that a near optimal condition can be restored, if the system is running abnormally or even in a suboptimal state. An automatic system provides process corrections from internal control mechanisms, whereas a manual system requires human intervention.

Sensor systems for a grinding process should be capable of detecting any malfunctions in the process with high reliability so that the production of substandard parts can be minimized. Some major quality issues in the grinding process are the occurrence of chatter vibration, grinding burn, and surface roughness deterioration. Quality problems have to be identified in order to maintain the desired workpiece quality.

11.1.3 Process Optimization

In addition to quality detection, another important task of the monitoring system is to provide useful information for optimizing the grinding process in terms of total grinding time or total grinding cost. Optimization of the process can possibly be achieved if degradation of the process behavior can be tracked by the monitoring system. The information obtained with a sensor system can also be used to establish databases as part of an intelligent system (Rowe et al. 1994; Tönshoff et al. 2002).

11.1.4 Grinding Wheel Wear

An important aspect of a grinding process is grinding wheel performance. The grinding wheel should be properly selected and conditioned to meet the requirements of the parts. Grinding wheel performance changes during the grinding process, which makes

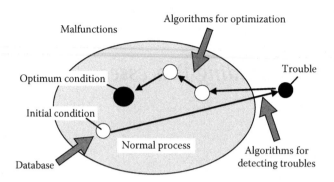

FIGURE 11.1
Role of a monitoring system for grinding.

it difficult to predict process behavior in advance. Changes occur in the grinding wheel shape leading to geometric workpiece errors. Changes also occur in the grinding wheel topography including distribution of grinding grits and their sharpness. These changes have various consequences for grinding forces, material removal rate, workpiece surface quality, and process stability. Conditioning of the grinding wheel surface is necessary before grinding starts. It becomes necessary as well after the wheel has reached the end of its redress life to restore the wheel configuration and the surface topography to the initial state. Therefore, sufficient sensor systems are required to minimize the additional machining time, to assure the desired grinding wheel topography is maintained and to minimize wasted abrasive material during conditioning.

11.2 Sensors for Monitoring Process Variables

11.2.1 Introduction

As with all manufacturing processes, ideally, the variables of greatest interest are measured directly as close to their origin as possible. Grinding processes are affected by a large number of input variables, which each influence the resulting output quantities. Brinksmeier proposed a systematic approach to distinguish between different types of quantities to describe a manufacturing process precisely (Brinksmeier 1991; Marinescu et al. 2004).

The most common sensors to be used in either the industrial or the research environment are for force, power, and acoustic emission (Byrne et al. 1995). Figure 11.2 shows the setup for the most popular integration of sensor systems in either surface or outer diameter grinding. Sensors mainly have to detect grinding performance during the period of intermittent contact between the grinding wheel and workpiece. Only during this limited interaction, can many process quantities can be detected.

11.2.2 Force Sensors

The first attempts to measure grinding forces go back to the early 1950s and were based on strain gauges. Although the system performed well to achieve substantial data on grinding, the most important disadvantage of this approach was the significant reduction of the

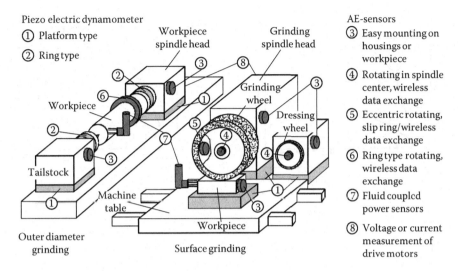

Piezo electric dynamometer
① Platform type
② Ring type

Workpiece spindle head

Grinding spindle head

AE-sensors
③ Easy mounting on housings or workpiece
④ Rotating in spindle center, wireless data exchange
⑤ Eccentric rotating, slip ring/wireless data exchange
⑥ Ring type rotating, wireless data exchange
⑦ Fluid coupled power sensors
⑧ Voltage or current measurement of drive motors

Workpiece

Grinding wheel

Dressing wheel

Tailstock

Machine table

Workpiece

Outer diameter grinding

Surface grinding

FIGURE 11.2
Possible mounting positions of force, AE, and power sensors.

total stiffness during grinding. Thus, research was done to develop alternative systems. With the introduction of piezoelectric quartz force transducers, a satisfactory solution was found. In Figure 11.2, different locations are shown for mounting a force platform. In surface grinding, the platform is most often mounted on the machine table to carry the workpiece. In internal diameter (ID) or external diameter (OD) grinding this solution is not available due to the rotation of the workpiece. In this case, either the whole grinding spindle head is mounted on a platform or the workpiece spindle head and sometimes also the tailstock are put on a platform (Karpuschewski 2001).

Figure 11.3 shows an example of force measurement with the spindle head on a platform during ID plunge grinding. In this case, the results are used to investigate the influence of coolant supply systems, while grinding case-hardened steel. The force measurements give a clear view that it is not possible to grind without coolant using the chosen grinding wheel due to wheel loading causing high normal and tangential forces. But it is also seen that there is a high potential for minimum quantity lubrication (MQL) with very constant force levels over the recorded material removal (Brunner 1998).

For OD grinding, it is also possible to use ring-type piezoelectric dynamometers. All three perpendicular force components can be measured, the sensors are mounted under preload behind the nonrotating center points. Dressing forces can also be monitored by the use of piezoelectric dynamometers, for example, the spindle head of a rotating dresser can be mounted on a platform. Beyond these general solutions, many special setups have been used for nonconventional grinding processes like ID cutoff grinding of silicon wafers or ID grinding of long small bores with rod-shaped tools.

Force measurements can also be used to get information about the surface integrity of a workpiece. The tangential force is the more important component, because the multiplication of tangential force and cutting speed results in the grinding power P_c, as shown in Figure 11.4 OD plunge grinding. If this grinding power is referred to the zone of contact the specific grinding power P_c'' can be calculated. Grinding power can be used to estimate the heat generation during grinding (Brinksmeier 1991; Karpuschewski 1995; Marinescu et al. 2004; Rowe 2014).

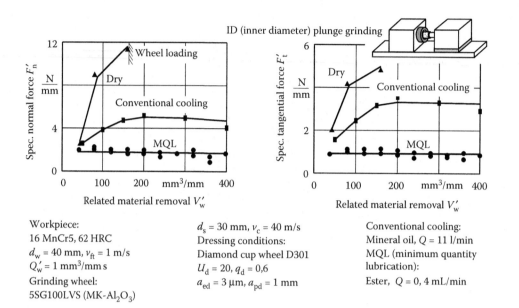

Workpiece:
16 MnCr5, 62 HRC
d_w = 40 mm, v_{ft} = 1 m/s
Q'_w = 1 mm³/mm s
Grinding wheel:
5SG100LVS (MK-Al₂O₃)

d_s = 30 mm, v_c = 40 m/s
Dressing conditions:
Diamond cup wheel D301
U_d = 20, q_d = 0,6
a_{ed} = 3 μm, a_{pd} = 1 mm

Conventional cooling:
Mineral oil, Q = 11 l/min
MQL (minimum quantity
lubrication):
Ester, Q = 0, 4 mL/min

FIGURE 11.3
Grinding force measurement with platform dynamometer. (Adapted from Brunner, G., 1998, Schleifen mit mikrokristallinem Aluminiumoxid, Dissertation, Universität Hannover.)

On the right-hand side of Figure 11.4, the residual stress change at the surface of a ground-hardened steel workpiece is schematically shown for increasing specific grinding power (Brinksmeier 1991). The effects of thermal and mechanical loads interact with each other. At the beginning only small thermally induced residual stresses due to external friction are likely to occur. With the beginning of plastic deformations, a steep increase of

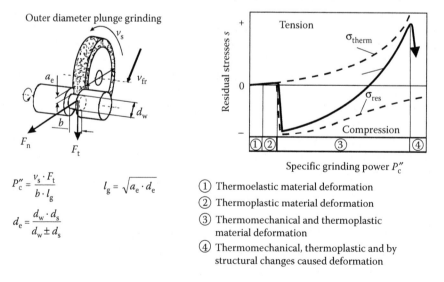

$$P''_c = \frac{v_s \cdot F_t}{b \cdot l_g} \qquad l_g = \sqrt{a_e \cdot d_e}$$

$$d_e = \frac{d_w \cdot d_s}{d_w \pm d_s}$$

① Thermoelastic material deformation
② Thermoplastic material deformation
③ Thermomechanical and thermoplastic material deformation
④ Thermomechanical, thermoplastic and by structural changes caused deformation

FIGURE 11.4
Residual stress determination depending on grinding power. (Adapted from Brinksmeier, E., 1991, *Prozeß- und Werkstückqualität in der Feinbearbeitung*, Habilitationsschrift, Universität Hannover.)

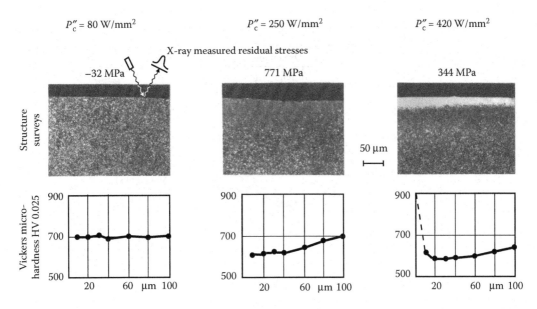

FIGURE 11.5
Influence of the specific grinding power on surface integrity of 16 MnCr5.

compressive residual stresses can be registered. With rising specific grinding power and thus higher temperatures in the contact zone, the mechanical influence decreases while the thermal load becomes dominant.

At very high levels of specific grinding power, structural changes might disturb the further tensile residual stress rise. Rehardening layers are likely to occur, drastically reducing tensile residual stresses (Karpuschewski 1995). Figure 11.5 shows representative structure surveys, Vickers microhardness depths and residual stress measurements of different plunge-cut ground workpieces made of case-hardened steel. The specific grinding power as the main characteristic was varied through the increase of the specific material removal rate Q'_w.

Brinksmeier analyzed a large variety of different grinding processes to establish an empirical model for the correlation between the specific grinding power based on force measurement and the x-ray calculated residual stress states (Brinksmeier 1991). Figure 11.6 shows so-called thermal transfer functions for different grinding operations on ball-bearing steel. The results reveal that it is possible to generate compressive residual stresses with any grinding operation as long as the specific grinding power is small enough. For higher values of P''_c, there is a clear tendency toward tensile residual stresses. The superior behavior of cubic boron nitride (CBN) grinding wheels compared to conventional abrasives is obvious, because this abrasive has a much better thermal conductivity and is thus able to remove more heat from the zone of contact.

Recent fundamental investigations of grinding efficiency and thermal damage have highlighted the importance of specific energy in grinding (Rowe and Jin 2001). Specific energy is the energy per unit volume of material removed usually quoted in joules/cubic millimeter. Specific energy can be calculated by dividing the grinding power by the material removal rate:

$$e_c = \frac{F_t v_s}{a_e v_w b}$$

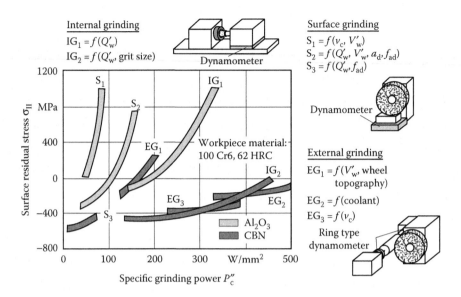

FIGURE 11.6
Set of different thermal transfer functions based on force measurement. (Adapted from Brinksmeier, E., 1991, Prozeß- und Werkstückqualität in der Feinbearbeitung, Habilitationsschrift, Universität Hannover.)

Specific energy is an inverse measure of grinding efficiency. Low specific energy represents high removal rate with low consumption of energy. Table 11.1 clearly demonstrates that low specific energy gives rise to lower temperatures than high specific energy. Table 11.1 shows a measurement where the lowest temperature was recorded at the highest removal rate. This is the opposite of normal expectation and is simply a result of low specific energy. For this reason, there is an increasing trend toward monitoring specific energy as a measure of the health of grinding processes. If the specific energy increases with time, it probably means the grinding wheel needs redressing.

TABLE 11.1

Effect of Specific Energy e_c and Removal Rate on Maximum Grinding Temperature

Depth of cut (mm)	0.407	0.98	0.92	0.96
Wheel diameter (mm)	173	173	173	170
Workspeed (m/s)	0.2	0.2	0.25	0.3
Contact length (mm)	8.37	13.0	12.6	12.8
Peclet number	45	70	85	103
Removal rate (mm²/s)	81	196	230	288
Total heat flux $q_t = P_c/b_w l_c$ (W/mm²)	165	200	238	248
Flux to chips q_{ch} (W/mm²)	79	122	149	184
Net heat flux q_t–q_{ch} (W/mm²)	85.8	78	89	64
Predicted T_{max} (wet °C)	238	224	249	172
Predicted T_{max} (dry °C)	1180	1290	1280	823
Measured T_{max} (msd °C)	1250	1350	1050	180
Specific energy (J/mm³)	17.0	13.3	13.0	11

Source: Adapted from Rowe, W. B., Jin, T., 2001, *Annals of the CIRP*, 50(1), 205–208.

Specific energy is conveniently computed from a grinding force sensor or from a power sensor. Power sensing is dealt with in the next section.

11.2.3 Power Measurement

The measurement of power consumption of a grinding spindle drive can be regarded as technically simple. But the evidence of this process quantity is definitely limited. The amount of power used for the material removal process is always only a fraction of the total power consumption. The grinding power is the net power measured after deducting the no-load machine power. The no-load grinding power includes the spindle power before grinding contact commences, the coolant supply power, and the power due to any other auxiliary motors. The grinding power is usually measured from the power supplied to the grinding spindle motor. Power monitoring is widely used in industrial applications by defining specific thresholds to avoid overload of the whole machine tool due to wheel wear or errors from operators or automatic handling systems. In grinding, power monitoring is a popular method to avoid thermal damage of the workpiece. The main reason is the easy installation without influencing the working space of the machine tool and relatively low costs. If the grinding power is used for more detailed prediction of actual grinding temperatures for research purposes, it is necessary to make a deduction for no-load power on the spindle. Different investigations show that the dynamic response of a power sensor at the main spindle is limited. Therefore, where forces sensors can be incorporated into the machine, more accurate determination of grinding power may be possible than is possible with a power sensor.

A typical result is shown in Figure 11.7 for a grinding process on spiral bevel ring gears, introducing a vitreous bond CBN grinding wheel for this complex operation. The cone-shaped grinding wheel with a metal core is fed to the workpiece made of case-hardened steel with a six-axis computer numerical control (CNC) grinding machine, the process is

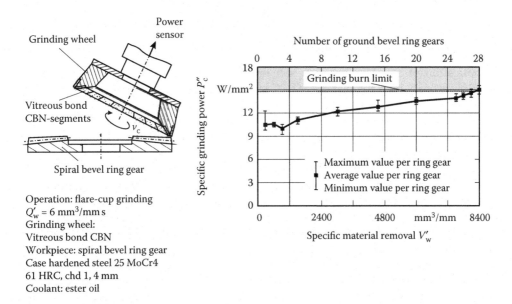

FIGURE 11.7
Power monitoring in spiral bevel gear grinding to avoid grinding burn.

called flare-cup grinding. Monitoring of the grinding power revealed a constant moderate increase over the material removal V_w''. At a specific material removal of 8100 mm^3/mm, which corresponds to the grinding of 27 ring gears, grinding burn was detected for the first time by nital etching. The micro- and macrogeometry of the twentieth workpiece was still within the tolerances, so the tool life criterion was the surface integrity state. After conditioning the grinding wheel with a diamond form roll, it was possible to continue the process. The grinding burn limit fixed by this test was proven in further succeeding investigations. For this type of medium or even large scale production in the automotive industry using grinding wheels with a long lifetime, this power monitoring is a very effective way to avoid thermal damage of the workpiece and also to get rid of the environmentally harmful etching process.

These results reveal that power monitoring can be a suitable sensor technique to avoid surface integrity changes during grinding. The most promising application is seen for superabrasives, because the slow wear increase of the grinding wheel can be clearly determined with this dynamically limited method.

11.2.4 Acceleration Sensors

In abrasive processes, the major application for acceleration sensors is related to balancing systems for grinding wheels. Large grinding wheels without a metal core may have a significant unbalance at the circumference. With the aid of acceleration sensors, the vibrations generated by this unbalance are monitored during the rotation of the grinding wheel at cutting speed. Different systems are in use to compensate this unbalance, for example, hydrocompensators using coolant to fill different chambers in the flange or mechanical balancing heads, which move small weights to specific positions. Although these systems are generally activated at the beginning of a shift, they are able to monitor the change of the balance state during grinding and can continuously compensate the unbalance. After balancing the grinding wheel, a dressing operation should be performed to eliminate wheel runout.

11.2.5 Acoustic Emission Systems

The application of acoustic emission sensors has become very popular in all kinds of machining process over the last decade. A large variety of sensors specially designed for monitoring purposes have been introduced on the market. They combine some of the most urgent requirements for sensor systems like relatively low costs, no negative influence on the stiffness of the machine tool, ease of mounting, and capability of transmitting signals from rotating parts.

First results on acoustic emissions were published in the fifties in tensile tests. Since then decades have passed, until this approach was first used to monitor manufacturing processes. The mechanisms leading to acoustic emission are mainly deformations through dislocations and distorted lattice planes, twin formation of polycrystalline structures, phase transitions, friction, crack formation, and propagation. Due to these different mechanisms, acoustic emission appears either as a burst type signal or as a continuous emission. The grinding process is characterized by the simultaneous contact of many different cutting edges randomly shaped with the workpiece surface. Every single contact of a grain generates a stress pulse in the workpiece. During operation, the properties of the single grain and their overall distribution on the circumference of the grinding wheel change due to the occurrence of wear. Thus, many different sources of acoustic emission have to be considered in the grinding process. The single pulse is a combination of the impact of the grain with the workpiece material and its fracture behavior, of wear of the individual abrasive as well as wear of the bond material. Even the structure of the workpiece, material may change due to thermal overload during grinding. A change from austenite to martensite structures in ferrous materials also generates acoustic emission, although the energy content is significantly lower compared to the other sources.

Different types of signal evaluation can be applied to the AE-sensor output. The most important quantities are root-mean-square value, raw acoustic emission signals, and frequency analysis. The time domain development of the root-mean-square value $U_{AE,RMS}$ contains essential information about the process condition (Inasaki 1991; Byrne et al. 1995). This value can be regarded as a physical quantity for the intensity of the acoustic signal. It is directly related to the load on the material, thus making this value attractive for any kind of monitoring. But it has to be regarded as an average statistical value because most often a low-pass filter is applied. If short transient effects like single grit contacts are to be revealed, the raw acoustic emission signal without any filtering is more attractive. Evaluation in the frequency domain is used to identify dominant patterns, which can be related to specific process conditions like chatter.

Possible positions of AE sensors in grinding are shown in Figure 11.2. The spindle drive units, the grinding wheel or the workpiece can be equipped with a sensor. In addition, fluid coupled sensors may be used without any direct mechanical contact to one of the mentioned components.

In Figure 11.8, the correlation between the surface roughness of a ground workpiece and the root-mean-square value of the AE signal is shown (Meyen 1991). A three-step outer diameter plunge grinding process with a conventional corundum grinding wheel was monitored. It is obvious that for a dressing overlap of $U_d = 2$, the coarse grinding wheel

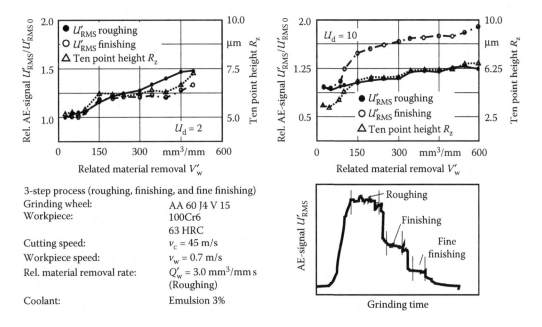

3-step process (roughing, finishing, and fine finishing)

Grinding wheel: AA 60 J4 V 15
Workpiece: 100Cr6
    63 HRC
Cutting speed: $v_c = 45$ m/s
Workpiece speed: $v_w = 0.7$ m/s
Rel. material removal rate: $Q'_w = 3.0$ mm³/mm s
    (Roughing)
Coolant: Emulsion 3%

FIGURE 11.8
Correlation between surface roughness and the AE-RMS signal. (Adapted from Meyen, H. P., 1991, Acoustic Emission (AE)—Mikroseismik im Schleifprozeß. Dissertation, RWTH Aachen.)

topography generated, leads to a high initial surface roughness of $R_z = 5$ µm. Due to continuous wear of the grains, the roughness increases as more material is removed. For the finer dressing overlap of $U_d = 10$, a smaller initial roughness with a significant increase can be seen for the first parts followed by a decreasing tendency. This tendency of the surface roughness is also represented by the AE signal. Higher dressing overlaps lead to more cutting edges thus resulting in a higher acoustic emission activity. The sensitivity of the fine finishing AE signal is higher, because the final roughness is mainly determined in this process step. Meyen has shown in many other tests that monitoring of the grinding process with acoustic emission is possible.

Acoustic emission signals can also be investigated in the frequency domain. Effects like wear or chatter vibration have different influences on the frequency spectrum, thus it may be possible to separate these effects. Figure 11.9 shows the result of a frequency analysis of the AE signal in outer diameter plunge grinding with a vitreous bond CBN grinding wheel (Wakuda et al. 1993). As a very special feature, the AE sensor is mounted in the grinding wheel core, and the signals are transferred via a slip ring to the evaluation computer, thus grinding as well as dressing operations can be monitored. The results reveal that no significant peak can be seen after dressing and first grinding tests. It is only after a long grinding time that specific frequency components emerge from the spectrum, which exhibit constantly rising power during the continuation of the test. The detected frequency is identical with the chatter frequency, which could be stated by additional measurements. The AE signals were used as input data into a neural network to automatically identify the occurrence of any chatter vibrations in grinding (Wakuda et al. 1993; Chen et al. 1996).

From the very first beginning of AE application in grinding, attempts were made to correlate the signal to the occurrence of grinding burn. The works of Klumpen and Saxler are directly related to the possibility of grinding burn detection with AE sensors

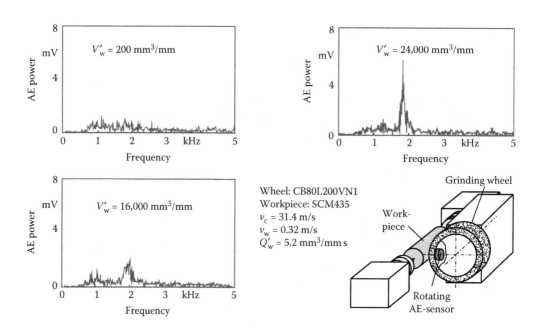

FIGURE 11.9
AE frequency analysis for chatter detection in grinding. (Adapted from Wakuda, M., Inasaki, I. et al., 1993, *Journal of Advanced Automation Technology*, 5(4), 179–184.)

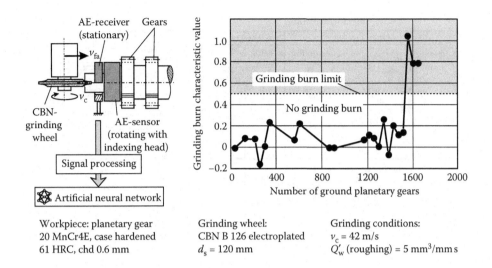

FIGURE 11.10
Grinding burn detection with acoustic emission. (Adapted from Saxler, W., 1997, Erkennung von Schleifbrand durch Schallemissionsanalyse. Dissertation, RWTH Aachen.)

(Klumpen 1994; Saxler 1997). They made a systematic approach to identify dominant influences on the AE signal during grinding.

One fundamental result was, that all process variations, which finally generate grinding burn, including increasing material removal rate or infeed or reduced coolant supply, lead to an increase of acoustic emission. Klumpen could only identify grinding burn by applying a frequency analysis of the AE signal to determine the inclination of the integral differences (Klumpen 1994). This must be regarded as a major disadvantage, because a frequency analysis is usually performed after grinding. This conclusion may change, of course, with the increasing availability of modern fast data signal processing computer devices. Saxler concentrated on the AE signal in the time domain.

The major result of the work by Klumpen (1994) is shown in Figure 11.10. Based on his investigations and theoretical considerations, he concludes that the AE sensor must be mounted on the workpiece to be most sensitive to grinding burn detection. This is of course a major drawback for practical applications. An industrial test was conducted during gear grinding of planetary gears with an electroplated CBN grinding wheel. The sensor was installed at the hydroexpansion clamping mandrel instead of one of a set of five gears. The sensor can rotate with the indexing head. The signals are wirelessly transferred to the stationary receiver. With the aid of artificial neural networks, he was able to achieve a dimensionless grinding burn characteristic value from the AE values of different frequency ranges in the time domain. Thus, an in-process detection of workpiece surface integrity changes became possible. The high efforts for training of the artificial neural network and the problems related to the sensor mounting at the workpiece side must be seen as limiting factors for a wider industrial application.

Results have clearly shown that AE systems can be regarded as suitable process quality sensors in grinding to monitor surface integrity changes. Acoustic emission techniques are rapidly developing for application in scientific research and it is therefore worthwhile to maintain an open interest in the subject as new sensor systems become available.

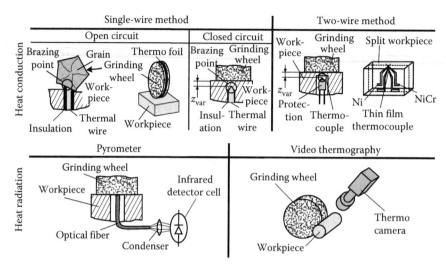

FIGURE 11.11
Temperature measurement systems in grinding.

11.2.6 Temperature Sensors

In any grinding process mechanical, thermal and even chemical effects are usually super-imposed in the zone of contact. Grinding generates a significant amount of heat, which if excessive may cause a deterioration of the dimensional accuracy of the workpiece, an undesirable change of the surface integrity or lead to increased wear of the wheel. Figure 11.11 shows the most popular temperature measurement devices. The preferred method for temperature measurement in grinding is most often the use of thermocouples. The second metal in a thermocouple can be the workpiece material itself, this setup is called the single-wire or single-pole method. A further distinction is made according to the type of insulation. A permanent insulation of the thin wire or foil from the workpiece by use of sheet mica is known as open circuit. The insulation is interrupted by the individual abrasive grains thus measurements can be repeated or process conditions varied until the wire is worn or damaged. Many authors used this setup. Also the grinding wheel can be equipped with the thin wire or a thermo foil, if the insulation properties of abrasive and bond material are sufficient. In the closed circuit type, a permanent contact of the thermal wire and the workpiece by welding or brazing is achieved. The most important advantage of this method is the possibility to measure temperatures at different distances from the zone of contact until the thermocouple is finally exposed at the surface. For the single-wire method, it is necessary to calibrate the thermocouple for every different workpiece material. This disadvantage may be overcome by the use of standard thermocouples, where the two different materials are assembled in a ready-for-use system with sufficient protection. A large variety of sizes and material combinations are available for a wide range of technical purposes.

With the two-wire method, it is possible to measure the temperatures at different distances from the zone of contact. This approach can be regarded as most popular for temperature measurement in grinding. One disadvantage of the double-pole technique is that the depth of the thermocouple junction is much larger than it is for the single-pole technique. This has the effect that the temperature reading is averaged over a greater depth below the surface in a region, where there is a steep temperature gradient.

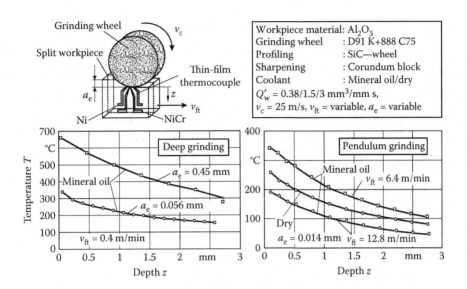

FIGURE 11.12

Grinding temperature measurement with thin-film thermocouples. (Adapted from Lierse, T., 1998, Mechanische und thermische Wirkungen beim Schleifen keramischer Werkstoffe. Dissertation, Universität Hannover.)

Recent advances in application of thin-film single-pole and double-pole thermocouples offer considerable advantages of accuracy and of giving a direct temperature reading at the contact zone (Marinescu et al. 2004; Batako et al. 2005).

Thin-film-grindable thermocouples are a special case of the single-wire and two-wire methods (Lierse 1998; Batako et al. 2005). An advantage of the thin-film method is an extremely small contact depth to resolve temperatures in a small area at the contact surface and the possibility to measure a temperature profile for every single test depending on the number of evaporated thermocouples in simultaneous use. Batako found that a thin but wide single-pole thermocouple greatly increases the probability of maintaining a continuous temperature signal throughout the passage through the grinding contact zone.

Temperature measurements using thin-film thermocouples are shown in Figure 11.12, in grinding of Al_2O_3 ceramic with a resin-bonded diamond grinding wheel (Lierse 1998). Obviously, the set grinding conditions have a significant influence on the generation of heat in the zone of contact. The heat penetration time is of major importance. In deep grinding with very small tangential feed speed, high temperatures are registered, whereas higher tangential feed speeds in pendulum grinding lead to a significant temperature reduction. As expected, the avoidance of coolant leads to higher temperatures compared to a use of mineral oil. But in any case, either for single or for two-wire methods, the major disadvantage is the high effort to carry out these measurements. Due to the necessity to install the thermocouple as close to the zone of contact as possible, it is always a technique where either grinding wheel or workpiece have to be specially prepared. Thus, all these methods are only used in fundamental research, an industrial use for monitoring is not possible due to the partial destruction of major components.

Besides these heat conduction-based methods, the second group of usable techniques is related to heat radiation. Infrared radiation techniques were used to investigate the temperature of grinding wheel and chips. By the use of a special infrared radiation pyrometer, with the radiation transmitted through optical fiber, it is even possible to measure the temperature of working grains of the grinding wheel just after cutting (Ueda et al. 1985).

Also the use of coolant was possible and could be evaluated. In any case, these radiation-based systems need a careful calibration, taking into account the properties of the material to be investigated, the optical fiber characteristics and the sensitivity of the detector cell. But again for most of the investigations a preparation of the workpiece was necessary as shown in Figure 11.11, bottom left.

A second heat radiation-based method uses thermography. For this type of measurement, the use of coolants is always a severe problem, because the initial radiation generated in the zone of contact is significantly reduced in the mist or direct flow of the coolant until it is detected in the camera. Thus, the major application of this technique was limited to dry machining. Brunner was able to use a high-speed video thermography system for OD grinding of steel to investigate the potential of dry or MQL grinding (Brunner 1998; Karpuschewski 2001).

11.3 Sensors for Monitoring the Grinding Wheel

11.3.1 Introduction

The grinding wheel state is of substantial importance for the quality of the grinding results. The wheel condition can be described by the characteristics of the grains. Wear can lead to flattening, breakage, and even pullout of whole grains. Moreover, the number of cutting edges and the ratio of active/passive grains are of importance. Also the bond of the grinding wheel is subject to wear. Due to its hardness and composition, the bond influences significantly the wear and variation of the grain distribution. Wheel loading, when it occurs, generates negative effects due to insufficient chip removal and coolant supply.

All these effects can be summarized as aspects of wheel topography, which changes during the wheel life between two dressing cycles. As a result, the diameter of the grinding wheel reduces with wear. In most cases, dressing cycles have to be carried out without any information about the actual wheel wear. Commonly, grinding wheels are dressed without reaching the end of acceptable wheel life in order to prevent workpiece damages, for example, workpiece burn. Figure 11.13 gives an overview of different geometrical quality features related to the redress life of grinding wheels. As a rule, the different types of wheel wear are divided into macroscopic and microscopic features. Many attempts have been made to describe the surface topography of a grinding wheel and to correlate the quantities to the result on the workpiece.

In Figure 11.14, suitable methods are introduced for dynamic measurement of the grinding wheel. Most of the systems are not able to detect all micro- and macrogeometrical quantities, but can only be used for special purposes.

11.3.2 Sensors for Macrogeometrical Quantities

The majority of sensors are capable of measuring macrogeometrical features. Any kind of mechanical contact of a sensor with the rotating grinding wheel causes serious problems, because the abrasives always tend to grind the material of the touching element. Only by realizing short touching pulses with small touching forces and by using a very hard tip material like tungsten carbide, it is possible to achieve satisfactory results. Another group of sensors for the measurement of grinding wheels is based on pneumatic systems. Although this method is, in principle, unable to detect microgeometrical features of a grinding wheel

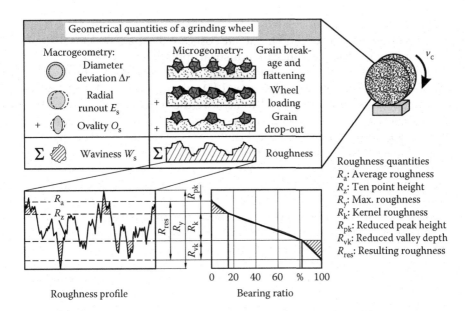

FIGURE 11.13
Geometrical quantities of a grinding wheel.

due to the nozzle diameter of 1 mm or more, the method is able to determine macrogeometry. A distinction should be made between systems that employ a compressed air supply or those that do not. The latter responds to airflow around the rotating grinding wheel. The results obtained reveal a dependence of the airflow on the distance of the sensor to the surface, on the circumferential speed and to some extent on the topography of the grinding wheel. The method with a compressed air supply is based on the nozzle-bounce plate principle, with the grinding wheel being the bounce plate. These systems are capable of

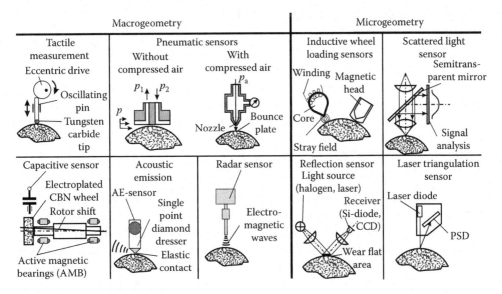

FIGURE 11.14
Sensors for grinding wheel topography measurement.

measuring the distance changes related to radial wear with a resolution of 0.2 μm. This feature, comparatively easy setup, and moderate costs are the main reasons that pneumatic sensors have already found acceptance in industrial application.

Another possibility to register the macrogeometry of a grinding wheel was reported by Westkämper and Klyk (1993). In high-speed ID grinding with CBN wheels, a spindle with active magnetic bearings was used to achieve the necessary circumferential speed of 200 m/s with small diameter wheels. These spindles have the opportunity to shift the rotor from rotation around the geometrical center axis to the main axis of inertia to compensate any unbalance. Especially if electroplated CBN wheels are used without the possibility of dressing, it is necessary to use balancing planes. Capacitive sensors have shown the best performance to measure runout of the abrasive layer of these grinding wheels at very high circumferential speed.

Also acoustic emission can be used to determine the macrogeometry of the grinding wheel. Oliveira proposed a system consisting of a single-point diamond dresser equipped with an AE sensor to detect the position of the grinding wheel surface (Oliveira et al. 1994). AE signals can be obtained without physical contact of the dresser and the wheel due to turbulence. In total, three different contact conditions can be distinguished including noncontact, elastic contact, and brittle contact.

Another principle used to determine radial wheel wear is based on a miniature radar sensor (Westkämper and Hoffmeister 1997). The radar technique is well known from speed as well as traffic control with a maximum accuracy in the centimeter range. The sensor used for grinding works on an interferometry principle. With an emitting frequency of 94 GHz and a wavelength of $\lambda = 3.18$ mm, this sensor has a measuring range of 1 mm and a resolution of 1 μm. Main advantages are the robustness against dust, mist, or coolant particles and the possibility to measure on any solid surface. The sensor was used in surface grinding of turbine blades with continuous dressing (CD). A control loop was established to detect and control the radial wear of the grinding wheel taking into account the infeed of the dressing wheel.

11.3.3 Sensors for Microgeometrical Quantities

The loading of a grinding wheel with conductive metallic particles is a special type of microgeometrical modification that can be detected using sensors based on inductive phenomena. The sensor consists of either a high permeability core or a winding. It is positioned a short distance from the surface. The metallic particles generate a change of impedance, which can be further processed to determine the state of wheel loading. Also a conventional magnetic tape recorder head may be used to detect the presence and relative size of ferrous particles in the surface layer of a grinding wheel. Due to the fact that only loading of particular metallic materials can be detected reliably, these sensors have not reached practical application.

The limitations of all techniques introduced so far, drive attention toward optical methods. Optical methods are promising because of their frequency range and independence of the surface material. A scattered light sensor was used to determine the reflected light from the grinding wheel surface by using CCD arrays. Gotou and Touge (1996) adopted the principle with a silicon diode as a receiver and using a laser source. Grinding wheels in wet grinding at 30 m/s could be measured. It was stated that the wear flat areas are registered by the output signal and that these areas change during grinding.

The most promising optical method is based on laser triangulation. Figure 11.15 shows the basic elements, which are a laser diode with 40 mW continuous wave power and a

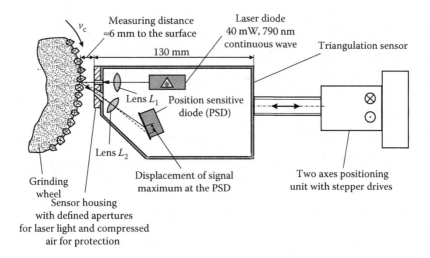

FIGURE 11.15
Measurement principle of a laser triangulation system. (Adapted from Tönshoff, H. K., Karpuschewski, B., Werner, F., 1993, Fast sensor systems for the diagnosis of grinding wheel and workpiece, *5th International Grinding Conference*, October 26–28, 1993, Cincinnati, Ohio MR93-369.)

position-sensitive detector (PSD) with an amplifier and two lenses (Tönshoff et al. 1993). The laser diode emits monochromatic laser light of 790 nm wavelength focused by lens L_1 on the grinding wheel surface. The scattered reflected light is collected by lens L_2 and focused on the PSD. If the distance to the sensor changes, the position of the reflected and focused light on the PSD also changes. The sensor is mounted relative to a two axes stepper drive unit moved in the normal direction to the grinding wheel surface and in the axial direction to make measurements of different traces on the grinding wheel circumference. This sensor system was intensively tested in a laboratory environment and in industrial application. For the determination of macrogeometrical quantities such as radial runout, no practical limitations exist. The maximum surface speed may even exceed 300 m/s.

Figure 11.16 shows the result of an investigation in outer diameter plunge grinding of ball-bearing steel with a corundum grinding wheel. The change of radial runout was measured as a function of material removed at 30 m/s at three different removal rates. For the smallest material removal rate, no change is detectable from the initial value after dressing. But for increasing material removal rates of $Q'_w = 1.0$ mm^3/mm s and 2.0 mm^3/mm s, the radial runout rises after a specific material removed. In the latter cases, the increasing radial runout leads to chatter vibrations with visible marks on the workpiece surface. Obviously, the system is capable of detecting significant macrogeometric changes due to wear of the grinding wheel.

For microgeometrical measurements the investigations revealed that the maximum speed of the grinding wheel should not exceed 20 m/s, based on hardware and software limitations. This means that for most applications the grinding wheel has to be slowed for the measurement. This major drawback limits the practical field of application. For conventional abrasives with relatively short dressing intervals, the use of this type of microgeometrical monitoring is not economic. The most interesting application for this sensor is seen in the monitoring of superabrasives, especially CBN grinding wheels. This sensor system was intensively tested during profile grinding of gears with an electroplated CBN grinding wheel (Regent 1999). The measurement was performed on the involute profile

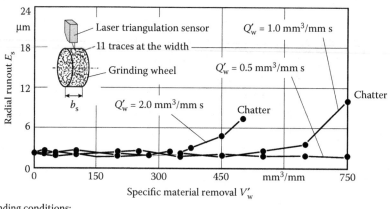

Grinding conditions:
Grinding wheel: EKW 80K5 V62   Multipoint diamond dresser
                     $q = 60$ Workpiece: 100 Cr6
                     $v_c = 30$ m/s $= v_{mea}$ ball bearing steel
Spark out time: 10 s

FIGURE 11.16
Optical macrogeometrical grinding wheel topography measurement.

of the grinding wheel making 10 traces, while changing the workpiece at a measurement speed of $v_{mea} = 10$ m/s.

Figure 11.17 (left) shows the setup for the investigation. On the right-hand side, an experimental result is presented. The measured quantity is the reduced peak height R_{pk} deduced from the bearing ratio curve, which can be used to describe the change of the grinding wheel topography at the grains tips.

As shown, the change in reduced peak height can be clearly correlated with the occurrence of grinding burn, which was confirmed by nital etching and succeeding metallographical and x-ray inspection. Although this result is very promising, some problems have to be taken into consideration. Measurements as well as simulations have revealed that it is not possible to correlate the sensor roughness results definitely to a specific wear pattern. In real applications, there are always several types of wear, for example, grain flattening and loading, that cannot be separated by the measuring quantities.

The examples shown for grinding wheel sensors reveal that the majority of systems are related to macrogeometrical features. Many attempts have been made to establish optical systems for the measurement of microgeometrical quantities. The overall limitation for these techniques will always be the hostile conditions in the working space of a grinding machine with coolant and process residues in direct contact with the object to be measured. In many cases, it is thus preferable to directly measure the manufactured workpiece itself.

11.4 Sensors for Monitoring the Workpiece

11.4.1 Introduction

Two essential quality aspects determine the result of a grinding process on the workpiece. On the one hand, the geometrical quality demands have to be fulfilled. These are

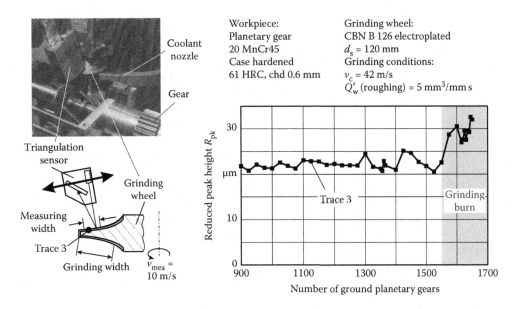

FIGURE 11.17
Grinding burn identification using a laser triangulation sensor. (Adapted from Regent, C., 1999, Prozeßsicherheit beim Schleifen. Dissertation, Universität Hannover.)

dimension, shape, and waviness as essential macrogeometrical quantities. The roughness condition is the main microgeometrical quality. But increasing attention is also paid to the surface integrity state of a ground workpiece due to its significant influence on the functional behavior. The physical properties are characterized by the change in hardness and residual stresses on the surface and in subsurface layers, by changes in the structure and the possible occurrence of cracks. All geometrical quantities can be determined by using laboratory reference measuring devices. For macrogeometrical properties, any kind of contacting systems are used, for example, 3D coordinate measuring machines, contour stylus instruments or gauges. Roughness measurement is usually performed with stylus instruments giving standardized values, but also optical systems are applied in some cases.

11.4.2 Contact-Based Workpiece Sensors for Macrogeometry

The determination of macrogeometrical properties of workpieces during manufacturing is the most common application of sensors in abrasive processes, especially grinding. For decades, contacting sensors have been in use to determine the dimensional change of workpieces during manufacturing. A variety of in-process gauges for most kinds of operation is available. In ID or OD grinding, measuring systems can either be comparator or be absolute measuring heads with the capability of automatic adjustment to different part diameters. The contacting tips are usually made of tungsten carbide combining the advantages of wear resistance, moderate costs, and sufficient frictional behavior. If constant access to the dimension of interest is possible during grinding, these gauges are often used as a signal feedback for adaptive control systems.

Figure 11.18 shows sizes maintained for successive parts ground on a CNC cylindrical grinding machine employing different control strategies (Liverton et al. 1993; Rowe 2014). Grinding without redressing after each part and without application of size gauging

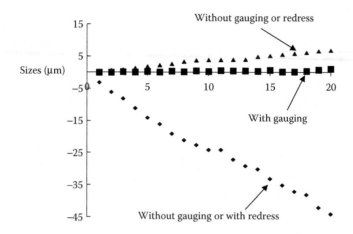

FIGURE 11.18
Sizes of successive parts in cylindrical plunge grinding: Wheel A465K5V30W at 33 m/s: Material hardened caststeel 19 mm diameter by 50 mm width. (Adapted from Liverton, J. et al., 1993, Adaptive control of cylindrical grinding—From development to commercialization, *Proceedings of Society of Manufacturing Engineers*, Dearborn, Paper MR93-370.)

equipment, it is seen that sizes steadily increase as the grinding wheel wears. Introducing a redress operation after each part causes sizes to progressively reduce due to wear of the dressing tool. With size gauging and automatic error compensation, sizes are accurately controlled to approximately 1 μm. The use of size gauging allows the grinding wheel to be retracted when final size is achieved for each part.

Grinding machines with CNC have over recent decades incorporated elements of intelligent control based on in-process monitoring. For example, the CNC system above was further extended to maintain high material removal rates by introducing adaptive feed-rate control and adaptive dwell control (Allanson et al. 1997). Adaptive feed-rate control automatically increases feed rate, if the grinding power drops during the redress cycle. In a stable grinding process, the grinding wheel initially wears more rapidly just after the redress operation. There is either an increase in power or more usually a decrease in power. Either way, corrective action is required. Maintaining a constant power level allows cycle time to be reduced.

Adaptive dwell control involves varying the dwell or spark-out period to ensure sufficient size accuracy is maintained with reduced cycle times. In a grinding cycle, there is a lag between programmed feed position and actual feed position due to machine deflections. The time lag can be identified by measuring the system time constant during the grinding cycle. A sufficient dwell period can then be introduced based on the system time constant to allow machine deflections to relax so that specified part size is approached. If the dwell is too short, size tolerances will not be achieved. If the dwell is too long, cycle time will be long and machining cost is increased. There is another effect of excessive dwell time and that is additional wear of the grinding wheel which reduces redress life.

Size control results with adaptive and nonadaptive systems are shown in Figure 11.19 (Rowe et al. 1994). The best accuracy is achieved with a cycle time of 60 s and a nonadaptive system. Cycle time can be reduced using an increased feed rate and a reduced cycle time of 30 s. However, it can be seen that the specified size has not been reduced and size variations are increased. Using adaptive feed-rate and dwell control, it was possible to achieve the size tolerance but with mean cycle time reduced to 37 s. Increasing the target power from 2.5 to 5 kW allowed mean cycle time to be further reduced to 22 s, although the error

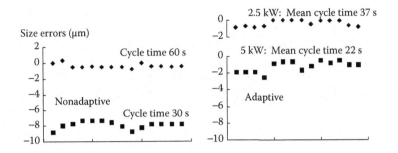

FIGURE 11.19
Size control using adaptive feed-rate and adaptive dwell control. A465K5V30W wheel. EN9 material. Plunge grinding width 50 mm. (Adapted from Rowe, W. B. et al., 1994, Keynote Paper, *Annals of the CIRP*, 43(2), 521–532.)

band was increased from approximately 1 to 2 µm. However, this accuracy was a considerable improvement on the nonadaptive system with a 30-s cycle time. The technique is further described by Rowe (2014).

The conventional technique for measuring round parts rotating around their rotational axis can be regarded as state of the art. The majority of automatically operating grinding machines are equipped with these systems. In a survey of contacting sensors for workpiece macrogeometry in Figure 11.20 (left), a more complex measurement setup is shown. Due to the development of new drives and control systems for grinding machines, a continuous path controlled grinding of crankshafts has now become possible (Tönshoff et al. 1998).

The crankshaft, Figure 11.20 (left), is clamped only once in the main axis of the journals. For machining the pins, the grinding wheel moves back and forth during rotation of the crankshaft around the main axis to generate a cylindrical surface on the pin. An in-process measurement device for the pin diameter has to follow this movement. A first prototype system was installed on a crankshaft grinding machine. The gauge was mounted to the grinding wheel head and moved back and forth together with the grinding wheel. A swivel joint allows height balancing.

The detection of waviness on the circumference of rotationally symmetric parts in grinding is more complex due to the demand for a significantly higher scanning frequency. Foth has developed a system with three contacting pins at nonconstant distances to detect the

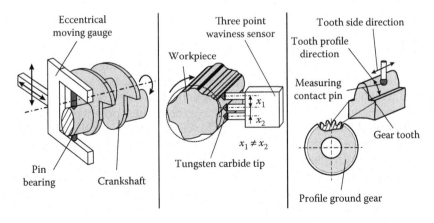

FIGURE 11.20
Contact sensor systems for workpiece macrogeometry.

development of waviness on workpieces during grinding as a result of, for example, regenerative chatter (Figure 11.20, middle; Foth 1989). Only by using this setup, it is possible to identify the real workpiece shape, taking into account the vibration of the workpiece center during rotation.

The last example of contact-based macrogeometry measurement in a machine tool is related to gear grinding, (Figure 11.20, right). Especially for manufacturing of small batch sizes or single components of high value, it is essential to fulfill the "first part good part" philosophy. For these reasons, several gear grinding machine tool builders decided to integrate an intelligent measuring head in their machines to be able to measure the characteristic quantities of a gear including, for example, flank modification, pitch, or root fillet. Usually a measurement is done after rough grinding, before the grinding wheel is changed or redressed for the finish operation. Sometimes also the initial state before grinding is checked to compensate large deviations resulting from distortions due to heat treatment. Of course the measurement can only be done if the manufacturing process is interrupted. But still the main advantage is a significant time saving. Any removal of the part from the grinding machine tool for checking on an additional gear measuring machine will take a longer time. Also the problem of precision losses due to rechucking is not valid, because the workpiece is rough machined, measured, and finished in the same setup. These arguments are generally true for any kind of high value parts with small batch sizes and complex grinding operations. Thus, it is not surprising that also in the field of aircraft engine manufacture, new radial grinding machines are equipped with the same kind of touch probe system.

11.4.3 Contact-Based Workpiece Sensors for Microgeometry

The determination of microgeometrical quantities on a moving workpiece by using contacting sensor systems is a challenging task. A permanent contact of any stylus to the surface is not possible, because the dynamic demands are much too high. Only intermittent contacts can be used to generate a signal, which should be proportional to the roughness. Saljé (1979) introduced a sensor based on a damped mass spring element. The surface of the fast moving workpiece stimulates self-oscillations of the sensing element, which are correlated to the roughness. Also rotating roughness sensors for OD grinding have been tested. But in the end due to limitations including wear and speed, the idea of contacting the surface for roughness measurement has not led to industrial success (Karpuschewski 2001).

11.4.4 Contact-Based Workpiece Sensors for Surface Integrity

The best way to investigate the influence of any cutting or grinding process on the physical properties of the machined workpiece would be to directly measure on the generated surface. But until now only very few sensors are available to meet this demand. In the following, two techniques will be explained, which have the highest potential for this purpose.

The principle of eddy-current measurement for crack detection is based on the fact that cracks at the workpiece surface will disturb the eddy-current lines, which are in the measuring area of a coil with alternating-current excitation. The feedback to the exciting field leads to changes of the impedance for coils with only one winding to a change of the signal voltage for sensors with two separated primary and secondary windings. All kinds of conductive materials can be tested. The penetration depth is determined by the excitation frequency. Conductivity as well as permeability of the workpiece can be investigated. In grinding, an eddy-current sensor was introduced to monitor the occurrence of cracks.

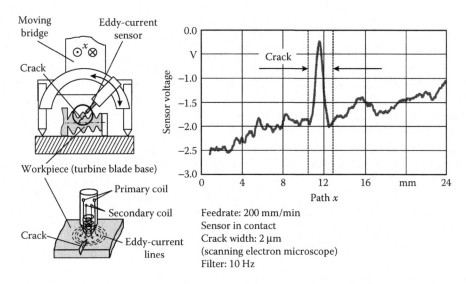

FIGURE 11.21
Eddy current crack detection after surface grinding of turbine blades. (Adapted from Lange, D., 1996, Sensoren zur Prozeßüberwachung und Qualitätsprüfung, 8. *Internationales Braunschweiger Feinbearbeitungskolloquium* 24–26. April 1996, p. 16/1–16/19.)

In Figure 11.21 (left), the setup for this eddy-current-based measurement is shown, used for the determination of cracks generated during profile surface grinding of turbine blade roots (Lange 1996). Figure 11.21 (right) shows the result of such a measurement. The crack was investigated afterward with the aid of a scanning electron microscope and had a width of 2 μm.

The eddy-current sensor could clearly determine this crack with a contact measurement. This size has to be regarded as the minimum resolution of the sensor. In any case, the sensor must be positioned in a perpendicular direction to the surface, because any tilting reduces the sensitivity. Thus, an additional shift option was implemented in the moving bridge. The results prove the suitability of eddy-current sensors for crack detection on turbine blade materials. Although the measurement speed was smaller than the grinding table speed, a check in the grinding machine may still be acceptable because of the high safety demand on these workpieces.

The second possibility to detect changes of the physical properties on machined surfaces of ferrous materials is based on micromagnetic techniques. Residual stresses, hardness values, and the structure in subsurface layers influence the magnetic domains of ferromagnetic materials. Any magnetization change can be measured with the so-called Barkhausen noise. The existence of compressive stress in ferromagnetic materials reduces the intensity of the Barkhausen noise, whereas tensile stresses increase the signal (Karpuschewski 1995). In addition to these stress-sensitive properties, also the hardness and structure state of the workpiece influence the Barkhausen noise. To separate the different material characteristics of a ground workpiece different quantities deduced from the Barkhausen noise signal must be taken into consideration. The most important quantities deduced from the signal are the maximum amplitude of the Barkhausen noise M_{max} and the coercivity H_{cM}. In any case, the measurement time is very short and amounts to only a few seconds, which is one of the major advantages of this technique. This so-called two-parameter micromagnetic setup was further improved by adding modules for the measurement of the incremental permeability, the harmonics of the exciting field and eddy current (Regent 1999). The major aim of this multiparameter system was to further separate the influence of the initial material properties from the changes due to machining operations.

A detailed investigation of the potential of the two-parameter micromagnetic approach was employed to characterize surface integrity states of workpieces with different heat treatments (Karpuschewski 1995). Afterward this technique was transferred to practical application. Figure 11.22 shows the results of a large industrial test on planetary gears ground with electroplated CBN grinding wheels (Regent 1999). It is essential to adapt the sensor geometry to the geometrical situation on the workpiece. In this case, the critical area to be tested was on the tooth flanks, thus the sensor had to be adapted to the module of the gear, left. On the right side, the results over the lifetime of an electroplated wheel are shown. The Barkhausen noise amplitude is corrected to consider slight changes of the excitation field. It can be seen that all gears where grinding burn was identified by nital etching were also recognized with the micromagnetic setup. But in addition gears with high $M_{max,corr.}$ values appear, which do not show any damage in nital etching. A possible explanation for this deviation is the different penetration depth of the methods. Nital etching gives only information about the very top layer of the workpiece. Any subsurface damage cannot be registered. Micromagnetic measurements can also reveal that damage depending on the frequency range.

The major challenge is to exactly identify the grinding burn limit. It has to be mentioned further that the measuring time required to scan all flanks of one gear is significantly higher than the grinding time. With intelligent strategies or increased number of sensors in parallel use this time can be shortened for a suitable random testing. A total automated measurement is possible. Thus, the very inaccurate and environmentally hazardous etching can be replaced by this technology (Karpuschewski 2001). A user has to be careful using the Barkhausen noise technique in that more severe burn cause a material phase

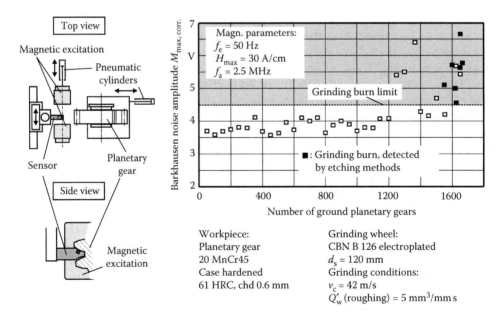

FIGURE 11.22
Micromagnetic surface integrity characterization of ground planetary gears. (Adapted from Regent, C., 1999, Prozeßsicherheit beim Schleifen. Dissertation, Universität Hannover.)

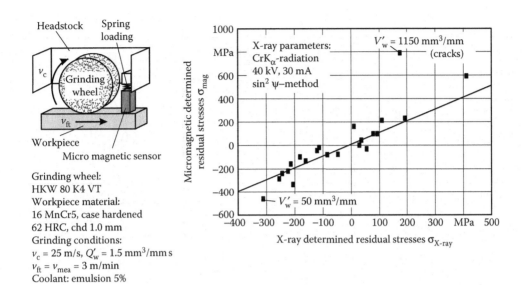

FIGURE 11.23
Micromagnetic in-process measurement of surface integrity during grinding. (Adapted from Regent, C., 1999, Prozeßsicherheit beim Schleifen. Dissertation, Universität Hannover.)

transformation reducing the Barkhausen noise level. Thus, when burn is encountered at first, the Barkhausen noise increases as shown in Figure 11.22. However, with higher temperatures and phase transformation, the Barkhausen noise level reduces again which could lead to the false conclusion that burn did not occur.

In the laboratory, the first tests for in-process measurements of surface integrity changes based on this micromagnetic sensing were conducted for outer diameter and surface grinding (Tönshoff et al. 1998; Regent 1999). In Figure 11.23, first results of this approach during surface grinding of steel are presented. The sensor with integrated excitation is moving along the surface behind the grinding wheel at the chosen table speed of 8 m/min. Permanent contact is assured by spring loading. The x-ray measurement is performed on one point of the ground surface, the micromagnetic result represents the average over the whole workpiece length. The deviations in the area of compressive residual stresses and low-tensile residual stresses are less than 100 MPa, and only in areas of significant damage with tensile stresses higher than 200 MPa, the deviations increase. This can be explained by the occurrence of cracks after grinding due to the high thermal load on the workpiece.

Although further investigations on the wear resistance of the sensor head, on long-term coolant influence, maximum workpiece speed, geometrical restrictions, and other parameters have to be conducted, this sensor seems to offer the possibility of in-process workpiece surface integrity measurement for the first time.

11.4.5 Noncontact-Based Workpiece Sensors

All the mentioned restrictions of contacting sensor systems on the workpiece surface gave a significant push to develop noncontact sensors. As for grinding wheels, again optical systems seem to have a high potential. In Figure 11.24, different optical systems as well as two other noncontacting sensor principles are introduced.

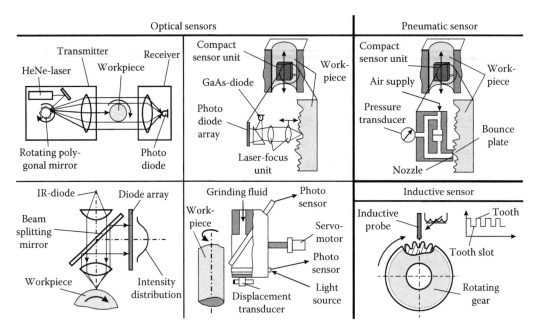

FIGURE 11.24
Noncontact sensor systems for workpiece quality characterization.

As a very fast optical system to measure macrogeometrical quantities, a laser scanner is shown. The scanner transmitter contains primarily the beam emitting HeNe laser, a rotating polygonal mirror and a collimating lens for paralleling the diffused laser beam. The setup of the scanner receiver contains a collective lens and a photo diode. The electronic evaluation unit counts the time, and the photo diode is covered by the shadow of the object. The diameter is a function of the speed of the polygonal mirror and the time the laser beam does not reach the covered photo diode. Conicity can be evaluated by an axial shifting of the workpiece. In principle, this optical measurement cannot be performed during the application of coolant. For a detailed workpiece characterization, a setup with a laser scanner outside of the working space of the grinding machine is preferred. A flexible measurement cell incorporating a laser scanner was introduced for the determination of macrogeometrical properties (Tönshoff et al. 1990). The system was able to automatically measure the desired quantities within the grinding cycle time, and the information was fed back to the grinding machine control unit. Noncontact laser diameter measuring systems are now commonly available commercially.

For the determination of micro- and macrogeometrical quantities, a different optical system has to be applied. The basis of a scattered light sensor for the measurement of both roughness and waviness is the angular deflection of nearly normal incident rays. The setup of a scattered light sensor is shown in Figure 11.24 (bottom left). A beam-splitting mirror guides the reflected light to an array of diodes. A commercially available system was introduced in the eighties (Brodtmann et al. 1984) and used in a wide range of tests. The optical roughness measurement quantity of this system is called scattering value S_N and deduced from the intensity distribution. In different tests, the scattered light sensor was directly mounted in the working space of the grinding machine to measure the workpiece roughness. A compressed air barrier protected the optical system. In all investigations, it was tried to establish a correlation between optical and stylus roughness measurements.

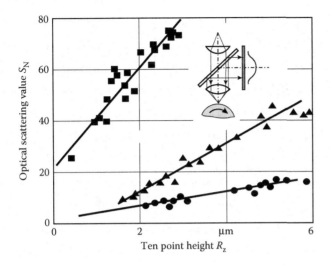

FIGURE 11.25
Different correlation curves for an optical scattered light sensor. (Adapted from König, W., Klumpen, T., 1993, Angepaßte Überwachungsstrategien und Sensorkonzepte—der Schlüssel für eine hohe Prozeßsicherheit, 7. *Internationales Braunschweiger Feinbearbeitungskolloquium*, 2–4 März 1993, p. 11.01–11.23; von See, M., 1989, Optimierung von Honprozessen auf der Basis von Modellversuchen und betrachtungen. Dissertation, TU Braunschweig.)

It was found to be possible to obtain a close relationship, while grinding or honing with constant process parameters (Figure 11.25; von See 1989; König and Klumpen 1993). This restriction is indispensable, because a change of input variables like dressing conditions or tool specification may lead to workpieces with the same stylus roughness values R_a or R_z but different optical scattering values S_N. If a quantitative roughness characterization referring to stylus values is demanded, a time-consuming calibration will be always necessary. As shown in Figure 11.25, the measuring direction has to be clearly defined to achieve the desired correlation. A second limitation is seen in the sensitivity of the system. The scattered light sensor is able to determine differences in high-quality surfaces, but for roughness states of 10-point height $R_z > 5.0$ μm, the scattering value S_N is reaching its saturation with a decreasing accuracy already starting at $R_z = 3.0$ μm [von See 1989]. Thus, some relevant grinding or honing operations cannot be supervised by this sensor system.

A different optical sensor is based on a laser diode (Figure 11.26, top middle; Westkämper and Kappmeyer 1992). The sensor is equipped with a gallium-arsenide diode, which is commonly used in a CD player. With a lens system, the beam is focused on the surface, and the reflected light is registered on an array of four photo diodes. This system can be used as an autofocus system, with the signal from the four diodes, the focus lens is moved until the best position for minimum diameter is reached. The correlation of the obtained optical average roughness $R_{a,opt}$ to the stylus reference measurement is shown in Figure 11.26 (left).

An almost linear dependence of the two different roughness quantities could be found. But this is much too slow to use the system for any in-process measurement. By using the focus-error signal of the four diodes without moving the lens, it is possible to increase the measurement speed significantly.

Another optical approach for in-process roughness, measurement is based on the use of optical fiber sensors (Inasaki 1985a). The workpiece surface is illuminated through fiber

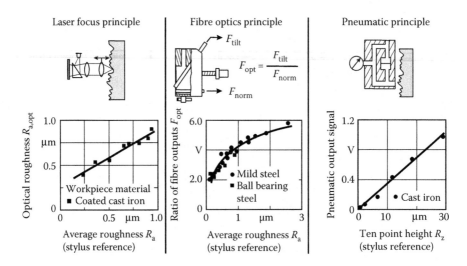

FIGURE 11.26
Correlation curves for different workpiece roughness sensors. (Adapted from Westkämper, E., Kappmeyer, G., 1992, Prozeßintegrierte Qualitätsprüfung beim Honen zylindrischer Bauteile, *Seminar des Sonderforschungsbereiches 326*, Universität Hannover und TU Braunschweig, pp. 41–58.)

optics and the intensity of the reflected light is detected and evaluated (Figure 11.26, middle). The latter setup was chosen to increase the sensitivity of the sensor system. The photo sensor in normal direction will register less intensity, whereas the inclined photo sensor will detect more intensity with larger light scattering due to increased roughness. The ratio of both photo sensors is related to roughness changes. A second advantage of the setup with two fiber optics, despite the increased sensitivity is the achieved independence of the workpiece material. Coolant flows around the whole sensor head to make measurement possible during grinding. It is essential to keep the coolant as clean as possible during operation, because the reflection conditions are definitely influenced by the filtering state of the fluid. This is the major drawback of the sensor system, because the coolant quality is not likely to be stable in production. Besides these mentioned systems, some other optical techniques for on-line measurement of surface topography have been proposed such as speckle patterns. Although the measurement speed may allow installing these systems in the production line surrounding, a use of speckle patterns for application as a sensor in the machine tool working space is not realistic.

In summary due to all the problems related to coolant supply, it must be stated that these conditions do not allow using optical systems during grinding as reliable and robust industrial sensors. Only optical sensor applications measuring in interruptions of coolant supply either in the working space of the machine tool or in the direct surrounding have gained importance in industrial production.

In addition to optical sensors, two other principles are used for noncontact workpiece characterization. A pneumatic sensor as shown in Figure 11.26 (top right) was designed and used for the measurement of cylinder surfaces. The measurement is based on the already-mentioned nozzle-bounce plate principle. A correlation to stylus measurements is possible (Figure 11.26, right). Main advantages of this system are the small size, the robustness against impurities and coolant and the fact that an area and not a trace is evaluated. So, in principle, any movement of the sensor during measurement is not necessary.

The last system to be introduced as a noncontact workpiece sensor is based on an inductive sensor. The sensor is used in gear grinding machines to identify the exact position of tooth and tooth slot at the circumference of the premachined and usually heat-treated gear (Figure 11.24, bottom right). The gear rotates at high speed and the signal obtained is evaluated in the control unit of the grinding machine. This signal is used to index the gear in relation to the grinding wheel to define its precise position (Karpuschewski 2001).

11.5 Sensors for Peripheral Systems

11.5.1 Introduction

Primary motion between tool and workpiece characterizes the grinding process, but also supporting processes and systems are of major importance. In this section, monitoring of the conditioning process and the coolant supply will be discussed.

11.5.2 Sensors for Monitoring of the Conditioning Process

The condition of the grinding wheel is a very decisive factor for satisfactory grinding results. Thus, the grinding wheel has to be prepared for grinding by using a suitable conditioning technology. The major problem in any conditioning operation is the possible difference between nominal and real conditioning infeed. There are four main reasons for these deviations. The unknown radial grinding wheel wear after removal of a workpiece material volume must be regarded as a significant factor. Also the changing relative position of grinding wheel and conditioning tool due to thermal expansion of machine components is relevant. As a third reason, infeed errors related to friction of the guideways and control accuracy have to be considered, although their influence is declining in modern grinding machines. The last reason to mention is the wear of the conditioning tool, which is of course dependent on the individual type of tool. The first wear effects for rotating dressers may be noticeable only after regular use for several weeks.

Due to the immense importance of the grinding wheel topography, the monitoring of the conditioning operation has been the subject for research over many decades. Already in the early eighties attempts were made to use an acoustic emission-based system for the monitoring of the dressing operation. At that time, the work was concentrated on dressing of conventional grinding wheels with a static single-point diamond dresser. It was possible to detect first contact of the dresser and the grinding wheel, and the AE intensity could be used to determine the real dressing infeed in dependence of dressing feed rate and grinding wheel speed. The dressing feed speed could be identified by the AE signal (Inasaki 1985b). In addition, it was stated that the AE signal reacts significantly faster to the first contact of the dressing tool and grinding wheel compared to monitoring by means of spindle power.

The limitation to straight cylindrical profiles was overcome by Meyen, who developed a system capable of detecting dressing errors on any complex grinding wheel profile (Figure 11.27; Meyen 1991). The strategy comprises the determination of a sliding average value with static and dynamic thresholds for every single dressing stroke. The different geometry elements are identified, and the currently measured AE signal is compared to the

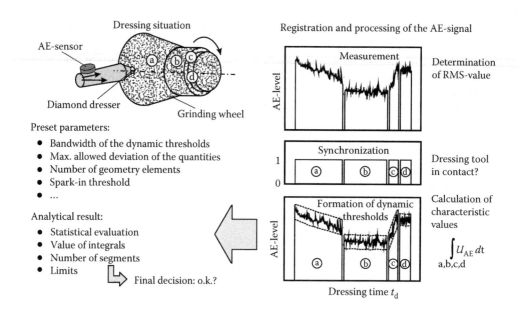

FIGURE 11.27
Dressing diagnosis for random grinding wheel profiles with AE signals. (Adapted from Meyen, H. P., 1991, Acoustic Emission (AE)—Mikroseismik im Schleifprozeß. Dissertation, RWTH Aachen.)

reference curve, which has to be defined in advance. With the calculation of further statistical quantities such as standard deviation or mean signal inclination, it is possible to identify the typical dressing errors in case of exceeding the thresholds.

As a consequent, next step AE systems were tested for conditioning operations of superabrasives such as CBN (e.g., Heuer 1992; Wakuda et al. 1993). The high hardness and wear resistance of these grinding wheels require a different conditioning strategy and monitoring accuracy compared to conventional abrasives.

The conditioning intervals due to the superior wear resistance can amount up to several hours. The dressing infeed should be limited to a range between 0.5 and 5 μm instead of 20–100 μm for conventional wheels in order to save wheel costs. Especially for vitreous-bonded CBN grinding wheels, it was proposed to use very small dressing infeeds more frequently in order to avoid an additional sharpening. This strategy known as "touch dressing" revealed the strong demand to establish reliable contact detection and a monitoring system for dressing of superabrasives. In most of the cases, rotating dressing tools are used. The schematic setup of a conditioning system with a rotary cup wheel, which is often used on internal grinding machines, is shown in Figure 11.28.

The conditioning cycle consists of four stages: fast approach, contact detection, defined infeed, and new initiation. Besides AE techniques, other methods have been tested. Heuer (1992) additionally investigated the possibility of using either the required power of the dressing tool spindle or a piezoelectric force measurement for monitoring. The latter technique was available, because a piezoelectric actuator was installed as a high-precision positioning system for the infeed of the dressing tool.

A further technique for contact detection was introduced (Tönshoff et al. 1995). The measurement of the rotational speed change of the high-frequency dressing spindle, which gives a maximum number of revolutions of 60,000 min⁻¹, was used to determine not only

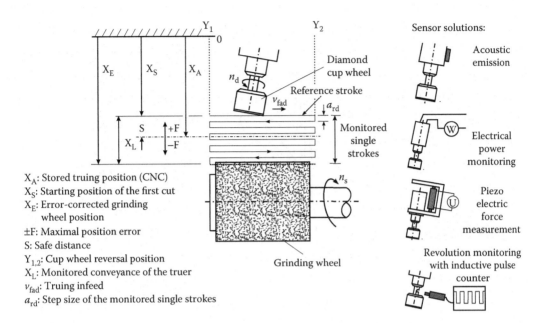

X$_A$: Stored truing position (CNC)
X$_S$: Starting position of the first cut
X$_E$: Error-corrected grinding
   wheel position
±F: Maximal position error
S: Safe distance
Y$_{1,2}$: Cup wheel reversal position
X$_L$: Monitored conveyance of the truer
v_{fad}: Truing infeed
a_{rd}: Step size of the monitored single strokes

FIGURE 11.28
Dressing monitoring with rotating diamond tools. (Adapted from Heuer, W., 1992, Außenrundschleifen mit kleinen keramisch gebundenen CBN-Schleifscheiben. Dissertation, Universität Hannover.)

the first contact, but also the whole dressing process. After contact detection of any of the mentioned systems, the conditioning program is continued until the desired number of strokes and infeed is reached. Depending on the type of system, it is possible to monitor the course of the signal over the whole width of the grinding wheel.

The use of AE sensors for contact detection of the conditioning dressing operation can be regarded as state of the art. Many different systems are available. New grinding machine tools with self-rotating conditioning tools are usually equipped with an AE system already in the delivery state.

11.5.3 Sensors for Coolant Supply Monitoring

In almost all grinding processes, coolant is used to reduce heat and to provide sufficient lubrication. These are the main functions of any coolant supply. Further requirements are the removal of chips and process residues from the workspace of the machine tool, the protection of surfaces and human compatibility. Modern coolant compositions also try to fulfill the contradictory demands of long-term stability and biological recycling capability.

With the wider use of superabrasives such as CBN and the possibility of high-speed grinding and high-efficiency deep grinding, more attention has been paid to coolant supply. Coolant pressure and flow rate measured with a simple flowmeter in the coolant supply tube before the nozzle are often specified. Different authors have also worked on the influence of different nozzle designs (e.g., Heuer 1992; Rowe 2014). In most cases, the influence of different supply options including conventional flood nozzles, shoe, spot jet, or spray nozzles or even internal supply through the grinding wheel is described by using the previously mentioned process quantities such as forces or temperature (Heinzel 1999).

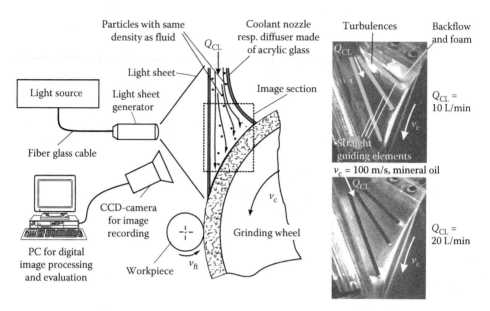

FIGURE 11.29
Flow behavior monitored by means of particle image velocimetry. (Adapted from Heinzel, C., 1999, Methoden zur Untersuchung und Optimierung der Kühlschmierung beim Schleifen. Dissertation, Universität Bremen.)

Environmental aspects of manufacturing also get significantly more attention these days leading to detailed investigation of coolant supply and the possibility to reduce or avoid coolants in grinding completely (Karpuschewski et al. 1997). Brinksmeier and Heinzel systematically investigated coolant-related influences in order to optimize relevant parameters and designs. A special flow visualization technique was used for the development of a suitable shoe nozzle design (Figure 11.29; Brinksmeier et al. 1999; Heinzel 1999). Tracer particles with almost the same density were added to the transparent fluid. Parts of the nozzle were made from acrylic glass and a CCD camera recorded the flow images perpendicular to the light sheet plane. Although only a qualitative result is available, this technique offers the possibility to systematically study and improve the design of coolant nozzles. As an example on the right side of Figure 11.29, the flow behavior of a nozzle with straight guiding elements is shown at two different flow rates.

The coolant was mineral oil and the grinding wheel rotated at a speed of 100 m/s. For the smaller flow rate of 10 L/min, inhomogeneous flow behavior was observed. Turbulence, backflow, and foam between top and center guiding elements and at the entry side of the grinding wheel were visible. A doubling of the flow rate led to steady flow behavior.

Besides this use of an optical monitoring method to optimize the design of coolant nozzles, a special sensor installation for pressure and force investigations was introduced (Heinzel 1999). The force measurement was achieved using a piezoelectric dynamometer. During grinding only the total normal force can be registered by this instrument. The idea is to separate the normal force component used for cutting, friction, and deformation from the component that is resulting for the buildup of hydrostatic pressure between the grinding wheel and workpiece and because of the impact of the coolant flow on the surfaces. For this purpose, an additional pressure sensor was integrated into the workpiece carrier. This made it possible to measure the pressure change over a grinding path through a

Grinding wheel: GRY B181 L 200 G
d_s = 400 mm, b_s = 7 mm, v_{ft} = 0.8 m/min,
v_c = 140 m/s
Coolant: emulsion 4%, Q_{cl} = 60 L/min, one nozzle

Effective width of pressure zone:
$\Delta y_{eff.}$ = 3.32 mm
Workpiece: 100Cr6, 62 HRC,
Length: 100 mm, width: 10 mm

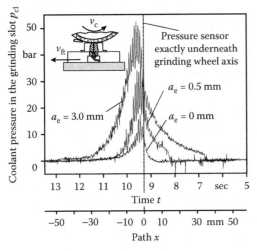

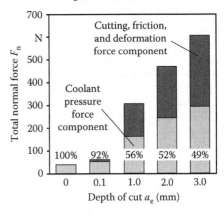

FIGURE 11.30

Coolant supply monitoring with pressure sensor and dynamometer. (Adapted from Heinzel, C., 1999, Methoden zur Untersuchung und Optimierung der Kühlschmierung beim Schleifen. Dissertation, Universität Bremen.)

bore in the workpiece. Results using this sensor configuration are shown in Figure 11.30 (Heinzel 1999).

The left figure shows the pressure measurements at different depths of cut. It is shown that with increasing infeed, the maximum of the pressure distribution is shifted in front of the contact zone, which can be explained by the geometry of the generated slot. Higher infeed leads to a geometrical boundary in front of the contact zone and results in a rise of the hydrodynamic pressure. If the measured pressure distribution is numerically integrated over the corresponding workpiece surface, the coolant pressure force component can be determined making some assumptions for the calculation.

The right side of Figure 11.30 shows results of this combined calculation and measurement. A pass with no infeed already leads to a normal force of 34 N generated entirely by the coolant pressure. With increasing depth of cut the amount of this force component is of course reduced. Almost half of the normal force is attributed to the coolant pressure, even under deep grinding conditions of a_e = 3 mm.

This method is suitable to investigate the influence of different coolant compositions. The efficiency of additives can be evaluated, if the coolant pressure force component is known and can be subtracted from the total normal force to emphasize the effect on the cutting, friction, and deformation component.

The use of special sensor systems for coolant supply investigations is a relative new field of research. First results have shown that these sensors contribute to a better understanding of the complex thermomechanical interaction in the zone of contact. Also direct industrial improvements such as coolant nozzle optimization or additive efficiency evaluations for grinding can be performed. Thus, a further improvement of the monitoring techniques is desirable.

References

Allanson, D. R., Rowe, W. B., Chen, X., Boyle, A., 1997, Automatic Dwell control in computer numerical control plunge grinding, *Proceedings of the Institution of Mechanical Engineers (London), Journal of the Engineering Manufacture, Part B*, 211, 565–575.

Batako, A. D., Rowe, W. B., Morgan, M. N., 2005, Temperature measurement in grinding, *International Journal of Machine Tools and Manufacture (Elsevier)*, 45(11), 1231–1245.

Brinksmeier, E., 1991, Prozeß- und Werkstückqualität in der Feinbearbeitung (Process and workpiece quality in finishing), Universität Hannover, Habilitationsschrift.

Brinksmeier, E., Heinzel, C., Wittmann, M., 1999, Friction, cooling and lubrication in grinding, *Annals of the CIRP*, 48(2), 581–598.

Brodtmann, R., Gast, T., Thurn, G., 1984, An optical instrument for measuring the surface roughness in production control, *Annals of the CIRP*, 33(1), 403–406.

Brunner, G., 1998. Schleifen mit mikrokristallinem Aluminiumoxid (Grinding with Microcrystalline Aluminum oxide). Dissertation, Universität Hannover.

Byrne, G., Dornfeld, D. A. et al., 1995, Tool condition monitoring (TCM): The status of research and industrial application, *Annals of the CIRP*, 44(2), 541–567.

Chen, X., Rowe, W. B., Li, Y., Mills, B., 1996, Grinding vibration detection using a neural network, *Proceedings of the Institution of Mechanical Engineers, London, Part B, Journal of the Engineering Manufacture*, 210, 349–352.

Foth, M., 1989, Erkennen und Mindern von Werkstückwelligkeiten während des Außenrundschleifens. Dissertation, Universität Hannover.

Gotou, E., Touge, M., 1996, Monitoring wear of abrasive grains, *Journal of the Materials Processing Technology*, 62, 408–414.

Heinzel, C., 1999, Methoden zur Untersuchung und Optimierung der Kühlschmierung beim Schleifen (Methods of investigating and optimizing coolant application in grinding). Dissertation, Universität Bremen.

Heuer, W., 1992, Außenrundschleifen mit kleinen keramisch gebundenen CBN-Schleifscheiben (External cylindrical grinding with small ceramic CBN grinding wheels). Dissertation, Universität Hannover.

Inasaki, I., 1985a. In-process measurement of surface roughness during cylindrical grinding process, *Precision Engineering*, 7(2), 73–76.

Inasaki, I., 1985b, Monitoring of dressing and grinding processes with acoustic emission signals, *Annals of the CIRP*, 34(1), 277–280.

Inasaki, I., 1991, Monitoring and optimization of internal grinding process, *Annals of the CIRP*, 40(1), 359–362.

Karpuschewski, B., 1995, Mikromagnetische Randzonenanalyse geschliffener einsatzgehärteter Bauteile. Dissertation, Universität Hannover.

Karpuschewski, B., 2001, Sensoren zur Prozeßüberwachung beim Spanen (Sensors for process supervision in machining), Universität Hannover, Habilitationsschrift.

Karpuschewski, B., Brunner, G., Falkenberg, Y., 1997, Strategien zur Reduzierung des Kühlschmierstoffverbrauchs beim Schleifen, Jahrbuch Schleifen (Strategies for reduction of coolant requirement in grinding), Honen, Läppen und Polieren, Vulkan-Verlag Essen, 58. Ausgabe, pp.146–158.

Klumpen, T., 1994. Acoustic Emission (AE) beim Schleifen, Grundlagen und Möglichkeiten der Schleifbranddetektion (Acoustic emission in grinding). Dissertation, RWTH Aachen.

König, W., Klumpen, T., 1993, Angepaßte Überwachungsstrategien und Sensorkonzepte— der Schlüssel für eine hohe Prozeßsicherheit, 7. *Internationales Braunschweiger Feinbearbeitungskolloquium*, 2–4 März 1993, p.11.01–11.23.

Lange, D., 1996, Sensoren zur Prozeßüberwachung und Qualitätsprüfung (Sensors for process monitoring and quality assessment), 8. *Internationales Braunschweiger Feinbearbeitungskolloquium* 24–26 April 1996, p.16/1–16/19.

Lierse, T., 1998, Mechanische und thermische Wirkungen beim Schleifen keramischer Werkstoffe (Mechanical and thermal actions in grinding ceramic workpieces). Dissertation, Universität Hannover.

Liverton, J., Rowe, W. B., Moruzzi, J. L., Thomas, D. A., Allanson, D. R., 1993, Adaptive control of cylindrical grinding—From development to commercialization, *Proceedings of Society of Manufacturing Engineers*, Dearborn, Paper MR93-370.

Marinescu, I., Rowe, W. B., Dimitrov, B., Inasaki, I., 2004, *Tribology of Abrasive Machining Processes*, William Andrew Publishing Norwich, New York.

Meyen, H. P., 1991, Acoustic Emission (AE)—Mikroseismik im Schleifprozeß (Acoustic emission—Micro-seismic action in grinding). Dissertation, RWTH Aachen.

Oliveira, G. J., Dornfeld, D. A., Winter, B., 1994, Dimensional characterization of grinding wheel surface through acoustic emission, *Annals of the CIRP*, 43,(1), 291–294.

Regent, C., 1999, Prozeßsicherheit beim Schleifen (Process assurance in grinding). Dissertation, Universität Hannover.

Rowe, W. B., 2014, *Principles of Modern Grinding Technology*, 2nd edition. Elsevier, Oxford.

Rowe, W. B., Jin, T., 2001, Temperatures in high efficiency deep grinding, *Annals of the CIRP*, 50(1), 205–208.

Rowe, W. B., Li, Y., Inasaki, I., Malkin, S., 1994, Applications of artificial intelligence in grinding, Keynote Paper, *Annals of the CIRP*, 43(2), 521–532.

Saljé, E., 1979, Roughness measuring device for controlling grinding processes, *Annals of the CIRP*, 28(1), 189–191.

Saxler, W., 1997, Erkennung von Schleifbrand durch Schallemissionsanalyse. Dissertation, RWTH Aachen.

Tönshoff, H. K., Brinksmeier, E., Karpuschewski, B., 1990, Information system for quality control in grinding, *4th International Grinding Conference*, October 9–11, 1990, Dearborn, Michigan, Paper MR90-503.

Tönshoff, H. K., Falkenberg, Y., Mohlfeld, A., 1995, Touch-dressing—Konditionieren von keramisch gebundenen CBN-Schleifscheiben (Conditioning of ceramic bond CBN grinding wheels), *IDR Industrie Diamanten Rundschau*, 1/1995, 43–48.

Tönshoff, H. K., Friemuth, T., Becker, J. C., 2002, Process monitoring in grinding, *Annals of the CIRP* 51(2), 551–571.

Tönshoff, H. K., Karpuschewski, B., Werner, F., 1993, Fast sensor systems for the diagnosis of grinding wheel and workpiece, *5th International Grinding Conference*, October 26–28, 1993, Cincinnati, Ohio MR93-369.

Tönshoff, H. K., Karpuschewski, B. et al., 1998, Grinding process achievements and their consequences on machine tools—Challenges and opportunities, *Annals of the CIRP* 47(2), 651–668.

Ueda, T., Hosokawa, A., Yamamoto, A., 1985, Studies on temperature of abrasive grains in grinding—Application of infrared radiation pyrometer, *Journal of Engineering for Industry, Transactions of the ASME* 107, 127–133.

von See, M., 1989, Optimierung von Honprozessen auf der Basis von Modellversuchen und betrachtungen. Dissertation, TU Braunschweig.

Wakuda, M., Inasaki, I. et al., 1993, Monitoring of the grinding process with an AE sensor integrated CBN wheel, *Journal of Advanced Automation Technology*, 5(4), 179–184.

Westkämper, E., Hoffmeister, H. -W., 1997, Prozeßintegrierte Qualitätsprüfung beim Profilschleifen hochbeanspruchter Triebwerksbauteile, Arbeits- und Ergebnisbericht 1995–97 des Sonderforschungsbereiches 326, Universität Hannover und TU Braunschweig 1997, pp. 345–401.

Westkämper, E., Kappmeyer, G., 1992, Prozeßintegrierte Qualitätsprüfung beim Honen zylindrischer Bauteile, Seminar des Sonderforschungsbereiches 326, Universität Hannover und TU Braunschweig, pp. 41–58.

Westkämper, E., Klyk, M., 1993, High-speed I.D. grinding with CBN wheels, *Production Engineering* I(1), 31–36.

12

Economics of Grinding

12.1 Introduction

Evaluation of grinding costs is key to achieving maximum profitability in grinding and it is important that costs are evaluated in a meaningful way. Old attitudes of treating abrasive cost, or even total perishable tooling costs, as a single measure of a process are completely misleading and unacceptable.

Models for evaluating "total grinding costs" came to the forefront in the early 1990s with the emergence of cubic boron nitride (CBN) for grinding automotive and aero-engine components. Abrasives costs with CBN at that time were often two or three times higher than with conventional abrasives but the reduction in labor costs and scrap produced far higher overall cost savings. Several models are available for costing. A detailed model including comparison for different abrasives and machines is given in Chapter 19 with respect to centerless and cylindrical grinding processes (Rowe et al. 2004; Rowe 2014). The following cost analysis gives a clear picture of all the costs involved in a decision involving the comparison of two proposed methods.

12.2 A Comparison Based on an Available Grinding Machine

12.2.1 Introduction

If similar machines are to be employed or the grinding machines are readily available for production, the problem of making cost comparisons between two processes is simplified. (English et al. 1991; Carius 1999) presented the items for evaluation of total manufacturing cost which are listed in Table 12.1.

The table shows the main process costs that enter into a comparison between two processes as required by a process engineer in decision-making ignoring the capital cost involved in setting up a grinding facility. It has to be emphasized that these costs depend on the efficiency of the process itself. In other words, it is necessary to carry out grinding trials to determine cycle time, number of parts per dress, and so on before the above costs can be established. An example of the investigation required in order to establish costs per part are given in Chapter 19.

An example of the implications and benefits of the costing approach are given here.

TABLE 12.1

Total Manufacturing Cost Calculation Worksheet

1	Total manpower overhead, cost/part	US$/part
2	Total wheel, cost/part	US$/part
3	Total wheel change, cost/part	US$/part
4	Total dresser roll, cost/part	US$/part
5	Total dresser roll change, cost/part	US$/part
6	Total maintenance labor, cost/part	US$/part
7	Total scrap, cost/part	US$/part
8	Total coolant, cost/part	US$/part
9	Total coolant filter, cost/part	US$/part
10	Total coolant disposal, cost/part	US$/part
11	Total inspection, cost/part	US$/part
	Total grinding cost	US$/part

12.2.2 Aero-Engine Shroud Grinding Example

The application illustrated in Figure 12.1 is the internal grinding of large aero-engine shroud assemblies for aircraft engines. Traditionally, the part was ground using seeded gel (SG) abrasive, but the SG process was replaced by a vitrified CBN wheel that reduced grinding cycle time by 50%. The greatest time saving was achieved by the elimination of the need to dress several times for each component ground. Relevant costs are given in Table 12.2.

Cost items 1–4 are tooling costs and illustrate that the switch to CBN abrasive created a negative impact on grinding costs of US$12.11. However, the increased productivity, items 5–8, had a positive impact on labor cost alone of US$179.44. Also, the quality improvement as a result of the very low wheel wear rate using the superabrasive yielded an even greater cost saving of $399 by the elimination of scrap. Finally, there was also a modest environmental benefit of $0.36 in terms of reduced coolant disposal costs. Converting from

FIGURE 12.1
Internal profile grinding of aero-engine turbine shroud assembly.

TABLE 12.2

Cost Calculation for Internal Profile Grinding of Aero-Engine
Shroud Assembly

		Alox	CBN	Impact of CBN
1	Wheel, cost/part	$10.00	$28.74	
2	Dresser roll, cost/part	$7.44	$3.15	
3	Coolant, cost/part	$5.56	$2.46	
4	Filter, cost/part	$2.00	$0.88	−$12.11
5	Labor/part	$247.77	$123.85	
6	Wheel change/part	$30.00	$0.62	
7	Diamond roll change/part	$0.20	$0.19	
8	Maintenance/part	$62.81	$36.75	+$179.44
9	Scrap	$440.48	$48.68	
10	Inspection, cost/part	$22.50	$14.40	+$399.90
11	Coolant disposal, cost/part	$0.65	$0.29	+$0.36
	Total grinding, costs/part	$829.48	$260.01	

a conventional wheel with a *G* ratio of 1 to a CBN wheel with a *G* ratio of 1000 effectively halved the amount of swarf in the coolant stream. Furthermore, the filtered swarf was much cleaner and therefore easier to recycle.

In summary, a change of process from traditional wisdom considering abrasive cost alone would have had a negative impact of $18.74/part. In fact, the change of process reduced total manufacturing cost by $569.47/part.

12.3 A Cost Comparison Including Capital Investment

12.3.1 Introduction

Total in-process costs may be only half the analysis. For new processes, consideration needs to be given to balance capital equipment costs against operating costs.

12.3.2 Automotive Camlobe Grinding Example

An example of this was the introduction in the late 1980s of CBN to grinding camshafts. After the initial installation, which was instigated because it was the only viable method of generating a particular profile on the camlobe, the abrasive cost was found to be 50% higher using CBN. However, the productivity was found to be about 30% higher than a typical alumina wheel-based process. This had a major impact on the number of grinders required on a later new installation even when the profile issues were not present.

An example is shown in Table 12.3. A new installation required six fewer grinders and each machine was actually less expensive. So even taking the most negative view of the process by considering abrasive cost only, for an increase in abrasive cost of $90 K/annum, a capital equipment cost saving of $5.3 M was achieved. Furthermore, since this was a relatively new technology, significant process costs were likely to be recuperated through future process optimization.

TABLE 12.3

Capital Equipment Investment versus Production Abrasive
Costs for a High-Production Camshaft Manufacturing Line
(Production Requirements: 750,000 Parts/Annum)

	Alox	CBN
Number of machines	22	16
Cost/machine	$750 K	$700 K
Capital equipment cost	$16.5 M	$11.2 M
Capital investment savings	$5.3 M	
Annual abrasive cost	$180 K	$270 K

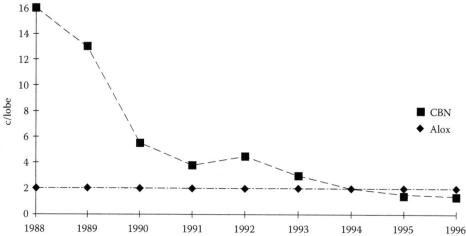

FIGURE 12.2
Abrasive cost improvement in camshaft lobe grinding 1988–1996. (From Renaud, W., Hitchiner, M., 1995, The development of camshaft lobe grinding with vitrified CBN. *SME Ist Intl. Machining & Grinding Conf.*, MR95-163.)

Dramatic process improvement was in fact proven to be the case. Figure 12.2 illustrates the improvement in abrasive cost alone over 8 years at the first CBN installation for grinding camshafts in North America.

12.4 Cost Comparison Including Tooling

12.4.1 Introduction

With better understanding of new technologies such as this it is often possible to push productivity considerably further. The example above based on grinder technology that is now 15 years old had a cycle time of about 6 min limited by burn. On the latest grinders with much more sophisticated computer numerical controls, higher wheel speeds, and faster linear motor technology, it is possible to grind camshafts up to 50% faster, albeit with higher abrasive cost initially. Capital equipment is very expensive and industry is trying

TABLE 12.4

Comparison of Capital Equipment Costs versus Tooling Costs for Various Production Rates

Camlobe Grinding of Automotive Camshafts			
Production requirement = 1,000,000 cams/annum			
Machine cost with gantry loading, installation, etc. = $1 M			
Production rates	• 10/h @ $0.25/camshaft		
	• 15/h @ $0.40/camshaft		
	• 20/h @ $0.60/camshaft		
Cycle Time	Grinders	Capital Cost	Tooling Cost
6 min	25	$25 M	$250 K/annum
4 min	17	$17 M	$400 K/annum
3 min	13	$13 M	$600 K/annum

to drive up their return on investment by limiting such expenditures. A project engineer must therefore weigh up carefully capital against process costs. The following example shows how more expensive tooling can bring down costs.

12.4.2 Effect of Tooling Costs in Camlobe Grinding

Consider the example of a camlobe grinding operation at a high-production automotive engine plant as shown in Table 12.4.

For an increase in tooling costs of $350 K per annum, a capital cost saving of $12 M can be achieved. Again past history would indicate that processing costs would be further reduced over the expected life of the grinding machines and would be a primary focus for future cost savings from both process optimization and competition between tooling suppliers.

12.5 Grinding as a Replacement for Other Processes

12.5.1 Introduction

With the rapid advances in metalworking technology, it is now necessary when selecting new machines to go back to the basic fundamentals of processing a part. An analysis of capital and processing costs may well indicate that the traditional metal-cutting process for producing a given part may not be the most cost-effective. Grinding has recently successfully replaced such processes as lapping and broaching, while grinding itself has been replaced on many occasions by hard turning. Machine tool builders are combining metal-cutting processes, converting machines from one process to another, or reducing costs by adopting commonality of machine components. At the same time, tighter environmental and quality demands, and the "just-in-time" approach to manufacturing, can make historic processing methods obsolete. Some examples are described below.

12.5.2 Fine Grinding as a Replacement for Lapping

Grinding of flat components by processes such as double disk grinding has traditionally been limited to achievable flatness tolerances of about a micron. To obtain flatnesses of the

TABLE 12.5

Comparative Example of Lapping and Fine-Grinding Costs

Input	Lapping Cost/Part	Grinding Cost/Part
Labor/machine cost	$0.60	$0.25
Abrasive costs	$0.08	$0.25
Plate cost (lapping)	$0.01	$0.00
Cleaning cost	$0.34	$0.00
Total	$1.03	$0.50

order of 0.2–0.6 µm (1–2 light bands) demanded for seals, fuel injection, automotive transmission, and pump components has required lapping. This is a slow, batch process carried out on large cast iron tables up to 3 m in diameter using free abrasive slurries. Lapping is very dirty and requires an expensive postlap cleaning process. However, the kinematics of some lappers is quite sophisticated as described later in this handbook. Machine tool builders have therefore combined lapping kinematics with fixed superabrasive grinding wheels to reduce cycle time by up to a factor of 10.

An example of a comparison of the cost of lapping and fine grinding is shown in Table 12.5 grinding PM steel parts on a Peter Wolters double-sided grinder with vitrified CBN pellet wheels (Hitchiner 2002).

The greater savings come not from the actual grinding process but from the elimination of the subsequent cleaning process required after lapping.

12.5.3 High-Speed Grinding with Electroplated CBN Wheels to Replace Turn Broaching

Traditionally, crankshaft journals are roughed from a forging using turn broaches. The majority of stock is present in the sidewall and undercut at the edges of the journal, where it takes a heavy toll on tool insert life. Insert resetting can take several hours using expensive specialized equipment. Current high-speed grinding technology using electroplated CBN wheels in oil coolant was able to process the part in half the time of a turn broach at comparable or lower tooling costs. A crank grinder costs the same as a turn broach effectively halving the capital equipment costs, while eliminating the labor and capital costs of tool insert resetting. See illustrative examples in Figure 12.3.

12.6 Multitasking Machines for Hard Turning with Grinding

Hard turning can remove stock much faster and with lower forces than regular grinding processes, but it cannot hold quite the tolerances and finishes of grinding. However, when combined in a single machine with a single chucking, the two processes can enhance each other. The photographs (Figure 12.4) show an operation to process two inner diameters and the faces of a hardened steel transmission gear component. The process was originally envisioned as being processed entirely by a single grinding wheel with an estimated cycle time of 4 min. The critical surfaces were the inner diameters. The problem, however, was the grinding of the top face due to quill deflection resulting in a cycle time of 10 min.

Turn broaching High-speed grinding with EPCBN wheels

FIGURE 12.3
Examples of turn broaching and grinding of crank pins. (Courtesy of Landis, Waynseboro, Pennsylvania, USA.)

FIGURE 12.4
Multitasking machine for grinding and hard-turning. (Courtesy of Campbell Grinders.)

However, the addition of a turning bar for hard turning the top face reduced the cycle time to 3 min to exceed end-user expectations while still maintaining very acceptable quality.

12.7 Summary

Close attention must be paid to the *entire* cost of a given manufacturing process. On very few occasions, does abrasive cost alone for a given operation govern the processing route. Continual advances in machine tool, wheel, and coolant technology together with ever-greater demands for productivity, quality, and environmental considerations demands that the manufacturing engineer review all options available each time an opportunity for new equipment arises. Careful consideration should be given to emerging technologies, global sourcing (if backed up with adequate local technical support), and fundamental university and corporate research.

References

Carius, A., 1999, Systems view of grinding costs reveals dramatic cost saving opportunities. *MAN*, 6.

English, W., Nolan, T. C., Ratterman, E., 1991, Hidden aspects of superabrasive economics. *Superabrasives 91, SME Conference Proceedings*, Chicago, IL.

Hitchiner, M., 2002, The growth of micro grinding technology using bonded superabrasive wheels. *IMTS 2002 SME Conference*, Chicago, IL, September 4–11, MR02-289.

Renaud, W., Hitchiner, M., 1995, The development of camshaft lobe grinding with vitrified CBN. *SME 1st International Machining & Grinding Conference*, September MR95-163.

Rowe, W. B., 2014, *Principles of Modern Grinding Technology*, 2nd edition. Elsevier, William Andrew Imprint, Oxford, UK and New York, USA.

Rowe, W. B., Ebbrell, S., Morgan, M. N., 2004, Process requirements for precision engineering. *CIRP of the Annals*, 53(1), 255–258.

Section II

Application of Grinding Processes

13

Grinding of Ductile Materials

13.1 Introduction

13.1.1 Grindability

Workpiece material composition and condition control all aspects of the grinding process. As ever greater demands are placed on productivity and quality, it becomes critical that the engineer understands the metallurgy of the materials being ground and their impact on the grinding process. Grindability is the term used to describe the ease of grinding a given workpiece material and is akin to the term machinability used in milling and turning. Machinability, a more familiar term to engineers, is usually judged by four criteria: tool life, tool forces and power consumption, surface quality (including roughness, integrity, and burrs), and chip form. Comparable criteria exist for grindability, namely G ratio, stock removal parameter Λ, surface quality (including roughness, surface residual stress, and burrs), and chip form. In broad terms, an easy to machine material is usually easy to grind especially when judged in terms of tool life and power.

13.1.2 Effect of Chip Form

However, major deviations can arise when the influence of the chip form is considered. Machining involves using a tool of defined geometry to produce uniform chips of the order of 100–1500 µm in thickness. By contrast, grinding is carried out with a random array of cutting point shapes, generally with a high-negative rake angle and random spacings and heights: chip thickness varies from <1 µm to no more than 50 µm even for the most aggressive of rough grinding. An easy to machine material would be judged more for the ease of handling and disposal of its chips. A material that produces long stringy chips would be given a poor rating as would one which produced a fine, discontinuous chip. The ideal chip would be a nicely broken chip of a half or full turn of the normal chip helix. In grinding, the greater factor is loading, or the imbedding of metal into the face of the wheel, especially from a long stringy chip, and a short discontinuous chip is preferred. Chip breakers can change machining chip form, while high-pressure coolant jets reduce built-up-edge on tool inserts. Similarly, optimized grind geometry, for example, small d_e values, combined with high-pressure coolant scrubbers can minimize loading. Under these conditions, grindability and machinability based on tool life or cutting energy or the rate of increase in cutting energy become more in step being based on the mechanical properties of the work material.

13.1.3 Chemical Reactivity

Chemical reactivity between a particular workpiece and abrasive material can affect these trends. Diamond is chemically reactive to most transition metals, cubic boron nitride (CBN) is reactive with titanium, while SiC is reactive to titanium, iron, or cobalt. These properties can reduce wheel life by $10–10^4$ times from expected values based on their relative mechanical properties. Small amounts of additives such as lead in steel or copper in porous powdered metal components will improve machinability by allowing a more continuous chip formation but in doing so reduce grindability due to increased wheel loading.

13.2 Cast Irons

Iron or "ferrous" alloys are the most ubiquitous metals in industry today. They include numerous classes of cast iron, carbon steels, tool, alloy, and stainless steels, as well as exotic materials for the aerospace, nuclear and medical industries. In their oldest and simplest forms, iron alloys depend on their carbon content and composition. Cast iron contains carbon in excess of the solubility limit in the austenite phase. It is used mainly to pour sand castings due to its excellent flow properties that make it ideal for complex cast shapes. It is most commonly found in automotive engine manufacture (blocks, manifolds, cranks, and small cams). Cast iron is generally very brittle compared to steel and incompatible with most forming operations such as drawing or forging (Figure 13.1).

The properties of cast iron depend heavily on how the carbon is distributed.

13.2.1 Gray Cast Iron

Gray cast iron is so called because it gives a gray fracture due to the presence of flake graphite. Its grindability is very good, in either the soft ferritic phase or the harder pearlitic

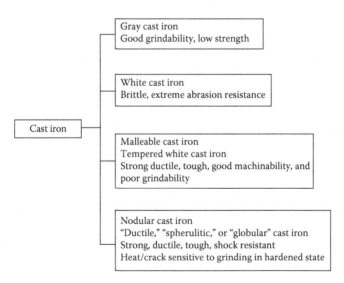

FIGURE 13.1
Classes and properties of cast iron.

phase, as the graphite disrupts the metal flow to produce very short chips. But it is relatively abrasive, especially in the pearlitic phase, requiring either a hard alox wheel grade or preferably the use of CBN.

13.2.2 White Cast Iron

White cast iron is an extremely hard alloy produced by rapid cooling of the casting to combine all the carbon with iron to form cementite. When chipped the fractured surface appears silvery white. White iron has no ductility and is thus incapable of handling bend or twist loads. It is used for its extreme abrasion resistance as rolls in mills or rock crushers. White iron is very heat sensitive and prone to cracking. It is usually ground with free-cutting silicon carbide abrasive wheels but recently white iron has begun to be ground with vitrified CBN in some applications.

13.2.3 Malleable Cast Iron

Malleable cast iron is produced by controlled annealing of white iron to produce round graphite nodules. The material is ductile, tough, and has good machinability, but grindability is lessened by a tendency for wheel loading.

13.2.4 Nodular Ductile Cast Iron

Nodular ductile cast iron again has the carbon in the form of round graphite nodules but now also alloyed with other numerous other elements such as silicon and magnesium. In this form, it is frequently used in automotive applications such as camshafts, crankshafts, and cylinder heads. The surface of the casting can be "chilled" by rapidly cooling it next to a metal chiller creating a hardened layer of white cast iron up to 55 HrC hardness under a softer, more ductile substrate. Again, a range of alloying elements is added to control the depth and hardness of this chilled layer. Chilled cast iron is particularly common in camshaft manufacture. Due to the highly abrasive nature of this layer combined with high productivity requirements and heat sensitive nature of the material, chilled cast iron for this type of precision application is ground almost exclusively with CBN.

13.3 Steels

Steels are differentiated from cast iron in part by the level of carbon. Cast iron has typically 1.7%–4.5% carbon, while steels have only 0.05%–1.5% carbon. The versatility of steel is evidenced by the enormous range of grades available selected for their processing, mechanical, electrical, magnetic, and corrosion resistance properties. The major families of steel types are listed in Figure 13.2.

13.3.1 Plain Carbon Steels

The simplest, and cheapest, steels are plain carbon steels having no other alloying element. As the carbon content is increased, the ductility is reduced while the maximum

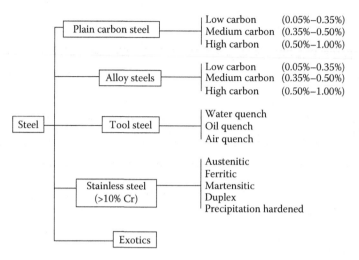

FIGURE 13.2
Major families of steel types.

hardness is increased up to about 63 HrC. Plain carbon steels are identified in the AISI or SAE four digit code system by 10xx, where the 10 identifies the material as plain steel; the values for xx indicates the carbon content where, for example, 1050 steel is 0.50% carbon (Figure 13.3).

The 1050 steel is the mainstay of many industries including the automotive and bearing. It is easy to grind using fused alox or ceramic abrasive wheels and able to take a good case-hardening depth without stress cracks. Case-hardening is achieved by a carburizing process to drive carbon into the surface layer followed by quenching. This results in a surface layer of depth 1 mm with a hardness value up to 63 HrC for maximum wear resistance on a softer more ductile core for toughness. Steels with higher carbon content, for example, 1080, are used where greater hardening depths might be required. Certain industries and applications such as internal grinding of small bearings and cam-lobe grinding of automotive camshafts have in part or almost totally converted to CBN and this will continue as machine, wheel, and processing technology improves. Even cylindrical grinding of 1050 steel in the soft state is now in production with vitrified CBN.

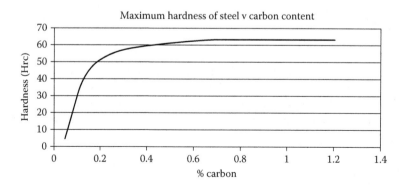

FIGURE 13.3
Effect of carbon content on maximum hardness of plain steel.

13.3.2 Alloy Steels

Alloy steels are carbon steels with additional alloying elements to enhance their physical properties or improve their machinability. Typical alloy steel types are given in Table 13.1 including the AISI codes. Physical property improvements include thorough hardenability, toughness, tensile strength, and wear resistance. Alloying elements adjust grain size that can adversely affect machinability or improve grindability. The alloying elements also form hard carbide particles in the grain structure which are highly abrasive and cause increased wheel grain wear. Martensitic structures, in particular, having high tensile strength and low elongation, produce short chips and cause high abrasion wear levels in the grinding wheel leading manufacturers toward increased usage of CBN.

13.3.3 Tool Steels

The impact of the carbide level becomes most extreme with tool steels, so named for their heat, abrasion, and shock resistance, dimensional stability in heat treatment and cutting ability. All of these contain one or more of the alloy elements chromium, tungsten, molybdenum, and vanadium and have special heat treatment processes to control grain structure (Table 13.2). The resultant material can have hardness values up to 67 HrC. The primary carbides of interest here are M_6C carbides of molybdenum and tungsten, and MC carbides from vanadium. Essentially, the carbides provide the wear resistance, while the softer metal matrix provides toughness. These carbide particles are very hard (Badger 2003).

TABLE 13.1

AISI Steel Grade Identification and Effect of Alloy Type

AISI Steel Grade Identification Using First Two Digits	
10xx/11xx	• Carbon only
13xx	• Manganese
23xx/25xx	• Nickel
31xx/33xx/303xx	• Nickel chromium
40xx/44xx	• Manganese molybdenum
41xx	• Chromium molybdenum
43xx/47xx/81xx/86xx/93xx/94xx/98xx	• Nickel chromium molybdenum
46xx/48xx	• Nickel molybdenum
50xx/51xx/501xx/511xx/521xx/514xx/515xx	• Chromium
61xx	• Chromium vanadium
92xx	• Silicon manganese
Impact of Alloy Elements	
Manganese	• Mechanical strength, improved heat treat depth
Nickel	• Improved toughness
Molybdenum	• Improved high-temperature hardness and corrosion resistance
Chromium	• Improved hardening and corrosion resistance
Vanadium	• Improved hardness and toughness, fine grain control
Silicon	• Improved tempered hardness

Note: xx identifies the amount of carbon present, e.g., xx = 50 is 0.50% carbon.

TABLE 13.2

Categories of Tool Steel

Category	Description
W	Water hardening
O	Oil hardening
A	Air hardening
D	Oil and air hardening (chromium)
S	Shock resistant
H	Hot working
M	High speed (molybdenum)
T	High speed (tungsten)
L & F	Special purpose
P	Mold making

Uses: Arbors, broaches, machine centers, chasers, cutting tools, dies, form and lathe tools, reamers, blades, rolls, taps, vise jaws, and wrenches.
Aerospace bearings (M50).

Tool steels are extremely difficult to grind with conventional wheels. The MC carbides exceed the hardness of alox grains (Table 13.3). Indeed in the 1960s prior to the introduction of CBN, some toolrooms would use resin-bonded wheels containing very friable diamond as an alternative to SiC or alox wheels. Interesting to note, it is not only due to the amount of carbide present but also due to the size of the carbide particles. This is of particular interest when considering the grinding of workpieces made using powder–metal (PM) technology.

The PM parts are pressed to shape in a dye then sintered in a furnace at temperatures around 2000°F (1100°C). The powder does not melt so it never fully densifies. Mechanical bonds formed during pressing transform into metallurgical bonds during sintering. A major advantage of using PM is the ability to produce complex shapes to near-net shape. Typically, the PM process can control lengths to within 0.003–0.005 inches (75–125 μm). Since the structures do not fully densify, they contain pores between the grains. Machining of PM parts is therefore quite difficult, because it consists of machining a series of interrupted cuts at the microscopic level resulting in high abrasive tool wear and chipping. The porosity also reduces heat conduction and drives more heat into the chip. All these factors which adversely affect machinability are a benefit for grindability, and PM technology, contrary to popular belief, may actually have increased the need for grinding as a precision finishing process.

TABLE 13.3

Comparison of the Hardness of Tool Steel Carbides to Abrasive Grains

Constituent	Hardness (Knoop)
Matrix	800
M_6C carbide	1600
MC carbide	2800
Alox grain	2100
CBN grain	4700

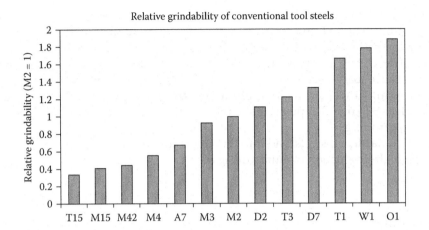

FIGURE 13.4
Relative grindability of conventionally produced tool steels.

Another benefit of PM technology is that it allows the control of the grain size of the carbides in tool steels. Badger (2003) reported that although the PM tool steel ASP2060 contained 15% MC carbide, it had only slightly poorer grindability based on G ratio than M2 steel containing only 2% carbide content. PM steels with comparable carbide content to those produced by conventional forging, and heat treatment processes had up to 10 times the grindability based on G ratio as shown in Figures 13.4 and 13.5.

It should be noted as a caveat that each PM application must be considered individually, as there are enormous variations in composition, particle hardness, apparent hardness, porosity, and pore fillers. It is not uncommon to have a material with an apparent hardness of only 30 HrC but a particle hardness of 60 HrC.

13.3.4 Stainless Steels

The last group of alloys is the stainless steels defined as steels containing more than 10% chromium. These are designed first and foremost to resist oxidation, where they

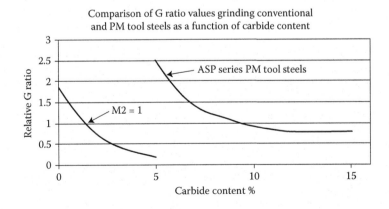

FIGURE 13.5
Comparison of the grindability of tools steels based on carbide content made by conventional and powder–metal methods.

are used in the food, medical, petroleum, and chemical industries. In addition, some precipitation-hardened stainless steels are designed to withstand high temperatures bringing them into the lower end of the range of heat-resistant alloys used in the aerospace industries. They are characterized by producing long chips when ground and a tendency to cause loading. They also will work harden producing high-grinding forces and wheel wear. Currently, most applications are ground using the more advanced ceramic such as SG- or TG-type grains or regular alox grains in continuous-dress creep-feed (CDCF) grinding.

Precipitation-hardened steels lead into the final and most difficult of material groups to grind in the high-production manufacturing arena.

13.4 Heat-Resistant Superalloys

Heat-resistant superalloys (HRSA) may be based on high levels of nickel, chromium, and/ or cobalt. In many cases, iron as an alloying element may be almost or entirely absent. These are the workhorses of the aerospace engine and land-based power generation industries. Their superior ability to retain strength and corrosion resistance at high temperatures allows their use in turbine vanes and blades, and afterburners that operate at temperatures up to 980°C (Figure 13.6).

13.4.1 Precipitation-Hardened Iron-Based Alloys

The lowest class of HRSA materials is the precipitation-hardened iron-based alloys discussed above.

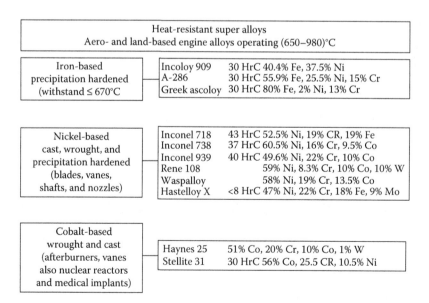

FIGURE 13.6
Major classes and examples of heat-resistant superalloys.

13.4.2 Nickel-Based Alloys

The second, and most prolific, group of alloys is the nickel-based superalloys. These are used extensively for blades, vanes, buckets, shrouds, shafts, nozzles, honeycomb, and rotors. The raw material may be supplied as a forging, casting, or as bar stock depending on the application. Blades and vanes, for examples, are supplied as castings, usually hollow to reduce weight. Because the grindability is so poor and hence grinding costs high, casting suppliers have improved their technology dramatically over the last decade to product near-net shapes and reduce grinding processes from ones requiring CDCF grinding to remove up to 10 mm of stock to finish grind processes removing as little as a tenth of this. Some blades are even produced as directionally solidified single crystals to maximize the strength in a particular direction. The material may also be supplied after various heat treatments including annealing, solution treatment, and aging. Many basic material grades may have several levels of heat treatment each of which significantly affects its grindability. Nevertheless, all nickel-based alloys share the following properties that make their grindability "difficult":

- High strength, including dynamic shear strength, is retained at the high temperatures seen in grinding
- Poor thermal conductivity
- Material structure contains hard carbide for wear resistance and crack arrest
- Material work hardens during chip formation
- Extreme sensitivity to heat damage; the so-called "white layer" is very detrimental to component strength in an industry where safety is paramount

The grindability of nickel-based alloys based on G ratio, and surface quality of chip form is perhaps a twentieth that of carbon steels or gray cast iron; based on stock removal parameter grindability is about a third. CBN and advanced ceramic grains are becoming increasingly dominant for grinding both nickel- and cobalt-based materials.

13.4.3 Cobalt-Based Alloys

The third group of alloys is cobalt based. These alloys display the best high-temperature corrosion resistance but have the lowest grindability and are the most expensive. They are, therefore, used in just the hottest part of the engine. However, one area where expense is not so important and where strength and corrosion resistance is critical is in the medical industry, where wrought and cast cobalt–chromium alloys are becoming common for implants such as knee and hip joints. They are strengthened by solid solution elements and the presence of carbides, and experience the same problems in grindability as the aerospace industry. Plated CBN and extruded ceramic grain are common for this type of application.

13.4.4 Titanium

Figure 13.7 gives typical values for the relative grindability of many of the material types described above. These numbers are dependent on advances in the abrasive and wheel bond technology, and on the type of application. With the growth of CBN technology, the grindability of hard steel and chilled iron has increased significantly. Similarly, recent

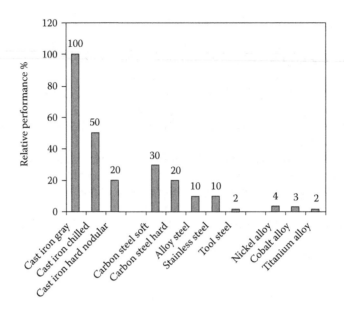

FIGURE 13.7
Relative grindability by material type.

developments in extruded ceramic grain technology may raise the grindability of nickel- and cobalt-based alloys based on *G* ratio and stock removal parameter from those shown. Titanium alloys are also included in Figure 13.7. Titanium is problematic because it is chemically reactive with CBN limiting the application of superabrasives; it is also reactive with alumina. Some success has been reported using plated diamond (Kumar 1990) to grind aerospace grade titanium alloys but most applications still use porous SiC wheels. On the other hand, the machinability of titanium remains better than nickel- or cobalt-based alloys based on chip form and surface quality, and in many cases, it should be considered as a better alternative to grinding.

References

Badger, J. A., 2003, Grindability of high-speed steel. *Abrasives Magazine*, December–March 2003, 16–19.

Kumar, K. V., 1990, Superabrasive grinding of titanium alloys. *SME Conference Proceedings Paper MR90-505*, October 1990, USA.

14

Grinding of Ceramics

14.1 Introduction

This chapter discusses factors affecting the grinding of the new generation of high-performance ceramics. Particular attention is drawn to effect on predressing time and surface quality. Electrolytic in-process dressing (ELID) grinding is introduced as a promising new process used for machining a range of very hard materials and, in particular, for machining ceramics.

14.1.1 Use of Ceramic Materials

Ceramic materials are used in engineering where there are demanding requirements for particular material properties. These may include requirements such as unusually high heat resistance, unusually high wear and corrosion resistance, low specific weight, or special electrical properties. Noticeable examples include ceramic bearings for high speeds in the machine tool industry and electronic components for the computer industry.

It is possible to engineer ceramic materials for high density and high strength. Developments in ceramic materials have greatly increased the range of engineering applications in recent decades. Zirconia, for example, is a modern material that is very tough apart from the obvious use for jewelry, and is finding increasing engineering use for wear resistance. Hard silicon nitride ceramics also offer a way to overcome the greatest disadvantage of traditional ceramics: significant brittleness. Ceramic–ceramic composites have further increased the range of properties that can be achieved.

14.1.2 Machining Hard Ceramics

Abrasive machining is practically the only successful method for machining these hard and brittle materials. Superabrasive grinding wheels with diamonds are generally employed due to the extreme hardness of some ceramics. The combination of high hardness and strength, however, causes considerable wear on the diamond abrasive itself with effect on the geometric accuracy of the grinding wheel, the protrusion height of individual grains (cutting edges), and the distribution pattern of active cutting edges. Therefore, careful preparation of the wheel prior to grinding is essential.

14.1.3 Wheel-Dressing Requirements

Surface conditioning of a grinding wheel generally includes the two requirements of truing and dressing. Dressing replaces worn (dulled) cutting edges with sharp edges by fracturing the worn grains and regenerates a wheel face loaded with swarf by removing a layer from the face with a dressing tool. Dressing a superabrasive wheel is much more difficult than dressing a conventional wheel and is not always possible. A special process known as ELID was introduced to facilitate the grinding of hard ceramics; this is discussed in more detail in this chapter.

The necessity of dressing increases costs in grinding. In addition, the loss of abrasive particles from the wheel surface is expensive due to the high cost of superabrasive wheels and the associated loss of wheel life. Consequently, it is very important to use superabrasive wheels effectively to minimize waste of the wheel as well as to reduce dressing time.

14.1.4 ELID Grinding

The new ELID technique has attracted special interest because it eliminates conventional dressing times while improving the quality of ground surfaces. In ELID grinding, the grinding wheel is connected to a positive electrode, and a negative electrode is mounted in proximity to the wheel with a gap of 0.1–0.3 mm. During predressing or grinding, a power source provides a pulsed direct current.

The gap is filled with an electrolyte grinding fluid to obtain an electrochemical action. The direct current pulse ionizes and removes the conductive bond material. As the action proceeds, the surface of the wheel becomes coated with a nonconducting layer such as iron oxide or hydroxide. The nonconducting layer has the effect of preventing excessive electrolytic action. The rate of electrolytic action is an essential aspect of process control to ensure adequate protrusion of the grains while avoiding excessive wheel wear.

14.1.5 Advantages of ELID

Advantages of ELID include the following:

- Elimination of dressing time by in-process dressing
- Elimination of the need for conventional dressing tools and hence reduced dressing costs
- Elimination of physical or mechanical damage to the dressed grains
- A significant reduction in normal grinding force with ELID grinding compared with conventional grinding
- Extremely smooth surfaces achieved by use of ultrafine abrasives
- High precision parts achieved with excellent process efficiency

14.2 Background on Ceramic Materials

14.2.1 History

The word ceramic can be traced back to the Greek term *keramos*, meaning "a potter" or "pottery." Keramos, in turn, is related to an older Sanskrit root meaning "to burn." Thus, the early Greeks used the term to mean, "burned stuff" or "burned earth" when referring to products obtained through the action of fire upon earthy materials (The American Ceramic Society; Marinescu et al. 2001).

14.2.2 Structure

The structure of ceramic crystals is among the most complex of all materials, containing various elements of different sizes. The bonding between these atoms is generally covalent (electron sharing, hence, strong bonds) and ionic (primary bonding between oppositely charged ions, thus, strong bonds). These bonds are much stronger than metallic bonds. Consequently, thermal and electrical resistance and hardness of ceramics may be significantly higher than those of metals (Inasaki 1998).

The properties of ceramic products are dependent on the chemical composition and atomic and microscale structure. Compositions of ceramic products vary widely, and both oxide and nonoxide materials are commonly used. In recent times, the composition, grain structure, and also the distribution and structure of porosity have been more carefully controlled to achieve greater product performance and reliability (Inasaki 1998).

14.2.3 Ceramic Groups

Ceramics are classified in the following subgroups (Marinescu et al. 2001):

- Oxides (alumina, zirconia, and partially stabilized zirconia)
- Carbides (tungsten and titanium used for cutting tools, silicon carbide used as abrasives in grinding wheels)
- Nitrides (cubic boron nitride, titanium nitride, and silicon nitride)
- Sialon (silicon nitride with various additions of aluminum oxide, yttrium oxide, and titanium carbide)
- Cermets (ceramics bonded with a metallic phase)
- Silicates are products of the reaction of silica with oxides of aluminum, magnesium, calcium, potassium, sodium, and iron such as clay, asbestos, mica, and silicate glasses

14.2.4 Ceramic Product Groups

Table 14.1 presents typical examples of products obtained by ceramic powder processing. Nowadays, ceramic–ceramic composites are used for many engineering applications. Development of ceramic–ceramic composites is largely due to improvements in microstructure of reinforced ceramics such as $Al_2O_3 + TiC$, $Al_2O_3 + ZrO_2$, $ZrO_2 + Al_2O_3$, and whisker-reinforced ceramics such as $Al_2O_3 + SiCw$, $Si_3N_4 + SiCw$, etc. Among these, $Al_2O_3 + ZrO_2$ and $ZrO_2 + Al_2O_3$ are referred to as toughened ceramics, which have widened the scope of application of oxide ceramics.

One of the main oxide ceramics is alumina, also called corundum. Emery is also a widely used oxide ceramic. It has high hardness and moderate strength. Aluminum oxide is almost totally manufactured synthetically in order to control quality.

Synthetic aluminum oxide, first made in 1893, is used for applications such as electrical and thermal insulation and as cutting tools and abrasives (Ruhle 1985).

Zirconia (zirconium oxide, ZrO_2) has high toughness, high resistance to thermal shock, wear, and corrosion; low thermal conductivity; and a low friction coefficient when paired with many materials (Green and Hannink 1989).

Zirconia-toughened alumina (ZTA) is a ceramic system obtained as a result of the transformation toughening of ZrO_2 in a constrained matrix of Al_2O_3. One field of application for

TABLE 14.1

Products Obtained by Ceramic Powder Processing

Group of Applications	Products Obtained by Ceramic Powder Processing
Electronics	Substrates, chip carriers, and electronic packaging
	Capacitors, inductors, resistors, and electrical insulation
	Transducers, electrodes, and igniters
Advanced structural materials	Cutting tools, wear-resistant inserts
	Engine components
	Resistant coatings
	Dental and orthopedic prostheses
	High-efficiency lamps
Chemical processing components	Ion exchange media
	Emission control components
	Catalyst supports
	Liquid and gas filters
Refractory structures	Refractory lining in furnaces
	Regenerators
	Crucibles
	Heating elements
Construction materials	Tile structural clay products
	Cement, concrete
Institutional and domestic products	Hotel china and dinnerware
	Bathroom fixtures
	Decorative fixtures and household items

Source: Malkin, S., 1989, *Grinding Technology. Theory and Applications of Machining with Abrasives.* Ellis Norwood Ltd., U.K. With permission.

ZTA due to excellent wear behavior is the application for cutting tools. In recent decades, ceramic cutting tools have made a strong impact in manufacturing.

14.2.5 Application of ZTA Ceramics

The enhanced strength and toughness of ZTA ceramics have made these materials more widely applicable and more productive than plain ceramics and cermets for machining steels and cast irons. ZTA ceramics can be used as cutting tool inserts for metal cutting applications (Sornakumar and Gopalakrishnan 1995).

Cutting inserts are made by sintering followed by grinding and lapping. During the sintering process, ceramics are subject to volumetric shrinkage and as a result cannot meet the requirements for form and size accuracy without further machining. To meet these demands, machining by grinding, lapping, honing, or polishing with diamonds is necessary.

The performance of ZTA cutting inserts is strongly influenced by the grinding process. The active faces of the inserts should be smooth, without cracks, and the cutting edges should be free of chips. ELID grinding can provide a solution to these requirements such that dimensional accuracy and surface quality are achieved and even surpassed with a considerable increase in productivity (Marinescu and Ohmori 1999; Ohmori and Li 2000a).

The ELID grinding technique was recently developed in Japan and there is an increased tendency to apply it for grinding hard and brittle materials (Bandyopadhyay 1996;

Ohmori and Li 2000a). However, there are still insufficiently explored areas that make industrialists hesitant in adopting the method. Very little experimentation on ELID grinding has been conducted in the United States until very recently. The need for further practical information provides the motivation for further research.

14.2.6 Grinding of Ceramics

The characteristics of grinding advanced ceramics are very different from the ones for metals and research is still ongoing for a more comprehensive understanding and a better control of grinding parameters. Figure 14.1 presents the factors affecting the grinding process (Marinescu and Ohmori 1999). Due to the fact that ceramics are brittle and have high hardness, they are best ground using diamond-grinding wheels (Marinescu et al. 2001).

14.3 Diamond Wheels for Grinding Ceramics

14.3.1 Type of Diamond Abrasive

Diamonds, used for grinding ceramics, are mainly synthetic diamonds. The synthesis process permits control of diamond characteristics generating either blocky grit shapes with very high-impact strength or friable grits with low-impact strength. Diamonds are also classified by grit size and grit-size distribution. The grit system can be used to control the desired working result within a wide range (Carius 2002).

The abrasive grains employed for ELID grinding usually consist of synthetic diamond. Diamond is the most efficient superabrasive material used for grinding all sorts of hard, nonmetallic materials. Depending on the materials to be ground, different strength and wear characteristics are required for the diamond used. For best economic performance, the abrasive grain formulation must be matched with a correct bond formulation.

The main property of diamond is its hardness (10 on Mohs hardness scale). Not only do different diamonds have appreciably different structures and mechanical properties, but also different parts of the same stone may differ appreciably. Diamonds as abrasive

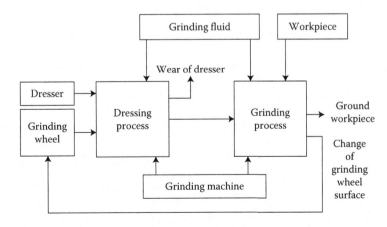

FIGURE 14.1
Factors affecting the grinding process. (From Marinescu, I., Ohmori, H., 1999, ELID grinding of ceramic materials. *Finishing of Advanced Ceramics, Ceramic Transactions*, 102. With permission.)

grist in grinding applications are used in monocrystalline or polycrystalline configuration. Essentially, these diamond types show different shatter behavior and therefore influence the grinding process characteristics significantly.

The physical characteristics of diamond are

- Material pure carbon (C)
- Atomic weight 12
- Melting point 3700 K

Diamond has the following properties:

- It behaves like a nonmetal
- It is an electrical insulator but conducts heat very well
- It is truly stable only at high pressures
- Crystals form in the area of pressure and temperature, where diamond is stable

For ELID grinding, the grinding wheel is normally a metal-bonded diamond wheel. This structure presents a very different challenge for dressing compared with a vitrified diamond wheel or a resin-bond diamond wheel or even an electroplated diamond wheel.

14.3.2 Types of Diamond Wheel

Basically, a grinding wheel has three components: abrasive, bond, and core. Figure 14.2 presents the three components of a superabrasive diamond or CBN wheel (Inasaki 1998).

In general application, diamond wheels can be resin, metal, or vitrified bond. Latest developments in the abrasive tool industry introduced new bond specifications, which are combinations of metal-resin or metal-vitrified bonding systems. These so-called hybrid bonds promise to merge the properties of different bonding systems.

Wheels can be classified according to the bond material as either (Marinescu et al. 2001) monolayer-electroplated wheels or multilayer wheels, which include resin bonded, vitrified bonded, and metal bonded.

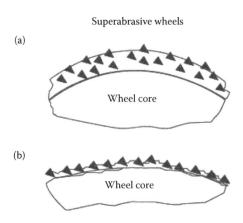

FIGURE 14.2
Abrasive system, bond system, and core system for a superabrasive wheel. (a) Impregnated rim and (b) electroplated single layer. (From Marinescu, I., Toenshoff, H., Inasaki, I., 2001, *Handbook of Ceramic Grinding and Polishing.* Noyes Publications, William Andrew Publishing, Norwich, NY. With permission.)

Resin bonds have been widely used for diamond wheels for many years and offer the particular feature of structural flexibility. Phenolic resins and their derivatives are still the most commonly used bond types. Resin-bonded wheels are relatively easy to use in a wide range of applications typically in the conventional grinding of cemented carbides.

Vitrified bond diamond wheels have steadily become a more important market segment of the diamond tool market for conventional grinding. They are relatively easy to profile and maintain their shape. A further advantage of this type of wheel is that porosity can be introduced into the bond in a controlled manner. This allows extra chip clearance volume on the wheel surface and space for the improved transport of coolant into the grinding zone (Marinescu et al. 2001).

Two basic techniques are available for the production of vitrified bond diamond wheels: hot pressing and cold pressing followed by free firing. Each method is designed for different application areas. Hot-pressed wheels are normally used for grinding diamond, as in the manufacture of both polycrystalline diamond cutting tools and natural diamond tools such as hardness-testing indenters. Cold-pressed tools, which have inbuilt porosity, are used in deep grinding of advanced ceramic materials and cemented carbides.

Metal bond wheels are used for applications such as grinding of glass, stone, ceramics, semiconductors, and plastics.

For ELID grinding, a cast-iron metal bond is normally used as this provides the necessary electrical conductivity. Figure 14.3 shows types of diamonds for metal-bonded wheels (Carius 2002). The strong, relatively hard, and inelastic metal bond requires diamond products with properties that are correctly matched to those of the bond. This type of tool is less tolerant of an unsuitable grit selection than other bond systems. The main feature of metal bonds is that the diamond is rigidly retained in the bond. Metal bonds provide very high abrasive resistance to the small detritus that occurs when machining short chipping materials. The wheel is manufactured by sintering at a high temperature.

The diamond layer in a wheel can be laid on different core materials. The diamond wheel is adapted for the specific machining operation. The nature of the core material is important in the machining of brittle ceramics and permits adaptation of the wheel with respect to vibration behavior and thermal conductivity (Marinescu et al. 2001).

When machining ceramic materials, it is important to minimize external vibration sources because of the brittle nature of these materials. The grinding wheel must be

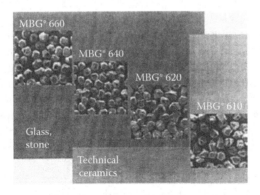

FIGURE 14.3
Diamonds for metal-bonded wheels. (From Carius, A., 2002, Characteristics of superabrasive diamond and CBN micron powders. GE Superabrasives, Superabrasive Certification Course, University of Toledo, May–October. With permission.)

prepared and conditioned to reduce runout error, clamping error, and grinding wheel unbalance. Keeping the weight of the grinding wheel low can help to minimize vibrations. Also, the use of machine-integrated dressing systems help to reduce the runout errors, as it is not necessary to shift the wheel from an external dressing unit into the machine.

14.3.3 Wheel Truing and Dressing

The actual cutting edges and points on abrasive grains at the wheel active surface are microcutting tools that interact with the workpiece material. The spatial distribution of abrasive grains over the wheel surface and their morphology comprise the grinding wheel topography.

The grinding wheel topography and the macroscopic wheel shape are initially generated by conditioning the wheel before grinding and periodically during the course of grinding. Wheel preparation generally includes truing and dressing. Truing usually refers to removal of material from the cutting surface of a grinding wheel so that the spinning wheel runs true with minimum runout from its macroscopic shape, although truing may also include profiling of the wheel to a particular shape (Marinescu et al. 2001).

Dressing is the process of conditioning of the wheel surface so as to achieve a certain grinding behavior. Generally, dressing is the process of conditioning worn grains on the surface of a grinding wheel in order to produce sharp new grains and truing out-of-round wheels. Dressing is necessary when excessive attritious wear dulls the wheel or when the wheel becomes loaded. Dulling of the wheel is known as glazing because of the shiny appearance of the wheel surface. Loading occurs when the pores on the surface of the wheel become filled or clogged with chips.

Pregrinding preparation of superabrasive wheels usually involves two distinct processes: truing and dressing. Most superabrasive wheels are trued and dressed, with the exception of electroplated wheels, which may only require "cleaning" or "touching up" with an abrasive stick. One popular truing method for diamond wheels utilizes a vitrified green (friable) silicon carbide grinding wheel mounted on a brake-controlled truing device. The truing wheel is operated as if it cylindrically traverse grinds the grinding wheel. The axes of the dressing wheel and grinding wheel are parallel to each other, as in Figure 14.4.

Other types of wheels (e.g., cup wheels) use a similar device. The depth increment a_b might be between 10 and 20 µm after each traverse across the wheel face with a cross-feed velocity v_c, corresponding to a lead of $s_b = 0.1$–0.2 mm per grinding wheel revolution.

Numerous other truing methods utilizing diamond tools, similar to those applied to conventional abrasive wheels, have been tried with diamond wheels, but with limited success. Diamond truing of diamond wheels removes wheel material much faster than silicon carbide truing, but there is a danger of damage both to the truing tool and to the wheel. Truing with single-point and multipoint diamond tools may also necessitate excessive wheel loss, and the truing tool wears out rapidly (Kalpakjian 1997).

After truing, dressing of superabrasive wheels is usually accomplished by in-feeding a fine-grained vitrified abrasive stick into the wheel surface either manually or with a holding device. Again, there is a danger of damage to the abrasive grains with excessive use of an abrasive stick.

The decisive factors for grinding behavior include grit type, grit size, and concentration, but also specification of the bond system and conditioning of the grinding layer. The interplay between wear processes on the grit and on the bond affects both grinding behavior and workpiece quality. Metal bond wheels in particular react strongly to differences in dressing (Malkin 1989).

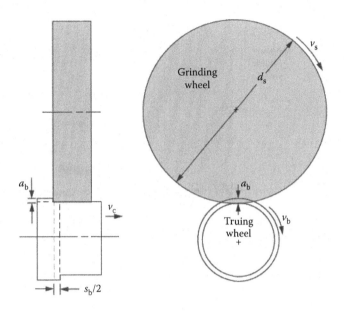

FIGURE 14.4
Brake-controlled truing arrangement for peripheral wheels. (From Kalpakjian, S., 1997, *Manufacturing Processes for Engineering Materials,* 3rd edition. Addison-Wesley. With permission.)

These operations for superabrasive wheels (truing and dressing) should be avoided as much as possible because they require production workflow to be interrupted. Some of the advantages of using superabrasives can be canceled if the dressing time is excessive. A further consideration is the loss of expensive abrasive particles from the wheel surface due to the dressing process. This can be important since initial superabrasive wheel cost may be several times the cost of conventional wheels. Consequently, it is very important for effective utilization of superabrasive wheels to avoid wasting wheel material due to dressing as well as to reduce dressing time.

14.4 Physics of Grinding Ceramics

The material-removal rates in grinding of ceramics vary widely depending on the application (Figure 14.5). With recent advances toward understanding of the mechanics of grinding ceramics, it is possible to achieve material-removal rates comparable to that in metal grinding. However, current practice achieves only about a tenth of these material-removal rates.

One model of chip formation is based on the Hertzian surface pressure of two bodies in contact with each other, where the stresses and deformations produce microcracks, which cause breakdown of the ceramic grains so that brittle material erosion takes place (Figures 14.6 and 14.7; Marinescu et al. 2001).

Ceramic materials are more brittle than metallic materials and show very little plastic deformation under load up to the point of fracture. For this reason, it might be expected that the mechanism of abrasive machining of ceramic materials would mainly involve

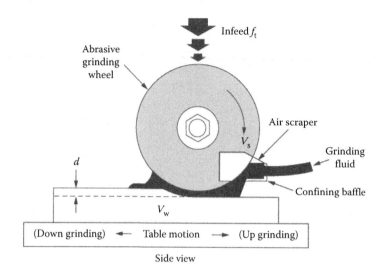

FIGURE 14.5
Schematic of surface grinding. (From Malkin, S., 1989, *Grinding Technology. Theory and Applications of Machining with Abrasives*. Ellis Norwood Ltd., U.K. With permission.)

brittle fracture. Microscopic examination of high-density polycrystalline alumina surfaces ground with diamond shows fractured areas that are consistent with a brittle fracture mechanism. However, evidence of plastic flow with striations along the grinding direction is also observed. This would suggest that both flow and fracture play an important role in the grinding process for ceramics (Bifano 1998).

Another model is based on the assumption that ceramic material is softened by higher local temperatures at the cutting points, thus becoming plastically deformable, and can, therefore, be machined in a similar manner to other materials. Ductile-regime grinding has been used to describe the material-removal mechanisms in grinding of ceramics under suitable conditions. A transition from brittle to ductile-mode material removal at smaller

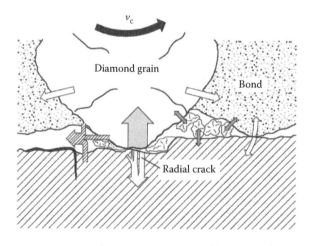

FIGURE 14.6
Surface contact. (From Uhlmann, E., 1998, Technological and ecological aspects of cooling lubrication during grinding of ultrahard materials. *Ultrahard Materials Technical Conference*, Conference Papers. With permission.)

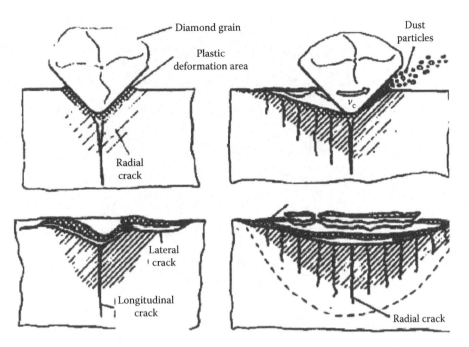

FIGURE 14.7
Phenomena of surface deformations for diamond on ceramic. (From Uhlmann, E., 1998, Technological and ecological aspects of cooling lubrication during grinding of ultrahard materials. *Ultrahard Materials Technical Conference*, Conference Papers. With permission.)

cutting depths can be argued purely from considerations of necessary material-removal energy. Specifically, for lower machining depths of cut, it can be shown that plastic flow is a more energetically favorable material-removal process than fracture. The limiting depth at which a brittle–ductile transition occurs is a function of the intrinsic material properties governing plastic deformation and fracture (Bifano 1998).

Ductile-mode grinding enhances surface quality, but is very slow and costly. One possible method of promoting ductile flow and achieving removal rates is by using high peripheral wheel speeds. Higher wheel speeds reduce the uncut chip thickness and result in a smaller force per grit, increased ductile flow, and decreased strength degradation. The reduction in surface fracture and apparent increase in flow may be associated with glassy-phase formation at elevated grinding temperatures (Bifano 1998; Bifano et al. 1998).

A combination of the two modes, brittle and ductile, depending on the type of ceramic and the machining conditions is probably involved in reality. The transition is decided by the size of plastically deformed layer caused by the interference between the material and the tool.

Figure 14.8a and b provides a comparison between ground surface finishes of silicon nitride in brittle-mode grinding and of zirconium oxide in ductile-mode grinding. Figure 14.8a shows the result of brittle-mode grinding (Ohmori and Li 2000b).

Brittle materials exhibit discontinuous failure phenomena. Ductile-mode grinding (Figure 14.8b) is superior to brittle-mode grinding with respect to both geometrical accuracy and surface integrity. The ductile regime grinding hypothesis of Bifano states that, for any material, if the dimensional scale of material removal is small enough, then plastic flow of the material will take place without fracture. The nature of the material does not affect ductile chip formation.

(a)

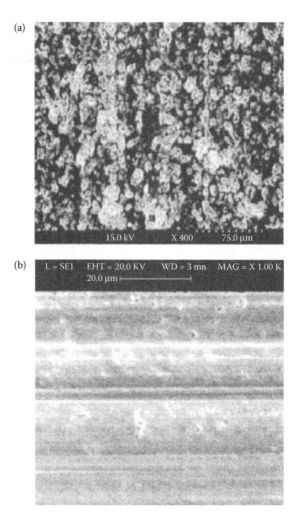

(b)

FIGURE 14.8
Grinding in brittle mode and ductile mode. (a) Brittle-mode grinding of Si_3N_4. (b) Ductile mode grinding of ZrO_2. (From Zhang, B. et al., 2003, *Journal of Materials Processing Technology*, 132, 353–364. With permission.)

The term semiductile grinding is used for the brittle–ductile transition process in grinding of ceramics. The transition from a brittle to a ductile mode during the machining of brittle materials is described in terms of energy balance between the strain energy and the surface energy. Another interpretation of ductile transition phenomena is based on cleavage fracture due to the presence of defects. The critical values of a cleavage and plastic deformation are affected by the density of defects/dislocations in the work material. Since the density of defects is not so large in brittle materials, the critical value of a fracture depends on size of the stress field (Ngoi and Sreejith 2000).

The mechanism of material removal with abrasive cutting edges on the wheel surface is basically the same as with cutting tools. The size of chips removed in grinding is, however, much smaller than the case of cutting providing better surface finish and machining accuracy. The order of chip thickness in grinding is far less than 0.1 mm, whereas it is larger than 0.1 mm in cutting.

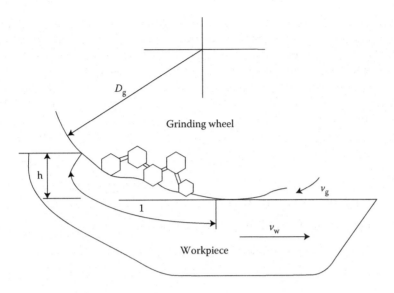

FIGURE 14.9
Parameters in surface grinding. (From Marinescu, I., Toenshoff, H., Inasaki, I., 2001, *Handbook of Ceramic Grinding and Polishing*. Noyes Publications, William Andrew Publishing, Norwich, NY. With permission.)

The average cross-sectional area of chips, A_{cu}, in grinding can be estimated as follows for surface grinding (Figure 14.9; Marinescu et al. 2001).

The material-removal rate is Q_w, where b is the grinding width, v_w is the workpiece speed, and a_e is the depth of cut.

$$Q_w = b \cdot v_w \cdot a_e \tag{14.1}$$

The number of chips produced per unit time N is given by

$$N = C \cdot b \cdot v_s \tag{14.2}$$

Assume that each cutting edge on the wheel surface produces a chip, where v_s is the grinding wheel surface speed and C is the number of cutting edges per unit area on the wheel surface. The average chip volume V_{cu} is

$$V_{cu} = \frac{Q_w}{N} = \frac{v_w \cdot a_e}{C \cdot v_s} \tag{14.3}$$

The chip length is given by

$$l_g = \sqrt{a_e \cdot d_s} \tag{14.4}$$

where d_s is the wheel diameter.

The average cross-sectional area of chips is (Green and Hannink 1989)

$$A_{cu} = \frac{V_{cu}}{l_g} = \frac{v_w}{C \cdot v_s} \cdot \sqrt{\frac{a_e}{d_s}} \tag{14.5}$$

A_{cu} is increased through increase of workspeed or depth of cut. This increases the cutting force acting on each cutting edge and results in breakage or dislodgement of the abrasive from the bond. This phenomenon is called "self-sharpening" (Marinescu et al. 2001). The surface roughness of the workpiece, however, deteriorates.

Conversely, A_{cu} is decreased through increase of wheel speed and attritious wear of cutting edges becomes more significant. In this case, heat generation increases while surface roughness becomes smaller.

14.5 ELID Grinding of Ceramics

14.5.1 Mechanism of ELID Grinding Technique

ELID is a new technique that uses electrolysis to remove bond material. The application of this dressing technique during the course of a grinding process allows chips and debris to be removed easily and allows grindability to be maintained over a long period of time. Various applications of ELID grinding have been developed. With ELID dressing, it was possible to realize mirror grinding with various types of systems using existing machine tools such as (Ohmori and Itoh 1998; Ohmori and Li 2000a)

- Surface grinders
- Rotary surface grinders
- Turning centers
- Infeed grinders
- Lap-grinding machines

The ELID system consists of a negative electrode, a brush, a power supply, a metal-bonded wheel, and an electrolytic grinding fluid (Figure 14.10).

The wheel is made the anode of the power supply through the application of a brush smoothly contacting the surface. The cathode is placed in proximity to the wheel surface. The power supply transforms an alternating current into direct current as necessary for electrolysis. Generally, ELID must be applied to a wheel, which has been trued to run concentric to the spindle axis. The shape of the wheel is profiled to the required geometry and dull abrasives are fractured. Before commencement of grinding, it is necessary to perform a predressing operation in order to expose diamond grains above the level of the bond. Truing of the grinding wheel is, therefore, required for the following reasons (Ohmori and Itoh 1998; Ohmori and Li 2000a):

- The wheel must be concentric to the spindle axis.
- The wheel profile must have the correct geometry.
- Dull abrasives need to be fractured.

After truing, an ELID predressing operation on the diamond wheel is necessary to

- Expose diamond grains above the bond level (sharpening)
- Provide chip clearance

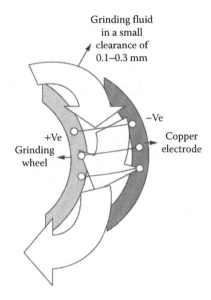

Grinding fluid in a small clearance of 0.1–0.3 mm

−Ve

+Ve

Copper electrode

Grinding wheel

FIGURE 14.10
The phenomenon of electrolysis in-between grinding wheel and a copper electrode.

Diamond grains protrude from the wheel surface after predressing of the wheel. The ELID process is performed in the following three steps (Marinescu and Ohmori 1999):

1. Precision truing of the grinding wheel carried out by the electrodischarge truing technique.
2. Predressing of the grinding wheel by electrolysis.
3. Grinding with ELID.

The cast-iron bond material of the grinding wheel is electrically conductive. Power is supplied to the metal bond through a brush that contacts the wheel. A copper electrode forms the cathode and a current is provided through the electrolyte in the gap between the anode and the cathode. Electrolysis occurs at the gap between the metal bond and the electrode. The cast iron is ionized into Fe^{2+} (Ohmori and Li 2000b). Further chemical reactions form the ionized Fe into hydroxides, and these substances change into oxides such as Fe_2O_3, which form an insulating layer. The electroconductivity of the wheel surface is reduced due to the growth of the insulating layer. After the oxide layer is formed and predressing is completed, ELID grinding can be performed (Ohmori and Itoh 1998).

Due to grinding, the oxide layer is removed when the wheel touches the workpiece. The thickness of the oxide layer is decreased and electroconductivity increases again. The metal bond is removed by electrolysis and transformed into a further oxide layer. The result is new sharp edges protruding from the surface of the wheel. This means that ELID is performed at the same time as grinding. The mechanism of ELID is shown in Figure 14.11.

14.5.2 Research Studies on ELID

ELID grinding technique was used for the first time in Japan at the RIKEN Laboratories. The pioneers of this technique are Dr. Hitoshi Ohmori and Prof. Takeo Nakagawa.

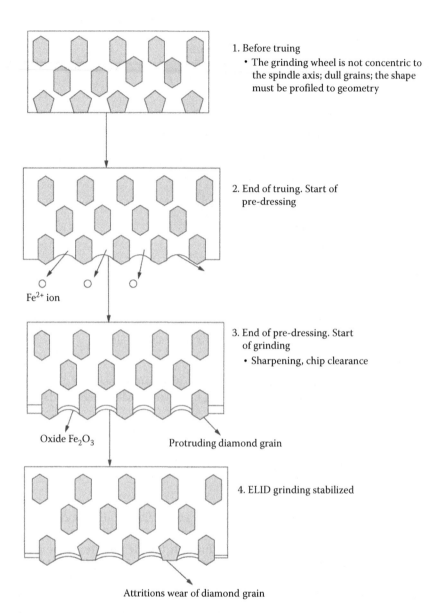

1. Before truing
   • The grinding wheel is not concentric to
   the spindle axis; dull grains; the shape
   must be profiled to geometry

2. End of truing. Start of
   pre-dressing

Fe^{2+} ion

3. End of pre-dressing. Start
   of grinding
   • Sharpening, chip clearance

Oxide Fe_2O_3 Protruding diamond grain

4. ELID grinding stabilized

Attritions wear of diamond grain

FIGURE 14.11
Mechanism of electrolytic in-process dressing.

Dr. Ohmori's new method is a result of the development over several years of the combined grinding and dressing technique for hard and brittle materials in an effort to effectively utilize cast-iron fiber-bonded wheels that are considered to be the toughest among
all metal-bonded wheels. Nakagawa found one solution for the efficient machining of the
ceramics is to use a cast-iron-bonded diamond-grinding wheel mounted on a rigid numerically controlled grinding machine. A cast-iron-bonded wheel is a metal-bonded wheel of
very high strength. It gives a low wear rate, and a high grinding ratio even at a low grinding speed (Green and Hannink 1989; Ohmori and Itoh 1998; Ohmori and Li 2000b). Ohmori
and Li (2000a), at the Institute of Physical and Chemical Research (RIKEN), successfully

applied the ELID to grind hard materials components and studied the possibility to apply this new technique for surface grinding, centerless grinding, and double-sided grinding. Efficient and precise ELID grinding was successfully carried out on Si_3N_4 and ZrO_2, titanium carbide, and sialon. Mirror surfaces were achieved using the ELID technique, the best surface roughness and precision being obtained with a #4000 diamond wheel using through-feed grinding. Marinescu and Ohmori (1999), from the Abrasive Micro-Machining Laboratory, Toledo, performed ELID centerless grinding on different types of ceramics at RIKEN Laboratories in collaboration with the ELID Grinding Project Team from Tokyo. The following conclusions were obtained:

- Stable, efficient, and precision grinding for ceramic components has been achieved by mirror grinding with ELID technique.
- Efficient and precision truing by the electrodischarge method for metal-bonded grinding wheels for centerless grinding can be achieved.
- Good ground surface roughness and accuracy are offered with #4000 metal-bonded grinding wheel by through-feed grinding for ZrO_2 optical fiber ferrules (Marinescu and Ohmori 1999).

Lee (2000), at Pusan National University, Korea, proposed a dressing system controlled by computer for ultra-precision grinding of STD-11, die steel. The superabrasive wheel is CBN 12000.

Bandyopadhyay (1996) performed ELID grinding experiments for Oak Ridge National Laboratory at RIKEN Laboratories under the supervision of Dr. H. Ohmori during two summers at RIKEN. Section I of his research addressed basic aspects of ELID grinding affecting the rate of material removal, the normal forces developed during grinding, and the conditions that improve the ratio of the material removed to that of abrasive consumed. Section II of the report addressed the effects of ELID grinding on the bending strength of silicon nitride.

Hong and Li (2000) studied theoretical aspects of ELID at the University of Rochester, New York. They studied the anodic metal matrix removal rate in ELID. The electric field or current density distribution around a diamond particle embedded in a metal anode during ELID was calculated for the two-dimensional case of a long diamond particle, without a particle and with some protrusion. The mathematical model and the assumptions are presented in two recent articles in the *Journal of Applied Physics*.

14.5.3 Summary on ELID Grinding

ELID grinding is a new technique and has given good results for machining hard silicon nitride and zirconia. This grinding technique is highly recommended for machining ceramics. However, up to this time, no published research could be found related to ELID grinding of ZTA ceramics.

The possibility of implementing this method in different types of grinding processes and for different concentrations of diamond in the grinding wheel has been demonstrated experimentally. Some characteristics related to surface finish are available. However, the influence of the grinding regime (rotational speed, depth of cut, and feed) on the surface roughness has been insufficiently explored.

A mathematical model of the anodic metal matrix removal rate in ELID has been developed. This theoretical approach, however, has not yet been applied to the modeling of the predressing time.

References

Bandyopadhyay, B. P., 1996, Application of electrolytic in process dressing for high efficiency grinding of ceramics parts. Oak Ridge National Laboratory report, Research activities 1995–1996, ORNL/SUB/96-SV716/1.

Bifano, T. G., 1998, Ductile regime grinding of brittle materials. PhD dissertation, North Carolina University.

Bifano, T. G., Dow, T. A., and Scattergood, R. O., 1998, Ductile-regime grinding: A new technology for machining brittle materials. *Intersociety Symposium on Machining of Advanced Ceramic Materials and Components*, ASME, New York.

Carius, A., 2002, Characteristics of superabrasive diamond and CBN micron powders. GE Superabrasives, Superabrasive Certification Course, University of Toledo, May–October.

Green, D. J., Hannink, R. H. J., 1989, *Transformation Toughening of Ceramics*, CRC Press, Boca Raton, FL.

Hong, C. J., Li, M., 2000, Anodic metal matrix removal rate in electrolytic in-process dressing. Part I: Two-dimensional modeling. *Journal of Applied Physics*, 87(6), 3151–3158. Part II: Protrusion Effect and Three-Dimensional Modeling, 3159–3164.

Inasaki, I., 1998, Fluid film in the grinding arc of contact. Contribution at January CIRP Meeting, Paris, 27–31.

Kalpakjian, S., 1997, *Manufacturing Processes for Engineering Materials,* 3rd edition. Addison-Wesley.

Lee, E. S., 2000, The effect of optimum in process electrolytic dressing in the ultra precision grinding of die steel by a superabrasive wheel. *International Journal of Advanced Manufacturing Technology,* 16, 814–821.

Malkin, S., 1989, *Grinding Technology. Theory and Applications of Machining with Abrasives.* Ellis Norwood Ltd., U.K.

Marinescu, I., Ohmori, H., 1999, ELID grinding of ceramic materials. *Finishing of Advanced Ceramics, Ceramic Transactions*, 102.

Marinescu, I., Toenshoff, H., Inasaki, I., 2001, *Handbook of Ceramic Grinding and Polishing.* Noyes Publications, William Andrew Publishing, Norwick, NY.

Ngoi, B. K. A., Sreejith, P. S., 2000, Ductile regime finish machining—A review. *International Journal of Advanced Manufacturing Technology*, 16, 547–550.

Ohmori, H., Itoh, N., 1998, Performances on mirror surface grinding with ELID for efficient fabrication of precision cylindrical components of hard materials. *International Journal of the Japan Society for Precision Engineering*, 32(5), 1451–1457.

Ohmori, H., Li, W., 2000a, Efficient and precision grinding of small hard and brittle cylindrical parts by the centerless grinding process combined with electro-discharge truing and electrolytic in-process dressing. *Journal of Materials Processing Technology*, 98, 321–327.

Ohmori, H., Li, W., 2000b, Fine cylindrical machining technique for hard material components by centerless grinding process combined with electro-discharge truing and electrolytic in-process dressing. *Finishing of Advanced Ceramics, Ceramic Transactions*.

Ruhle, M., 1985, Ceramic microstructures and properties. *Journal of Vacuum Science & Technology*, A3, 749–756.

Sornakumar, T., Gopalakrishnan, M. V., 1995, Development of alumina and Ce-TTZ ceramic-ceramic composite (ZTA) cutting tool. *International Journal of Refractory Metals and Hard Materials*, 13, 375–378.

Uhlmann, E., 1998, Technological and ecological aspects of cooling lubrication during grinding of ultrahard materials. *Ultrahard Materials Technical Conference*, Conference Papers.

Zhang, B., Bheng, X. L., Tokura, H., Yoshikawa, H. M., 2003, Grinding induced damage in ceramics. *Journal of Materials Processing Technology*, 132, 353–364.

15

Grinding Machine Technology

A detailed discussion of machine tool design is beyond the scope of this book. The following sections highlight some important features in the design and construction of grinding machines and their elements. For more details, there are publications dedicated to general field, such as Slocum (1992) and for grinding machine principles, Rowe (2014).

15.1 The Machine Base

15.1.1 Introduction

The machine base of a grinder is the most fundamental element on which all the active and passive assemblies are carried. It must be rigid yet deform elastically, must inhibit vibration, survive in a hostile shop environment, and be thermally stable or at least have predictable thermal properties. Construction is usually based either on a casting or weldments, and selection is generally based on cost and/or application (Figure 15.1).

15.1.2 Cast-Iron Bases

Class 40 cast iron (40,000 psi tensile strength) is the traditional material and still viewed by many original equipment manufacturers (OEMs) as the best material for a machine tool base. Castings have a good stiffness to weight ratio and good damping qualities. The material is inexpensive and easy to mold in required shapes to include coolant ways and run-offs, hydraulic and electrical conduit holes and honeycombing to further improve weight/stiffness ratio.

Problems with castings are first that they require a pattern and second they must be stress relieved both of which can be very expensive for larger forms. High-production grinders produced in volume and/or designed to a common base can absorb these costs. Custom or one-off grinder manufacturers are usually forced to use less expensive alternatives.

15.1.3 Reuse of Cast Bases

The desire for stable cast-iron bases has led to a whole industry of machine tool builders who are referred to as "remanufacturers." These builders will take an old machine tool, strip it down to the cast-iron base that has become extremely stable over its lifetime, and essentially reassemble the grinder using all the latest components of current grinding technology. This should not be confused with "retrofits" of electrical, hydraulic or lubrication updates to a machine, or a "rebuild" which is the reworking of particular parts of a grinder to bring it up, at best, to the original OEM specification.

FIGURE 15.1
Goldcrown centerless grinder base casting; side and underside views. (Courtesy of Landis Grinders.)

15.1.4 Welded Bases

Where cast iron is impracticable from a cost point of view several alternatives are used normally based around welded steel. This material has a higher modulus and strength than cast iron so machine tool builders will weld steel sections together with ribbing to provide additional stiffness. Weld joints add some vibration damping and each additional rib weld will also add mass where required. Nevertheless, simple steel weldments have considerably less damping ability than cast iron. This can be improved slightly by pumping coolant through ducting which also aids thermal stability.

15.1.5 Damping in Machine Tools

Most machine tools are lightly damped (Tobias 1965). Typically, the damping ratio is approximately 0.05–0.15. Damping in a machine tool derives mainly from sliding joints and bearings. Significant damping also arises from screwed and bolted joints. The reason that damping arises from joints is because dissipation of vibration energy requires irreversible strains to take place on a continual basis. Irreversible strains or friction, as it is more commonly known occurs through plastic and viscous shear mechanisms and not through elastic deflections from which the vibration energy is recovered. Frictional dissipation occurs mainly in joints and only in a minor way by deflections within the body of a casting. It is for this reason that the main structural elements provide only a minor contribution to the total damping of a machine. In order to introduce damping into a structure, it is necessary to engineer the distribution of mass, structural stiffness, friction and movement within the design.

15.1.6 Large Mass Bases

One approach to design of a machine base is to make it truly massive. An example is the use of a solid granite base described below. This makes the machine base very rigid and minimizes distortion of the base when subjected to a rocking motion. Most machines exhibit rocking behavior in the frequency range 20–35 Hz as described by Tobias (1965). The rocking motion has to be minimized by careful attention to the foundation and mounting on the foundation. From a vibration viewpoint, a massive base reduces the rocking frequency to a low value. This has the advantage that it tends to isolate the rocking vibration from the higher frequency modes of the machine. A massive and rigid base therefore inhibits vibration of the whole machine assembly rather than contributing toward it.

15.1.7 Tuned Mass Dampers

A particular vibration can be substantially reduced by addition of a spring-mounted mass tuned to have the same resonant frequency as the unwanted vibration. The additional mass required can be less than one-fifth the size of the mass requiring to be damped (Den Hartog 1956). For example, if the rocking frequency of the base is of particular concern, a mass suspended on springs within the base and tuned to the appropriate frequency will greatly reduce the amplitude of vibration for a relatively low cost. This principle was employed in a 508 m high tower in Taipei to eliminate sway of the building. A large steel ball weighing approximately 600 ton was suspended internally at the top of the tower.

15.1.8 Composite Material Bases

Builders have sought alternative means of providing improved damping. One method is the use of a polymer matrix composite made using crushed concrete, granite, or quartz with trade names such as Granitan S103® (Studer), Mineralit® (Emag 1998), and Micro-Granite (Elb 1997). The materials have significantly greater damping characteristics than steel or even cast iron see Figure 15.2. They can be cast into almost any shape, do not require aging or annealing, have a third the density of cast iron, and can support rails and slideways if inserts are used to anchor them. They are used either as fillers in weldment structures or as monolithic bases (Drake 2000). In the latter case, foam cores may be added to improve the weight/strength ratio. With appropriate design, a monolithic polymer structure, particularly for a low load-bearing application, can have the same stiffness as cast iron but much greater damping. Perhaps the optimum use of polymers, however, is as used in the Viking centerless grinder (Viking 1998), where epoxy-granite is used as a filler material in a nodular cast-iron base thus gaining the best of both worlds.

It should be noted when designing machine bases that polymer matrices are not as strong as cast iron and have a much lower thermal conductivity. Arnone (1997) provides the following data for one particular grade of polymer concrete.

An example of a company that has made significant effort to improve upon a standard weldment is Weldon Machine (York, Pennsylvania; Weldon 1991) (Figure 15.3). This

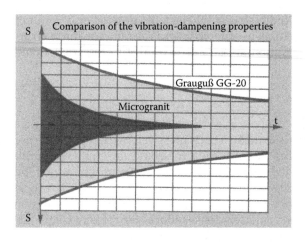 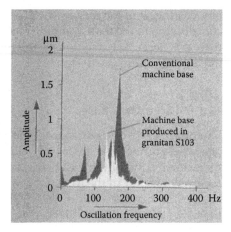

FIGURE 15.2
Damping improvements of polymer matrices relative to cast iron. (From Elb, 1997, *Micro-Cut A Production Machine for Precision- Profile- & Creepfeed Grinding.* Elb-Schliff Werkzeugmaschinen GmbH, Frankfurt, Germany. Trade Brochure, and Studer, n.d. *All about S21 Lean CNC.* Studer Schleifring Group. Trade Brochure. With permission.)

machine tool builder makes both standard cylindrical grinders and custom machines with unique base shapes. In earlier efforts to improve damping, they used expansive concrete as filler in a welded steel base.

This proved quite effective but added significantly to the weight of the base. More critically the thermal expansion mismatch could cause structural bowing. Weldon therefore worked jointly with Slocum of MIT to develop a more thermally stable system based on internal viscous damping (Hallum 1994; Weldon 2 1994). The "Shear Damper®" base consists of a series of stiff steel tubes wrapped with a highly viscous polymer tape. The tubes are suspended within the base weldment leaving a 3- to 5-mm gap with the walls. This gap is injected with an epoxy material constraining the viscous layer such that its only movement is in shear that dissipates vibration energy. The beauty of this method is that

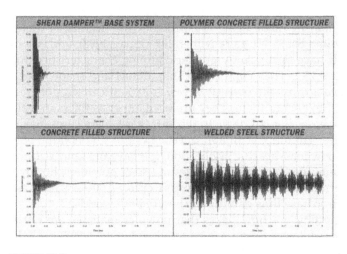 

FIGURE 15.3
Frequency response of various Weldon machine base constructions. (From Weldon 2, 1994, *We've Struck Gold!* Weldon Machine Tool, York, USA. Trade Brochure.)

it decouples the damping and stiffness functions of the structure and is broad frequency based unlike, for example, a tuned mass damper. The internal tubes can also act as conduits for coolant or other fluids to control the base temperature.

15.1.9 Granite Bases

For grinders where thermal stability is absolutely critical an alternative approach is to use solid granite as used for inspection tables. Tschudin reports using Granitline® a natural quarried granite for the bases of high-precision centerless grinders. Buderus (1998) also reports using a natural granite bed for its combination grinding/hard-turning centers that has both higher static stiffness and better damping characteristics than a typical cast-iron base. The use of solid granite illustrates the application of the massive base principle discussed above.

15.2 Foundations

Floor construction is a critical consideration when installing a new grinder. It must first and foremost be capable of supporting the weight without deforming. It should also absorb shock and isolate the machine from adjacent machine noise. The principles of vibration isolation of machines are described in detail by Den Hartog (1956). For light machines, a sealed concrete floor 150–200 mm thick is usually sufficient in combination with elastic pads on the machine. Larger grinders will require an independent thick cast slab that may have to be isolated from the factory floor by shock-isolation elements. Since these slabs are made of concrete which are slow to react to changes in temperature and therefore liable to warping, good temperature control of the factory is recommended. Many grinder bases, such as the one in Figure 15.4, are designed with 3-point floor support that allows,

FIGURE 15.4
Construction steps in a polymer concrete-filled steel base weldment. (Courtesy of Weldon Solutions.)

assuming the base has sufficient stiffness, for leveling of the grinder on uneven floors. However, with care of installation, a 5-point mount is actually more desirable in eliminating vibration modes especially from rocking or reversal of machine axes on machines such as reciprocating surface grinders (Yoshida 2000).

A machine table that reciprocates creates substantial inertia forces during the reversal of motion. A reciprocating surface grinder unless anchored to the foundation, tends to "walk" across the floor. This can be avoided by using anchor bolts sunk into the foundation. Provision must be included for leveling the machine on the foundations and also for vibration damping.

15.3 Guideways

15.3.1 Introduction

Guideways are the elements or surfaces that carry and guide the moving elements such as workpiece holder/drives and wheel and dresser heads. Guideways can be divided into two categories—sliding and rolling element. Guideways of either type are often known simply as slideways. Their selection is dependent on many factors such as speed, acceleration and range of motion, accuracy and repeatability, stiffness, damping characteristics, thermal stability, weight, load capacity, slip-stick characteristics, ease of manufacturing, and cost including support equipment, for example, hydraulic pumps, chillers, filters, etc. In many cases, the choice is also dependent on which machine axis is involved and it is therefore opportune at this point to define the primary machine axes as defined by International Machine Tool Standards (Smith 1993), and some examples for the more common grinder types.

15.3.2 Definition of Axes

The six degrees of motion possible are shown in Figure 15.5.

X axis—The horizontal axis of motion parallel to the work-holding surface

Y axis—The axis of motion perpendicular to both the X and Z axes

Z axis—The axis of motion parallel to the principal spindle of the machine

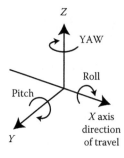

Pitch—rotation about a horizontal axis perpendicular to the direction of motion

Roll—rotation about the direction of motion

Yaw—rotation about a vertical axis perpendicular to the direction of motion

FIGURE 15.5
Axis error definitions in machine tool technology.

A axis—The axis of rotary motion of a machine tool member about the *X* axis

B axis—The axis of rotary motion of a machine tool about the *Y* axis

C axis—The axis of rotary motion of a machine tool about the *Z* axis

Additional linear axes follow sequentially backward through the alphabet, for example, dressing axes tend to be *U* and *V*, while additionally rotary axes follow sequentially forward through the alphabet. Axes carrying in-line probes are also given letters generally out of sequence. After the three principal axes, labeling varies from one machine tool builder to another.

Misalignments in a system lead to the following types of error in slide motion.

15.4 Slideway Configurations

15.4.1 Introduction

Slideways traditionally refer to a range of linear sliding contact guideways based on square, *T*, or Flat and Vee cross sections. In recent decades, the advent of numerical control (NC) and computer numerical control (CNC) machines has led to rolling element slideways for ease of positional control.

15.4.2 Flat and Vee Way

The simplest load-bearing arrangement is the "Flat and Vee" slideway. This uses gravity to preload the bearing configuration. It is considered by many machine tool builders as the optimum configuration for horizontal axes on large grinders under heavy but even downward loading. The path accuracy can be <1 μm and wear of the slide is self-compensating (Waldrich Siegen 1996). Under uneven or changing load, it is prone to yaw errors. It is most commonly seen still on large surface grinders, for example, bed grinding or on cylindrical roll grinders that handle workpieces up to 10 tons or more. Figure 15.6 shows a photograph of the Flat and Vee way on a Favretto slideway grinder. The bearing surfaces are hand scraped for good bearing contact and oil retention. The upper and lower surfaces are kept lubricated by oilers that consist of a series of rollers semisubmerged in small oil reservoirs. The method is simple and eliminates the need for hydraulic pumps and plumbing.

Alternatively, oil is supplied under pressure from an external pump and fed through oil lubrication channels such as those shown on the upper slide portion of the *X* axis of a Mattison surface grinder (Figure 15.7).

15.4.3 Double Vee Slideway

A variation on the Flat and Vee is the Double Vee slideway. This again depends on gravity for its preload and is self-compensating for wear. However, it can handle changes in the load position with much less yaw error. The major difficulty with this slideway design is in the manufacture, especially over any significant length. For this reason, it is most often used on short stroke axes such as the *Z* axis on small manual and NC surface grinders, for example, Acer or as the example in Figure 15.8 (Kent n.d.).

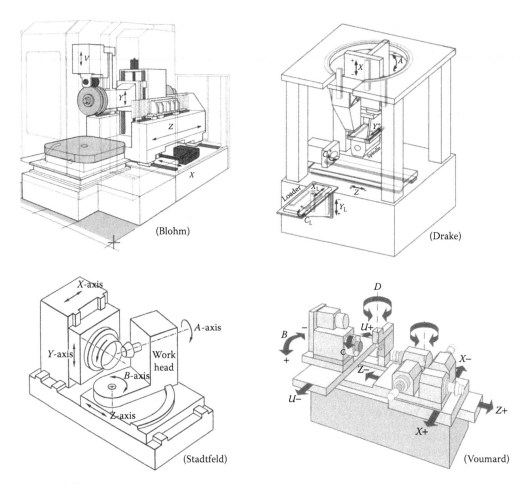

FIGURE 15.6
Examples of machine axes.

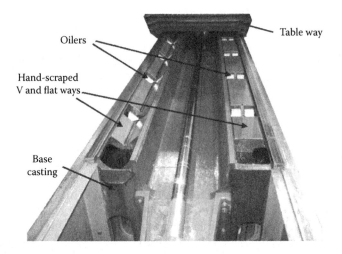

FIGURE 15.7
Example of flat and vee sideways.

FIGURE 15.8
Example of double vee slideways. (From Kent, n.d. *Precision Surface Grinder KGS-250 Series*. Kent Industrial Co, Taiwan. Trade Brochure. With permission.)

15.4.4 Dovetail Slideway

Other designs of slideway cross sections do not have the luxury of using gravity for preload and have to incorporate some form of constraint or box. The standard Tee-shaped slideway is shown in Figure 15.9 (Slocum 1992).

Preloading must now be achieved by gibs that are held in place by torqued setscrews. This leads to problems with backlash over time as the gibs wear. This particular design is more common in machining centers and mills where it can provide excellent damping at relatively low cost.

Plain constrained slideways offer probably the greatest stiffness and damping of any guideway system. It is perhaps not surprising that the Viking centerless grinder, which was designed specifically around maximizing its stiffness at 3 m lbf/in., uses dovetail plain slideways.

Koyo makes use of a square slide version in order to obtain very high stiffness within a confined space for a slide to hold an upper vertical spindle for its superabrasive double disk grinder (Figure 15.10; Koyo n.d.). The sliding surface is hardened chrome, while the slideway is coated with Turcite® (see below).

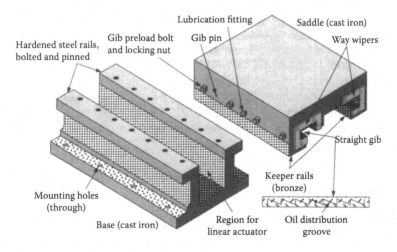

FIGURE 15.9
Example of dovetail slideways. (From Slocum, A. H., 1992, *Precision Machine Design*. SME. With permission.)

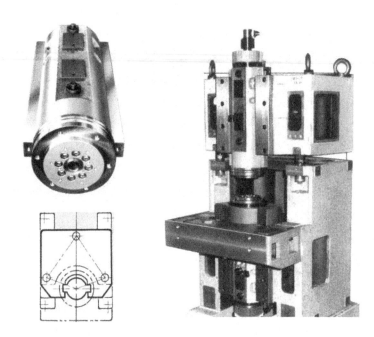

FIGURE 15.10
Koyo machine KVD series vertical spindle surface grinder.

15.4.5 Plain Slideway Materials

Slideways are traditionally made of cast-iron sliding on cast iron making use of the graphite present as lubricant or on porous brass, especially where gibs are applied. The brass is designed to wear preferentially while providing good lubrication.

The major limitation of these combinations is static friction or stick-slip that can create drive problems with servo errors. One way to reduce this is to use PTFE-based coatings with bronze fillers such as Turcite and Roulon®. These have been used extensively, for example, Koyo, Kent, Chevalier 2000, Studer. Under load the material has a static friction coefficient within 20% of its dynamic friction coefficient of 0.015 (Slocum 1992).

Alternative approaches include hydrostatic slideways and rolling element slideways. The use of simple ball-bearing ways as in the example of the X axis slide on a Kent manual surface grinder Figure 15.11, or preloaded nonrecirculating needle rollers (Figure 15.12), as used by, for example, Fortuna in camlobe grinders in the 1980s and by Mitsui NC thread grinders at about the same time (THK n.d.). These guideway systems have less stiffness and damping than plain slideways while still having nonzero stick-slip. Rolling element slideways are discussed in more detail below after the discussion of hydrostatic slideways.

15.5 Hydrostatic Slideways

In high-precision, high-production grinders, rolling element slideways tend to have been replaced by advances in hydrostatic bearing design unless there are severe space limitations.

FIGURE 15.11
Example of rolling slideway machine. (From Kent, n.d. *Precision Surface Grinder KGS-250 Series*. Kent Industrial Co, Taiwan Trade Brochure. With permission.)

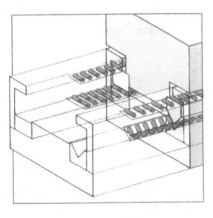 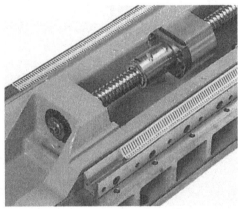

FIGURE 15.12
Example of rolling slideway machines. (From Fortuna, 1991, *Automated Camshaft Grinding*. Fortuna-Werke, Stuttgart, Germany, Trade Brochure With permission.)

15.5.1 Hydrostatic Bearing Principle

A hydrostatic bearing is noncontact in that the slide is supported by a film of high-pressure oil of the order of 10 micron thick. Stick-slip is therefore eliminated. It functions by producing a continuous flow of high-pressure oil through and out of the bearing, the flowrate being controlled by restrictors such as small orifices or capillaries. The control restrictors are necessary to allow the slideway to cope with varying applied loads and maintain a near constant film thickness. Throughout the bearing surface small pockets or pads are machined to create areas of high pressure fed by these restrictors and from which the oil flows to the rest of the bearing surfaces before exiting the bearing. The oil is then collected, filtered, repressurized and recirculated. Information on the operating principles and design of hydrostatic slideways is given by Rowe (1983, 2012).

15.5.2 Plane-Pad Hydrostatic Slideway Configurations

As shown in the examples, Figures 15.13 through 15.16, opposed pad hydrostatic ways are used extensively in high-production grinders both in Flat and Vee type configurations, constrained Flat and Vee configurations to handle side force, and square and *T* constrained ways for vertical and high-load normal applications. Figures 15.17 and 15.18 show round slideways.

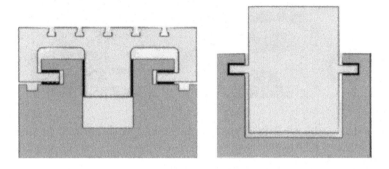

FIGURE 15.13
Magerle MFP surface profile grinder horizontal *X* axis and vertical *Y* axis hydrostatic bearing slideway. (From Magerle, n.d. *Magerle Grinding Systems.* Schleifring-Gruppe, Uster Switzerland. Trade Brochure. With permission.)

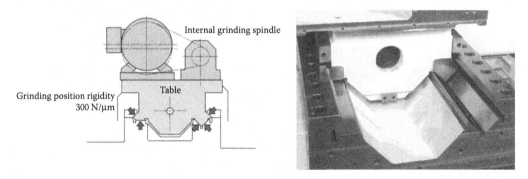

FIGURE 15.14
Five-way restrained Z axis hydrostatic slide on an Okuma GI 10 internal grinder. (From Okuma 1 2000, GI-10N CNC Internal grinder for mass production. Okuma Corp, Aichi, Japan, Trade Brochure.)

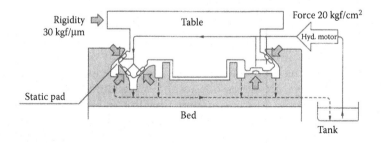

FIGURE 15.15
Five-way restrained *X* axis hydrostatic slide on an Okuma GC-super33 high-speed cam grinder. (From Okuma 2 1998, *GC-Super33 Triple-Speed Cam Grinder.* Okuma Corp, Aichi, Japan, Trade Brochure.)

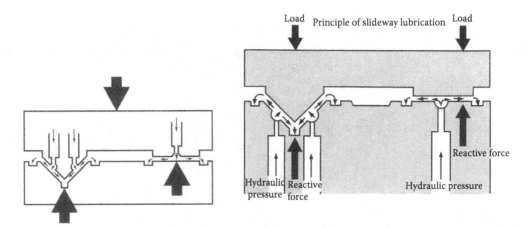

FIGURE 15.16
Cylindrical grinder examples of the use of hydrostatic flat and vee ways. (From Shigiya 2, n.d. *Cylindrical Grinders—General Catalog*. Shigiya Machinery Works, Hiroshima, Japan, Trade Brochure; and Toyoda, 1994, *GL3A GL3P CNC Cylindrical Grinder*. Toyoda Machine Works, Japan, Trade Brochure. With permission.)

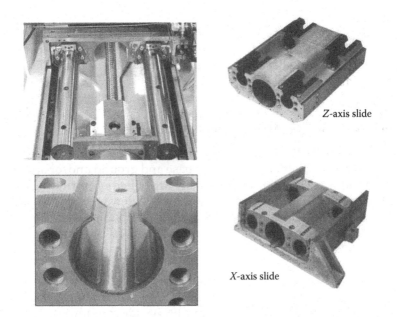

FIGURE 15.17
Bryant UL2 hydrostatic round slideway designs.

15.5.3 Plane-Pad Hydrostatic Flowrate

For a hydrostatic bearing, the pad pressure is related to the flowrate by (Rowe 1983, 2014):

$$q = Ph^3 \frac{B}{\eta}$$

where q is the flowrate, h is the film thickness, η is the film viscosity, and B is the flow factor (constant).

FIGURE 15.18
Danobat round hydrostatic slideways for the internal grinding Z axis on the 585 Universal Grinder.

Due to the cube relationship between flowrate and film thickness, slideway performance is very sensitive to the accuracy of manufacture. Also the viscosity depends on oil temperature and hence flowrate depends on temperature. Chillers are required and oil must be filtered to <3 μm.

15.5.4 Hydrostatic Slideway Materials and Manufacture

Generating the necessary tolerance on any slideway system can be expensive and time consuming. The ways are either cast into the base and ground and/or are scraped, or the base is ground and ways bought "off-the-shelf" are bolted on. Grinding a base requires a significant investment in capital equipment for a slideway grinder, while scraping is particularly skilled and labor intensive and may still be necessary for preparing the base. Low friction coatings are added in many cases to the surface as insurance in case of a loss of hydrostatic pressure. The ways can still act as plain bearings to avoid damage.

Alloys such as Moglice® have been developed which are low friction castable materials that requires only a master to produce a finished surface. An example of its use is on the X and Z axes of the Bryant UL2 grinders (Bryant 1995) using roundway hydrostatic designs (Figure 15.17).

15.5.5 Round Hydrostatic Slideways

Round hydrostatic slideway designs have been available since the 1960s but it was the use of replication methods to control film thickness that resulted in slides that could move accurately and smoothly at up to 46 m/min (1800 in./min). Danobat (1991) has also used round hydrostatic bearings in the Z-axis internal grinding slideway of their 585 Universal grinder. Round slides require close control to ensure precise parallel assembly of the slideways.

15.5.6 Diaphragm-Controlled Hydrostatic Slideways

A simple double diaphragm valve design was developed by Rowe for hydrostatic slideways and for hydrostatic journal bearings. Designs of this type have been successfully implemented in various machine tools and give very high bearing stiffness (Rowe 1969, 1970, 2012).

The pressures at the individual pads can nowadays be controlled using ultrasensitive pressure membrane monitors and computer algorithms that regulate and optimize the flow and pressures. The method both increases the stiffness and reduces the oil demand of the system: this in turn reduces heat generation. The approach has been reported to be common on cylindrical, gear, and crankshaft grinders in Europe (Anon 1999).

15.5.7 Self-Compensating Hydrostatic Slideways

In addition to the shear damping base design, Slocum (1992) and Weldon (Hallum 1994) developed a self-compensating hydrostatic bearing design for use on their Model 1632 cylindrical grinder.

A gear pump provides vibration-free fluid delivery to the self-compensating pads. The fluid enters the pad through ports at the edge of each and flows to collector grooves at the center that act as a feedback source for the bearing pads. There are two travel paths for the fluid one from the lower self-compensating pad to the upper bearing pad and a similar one from the upper self-compensating pad to the lower bearing pad. As a load is applied over a compensation pad, the pad restricts fluid to the opposing bearing pad and hence directs additional fluid to the bearing pad opposing the load. The beauty of this system is it insensitive to manufacturing errors in the bearing gap and insensitive to dirt because there are no small diameter passages. Typical port size is 3 mm. It even permits the use of low-viscosity water-based coolants as the fluid medium.

This type of slideway supplied by New Way® Machine Components Inc. has also been incorporated in machine design by Torrington's Advance Machinery Center (Sotiropoulos 1998).

15.6 Recirculating Rolling Element Slideways

Slideways employing recirculating rolling elements were the salvation of the machining center manufacturers as a means of reducing manufacturing costs while increasing speed for high-precision circular interpolation with CNCs. They also virtually eliminated stick-slip ($\mu = 0.001$–0.003) and accompanying loss of motion, without the need for expensive hydrostatic way designs. They were light and allowed much faster acceleration and deceleration than plane sliding bearing ways with less energy and build up of heat. Plane sliding bearings are limited to 20–25 m/min, while rolling element bearing designs can run up to and over 120 m/min.

A grinder axis with rolling element slideways will usually consist of two guideways or rails and four bearing blocks to carry the spindle or work table. The rails come preground with straightness accuracy down to 1 μm or less which are bolted directly to the machine bed. This can be facilitated without scraping or grinding of the bed surface by using 3-point laser aligning of the rails and then filling underneath with a rigid epoxy. The approach creates 100% contact for good damping and stiffness although the process can be lengthy.

There are two types of rolling element—ball and roller. Regardless of type, the load is carried by two or more channels filled with bearing elements that circulate through them. Ball-bearing slideways can be made more accurately, are slightly less expensive, and can handle faster velocity, acceleration, and deceleration rates than roller slideways. Rollers, on the other hand, have better damping and stiffness characteristics, and can handle twice the load for the same-sized bearing package. Rollers, however, are more sensitive to alignment errors, and with problems with skew caused by the end of the rollers getting pinched such that only the corners bear the load. For this reason, they are often crowned but this reduces the length of contact and hence level of damping.

Rolling element slideways are much more sensitive to crashes than plane slideways and provide less damping especially in the direction of motion where the damping is almost nonexistent. In some cases, additional plane-bearing elements may be added to compensate. Also, the bearings are preloaded, by making the balls or rollers slightly larger than the channel, such that they distort increasing their contact area and hence damping capacity. The slideways are also very sensitive to contamination and must be kept clean of grinding debris or coolant. This mandates using wiper seals on the sliding trucks and slideway covers.

Truck spacing along the length of a rail should be at least twice the width between the rails and never less than 1:1. For additional stiffness, four or more rails may be used as shown in Figure 15.19. This shows the Y-axis guideway for a 700 series Campbell vertical grinding center with four rails together with a stacked two-rail X-axis slides. Note also the oil lubrication lines for each truck (Figure 15.20).

The three most common suppliers of rolling element slides appear to be Schneeberger, THK, and INA bearings based on grinders made in or entering the U.S. market.

Schneeberger supply roller-bearing slides with two sets of opposed, crowned rollers in flat raceways. The geometry of the forces is such that the resultant points far outside the rail for greater stiffness (Figures 15.21 and 15.22).

INA uses three tracks of ball bearings with contact angles optimized to carry compressive and side loads.

THK provides primarily recirculating-ball rolling elements because of the greater capacity rolling balls to cope with misalignment. In some cases, flexible plastic retainer rings

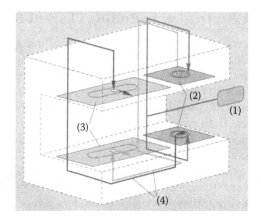

FIGURE 15.19
Opposed pad hydrostatic slideway bearings. (After Hallum, D. L., 1994, A machine tool built from mathematics. *American Machinist*, October.)

FIGURE 15.20
Campbell 700 series vertical grinding center during construction.

keep the balls separated to reduce friction, increase damping, and aid lubrication by carrying the grease.

The selection of a particular type of slide is often dependent on the personal choice of the individual design engineer based more on delivery, price, and technical support. Roller bearings, with their higher load capacity and slightly greater resistance to shock, appear to dominate grinders for creep-feed grinding, while ball bearings are used more for high speed on hybrid machining/grinding centers.

The slides must be kept well lubricated with a thick oil film to prevent fretting. Grease or slideway oil is used. The simplest greasing method is a manual grease gun as part of standard planned maintenance practice. However, automated grease or forced oil lubrication from a central pump is perhaps more reliable. Grease is typically used where there are a lot of short fast movements such as oscillation in internal grinding where an oil film could break down too quickly. Be it roller or ball, the slideways should be run over their

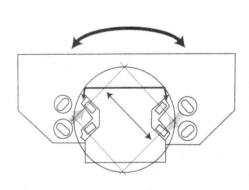
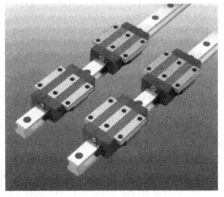

FIGURE 15.21
Example of recirculating rolling element slideways. (From Schneeberger.)

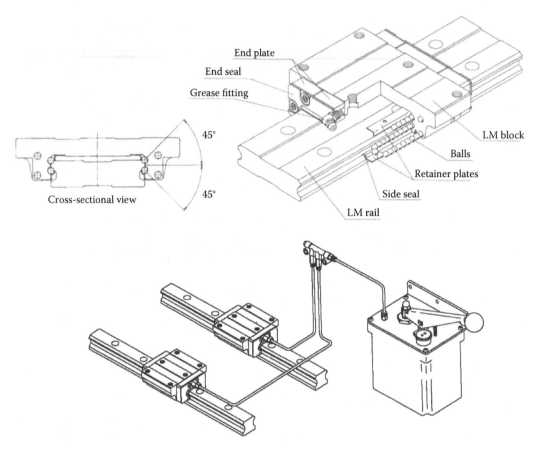

FIGURE 15.22
Examples of recirculating rolling element slideway.

full length every so often to ensure a good oil film is maintained. Under normal operations such a system should need lubrication only every 100 km of operation.

It is also recommended that the rails be flash chromed due to coolant, even with the best of slideway covers, to prevent corrosion. This is a treatment provided by the rail manufacturer.

15.7 Linear Axis Drives and Motion Control

15.7.1 Introduction

Motion control is the technology required to drive the carriages on the machine slideways and ensure that the motion is correct. Glancing through any used machinery brochure, it is easy to understand the changes that have occurred in the design of machine tools over the last 25 years. In no area is this more apparent than in linear axis motion control. Prior to about 1980, nearly all machine tools were hydraulically driven; today more than 80% of new machines are driven by servo motors with ball screws. The driving factors for this have been cost, advances in motor, encoder, and machine control technologies, reduction in the number of mechanical machine components for easier maintenance, and the desire to eliminate hydraulics from machines for heat and vibration reasons.

15.7.2 Hydraulic Drives

A hydraulic drive consists of a cylinder with a piston and rod that slides in and out of it. The force produced is the product of the applied pressure on the piston and the area over which it acts. Manual or servo valves control velocity and position by controlling the flow of fluid in and out of the cylinder. The cylinder can be either single action, often used in opposed pairs, or double action. Hydraulic systems usually offer only medium accuracy and are now most often used in surface grinders with a large range of travel such as in a slideway grinder.

15.7.3 Electrohydraulic Drives

For precision applications requiring only a short range of motion, such as some of the dresser infeed examples illustrated in the previous chapter, electrohydraulic drives are used. These consist of a sealed system with two bellows; a small diameter, long stroke master connected to a large diameter short stroke slave. A motor-driven screw compresses the slave bellows and actuates motion of the master bellows. The resolution of the screw motion is increased by the ratio of the master/slave bellows diameter.

15.7.4 AC Servo and Ball Screw Drives

Ball screw and AC servo-drives are the basis of most of the current grinders on the market today and led to the elimination of hydraulics. They consist of several components namely the recirculating-ball leadscrew, or ball screw, ball nut, motor, and positional/speed monitoring encoders as illustrated in Figure 15.23.

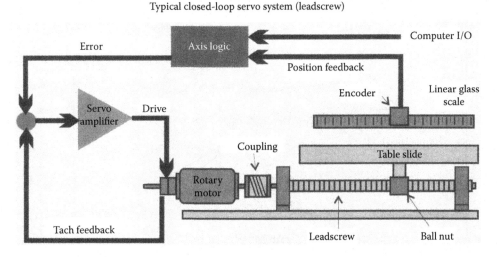

FIGURE 15.23
Typical closed-loop servo-drive ball screw system.

15.8 Elements of AC Servo-Drive Ball Screw Systems

15.8.1 Ball Screw

A ball screw (or recirculating-ball leadscrew) is at the heart of a controlled linear motion system. It is a precision ground hardened steel or stainless steel threaded shaft on which a ball nut rides with rolling contact to provide a positive, high efficiency of transmission, and low friction. The ball screw is manufactured by turning, heat treating, and then finish grinding the thread. In the process of finish grinding, the shaft will "unwind" developing a lead error both cyclic and accumulative. Some of this is compensated in the grinding process but nevertheless there is a residual error.

The class of ball screw is governed by a combination of accumulated lead error and cyclic variation. For a nominal 1 m length, a C0 class would have <8 μm accumulated error and <6 μm variation, while a C3 class would have <21 μm accumulated error and <15 μm of variation. Accumulated error can be programmed into modern CNCs as an offset to correct for the error (Figure 15.24).

Lower quality ball screws are made by rolling the form then heat treating. They are sometimes used on low-cost surface grinders, axes requiring noncritical accuracies, or for materials handling and peripherals. Typical ball-screw arrangements are shown in Figure 15.25.

Mounting of the ball screw is somewhat dependent on the speed of operation and personal preference. For grinders, the end coupled to the servo motor is always held fixed using a pair of angular contact bearings. The other end may also be fixed with a similar bearing arrangement, or just supported with a plane bearing. Fixing the bearing at both ends provides the greatest stiffness and raises the critical speed at which vibration is induced and

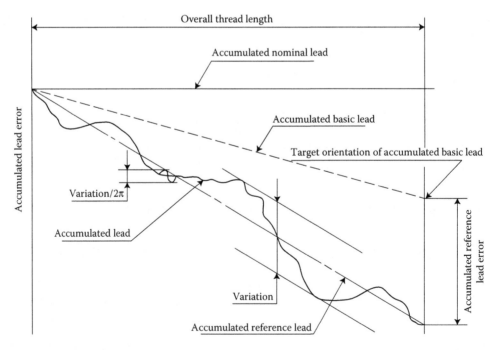

FIGURE 15.24
Ball screw lead errors. (After THK 2 1995, THK LM System Ball Screws Catalog #75-1BE. With permission.)

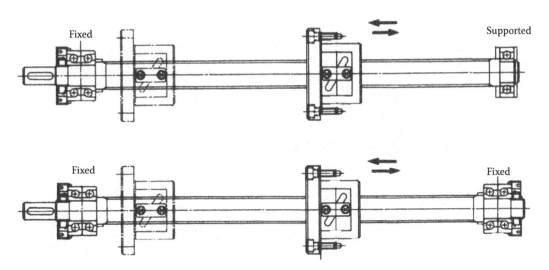

FIGURE 15.25
Typical ball screw assemblies.

is used especially for higher speed requirements. A single fixed end allows for thermal expansion without distortion of the screw. However, heat is also expected to be a greater issue with higher speed. For this reason, screws fixed at both ends for higher speeds are often preloaded axially to allow for some expected thermal growth. Hollow screws are also available through which coolant can be pumped to control thermal expansion.

15.8.2 Ball Nut

A ball nut is the other mechanical part of the ball screw system. Somewhat akin to linear ball-slide systems, ball nuts consist of a nut within which is a recirculating helical track for the bearing balls. There are a number of designs available with varying number of turns and internal or external recirculation of the balls, or with 1 or 2 starting points. The selection usually resolves around a compromise of friction, noise, damping, etc. As bearing balls are easily made with high precision, most accuracy problems usually lie in the shaft. Light preload is achieved by the use of slightly oversized balls. Medium to heavy preload to eliminate backlash is achieved by the use of two nuts with a spacer between them (Figure 15.26; THK 2 1995).

The speed capability of a ball screw system is related to its DN value, the ball diameter in mm × speed in rpm. Values range up to 120,000, which cover the highest speed required of most classes of grinder. The current limits of ball screws for motion accuracy suitable for grinders are about 40 m/min, although some hybrid machining/grinding centers can achieve 90 m/min and 1g acceleration. Most standard grinders will use ball screws with a pitch of about 5–10 mm and a diameter of at least 35 mm. Higher speed hybrid machining/grinding centers use pitches from 20 to 30 mm.

Special attention should be paid to the mount for the carrier plate to the ball nut and the supporting journal as any misalignment, in particular pitch errors, will reduce ball screw life dramatically. Also, remember when switching from a plane slideway design to a truck and way design where there is a vertical component under gravity that the full load is now being borne by the nut journal without aid from friction in the slide. The ball nut/ball screw contact surface can represent >50% of the total friction of the motion resistance.

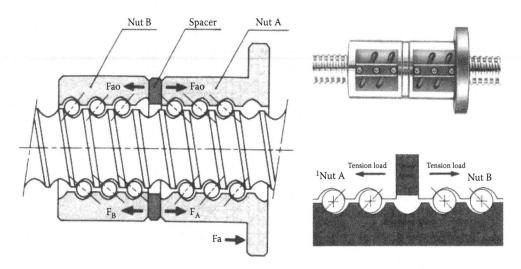

FIGURE 15.26
Preloaded ball nut assembly required for synchrobelt to keep motor below surface. (From THK 2, 1995, THK LM System Ball Screws Catalog #75-1BE. With permission.)

In certain applications such as camlobe grinding where inertia, acceleration, and jerk are key factors to profile accuracy, OEMs such as Landis (Waynseboro, Virginia) developed hydrostatic nuts in the early 1990s to reduce friction and eliminate backlash. Instead of metallic contact, the force is transmitted by an oil film between the male screw and the female formed nut. Backlash is reduced to <0.1 µm. The performance of this system has only recently been superseded by the use of linear motors.

15.8.3 AC Servo Motors

Figure 15.27 shows an ac servo-drive motor mounted in a ball-screw drive system. AC servo motors are a class of induction motor modified for servo operation. It is an asynchronous motor consisting of a stationary AC transformer (stator) and a rotating shorted secondary circuit (rotor) that carries an induced secondary current. The rotor consists of laminated iron cores with slots for conductors. Torque is produced by the interaction of the moving magnetic field and the induced current in the shorted conductors. In a simple induction motor, the speed at which the magnetic field rotates is the synchronous speed of the motor and is determined by the number of poles in the stator and the frequency of the power supply. For servo-motor operation, the stator is wound with two phases (or four) at right angles. The first winding has a fixed voltage supply, while the second has an adjustable voltage controlled by a servo amplifier. The great attraction of the motor is its accurate and rapid response characteristics. The motors are available in fractional and integral horsepower sizes and are used in a closed-loop control system in which energy is the control variable. In order to monitor this, the controller must be able to constantly monitor velocity and position. This is achieved through the use of encoders.

15.8.4 Encoders

Encoders monitor position and rotation and their derivatives, speed, acceleration, etc. The leading manufacturer is Heidenhain, on whose literature most of this discussion is based

FIGURE 15.27
An AC servo-drive motor in a ball screw drive system. (Note: The belt drive between the servo motor encoder and the ball screw connection to gain additional table travel. Belts can be the cause of increased maintenance.)

(Ernst 1998). The simplest and most common are rotary encoders mounted directly on the back of the servo motor to monitor the rotation position. The encoder functions on the Moiré principle. The system uses a glass scale or disk because of its transparency and low thermal expansion. The scale has a grating, produced by photolithography, of fine opaque chromium lines of thickness C/2 and spacing C.

The grating is scanned using a reticule with four windows illuminated by collimated light source. Each window has similar grating lines parallel to the scale. The light passing through each window and the scale is measured by its own photodiode behind.

As the scale moves relative to the windows a sinusoidal light intensity is produced which is converted to an electrical current. The signal from the four windows is each phase shifted 90°, signals (a)–(d) in Figure 15.28. Signal (e) is a reference. Signal (a) and (c), and (b) and (d) are then combined to produce the balanced amplitude signals S1 (f) and S2 (g) which are converted to square wave signals (h) and (i). Combination of (h) and (i) gives a final pulse with a resolution of a quarter of the grating spacing. Further resolution can be obtained via an interpolation circuit. The output is typically 1 V peak to peak.

Rotary encoders can have up to 5000 radial lines giving 20,000 measuring points per revolution. For a ball screw pitch of 10 mm, this gives a resolution of 0.5 μm. Better resolution can be obtained with interpolation. Rotary encoders function at signal frequencies of 160–400 kHz. At 400 kHz with 5000 lines, the maximum table speed allowed corresponds to 400,000/5000 = 80 rev/s giving 48 m/min. This speed is acceptable for all normal grinding operations and even some speed-stroke grinding applications. Gratings with up to 36,000 lines are available for purely angular positioning. Rotary encoders are also available that have several graduated disks linked by a mechanical transmission that allows them to resolve the number of spindle revolutions and therefore the absolute slide position. This negates the need for limit and reference switches but at some loss of resolution.

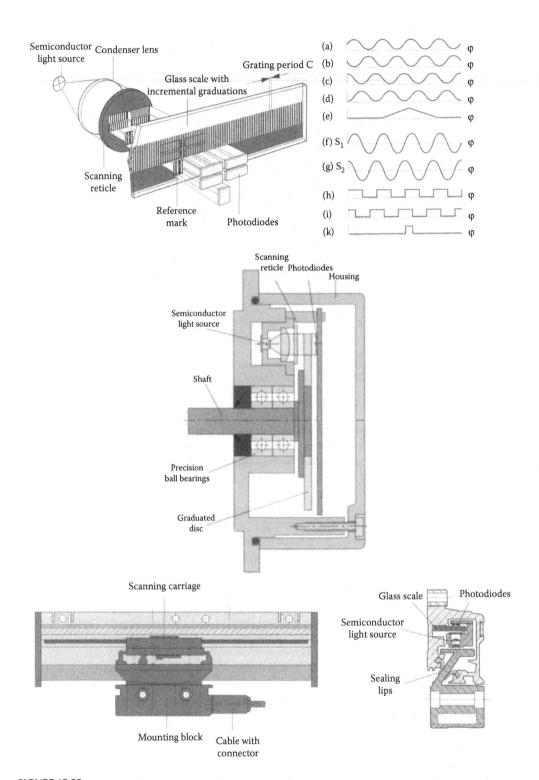

FIGURE 15.28
Examples of linear and rotary encoders. (From Ernst, A., 1998, Digital linear and angular metrology. *Verlag Moderne Industrie*. With permission.)

Rotary encoders are as good if not better at giving feedback on velocity as a tachometer but they are sometimes too remote from the point of grinding to accurately know the position of a slide where it matters. The machine is too sensitive to thermal variations especially for the resolution required for some high-precision applications or the use of CBN. For this reason, glass linear scales are placed next to the slides for positional reading with feedback direct to the PMAC. In many cases, if the material for the scale carrier has the same expansion as the workpiece, the thermal effects of the scale and workpiece compensate each other. A standard linear encoder with a period of 20 μm has accuracy of the order of 3 μm/m. As such, this allows measuring steps of 0.5–1 μm.

The units come "sealed" but nevertheless they should be carefully guarded from coolant and contamination. It is common to mount them under the same slideway guards used to protect the linear guides.

15.8.5 Resolvers

Machines built prior to about 1990 may have a resolver, in place of an encoder, based on the inductive measuring principle or "Inductosyn." In this case, a scale and slider consist of two staggered or zig-zag conductive strips. An alternating current is passed through the slider and the inductive current measured in the scale circuit. Resolvers of this type are much less accurate than modern encoders and more sensitive to temperature changes.

15.9 Linear Motor Drive Systems

15.9.1 Introduction

Linear motor systems started to make an impact on machine tools in the late 1990s for specialist applications such as CNC crank-pin grinding and camlobe grinding (Landis 1999).

More recently several OEMs such as Danobat (Deba, Spain) have designed machines for speed-stroke grinding with speeds up to 250 m/min and acceleration/deceleration of 5g. These are based on the superior speed, reduced friction and inertia obtainable with linear motors.

Linear motors have many advantages. They consist of a single moving part eliminating the ball screw with its inherent backlash, pitch errors, and compliance. That combined with low inertia and zero friction gives a much faster response time and higher accel/decel rates. The reliability is far greater than a ball screw while requiring much less maintenance. The system can also run either at low speeds for creep-feed grinding or extremely high speeds for speed-stroke grinding all with submicron accuracy. In conjunction with direct control through the PMAC via a suitable linear encoder, it is possible to control, position, velocity, and acceleration which is key for cylindrical grinding of nonround parts.

Linear motors do have disadvantages. First and foremost is cost; the price of a linear motor is still about five times higher for a given power than servo-drive systems. Also the motors generate a lot of heat especially at high speed which must be dissipated through internal cooling. The magnets are very high strength and will attract metal swarf unless well guarded or sealed. Linear slides are not commonly used on vertical axes because a counter balance or brake must be supplied in the event of a power failure. In fact, even for horizontal axes, a capacitor or auxiliary power supply is normally provided to control the axes in the event of a power loss.

Linear motors like most technologies had initial teething problems. The slider windings are embedded in epoxy which is prone to attack by water vapor which allows water to gradually eat its way into the windings and eventually short them. Where linear motors have been applied, the closest attention has been paid to slideway guarding. Landis for example tested their first linear motor-driven grinder in its laboratory for over 2 years running production under a range of conditions prior to offering linear motors commercially.

Linear motors are still a niche drive for very specific machine movement requirements. If such machines are under consideration, it is recommended that the potential buyer confirm with the machine tool builder that they have a successful track record of at least 2–3 years.

15.9.2 A Linear Motor System

Figure 15.29 shows a typical closed-loop linear motor drive system. Figure 15.30 shows the elements of a system.

An AC linear induction servo motor is essentially a rotary motor that has been laid out flat. It consists of a track containing rare earth (e.g., samarium/cobalt) high-density permanent magnetic strips embedded in epoxy and fixed to the machine base, and a slider assembly made up of a laminated steel structure with conductors wound in transverse slots. The slider is supported by a pair of linear bearings; ideally hydrostatic bearings. Thrust is developed by the AC current in the motor conductors interacting with the permanent magnetic field.

15.9.3 Laser Interferometer Encoders for Linear Motor Drives

Linear motors demand fast positional monitoring response made possible by using linear encoders based on laser interferometry. The simplest form using a homodyne laser is illustrated below in Figure 15.31. A laser beam L is split by a neutral beam splitter N into two beams. The measuring beam strikes the measuring reflector MR which is returned in a parallel path where it passes through a λ/4 reticule V, where a portion of the beam is delayed to create a circularly polarized wave. The reference beam is reflected by the

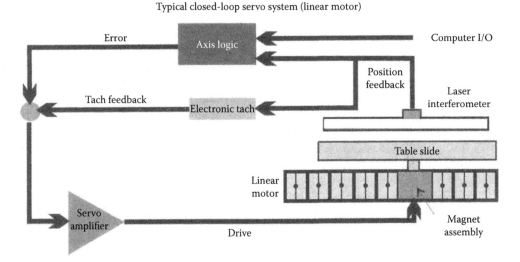

FIGURE 15.29
A closed-loop control system using a linear motor drive.

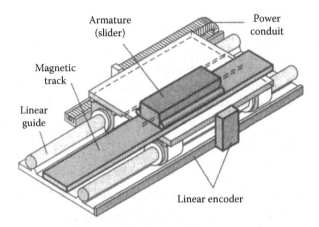

FIGURE 15.30
Elements of a linear motor drive system.

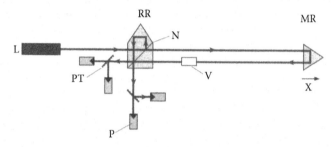

FIGURE 15.31
Principle of the laser interferometer.

reference reflector such that the two beams meet again at the neutral beam splitter and interfere. Further polarizing beam splitters PT put the interfering waves into quadrature measured by the four photodiodes P resulting in two electrical signals phase shifted by 90°. These signals are then converted into two square wave signals with eight cross-over points per wavelength of light giving an accuracy of about 0.08 μm. An interpolation circuit can increase sensitivity by a further order of magnitude.

Heterodyne laser systems are also available that use two almost identical wavelengths. Compensation is also possible for temperature, humidity, and air pressure (Heidenhain). Fortunately, the clean conditions required for linear motor operation also make conditions unusually suitable, for a grinder application, for laser interferometers.

15.10 Spindle Motors and Grinding Wheel Drives

15.10.1 Introduction

Spindle design and selection can be categorized by the type of wheel and operation into:

1. Large wheel applications using conventional wheels
2. Small wheel applications limited by burst speed and dynamic issues in grinding

3. High-speed grinding

4. Specialty applications

In many cases, one area overlaps with another especially where a machine tool builder either standardizes components from one model to another or offers flexibility from upgrading the machine design to handle more than one of the above categories.

15.11 Drive Arrangements for Large Conventional Wheels

15.11.1 Rolling Element Spindle Bearings for Large Wheels

Rolling bearings have increased in accuracy and reliability over recent decades as closer control has been achieved of materials and manufacturing tolerances. This has allowed rolling bearings to replace hydrodynamic bearings as the choice for many larger wheel machines to achieve relatively cool running. Rolling bearings have always been more common for smaller high-speed spindles. However, rolling bearings are still subject to wear and eventually need to be replaced. For machines involved in continuous production, it is important to have a program of planned maintenance and replacement in place.

If a machine tool with rolling bearings is to be moved or transported, it is absolutely essential to seek expert advice on care of the bearings. The bearings must be carefully supported and locked to prevent transmission of vibration through the bearings during transport or handling. Failure to heed this warning will lead to brinelling (indentation of the bearing tracks) and possible complete destruction of the bearings. Subsequent grinding performance will show vibration marks on the workpieces and the bearings will fail after a short running period.

The commonest grinder in the field uses a conventional grinding wheel of 7 in diameter or larger limited to wheel surface speeds between 6500 and 8500 ft/min. This leads to a wheel-spindle speed requirement <4600 rpm, a maximum spindle power requirement of about 10 hp/in. of wheel width, and a maximum runout specification of 2 microns (80 micron in.).

Most spindles consist of a cartridge design with the wheel arbor supported by a minimum of two pairs of ABEC seven or nine angular contact ball bearings with grease packed lubrication. Forced oil lubrication may be used for higher loads or speeds. The spindle is driven by either a simple DC brush or AC synchronous fixed speed motor or, for variable speed, a synchronous multipole induction motor. For reasons of space restraints, the spindle is usually coupled to the motor by a V-belt drive in a back to back configuration as shown in the example in Figure 15.32. The design is well established and robust but is prone to problems of belt slippage and balance issues and requires regular maintenance.

For higher load and stiffness requirements such as centerless grinding, the number of angular contact bearings are increased, as in the example in Figure 15.32 (Royal Master 1996). In addition, when increasing the number of bearing pairs, some are replaced by roller bearings for increased radial stiffness while maintaining angular contact bearings for axial thrust stiffness as shown in the example from Micron (Figure 15.33).

Roller bearings provide better stiffness and damping than ball bearings. They are also less sensitive to temperature than other bearing systems to be discussed below. Another

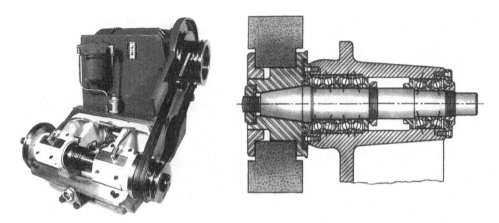

Jones and Shipman cylindrical   Centerless (Royal Master 1996)

FIGURE 15.32
Typical wheel-spindle arrangement for large conventional wheels. Cylindrical. (From Jones & Shipman. With permission.) Centerlesss. (From Royal Master, 1996. *TG12X4 Centerless Grinder*. Royal Master Grinders Inc, Oakland, USA. Trade Brochure. With permission.)

important factor, especially then buying imported machines, is that ball and roller bearings are readily available locally and easy to maintain and service.

Hollow roller bearings have proved of particular interest, certainly in the United States, to rebuilders of surface and centerless grinders. Sold under the trade name Holo-Rol® (ZRB 1998), these are high-stiffness roller bearings, where each roller is a tube. The hollow rollers act like shock absorbers and reduce vibration without loss of stiffness. Bearings for use in, for example, the Cincinnati 220-8 centerless grinder each have a stiffness of 4.47 m lbf/in., four times that of a comparably sized ball-bearing pair.

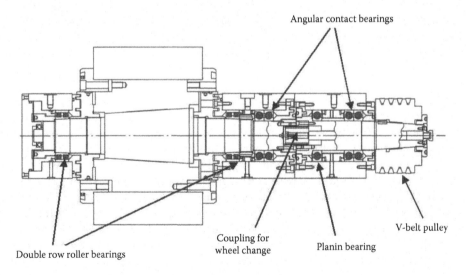

FIGURE 15.33
Heavy-duty wheel-spindle arrangement with V-belt drive. (From Micron, n.d. *Twin Grip Centerless Grinder* Micron Machinery Co Ltd., Yamagata, Japan, Trade Brochure. With permission.)

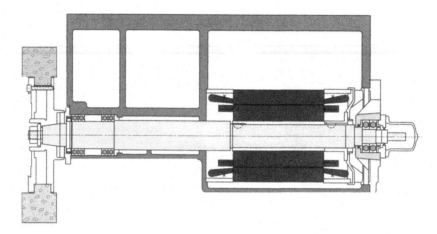

FIGURE 15.34
Wheel-spindle arrangement including an integral motor. (From Mattison, n.d. *Horizontal Surface Grinders*, Mattison, Rockford, USA. With permission.)

Where possible belt-drive designs have been replaced by direct drive motors in which the motor shaft is also the spindle shaft. This dramatically reduces vibration and maintenance issues (Figure 15.34). However, it is an essential requirement that an integral motor is capable of running at the required speed range for the range of wheel sizes to be employed.

15.11.2 Hydrodynamic Spindle Bearings for Large Wheels

To improve damping characteristics, part finish, and wheel roundness, a number of Japanese machine tool builders in particular have gone back to the use of hydrodynamic spindle bearings (Okuma 3 1993; Kondo 1998; Shigiya 2 n.d.; Toyo n.d.). Hydrodynamic bearings were popular in early grinding machines due to their reliability, good accuracy, and long life. The long life of a properly designed hydrodynamic bearing is due to the fact that the bearing is only subject to wear on start-up before the oil film is established and on stopping. However, hydrodynamic bearings create a significant quantity of heat leading to elevated machine temperatures. It is therefore important to allow machines with hydrodynamic bearings to warm up for a few hours to achieve close size tolerances in continuous production runs.

A modern hydrodynamic bearing is not a simple sleeve bearing but relies on the creation of a wedge by having two surfaces rotating relative to each other with different radii of curvature. Between the two surfaces is low-viscosity spindle oil. Adjacent to each surface, this fluid does not move relative to that particular surface. Consequently, a certain amount of fluid will be constantly dragged into the wedge. The amount entering will be greater than the amount permitted to leave because of the wedge shape resulting in a tendency to try and compress the fluid creating a pressure in the film between the surfaces. The pressure gradient that exists is such as to cause equal quantities to pass each point in the bearing. A pressure gradient created is illustrated in Figure 13.35 (Dayton 1949). The point of highest pressure may be up to four times the average pressure over the bearing.

The hydrodynamic bearing design as adopted for a grinding spindle is best described by Kondo (1998). The bearing is made of hardened steel lined with Babbitt metal and consists of three bearing surfaces and three elastic areas. The areas of elastic deformation provide adequate passage for the flow of fluid through the bearing with good temperature control.

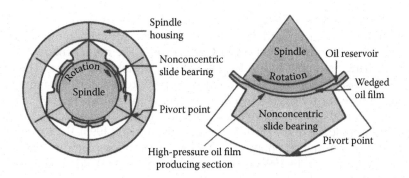

FIGURE 15.35
A multiwedge hydrodynamic bearing arrangement. (From Okuma 3, n.d. *GP-34/44N Plain Cylindrical Grinders* Okuma Corp, Aichi, Japan, Trade Brochure. With permission.)

The clearance gap is 8–12 μm and generates a wedge dynamic pressure of 80 kgf/cm². Kondo claims a precision of 1 μm.

The shape of the bearing surfaces is better illustrated by Okuma 3 (1993), as shown in Figure 15.35. The hydrodynamic bearing illustrated is to provide radial support. There is also a wedge-shaped hydrodynamics thrust bearing near the wheel to minimize axial deflection.

This is perhaps more clearly illustrated in Figure 15.36 which is a bearing arrangement for a Toyo outer diameter (o.d.) grinder for bearing inner races. The runout claimed by machine tool builders for hydrodynamic spindle bearing designs range from <1 to 0.2 μm. Hydrodynamic spindles are also supplied by Danobat (Danobat 2 n.d.), Kellenberger (n.d.), Monza (1998), and Weldon (Weldon 3 1998) among others. One feature in common to most is their use in cylindrical or centerless grinders in a very competitive market, where roundness, finish, and chatter elimination are critical. Hydrodynamic bearings do not require a lot of the ancillary equipment and are lower cost than the alternative, hydrostatic bearings.

15.11.3 Hydrostatic Spindle Bearings for Large Wheels

For applications demanding the greatest stiffness, damping, and power handling requirements with minimal runout many machine tool builders choose hydrostatic bearings.

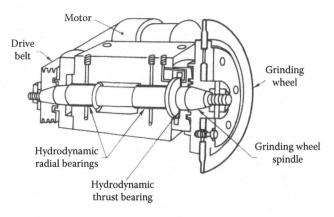

FIGURE 15.36
A hydrodynamic grinding wheel bearing arrangement.

A hybrid hydrostatic bearing can provide greater load support than many hydrodynamic bearings of the same diameter and can withstand crashes better than other bearing types; a significant issue when using steel-cored CBN wheels! The net stiffness of a hydrostatic spindle and bearing assembly will often exceed that of any other type. The other major advantage of hydrostatic bearings is that, when properly tuned, the averaging effect of the bearing will improve running truth by over an order of magnitude relative to the accuracy at which the individual components were machined (Aleyaasin et al. 2000). Radial errors of <0.5 μm throughout the life of the spindle are typical (Aronson 1995). For this reason, hydrostatics have become common, if not the standard, for applications such as camshafts (Landis 1996a) and crankshafts (Landis 1996b), roll grinding (Elgin n.d.; Waldrich Siegen 1996), some centerless grinders (Koyo 2 n.d.), and several standard cylindrical grinders (Paragon n.d.).

Rotary hydrostatic bearings are built in a similar fashion to their linear counterparts with metered fluid under pressure fed to a series of bearing pads. There are usually two radial or journal bearings with four opposed pads in each and either two axial thrust bearings or an axial thrust and additional oil-filled chamber behind the journal bearing and pressurized by it (Montusiewicz and Osyczka 1997) (Figure 15.37).

At rest the bearing is purely hydrostatic but as the spindle rotates fluid will be dragged into the narrow space between the pads generating additional hydrodynamic pressure. While some drag is inevitable and provides additional support at lower speeds, it can lead to instability and excessive heat at higher speed unless flowrate, pad widths, and depths are selected carefully. The principles of hybrid hydrostatic/hydrodynamic bearings were developed by Rowe (1983, 1988, 2012), who provided design charts for load-bearing capacity and criteria for stable operation. Recessed and plain hybrid hydrostatic designs have been established (Rowe et al. 1977, 1982; Koshal and Rowe 1980; Rowe and Koshal 1980; Ives and Rowe 1987).

Several Japanese machine tool builders design spindles specifically to incorporate the effect into "hybrid" hydrostatic spindles (Toyoda 1996; NTC 1999). Whereas a pure hydrostatic spindle might need 1000 psi oil pressure, a hybrid bearing might need only 300 psi for the same stiffness. The disadvantage is that the center of rotation will change with spindle speed and oil viscosity (Aronson 1995).

The major disadvantages of hydrostatic bearings in general are in the manufacturing and ancillary equipment requirements and their associated costs. The bearings require

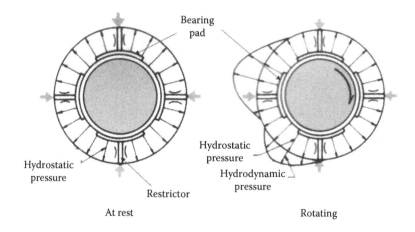

FIGURE 15.37
Example of hydrostatic bearing spindle. (After Toyoda, 1994, *GL3A GL3P CNC Cylindrical Grinder*. Toyoda Machine Works, Japan, Trade Brochure. With permission.)

an oil pump, collection, and recirculation system with filtration to <5 μm, and a chiller to hold the oil to ambient temperature. Hydrostatic spindles tend to be limited to about 750,000 DN due to heat generated by fluid shear, although bearing temperatures can be kept low by optimization of flowrate to remove unwanted heat (Rowe et al. 1970; Rowe 2012).

15.12 Drive Arrangements for Small Wheel Spindles

15.12.1 Introduction

The problems with spindles designed for small wheels is somewhat different to those for larger wheels in that nearly all wheel bond types, including most superabrasives, are limited by bond strength to operating in the speed range of 35–60 m/s. Taking 100 mm (4″) diameter as the cut-off for this discussion, the limitations are in flexibility to handle a wide range of wheel diameters and maximum wheel speeds for very small wheels as those used in, for example, the fuel injection industry. The operating range under consideration is therefore 6000–200,000 rpm.

15.12.2 Rolling Bearing Spindles with Belt Drive for Small Wheels

The simplest and oldest designs of spindles are grease packed, angular contact bearing spindles, belt-driven similar to the larger cylindrical wheel counterpart discussed above except the bearings are always at least ABEC 9 or 11 and the belts are multiribbed. Spindle nose runout is maintained at 2 μm or better. Grease-packed bearings are capable of running up to about 30,000 rpm (Bryant 1986) or 500,000 DN (Nakamura 1996). Similarly, belt drives are available up to about 30,000 rpm (Bryant 1986), although the preferred limit is nearer 10,000 rpm (Aronson 1998). With variable-speed control, this arrangement provides a compact flexible arrangement for dedicated internal and universal grinders performing internal and face grinding with conventional wheels in sizes in the range of 4 in down to 1 or 2 in. in diameter.

15.13 High-Speed Spindles for Small Wheels

For speeds faster than 30,000 rpm, various design considerations need to be incorporated.

15.13.1 Direct Drive Motors

Use of direct drive AC asynchronous induction motors with frequency converters are mostly used to eliminate the need for belts. The drive will usually have a closed-loop tachometer, thermistor monitored temperature control, and overload protection (Precise n.d.). Spindle horsepower curves need to be closely married to the rpm requirements as each drive will have a particular operating range in terms of rpm and torque. Controls are now available that can extend the constant horsepower range over a much wider speed range. Bryant (1992, 3 *High Speed Range Motorized Spindles*. Bryant Grinder Corp, Springfield,

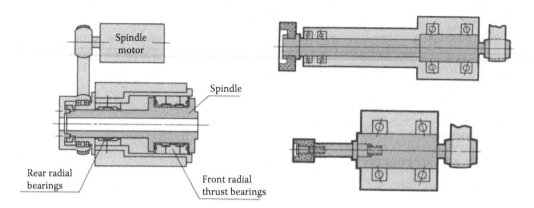

FIGURE 15.38
Examples of internal grinding spindles. (From Danobat 2, n.d. *R1-100-CNC Universal and Cylindrical Grinders.* Danobat, Elgoibar, Spain, Trade Brochure. With permission.)

Vermont. Private communication 1992.), for example, provides just two spindles to cover the entire range from 8000 to 70,000 rpm (Figure 15.38).

15.13.2 Dynamic Balancing of High-Speed Spindles

Dynamic balancing of all rotating components and in situ balancing to better than 1.2 mm/s displacement is critical to spindle life and grind quality (Gamfior n.d.).

15.13.3 Oil-Mist Lubrication for High-Speed Spindles

Oil-mist lubrication to replace grease-packed bearings can increase the DN value of steel ball bearings from 500,000 to 1,000,000. Oil-mist lubrication requires an oil-mist source and a compressed air source (Figure 15.39). The air needs to be carefully dried to prevent

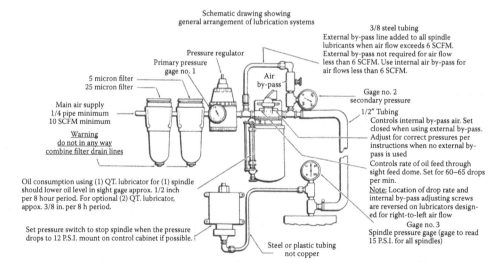

FIGURE 15.39
Oil-mist lubrication system. (After Bryant, 1986, 2 *Standard Wheelheads High-Frequency, Belt-Driven, Accessories High Speed Spindle Systems.* Bryant Grinder Corp, Springfield, Vermont, USA, Trade Brochure. With permission.)

rusting of the spindle and must be filtered to 5 μm or better to prevent contamination of the bearings. The oil-mist at a pressure of about 1 bar provides positive air displacement to the unit which coupled with a labyrinth seal prevents coolant ingression. Also note, even with the current oil-mist systems, when the spindles are run over extended periods of time a long length of pipe connection should be used (>1 m) to facilitate air–oil mixing.

15.13.4 Adjustment of Bearing Preload for High-Speed Spindles

Bearing preload needs to be carefully controlled and understood. Insufficient preload leads to low stiffness, while excessive preload leads to higher temperature rise and even seizure. As rotational speed increases, several factors cause an increase in preload

- Centrifugal force from the balls
- Hydrodynamic pressure from the oil lubricant
- Thermal expansion from temperature differences due to frictional heat generation causes a temperature gradient between the inner and outer ring
- Thermal expansion of the balls
- Expansion of the inner ring due to centrifugal force.

In addition to affecting preload, thermal effects cause expansion of the spindle arbor leading to axial positional error. To minimize preload effects, several options are available to the machine tool builder as follows.

15.13.5 Use of Ceramic Balls for High-Speed Spindles

The first option is the use of hybrid ceramic (silicon nitride) ball bearings. Ceramic balls have a lower density than steel to reduce centrifugal forces; they have lower friction coefficient to reduce frictional heating and lower thermal expansion to minimize the effects of heating. Ceramic is also chemically inert and allows the oil film to work better and for longer. In many cases, the attraction of ceramic balls is to allow grease-packed lubrication at higher DN values. Hybrid ceramic bearings can be rated for 0.8–1.4 m DN when grease packed, or to 2–3 m DN with oil-mist lubrication (Aronson 1994, 1996; Nakamura 1996).

15.13.6 Liquid Cooled High-Speed Spindles

The second option is to liquid cool the spindle by pumping oil through the housing and even injecting cooled oil into the bearings either as a metered oil–air mixture based on monitoring spindle characteristics (i.e., with greater control than a standard oil-mist system) or as a flood jet of oil. The oil–air approach can play a major role in extracting heat from the whole spindle as well as maintaining a positive pressure to the bearings to prevent contamination.

15.13.7 Floating Rear Bearing for High-Speed Spindles

The third option is to fix the front bearings but allow the rear bearing housing to float in a fluid-film bearing. This allows the radial preload to remain near constant, while driving all the axial thermal expansion backward and away from the wheel reference position.

15.14 Spindles for High-Speed Grinding

15.14.1 Introduction

Spindle design is not the limiting factor in most high-speed grinding opportunities. As discussed in earlier chapters, wheel speed limitations, dressing, burn, and a host of other factors normally prevent higher speeds before there is a spindle limitation. This is all too clear in, for example, high-speed cylindrical grinding. Current top grinding speeds with CBN wheels is about 200 m/s in actual production. Most of this is focused at the camshaft industry. It may be surprising but there are probably more grinders in countries such as India and Taiwan running at 160 m/s or greater than that are in the United States. This is because car and scooter engines in those countries use cast-iron cams, whereas U.S. cars require hardened steel cams because of the engine power.

15.14.2 Spindles Bearings for High-Speed Grinding of Hardened Steel

Hardened steel is much more sensitive to burn than cast iron is and has been limited to grinding speeds under 100 m/s. Even at 200 m/s, a 380-mm wheel rotates at 10,000 rpm which is within the limits of hydrostatic, hydrodynamic, or ceramic hybrid ball-bearing spindles. High-wheel speed is used to increase stock removal rates and therefore high-load capacity is to be expected. For this reason, most high-speed grinders use hydrostatic spindles, although Okuma (1998) has optimized its 59 kW hydrodynamic spindle to run 350 mm wheels up to 11,000 rpm. The highly efficient and successful Junker Quickpoint® contour grinding process uses an air-cooled grease-packed rolling bearing spindle for speeds up to 8000 rpm or 140 m/s (Junker 1996). In some cases, it is even belt-driven (Junker 1992).

15.14.3 Spindle Bearings for HEDG

Grinders for HEDG grinding and superabrasive high-speed grinding with plated wheels have been influenced strongly by machining center design. In many cases, the grinders are, in fact, a multipurpose machining/grinding/drilling center. The wheels are designed around sizes of 5 in. up to 12 in. running at up to 120 m/s. Numerous machine tool companies fall into this category with either a grinding background (e.g., Campbell, Magerle, Huffman, Edgetek, and Gendron) or a machining background (e.g., Makino, Mori Seiki, and Hitachi Seiki). The spindle requirement is therefore both speed and flexibility to operate over a range of speeds. Most machine builders use liquid cooled, ceramic hybrid bearings with AC servo motors operating at speeds 5000 rpm up to 20,000 rpm. Use of ceramic hybrid bearings allows for larger diameter shafts of 100 mm or greater and therefore enhanced stiffness. A similar approach has been taken by the manufacturers of high-production tool grinders that use superabrasive wheel packs in the 4–12 in. diameter range.

15.14.4 Spindle Cooling for High-Speed Grinding

It is interesting to note the attention paid by a number of the machine tool builders to how the spindle is cooled. One concern with a demand for flexibility is that to achieve the necessary preload at high-rotational speeds may mean a loss of preload at slower speeds. Makino (n.d.) reduces this by injecting chilled oil through the center of the shaft where centrifugal force then injects the oil from the inner to the outer bearing race. The oil is then injected through the housing setting up a temperature gradient from the inner to the outer race and between the bearing and the housing. This temperature difference combined with the efficiency of core cooling provides increased thermal stability at higher speeds and allows the bearings to be amply preloaded at rest.

15.14.5 Spindle Bearings for Very Small High-Precision High-Speed Wheels

It is small diameter wheels that have pushed the limits of spindle design. The fuel injection industry, for example, especially for diesel engines, requires bores as small as 2 mm to be ground to extremely high precision. Current spindle technology is limited to about 200,000 rpm which is only 21 m/s for a 2 mm wheel. The greatest problems at these speeds are wheel and arbor imbalance, and resonance problems and lack of stiffness of the arbor. Spindles running up to 150,000 rpm are more typical of production currently and these are dedicated to a small range of wheel diameters.

15.14.6 Active Magnetic Bearings for High-Speed Wheels

One possible solution to improve stiffness and flexibility in the use of small- and medium-sized wheels is the development of active magnetic bearing (AMB) spindles. AMB spindles have been available since the 1980s (Moritomo and Ota 1986) and a few are in use, certainly in the United States, although their acceptance has been slow.

Active magnetic bearing spindles use a magnetic field to support the rotor. Positional sensors and closed-loop monitoring to a controller constantly controls its position in the field. Originally developed for defense and space applications, AMB spindles have been used at speeds up to 1,000,000 rpm in., for example, centrifuges. For spindle applications, the rotor is levitated by two radial bearings and at least one thrust bearing (Wang et al. 2001). AMBs are offered at speeds up to 200,000 rpm (IBAG 1996). Catch bearings are provided in the event of a power loss. One of the big attractions of these spindles is the ability to provide a high-power output over a very large range of spindle speeds which in turn allows flexibility in wheel diameters when grinding a family of different-sized parts. As with any new technology, there is a price barrier and learning curve to be overcome before AMB spindles become generally accepted. Certainly, the avoidance of oil-mist lubrication with its associated environmental issues will be an attraction.

15.15 Miscellaneous Wheel Spindles and Drives

15.15.1 Hydraulic Spindle Drives

Although much of the interest is in new and emerging technologies, many machines remain in the field built 30 or more years ago. Very often aspects of an older technology will linger because it has proved such a practical or reliable method. A case in point is the use of gerotor hydraulic motors for dressing spindles discussed in the previous chapter. Hydraulic motors were common wheel drives for applications such as centerless grinding. The motors were compact and had good torque at low rpm. Lidkoping (1998) as of 1998 still offered a state of the art grinder suitable for vitrified CBN wheels with hydraulic piston pump drives on both its grinding and regulating wheels. The hydraulic drive was claimed to give an exact control of wheel velocity.

15.15.2 Air Motors and Bearings

The other drive still seen is based on air motors and bearings. Air bearings have very low friction and are therefore used for very high speeds. The limitation with air is the load-carrying capacity of the air film. The supply pressures that can be employed are much lower than can be employed with oil. This limits the bearing pressures that can be generated (Rowe 1967, 2012).

The most common use for these is in jig grinding. The process is a very low-stock removal process, often carried out dry, with small superabrasive wheels. The workpieces can have deep bores to be ground leading to long, flexible wheel shanks. The torque requirements for the motor are therefore low. Moore jig grinders are available with a complete range of air turbine and vane driven spindles from 9000 to 175,000 rpm. Similarly Hauser provides an air spindle for operations in the range of 80,000–160,000 rpm although it also provides direct drive electric options for slower speeds.

Air bearings in spindles are seen in ultraprecision grinders for specialist applications such as carbide grinding for the die industries with tolerances of <0.2 μm and surface finishes of 0.025 Ra (Pride n.d.). Many of the grinder manufacturers are focused on the electrooptical and semiconductor businesses and as such are outside the scope of this book but, on occasion, this type of machine is found suitable for ultrahigh-precision prototype work.

A case in point is Precitech who manufacture machines for single-point diamond turning or grinding of aspheric surfaces for lenses. The machines are equipped with aerostatic bearing, air-driven spindles for speeds up to 60,000 rpm, although the motor is replaced with a DC-servo motor for speeds around 10,000 rpm. With laser interferometer feedback the machine positional resolution is <10 nm indicative of the type of low-stock removal, high-precision applications where air spindles and bearings are found.

15.16 Rotary Dressing Systems

Rotary dressing heads and spindles have become a key feature of modern grinding systems particularly for application with superabrasive wheels. Rotary dressing units are often purchased separately from the basic machine tool. The subject has previously been rather overlooked in books on grinding technology. The following section, therefore, attempts to provide a picture of the overall range of provision and application.

Rotary dressing drives can be basically categorized by their source of power.

15.16.1 Pneumatic Drives

Pneumatic spindles are relatively inexpensive, can run on shop air of 6–7 bar (90 psi), and do not suffer from heat problems that would cause significant thermal movement. However, they have low torque or power (or more specifically a low power density) and are noisy. Larger motors suitable for form roll dressing can require 3 m^3/min (100 ft^3/min) of air. In general, their use is limited to internal grinding often in cross-axis dress mode, although even here they are being superseded by electric motors. Where used in uniaxial dress mode, they tend to be equipped with small diameter cup dressers to limit torque requirements. Speed selection is made by adjustment of air flow.

15.16.2 Hydraulic Drives

Hydraulic drives generally use gerotor technology to provide a compact high-power/high-torque motor that has made this style of dresser drive the mainstay of form roll dressing applications for conventional wheels for the last 30 years. Hydraulic spindles suffer from serious thermal problems as they get hot after only a few minutes, although this can be mitigated by pumping coolant through the housing and/or flooding the exterior of the housing with coolant. Nevertheless, they are not recommended for running continuously for more than 15 min. The other problem with hydraulic motors is that they do not react well to being driven, that is, to be used in a uniaxial (codirectional as opposed to counter-directional) dress mode, as the motors have a tendency to cavitate.

Hydraulic spindles require of the order of 500 psi and between 5 and 10 gpm flow which should ideally be from a separate closed-loop system filtered to 10 µm. The pressure line from the pump to the spindle should be as short as possible and a minimum of 10 mm (approx. 3/8 in) diameter. Also, the exhaust hose from the motor should never be smaller than the inlet. Speed adjustment is made by use of a flow control in line on the exhaust side of the motor. The hydraulics can also provide pressure and lubrication to sleeve bearings used in many of these older spindle designs and as a means of actuating retractable centers for quick roll changes.

In the last 10 years, there has been a trend to eliminate hydraulics on many new grinders which has pushed dresser spindle technology away from hydraulic drives and toward electric drives.

15.16.3 Electric Drives

Electric motors can be subdivided into three categories; traditional AC motors, AC servo motors with high-frequency drives popular since the early 1990s, and most recently DC brushless servo motors with high-frequency drives. The trend in these motors is to go to higher and higher power densities so as to be as compact as possible and hence give the maximum flexibility both to be fitted into smaller grinder footprints and to provide the additional power required for creep feed and vitrified CBN dressing.

There are numerous advantages to electric drives including the ability to run unidirectional (i.e., in an over-run situation, where the grinding wheel tries to drive the dressing tool) a high-power and power density, simple speed control, and cleanliness. Motor guarding is more of an issue to prevent coolant access to the windings but this is readily achieved with good rotating seals and/or positive air purging.

The enormous changes that have occurred in the last few years are exemplified in the photographs in Figure 15.40. These show two dressers from Wheel Dressing Inc. (Saint-Gobain Abrasives). The spindle to the left is a model ACI pneumatic spindle designed and in common use for over 25 years specifically for Heald internal grinders; it has a power rating of about 0.05 kW at 20,000 rpm. The spindle to the right is a model VS1 DC brushless electric dresser with a peak power rating of 0.80 kW at 20,000 rpm. Both photographs are at the same magnification.

ACI air spindle　　　　　　　　DC brushless servo spindle

FIGURE 15.40
Advances in dresser spindle motor technology.

15.17 Power and Stiffness Requirements for Rotary Dressers

A review of some commonly available dresser spindles sold to machine tool builders and as retrofits for grinders previously using single-point dressing gives a good indication of the power requirements and hence stiffness. Table 15.1 lists the capacity of various Wheel Dressing and Norton (Divisions of Saint-Gobain Abrasives) dressing spindles.

The first observation is that the dressers for roll applications tend to operate at speeds around 7–8 m/s, which for conventional applications give a crush ratio of the order of –0.2. It is also possible to extrapolate guidelines for dresser power requirements for traverse and plunge rotary dressers. It should be noted that spindles handling up to 300 mm wide rolls will in fact have a series of rolls and spacers and so the diamond contact width will be considerably less. Based on this, one can estimate approximately 50 W/cm for counter-directional plunge dressing of conventional wheels which is four to five times higher than that predicted for a flat roll from Tables 15.2 and 15.3 but in line with the expected higher power required for profiles and necessary safety factor.

Based on this and additional data from the field, the recommendations for power requirements are made and summarized in Table 15.2.

Taking these data, estimates can now be made for spindle stiffness requirements and the limits thereof. For example, a 10-cm wide roll dressing a CBN wheel might draw up to 1.5 kW running at 25 m/s. This would generate a normal force, based on $F_t/F_n = 0.2$, of up to 300 N. If deflection is to be kept under 2 μm then the system stiffness must be greater than

TABLE 15.1

ITW Philadelphia's Polymer Concrete Compared with Class 40 Cast Iron

Material Property	ITW Concrete	Class 40 Cast Iron
Compressive strength (PSI)	20,000	130,000
Tensile strength (PSI)	2000	40,000
Compressive modulus (PSI)	4.2E6	15.0E6
Coefficient of thermal expansion (IN/INF)	6.8E-6	6.7E-6
Thermal conductivity (BTU/FT-HR-F)	91.2	1300

TABLE 15.2

Capacities of Common Dresser Spindle Designs

Manufacturer	Model	Drive	rpm (max)	Roll Width	Roll Diameter	Roll Speed	Power	Application
WDI	ACI	Air	21,000	N/A	63.5 mm	34 m/s	0.05 kW	Traverse
WDI	HCI	Hydraulic	11,500	N/A	28.5 mm	7.5 m/s	0.10 kW	Traverse
WDI	HO 5/8	Hydraulic	3600	N/A	70.0 mm	7.5 m/s	0.35 kW	Traverse
WDI	ECI	Hydraulic	4000	25 mm	76.2 mm	7.8 m/s	0.18 kW	Plunge
WDI	HO	Hydraulic	3600	100 mm	76.2 mm	7.0 m/s	0.35 kW	Plunge
WDI	HI	Hydraulic	2400	75 mm	100 mm	6.2 m/s	0.45 kW	Plunge
WDI	DFW-AHO	Air	4000	100 mm	76.2 mm	7.8 m/s	0.30 kW	Plunge
WDI	HHD	Hydraulic	2300	300 mm	120 mm	7.1 m/s	0.70 kW	Plunge
WDI	DFW-AC	Air	2400	300 mm	125 mm	7.8 m/s	1.10 kW	Plunge
WDI	DFW-N3	Electric	3600	300 mm	120 mm	11.1 m/s	0.75 kW	Plunge
Norton	AXH-1464	Hydraulic	20,000	N/A	76.2 mm	39 m/s	0.33 kW	Traverse
Norton	AXH-1440	Hydraulic	4200	N/A	76.2 mm	8 m/s	0.48 kW	Traverse
Norton	AXH-1416	Hydraulic	4200	31.75 mm	152 mm	16 m/s	0.48 kW	Plunge
Norton	AXH-1418	Hydraulic	7450	31.75 mm	152 mm	29 m/s	0.33 kW	Plunge

N/A: Not available.

150 N/µm or 1.2 m lbf/in. Precision dressing spindles for form roll applications have maximum stiffness values in the range of 1–2 m lbf/in. due to the nature of the bearing designs available and required arbor diameters. However, the sum of the remaining machine components is rarely 1 m lbf/in. except on a small number of specifically designed grinders.

For crush forming, TVMK offer a hydraulic unit with a power of only 25 W/cm maximum but a torque of 10.8 Nm rated for rolls up to 35 mm wide. Since most crush-forming operations occur at roll speeds of 1–2 m/s, the normal force created would be expected to be as high as a 30 N/cm roll width. This is 50% higher than those for diamond roll dressing conventional wheels with a +0.8 crush ratio, and comparable to the highest dressing forces for CBN wheels indicated above. The advantage of the hydraulic drive is to provide the necessary high torque at low-rotational speeds. The results also illustrate why crush forming is uncommon and limited to narrow forms for applications such as thread grinding. Most standard grinders do not have the stiffness capability. Also note, crush forming

TABLE 15.3

Power Requirements for Dresser Spindles by Abrasive Type and Dress Application

Abrasive	Crush Ratio	Application	Spindle Power
Alox	Cross axis	Traverse—internal grind	50 W
Alox	(–ve)	Traverse—finish grind	100 W
Alox	(–ve)	Traverse—rough grind	250 W
Alox	(–ve)	Plunge	50 W/cm roll width
Alox	(+ve)	Plunge	100 W/cm roll width
CBN	(+ve)	Traverse	150 W
CBN	(–ve)	Plunge	75 W/cm roll width
CBN	(+ve)	Plunge	150 W/cm roll width

has a form accuracy of only about 25–50 μm, which is an order of magnitude less accurate than diamond roll dressing.

15.18 Rotary Dressing Spindle Examples

15.18.1 Introduction

Diamond roll dressing was introduced in the 1960s and has been growing in usage until it now represents the commonest dressing method for the majority of profiled and flat grinding applications, and virtually all vitrified CBN applications. Nevertheless, it often remains just an option on new grinders still designed primarily for stationary dressers or a later requirement for retrofitting to existing grinders. Conversion of machines to rotary dressing may require machine modifications including mounting plates, guarding changes, or even additional axes of motion. These axes may be used as controlled infeed motion for the dressing operation or merely to move into a fixed stop and then after dressing retract in order to clear the dresser from the grind area. Some examples of both dressing spindles and the accompanying infeed systems are illustrated in the proceeding pages [Courtesy Wheel Dressing (Saint-Gobain Abrasives)].

15.18.2 DFW-ACI Air-Driven Spindle

Figure 15.40 shows an impellor rotor driven spindle with ABEC 7 twin ball bearings for radial stiffness and ABEC 7 thrust bearing for axial stiffness. Note a muffler is required to reduce noise level with air operation. The unit requires dry shop air lubricated and filtered to 5 μm. It was originally designed for Heald internal grinders and is most effective on alox wheels but has been used successfully on small vitrified CBN internal wheels. Since air is compressible, a larger cylinder is required to achieve the same drive stiffness as with a hydraulic cylinder.

15.18.3 ECI Hydraulic Spindle

Figure 15.41 shows a cantilevered hydraulic powered dresser with three pressure lubricated sleeve bearings for radial stiffness and damping and two ABEC 7 thrust bearings for axial stiffness. Used with up to 3″ diameter by 1″ wide rolls typically for traverse dress operations or narrow plunge forms. This hydraulically driven spindle is rated for up to 7000 rpm, although generally runs at 4000 rpm.

15.18.4 DFW-HI Heavy-Duty Hydraulic Spindle for Internal Grinders

This unit shown in Figure 15.42 is a cantilevered hydraulic powered dresser with three sleeve bearings for radial stiffness and two Torrington thrust bearings for axial stiffness. It is a unit designed for heavy-duty plunge dressing applications on internal grinders. Rolls can be counter bored to wrap around the spindle to accommodate up to 3″ roll widths. This type of layout reduces the cantilever overhang and reduces deflections.

FIGURE 15.41
EC1 Hydraulic spindle.

FIGURE 15.42
DFW-HI hydraulic dresser spindle.

15.18.5 DFW-HO 5/8 Heavy-Duty Hydraulic Spindle Typically Used for Centerless Wheels

This unit shown in Figure 15.43 is an outboard supported heavy-duty traversing spindle with zero end play. The outboard support greatly increases the stiffness of the design compared with a cantilever layout. The unit has three sleeve bearings and four thrust bearings with coolant porting for thermal stability. The roll is replaced by removing the end cap, where the diamond roll is bolted to a face plate. The dresser is typically used for flat form dressing centerless wheels mounted cross axis or uniaxially to dress forms and radii. Centerless wheels are wide compared to wheels for most other grinding processes and therefore require a heavy-duty spindle.

FIGURE 15.43
HO 5/8 hydraulic dresser spindle.

15.18.6 DFW-HO Variable-Speed Hydraulic Dresser

The unit shown in Figure 15.44 is a variable speed, outboard supported light-duty plunge dresser with pressure lubricated sleeve and thrust bearings with easy removal of the dresser shaft from the housing for roll replacement. Coolant porting is supplied custom to each unit for the dressing process.

15.18.7 DFW-HHD Hydraulic Heavy-Duty Plunge Dresser

Figure 15.45 shows the heavy-duty hydraulic plunge dresser with sleeve and thrust bearings as in the previous example. Custom coolant porting is provided, and easy spindle removal is possible from the housing for roll changes. This unit is typically used for dressing multiple wheel assemblies.

FIGURE 15.44
DFW-HO hydraulic dresser spindle.

FIGURE 15.45
DFW-HHD heavy-duty plunge hydraulic dresser spindle.

15.18.8 DFW-HTG Heavy-Duty Hydraulic Spindle

The unit shown in Figure 15.46 is a comparable spindle to the DFW-HHD but using two pairs of ABEC 7 angular contact ball bearings to allow the spindle to be used with nonhydraulic drive systems. The housing casting can be custom made to accommodate up to 500 mm shaft width.

15.18.9 DFW-NTG Belt-Drive Spindle

The classic electric, belt-driven dresser shown in Figures 15.47 and 15.48 has an ABEC 7 ball-bearing arrangement for table mounted surface grinders such as creep feed. Air–oil mist porting is required to the bearings, while custom porting is provided for dressing. A proximity sensor is often placed on the pulley to confirm spindle rotation. The spindle design can handle up to a 375-mm roll assembly at a fixed speed of up to 3600 rpm with speed adjustment through changes to the pulley ratio. For the latest grinders with

FIGURE 15.46
DFW-HTG plunge hydraulic dresser spindle.

FIGURE 15.47
NTG plunge electric dresser for creep-feed grinding.

FIGURE 15.48
TG77 upgraded plunge electric dresser.

high-wheel-spindle power and coolant delivery systems, the standard electric motor has been replaced by custom AC servo or DC servo motors for variable-speed control and better motor sealing, as shown in Figure 15.49.

15.18.10 DFW-VF44 AC Servo High-Frequency Spindle

The AC servo high-frequency motor shown in Figure 15.49 has a peak power of 0.375 kW at 5500 rpm and a base frequency of 92 Hz. This style of dresser has been the standard for the United States for traverse dressing of vitrified CBN wheels for cylindrical grinding. It is now often fitted with an internal acoustic sensor for touch dressing.

15.18.11 DFS-VS8 DC Servo Variable-Speed Dresser

Figure 15.50 shows a DC brushless servo variable-speed dresser with constant torque and a peak power of 1.3 kW at 6000 rpm. The spindle is supported on three sets of angular contact bearings including (TG) outboard support. The 125 mm diameter diamond roll has

FIGURE 15.49
DFW VF44 AC servo HF electric traverse dresser spindle.

FIGURE 15.50
DC brushless servo electric drive dresser spindle.

a 35-mm wheel width capacity. It is targeted at heavy-duty traverse applications and form plunge applications with narrow CBN and engineered ceramic (SG) wheels. Figure 15.51 shows the unit mounted in a grinding machine for a dressing operation on an SG wheel.

15.19 Dressing Infeed Systems

15.19.1 Introduction

The key to any infeed system is high stiffness and damping, repeatability, and ease of maintenance. The units described below (Courtesy Wheel Dressing Division of Saint-Gobain Abrasives) have proven effective over numerous years in the field. There are a variety of designs combining dresser spindle units and infeed systems varying according to the application and particular features required for the particular application. The following range is described in chronological order from older hydraulic-mechanical compensator designs through DC stepper motors to direct drive AC servos.

FIGURE 15.51
A postmounted DC brushless servo electric drive dresser spindle mounted in a grinder dressing on seeded gel wheels.

15.19.2 DFW-SB-HO Single, Hydraulically Driven Carrier

Figure 15.52 shows a single hydraulically oil-lubricated, roundway infeed carrier with mechanical compensation for plunge-roll dressing conventional wheels on small cylindrical grinders. Chromed hardened steel roundways allow more efficient guarding from coolant and abrasive ingression. They also provide excellent damping characteristics. The example shown is equipped with a DFW-HO hydraulic spindle.

15.19.3 Mini Double-Cylinder Infeed

A mini double-barrel hydraulic-mechanical compensator for medium-sized grinders is shown in Figure 15.53. The system is equipped with a DFW-HO hydraulic spindle.

FIGURE 15.52
DFW-SB-HO hydraulic-driven infeed system.

FIGURE 15.53
Mini hydraulic-mechanical compensated infeed mechanism.

15.19.4 Double-Barrel Infeed Carrier

The double-barrel infeed carrier in Figure 15.54 has a hydraulic-mechanical compensator. It is suitable for plunge form dressing on machines such as centerless grinders.

Figure 15.55 shows an example of a hydraulic-mechanical compensator infeed system with a mounting casting attached directly to the existing platform of a single-point traverse diamond on a Cincinnati centerless grinder.

15.19.5 Double-Barrel Plunge-Form Dresser DFW-GA-H

Figure 15.56 shows a double-barrel infeed carrier with hydraulic-mechanical compensator for plunge form dressing used on a Cincinnati 480 30°-plunge centerless grinder. It replaced a hydrostatic single-point dresser unit and was equipped with a DFW-HHD heavy-duty hydraulic spindle.

FIGURE 15.54
Hydraulic-mechanical compensator infeed system.

FIGURE 15.55
Hydraulic-mechanical compensator infeed system for centerless grinder.

FIGURE 15.56
Hydraulic-mechanical compensator infeed system for centerless grinder.

15.19.6 Triple Barrel Infeed Carrier with Hydraulic-Mechanical Compensator

Figure 15.57 shows a triple-barrel infeed carrier for use on large centerless and multi-wheel grinders. Custom designs can take up to 20″ wide diamond roll assemblies. Note the series of coolant nozzles plumbed through the housing common to all the systems described.

15.19.7 DFW-CG-SM-HC Stepping Motor Carrier

Figure 15.58 shows a stepping motor and ball screw infeed compensated double-barrel carrier designed for the SGE-195 bearing grinder. It is equipped with a DFW-HC spindle for quick diamond roll changes.

FIGURE 15.57
Triple-barrel plunge dresser.

FIGURE 15.58
Stepper motor/ball screw double-barrel infeed system.

15.19.8 Stepping Motor Carrier for a Cylindrical Grinder

Figure 15.59 shows a stepping motor and ball screw infeed-compensation double-barrel carrier suitable for a cylindrical grinder. The carrier is equipped with a DFW-HHD hydraulic spindle.

15.19.9 Combination Stepper Motor and DC Traverse Motor

Figure 15.60 shows a traverse diamond roll unit with a small double-barrel stepper motor infeed and a DC motor cross-slide and DFW-HTG spindle. This carrier is designed for traverse dressing of CBN and engineered ceramic wheels.

FIGURE 15.59
Stepper motor/ball screw infeed system for plunge cylindrical grinder.

FIGURE 15.60
Two-axis combination stepper/DC motor traverse dressing system.

15.19.10 FW-BS-E Plunge-Roll Infeed System for a Creep-Feed Grinder

Figure 15.61 shows a plunge-roll infeed system for a Brown & Sharpe (Jones & Shipman) 1236 Hi-Tech creep-feed grinder. The dresser is mounted over the wheel on the spindle head. The unit has a servo-motor-driven triple-barrel infeed and electric belt-driven spindle assembly. It is equipped with TG angular contact bearing spindle arrangement.

FIGURE 15.61
Servo motor-driven plunge dresser for creep-feed grinding.

15.19.11 Servo-motor Infeed and Double-Barrel Carrier Dresser

Figure 15.62 shows a rebuilt Cincinnati 220-8 centerless grinder with a servo-motor infeed and double-barrel carrier dresser unit. The spindle has an 8-in. wide roll capacity.

15.19.12 Two-Axis CNC Profile Dresser

Figure 15.63 shows a two-axis CNC dresser for dressing the o.d., radii, and flanks profile for grinding diesel cranks on a remanufactured Landis 4R pin grinder. It is a programmable controller with linear and circular interpolation. The axes are direct drive servo motor/ball screw on linear slides to eliminate coupling misalignment and backlash.

Finally, Figures 15.64 and 15.65 provide illustrations of plunge-roll dresser spindles in situ on two typical grinding applications.

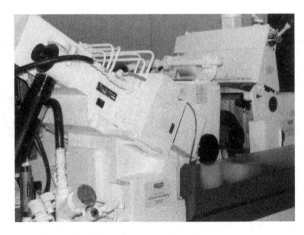

FIGURE 15.62
Rebuilt Cincinnati 220-8 centerless grinder with servo-driven plunge rotary diamond dressing.

FIGURE 15.63
Pictures of a Landis 4R crank grinder retrofitted with a two-axis CNC dresser for radius profiling.

FIGURE 15.64
Example of dresser mounted in angle-approach for a cylindrical grinder dressing alox wheels. (Courtesy of Landis Grinders.)

FIGURE 15.65
Example of dresser mounted in rear of Landis 14RE centerless grinder. (Courtesy of Landis Grinder.)

FIGURE 15.66
Examples of online and offline CNC profiling dressers. (From Normac Grinders. With permission.)

Generation of radii for the bearing industry is still often a problem using circular interpolation of two axes due to the limitations in encoder/ball screw/control resolution. In many cases, the smoothest radial form is still generated by swiveling about a fixed radius point. Most grinders in the field are equipped with single-point devices dressing alox or SG-type abrasive wheels, but Nevue (1993) demonstrated these could be readily retrofitted to small rotary devices for dressing CBN wheels.

For the most complex forms, especially in the gear industry, companies such as Normac (Arden, California) have specialized in compact multiaxis units with accompanying software that can generate and interpolate gear tooth profiles direct from print dimensions. The dresser can be either mounted on the grinder or the wheels dressed offline on a separate machine (Figure 15.66). Offline dressing is common in the tool, thread, and gear grinding industries. The equipment provided for this range from manual grinders, such as the Junker ARJ 250 example in Figure 15.67 which is designed to generate chamfers, bevels, radii, and other simple shapes, to fully programmable rotary dresser systems with

FIGURE 15.67
ARJ 250 offline dresser. (Courtesy of Max Engineering, Howell, Michigan.)

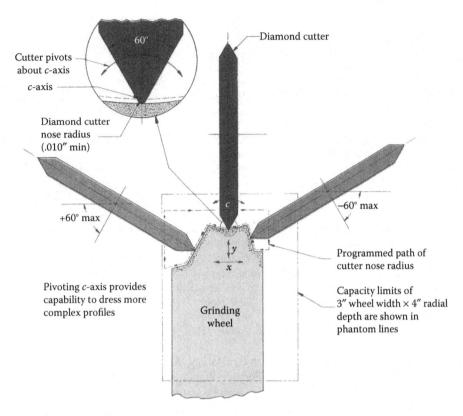

FIGURE 15.68
Orion engineering dresser (From Wendt Boart. With permission) and profiling method.

control of *X*, *Y*, and *C* axes in order to maintain the dresser perpendicular to the wheel face for very complex profiles (Figure 15.68; Orion 1993). Again, software development is the most critical factor in this technology especially where interpolation of involute forms is involved (Kelly and Smith 1993).

References

Acer, *High Precision Manual/Automatic Surface Grinder Supra Series*. Acer Group, Taiwan. Trade brochure, n.d.

Aleyaasin, M., Whalley, R., Ebrahimi, M., 2000, Error correction in hydrostatic spindles by optimal bearing tuning. *International Journal of Machine Tools and Manufacture*, 40, 809–822.

Anon, 1999, Hydrostatics ready for prime-time machining. *Tooling & Production*, January, 29–30.

Arnone, M., 1997, Learning the discipline of precision machining—Pt1 machine tool construction, motion control. *Manufacturing Engineering*, June, 58–64.

Aronson, R. B., 1994, Machine tool 101: Part 3. Spindles and motors. *Manufacturing Engineering*, March, 49–54.

Aronson, R. B., 1995, Machine tool 102: Spindles. *Manufacturing Engineering*, March, 99–107, using Landis as reference source.

Aronson, R. B., 1996, Spotlight on spindles. *Manufacturing Engineering*, June, 59–66.

Aronson, R. B., 1998, Spindles for 2000 and beyond. *Manufacturing Engineering*, 7, 106–113.

Bryant, 1995, *The Ultraline UL2 High Speed Grinding Machine* Bryant Grinder Corp. Springfield, Vermont. Trade brochure.

Bryant, 1986, 2 *Standard Wheelheads High-Frequency, Belt-Driven, Accessories High Speed Spindle Systems*. Bryant Grinder Corp, Springfield, Vermont. Trade brochure.

Buderus, 1998, *Buderus CNC 335 Machining Center—For Machining Hardened Stock*. Buderus Schleiftechnik Asslar, Germany.

Chevalier, 2000, *FSG-H818CNC, B8818CNC, C1224CNC, High Precision CNC Profile Grinder* Chevalier Machinery, Santa Fe Springs, California. Trade brochure.

Danobat, 1991, *CNC 585 High-Production and Universal Cylindrical Internal Grinding Machines* Danobat, Elgoibar, Spain. Trade brochure.

Danobat 2, n.d. *R1-100-CNC Universal and Cylindrical Grinders*. Danobat, Elgoibar, Spain. Trade brochure.

Dayton, R. W., 1949, Sleeve bearing materials. *American Society for Metals*, 5–6.

Den Hartog, 1956, *Mechanical Vibrations*. 4th edition. McGraw-Hill.

Drake, 2000, *GS: F Flute Grinder* Drake Manufacturing. Akron, Ohio. Trade brochure.

Elb, 1997, *Micro-Cut A Production Machine for Precision- Profile- & Creepfeed Grinding*. Elb-Schliff Werkzeugmaschinen GmbH, Frankfurt, Germany. Trade brochure.

Elgin, n.d. *Elgin Traveling Head and Traveling Table Heavy Duty Roll Grinders*. S&S Machinery, New York. Trade brochure.

EMAG, 1998, *VSC We Also Turn the Drilling Process on its Head*. Emag Maschinenfabrik GmbH, Salach, Germany Trade brochure.

Ernst, A., 1998, *Digital Linear and Angular Metrology*. Verlag Moderne Industrie.

Fortuna, 1991, *Automated Camshaft Grinding*. Fortuna-Werke, Stuttgart, Germany. Trade brochure.

Gamfior, n.d. *Use and Installation Handbook for Spindles and High-Frequency Spindles*. Gamfior S.p.a. Torino, Italy User Manual.

Hallum, D. L., 1994, A machine tool built from mathematics. *American Machinist*, October.

IBAG, 1996, *IMTS '96 debut—IBAG Introduces a New Generation of High Speed Machining with Active Magnetic Bearing Spindles*. IBAG NA Div of Burnco, Milford, Connecticut. Trade brochure.

Ives, D., Rowe, W. B., 1987, The effect of multiple supply sources on the performance of heavily loaded pressurized high-speed journal bearings. *Paper C199/87, Proceedings of the Institution of Mechanical Engineers (London), Tribology Conference*, July, London.

Junker, 1992, A new era in the field of o.d. grinding Erwin Junker Maschinenfabrik GmbH, Nordrach, Germany. Trade brochure.

Junker, 1996, *Quickpoint CNC o.d. Single Point Grinding*. Erwin Junker Maschinenfabrik GmbH, Nordrach, Germany. Trade brochure 1989.

Kellenberger, n.d., *Kel-Universal Grinding Machien for the Highest Demands*. Kellenberger Inc, Emslford, New York. Trade brochure.

Kelly, P. W., Smith, W. C., 1993, CNC grinding wheel dressing with emphasis on vitrified CBN. *5th International Grinding Conference*, October 26, Cincinnati, Ohio, SME.

Kent, n.d., *Precision Surface Grinder KGS-250 Series* Kent Industrial Co, Taiwan. Trade brochure.

Kondo, 1998, *Machine Guide* Kondo Machine Works Co. Ltd. Toyohashi City, Japan. Trade brochure.

Koyo, n.d., *KVD Series Vertical Spindle Surface Grinder* Koyo Machine Ind, Osaka, Japan, Cat # KVD-9512ET. Trade brochure.

Koyo 2, n.d., *Centerless Grinder KC Series* Koyo Machine Ind, Osaka, Japan, Cat # KVD-H608ET. Trade brochure.

Koshal, D., Rowe, W. B., 1980, Fluid film journal bearings operating in a hybrid mode. *Transactions of the ASME, Journal of Lubrication Technology*, 103, 558–572 (Part 1: Theory, Part 2: Experiments).

Landis 2, 1996a, *Landis® 3L Twin Wheelhead CNC Grinder for Camshafts and Crankshafts*. Landis Waynesboro, Virginia. Trade brochure.

Landis 3, 1996b, *Landis® 2SE CNC Plain and Eccentric Diameter Shaft Grinders*. Landis Waynesboro, Virginia. Trade brochure.

Landis, 1999, *Linear Motor Wheelfeeds Deliver High-Precision Camlobe and Crankpin Contour Grinding.* Wolftracks 6/1, 26. UNOVA, Cincinnati, Ohio.

Lidkoping, 1998, *CG 300 High-Output Precision Centerless Grinding.* Lidkoping Machine Tools AB, Lidkoping, Sweden. Trade brochure #1677, issue 1.

Magerle, n.d., *Magerle Grinding Systems.* Schleifring-Gruppe, Uster Switzerland. Trade brochure.

Makino, n.d., *A77 Horizontal Machining Center.* Makino Milling Machine Co. Ltd., Tokyo, Japan. Trade brochure.

Mattison, n.d., *Horizontal Surface Grinders* Mattison, Rockford, Illinois.

Micron, n.d., *Twin Grip Centerless Grinder* Micron Machinery Co. Ltd., Yamagata, Japan Trade brochure.

Monza, 1998, *Centerless Grinding Machines.* Officine monzeni spa, Italy. Trade brochure.

Montusiewicz, J., Osyczka, A., 1997, Computer aided optimum design of machine tool spindle systems with hydrostatic bearings. *Proceedings of the Institution of Mechanical Engineers*, 211B, 43–51 (spindle design based on Landis-Gendron design).

Moritomo, S., Ota, M., 1986, Present high speed machine tool spindles. *Bulletin of JSPE*, 20/1, 1–6.

Nakamura, S., 1996, High-speed spindles for machine tools. *IInternational journal of JSPE*, 30/4, 291–294.

Nevue, S. R., 1993, Design and development of a diamond disk form truer for CBN raceway grinding. MSc thesis, University of Connecticut.

NTC, 1999, *Machine Tools for Automotive Industry.* Nippei Toyama Corp, Tokyo, Japan. Trade brochure.

Okuma 1 2000, GI-10N CNC Internal grinder for mass production. Okuma Corp, Aichi, Japan. Trade brochure.

Okuma 2 1998, *GC-Super33 Triple-Speed Cam Grinder.* Okuma Corp, Aichi, Japan. Trade brochure.

Okuma 3, n.d., *GP-34/44N Plain Cylindrical Grinders.* Okuma Corp, Aichi, Japan. Trade brochure.

Orion, 1993, *Grinding Wheel Profile Dresser CNC 3 Axes.* Orion Engineering Co. Trade brochure.

Paragon, n.d., *Universal Cylindrical Grinding Machine.* Rong Kuang Machinery Co, Taichung, Taiwan. Trade brochure.

Precise, n.d., *Precise High Speed Spindle Systems.* The Precise Corporation, Racine, Wisconsin. Trade brochure.

Precitech, 1998, *Precitech Precision* Precitech Precision, Keene, New Hampshire. Trade Brochures, reference Nanoform® series of machines.

Pride, n.d., *Ultra Precision Grinders.* Pride Industries Inc., Champlin, Minnesota. Trade brochure.

Rowe, W. B., 1967, Experience with four types of grinding machine spindle. In: *Advances in Machine Tool Design and Research*, Tobias and Koenigsberger (eds.). Pergamon Press, Oxford and New York. *Proceedings of the 8th International MTDR Conference.* September, Birmingham University.

Rowe, W. B., 1969, Hydrostatic bearings. UK Patent 1 170 602. (Application 2 2072/66, 18 May 1966).

Rowe, W. B., 1970, Diaphragm valves for controlling opposed pad hydrostatic bearings. *Proceedings of the Tribology Convention, Brighton, 1970 and Proceedings of the Institution of Mechanical Engineers*, 184 (Pt-3L), 1–9.

Rowe, W. B., 1983, *Hydrostatic and Hybrid Bearing Design.* Butterworths, Boston.

Rowe, W. B., 1988, Advances in hydrostatic & hybrid bearing technology IME tribology group—Donald Julius Groen Prize Lecture 1988. *Proceedings of the Institution of Mechanical Engineers*, London.

Rowe, W. B., 2012, *Hydrostatic, Aerostatic, and Hybrid Bearing Design.* Butterworth Heinemann imprint of Elsevier, Oxford UK and Waltham, Massachusetts.

Rowe, W. B., 2014, *Principles of Modern Grinding Machine Technology*, 2nd edition. Elsevier.

Rowe, W. B., Koshal, D., 1980, A new basis for the optimization of hybrid journal bearings. *Wear*, 64(3), 115–131.

Rowe, W. B., Koshal, D., Stout, K., 1977, Investigation of recessed and slot-entry journal bearings for hybrid hydrodynamic and hydrostatic operation. *Wear*, 43(1), 55–70.

Rowe, W. B., O'Donoghue, J. P., Cameron, A., 1970, Optimization of externally pressurized bearings for minimum power and low temperature rise. *Tribology International*, August, 153–157.

Rowe, W. B., Xu, S. X., Chong, F. S., Weston, W., 1982, Hybrid journal bearings with particular reference to hole-entry configurations. *Tribology International*, 15(6), 339–348.

Royal Master, 1996. *TG12X4 Centerless Grinder.* Royal Master Grinders Inc, Oakland, California. Trade brochure.

Shigiya, n.d., *GPS-30 CNC Cylindrical Grinders.* Shigiya Machinery Works, Hiroshima, Japan. Trade brochure.

Shigiya 2, n.d., *Cylindrical Grinders—General Catalog.* Shigiya Machinery Works, Hiroshima, Japan. Trade brochure.

Slocum, A. H., 1992, *Precision Machine Design.* SME.

Smith, G. T., 1993, *CNC Machining Technology 3 Part Programming Techniques.* Springer Verlag, 101–124.

Sotiropoulos, N., 1998, The evolution of grinding machine designs and a look into the next generation. *Abrasives Magazine,* June/July, 8–10.

Studer, n.d., *All about S21 lean CNC* Studer Schleifring Group. Trade brochure.

THK, n.d., LM System—Linear Motion Systems. Catalog #75EA THK, Tokyo, Japan. Trade brochure.

THK 2, 1995, THK LM System Ball Screws. Catalog #75-1BE.

Tobias, S. A., 1965, *Machine Tool Vibration.* Blackie & Son, Glasgow and Bombay.

Toyo, n.d., *T-235 CNC External Grinding Machine.* Toyo Advanced Technologies Co. Ltd., Hiroshima, Japan. Trade brochure.

Toyoda, 1994, *GL3A GL3P CNC Cylindrical Grinder.* Toyoda Machine Works, Japan. Trade brochure.

Toyoda 2 1996, *GE3 CNC Cylindrical Grinder.* Toyoda Machine Works, Japan. Trade brochure.

Tschudin, n.d. Innovative centerless-grinding machines. Tschudin Systeme, Grenchen, Switzerland. Trade brochure.

Viking, 1998, Design news honors Milacron Chief Engineer. *Wolftracks* 5(2), 12–16.

Waldrich Siegen, 1996, *Worldwide Partners for Intelligent Manufacturing Solutions.* Waldrich Siegen, Burbach, Germany. Trade brochure.

Wang, X. P. et al., 2001, Machine tool spindles and active magnetic bearings. *Key Engineering Materials,* 202–203, 465–468.

Weldon, 1991, *Innovative CNC Grinding Technology.* Weldon Machine Tool, York, Pennsylvania. Trade brochure.

Weldon 2, 1994, *We've struck Gold!* Weldon Machine Tool, York, Pennsylvania. Trade brochure. (Note—Shear Damper® is a trademark of Aesop Inc.).

Weldon 3, 1998, *P175 CNC Punch Grinder* Weldon Machine Tool, York, Pennsylvania. Trade brochure (developed jointly with Tsugami, Japan).

Yoshida, K., 2000, Effects of mounting numbers of surface grinding machines on their rocking mode vibrations. *Abrasives Magazine,* June/July, 21–23.

ZRB, 1998, *ZRB Bearings* Harwinton, Connecticut. Trade catalog.

Additional References

Blohm Motion #23, March 1999, *Blohm Profimat RT p31.* Schleifring. Blohm, Hamburg, Germany. Trade brochure.

Goldcrown, 1993, *GC Series—Precision Centerless Grinding Machines.* Goldcrown Machinery (tradename for of Landis Machine, Waynesboro, Virginia). Trade Brochure.

Hauser, n.d. *Hauser—Your partner in grinding* Usach Technologies Elgin, Illinois. Trade brochure.

Junker, 1997, *ARJ 250 Dressing Machine.* Erwin Junker. Trade brochure.

Moore, 1996, *Moore Precision jig Grinding Wheels 1996 Edition.* Moore Tool Company, Bridgeport, Connecticut. Trade catalog.

Normac, 1998, *Profilers—CNC Grinding Wheel Profiles & Profiling Centers.* Normac Precision Grinding Machines. Trade brochure.

Normac, 2, 2000, *CBN-465 Super-Abrasive Wheel Profiling Center.* Trade advert.

Voumard, 2001, *Grinding Machines.* Voumard Machines CO SA, La Chaux-de-Fonds, Switzerland. Trade brochure.

16

Surface Grinding

16.1 Types of Surface Grinding Process

Terms for grinding were incorporated in an international dictionary for material removal processes by CIRP after agreement between representatives of a number of countries including from the United States, European Union, China, Japan, and Asia (CIRP 2005). The dictionary gives terms for material removal processes in French, German, and English and largely incorporates the terms employed in the German standard.

In this handbook, the terms employed are closely aligned with the CIRP dictionary. It allows some flexibility between the general terms derived from general cutting terminology and the more specific terms generally used in grinding. The symbols in Figure 16.1 are as defined by the CIRP dictionary which gives the commonly used symbols for wheel speed, work speed, depth of cut, geometric contact length, and equivalent wheel diameter.

Surface grinding processes are classified in Germany according to DIN 8589-11 in terms of the predominantly active grinding wheel surface position and of the table feed motion type. In the case of peripheral grinding, the grinding spindle is parallel to the workpiece surface to be machined. The workpiece material is mainly cut with the circumferential surface of the grinding wheel. In the case of face grinding with axial feed, in contrast, the grinding spindle is vertical to the workpiece surface. In this process, the workpiece material is sometimes mainly cut with the face side of the grinding wheel. The table feed motion can be translational or rotary. Figure 16.2 shows the classification of surface grinding processes according to DIN 8589-11 (DIN 8589). The reader will find that alternate symbols are used for some terms. Three examples are as follows.

Parameter	Acceptable Symbols
Grinding wheel speed	v_s or v_c
Work speed	v_w or v_{ft}
Width of grinding	b_w or a_p or $b_{s,eff}$

16.2 Basics of Reciprocating Grinding

According to DIN 8589-11, reciprocating grinding is a peripheral longitudinal grinding process with a back and forth feed motion, in which the feed motion takes place gradually in small steps with a relatively high feed rate (DIN 8589).

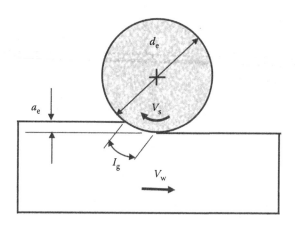

FIGURE 16.1
Wheel speed, work speed, depth of cut, equivalent grinding wheel diameter, and geometric contact length in grinding. (After CIRP, 2005, *Dictionary of Production Engineering II, Material Removal Processes*. Springer-Verlag. With permission.)

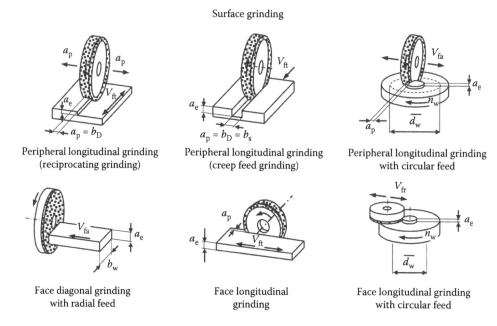

FIGURE 16.2
Classification of surface grinding processes. (After DIN 8589, Teil 11, Entwurf (05.2002): Fertigungsverfahren Spanen; Schleifen mit rotierendem Werkzeug; Einordnung, Unterteilung, Begriffe. Berlin, Beuth.)

16.2.1 Process Characterization

Reciprocating grinding is used for generating plain surfaces of usually large lateral dimensions. The grinding spindle is parallel to the workpiece surface to be machined, the workpiece material being cut mainly with the circumferential side of the grinding wheel. The grinding wheel is fed orthogonally to the workpiece surface by the amount a_p relatively to the workpiece (see Figure 16.3). This direction is in the axial direction with respect to the wheel spindle.

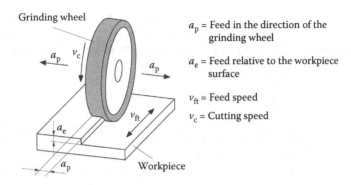

a_p = Feed in the direction of the grinding wheel

a_e = Feed relative to the workpiece surface

v_{ft} = Feed speed

v_c = Cutting speed

FIGURE 16.3
Longitudinal peripheral grinding.

16.2.1.1 Real Depth of Cut

The real depth of cut a_e is the feed relative to the workpiece surface. It is important to recognize that the real depth of cut is not the same as the set depth of cut. This is due to deflections of the grinding wheel, of the machine and due to wheel wear. Grinding performance should always be related to the real depth of cut, otherwise the results will depend very strongly on the particular grinding wheel, and the particular grinding machine setup.

In the case of reciprocating grinding, material removal typically takes place at low infeed and a high feed rate with a high number of passes with changing tool engagement and at constant peripheral speed and rotational direction of the grinding wheel. The alternation of up and down grinding is inevitable.

16.2.1.2 Speed Ratio

As a consequence, the sign of the speed ratio alternates too, due to directional change of feed rate v_{ft} during grinding with synchronous and counter rotation.

$$q = \frac{v_c}{v_{ft}} \tag{16.1}$$

16.2.1.3 Specific Removal Rate

The specific material removal rate Q'_w for reciprocating grinding is calculated from the product of infeed a_e and feed rate v_{ft}:

$$Q'_w = a_e \cdot v_{ft} \tag{16.2}$$

16.2.1.4 Upcut and Downcut Grinding

Differences can be traced back to the progression of cutting edge engagement. In the case of down grinding, the uncut chip thickness increases directly to the maximum value as the cutting edge is engaged in the material and continuously drops after the cutting edge exits the workpiece. In the case of up grinding, cutting edge engagement starts nearly tangentially to the already machined workpiece surface; the uncut chip thickness continuously increases to the maximum value until the cutting edge exits the workpiece. This has consequences for grinding forces, lubrication and wheel wear particular for deep cuts. These effects are discussed below under creep grinding.

16.2.1.5 Nonproductive Time

During machining, there is a considerable amount of nonproductive time. This depends on table speed and acceleration and the nongrinding time can be a multiple of the actual grinding time, that is, the contact time between grinding wheel and workpiece, due to the grinding wheel running beyond the workpiece, and due to the reversing table motion.

16.2.2 Influences of Grinding Parameters on Grinding Performance

Kinematic parameters and process parameters, as well as the work results are used for surface grinding process assessment. Table 16.1 shows the relevant assessment values.

16.2.2.1 The Influence of Cutting Speed (Wheel Speed)

In the case of reciprocating grinding, the contact length between workpiece and wheel is small and the grinding forces are relatively small. There are only a few cutting edges engaged at the same time, the force on individual cutting edges is accordingly high. By increasing the cutting speed at constant feed rate, the average chip thickness and length are reduced, making the total grinding forces decrease. The thermal stress of tool and workpiece, however, grows if the cutting speed is increased.

16.2.2.2 The Influence of Feed Rate (Work Speed)

Alongside infeed (i.e., depth of cut), the feed rate considerably affects the total machining time in reciprocating grinding. Increasing the feed rate results in higher average

TABLE 16.1

Kinematic and Process Parameters and Work Result in the Surface Grinding Process

Kinematic parameters	Effective cutting edge number N_{SkinA}
	Present cutting edge number N_{sact}
	Average chip thickness $\bar{h}_{\text{cu}}$
	Average chip length l_{c}
	Average material removal rate Q_{w}
	Blunting coefficient ϵ_{a}
Process parameters and work result	Surface roughness R_{z}
	Single grain force f_{g}
	Grinding force F_{c}
	Contact zone temperature ϑ_{K}
	Workpiece fringe area temperature ϑ_{z}
	Edge wear area A_{sw}
	Radial wear Δr_{s}

Source: Uhlmann, E. G., 1994a, *Tiefschleifen hochfester keramischer Werkstoffe,* Produktionstechnik. Berlin, Forschungsberichte für die Praxis, Band 129, Carl Hanser Verlag München, Wien.

chip thickness and length, and thus in an increase of grinding forces and thermal stress. A repeated start of the grinding wheel to the workpiece and the punctual impact stresses linked with it cause an increase in grinding wheel wear. The surface roughness of the workpiece grows with increasing feed rate. In metal working, feed rates in the range of 25–30 m/min are accepted (Spur and Stöefele 1980).

16.2.2.3 The Influence of Infeed

Infeed and feed rate (work speed) crucially determine the total machining time in reciprocating grinding. An increase of infeed increases grinding forces and thermal stress. The workpiece surface roughness decreases with a higher number of engaged cutting edges. In low removal rate grinding, maximum infeeds are recommended in the range of 5–10 μm (Spur and Stöefele 1980). Larger infeeds can be achieved in creep grinding and high-removal rate grinding.

16.2.2.4 The Influence of the Interrupted Cut

As a result of alternating up- and downcut grinding, and of frequent approaches to the workpiece, high alternating and impact stresses occur at the grinding wheel in reciprocating grinding. Appropriate grinding wheels and bond materials must be chosen to cope with the additional wear caused by this effect. The abrasive grain must be suitable for alternating and impact stresses and the bond must be capable of holding the abrasive grains firmly in the matrix, even under these conditions. If the bond does not meet this requirement, there will be increased grinding wheel wear and poor surface quality of the workpiece. With higher process forces and temperatures, the risk of displacements between tool and workpiece increases, having negative effects on the work result and on the wheel wear.

16.2.2.5 Reciprocating Grinding Without CF

Reciprocating grinding without cross feed is employed either for slot grinding or for generating a profile with a preformed wheel. The dresser is usually table mounted and for a conventional wheel may consist either of a simple single point or form block at the end of the slide or of a driven spindle with a form diamond roll. This latter approach is quite common for burn sensitive components such as nickel-based aerospace components on older grinders using both conventional and plated cubic boron nitride (CBN) wheels.

16.2.2.6 Multiple Small Parts

Care should be taken when laying out parts on the table especially when large number of small parts are ground together. Where possible, the parts should be butted together. Gaps between parts will cause drop-off of the wheel leading to vibration, force variations, and the other negative effects of an interrupted grind. For example, Figure 16.4 shows the impact of spacing when grinding tungsten carbide with resin diamond wheels (Hughes and Dean n.d.). Wheel life for a given wheel grade could be varied by a factor 3 depending on the spacing between parts. In the example shown, grit retention was improved by using metal coated diamond.

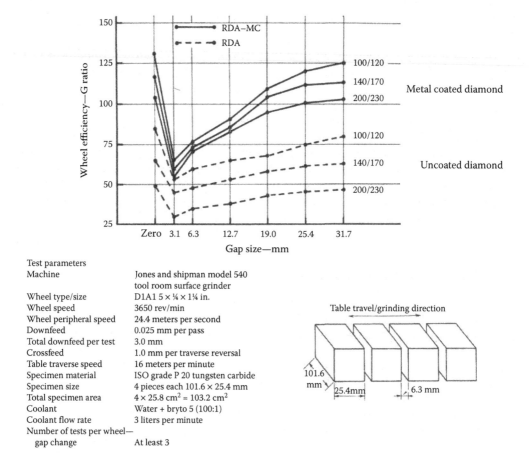

Test parameters
Machine   Jones and shipman model 540
                               tool room surface grinder
Wheel type/size   D1A1 5 × ¼ × 1¼ in.
Wheel speed   3650 rev/min
Wheel peripheral speed   24.4 meters per second
Downfeed   0.025 mm per pass
Total downfeed per test   3.0 mm
Crossfeed   1.0 mm per traverse reversal
Table traverse speed   16 meters per minute
Specimen material   ISO grade P 20 tungsten carbide
Specimen size   4 pieces each 101.6 × 25.4 mm
Total specimen area $4 \times 25.8 \text{ cm}^2 = 103.2 \text{ cm}^2$
Coolant   Water + bryto 5 (100:1)
Coolant flow rate   3 liters per minute
Number of tests per wheel—
   gap change   At least 3

FIGURE 16.4
The effect of part spacing on G-ratio.

16.2.3 Economics

Low thermal, static, and dynamic stresses are advantages of reciprocating grinding, in contrast to creep feed (CF) grinding. They allow for a more simple machine design in terms of drive capacity, rigidity, and ancillary units, thus accounting for lower purchase and operating costs. The generally high total machining times are a disadvantage for the workpiece, as well as higher surface roughness of the workpiece (being usually a functional surface), in contrast to CF grinding. Considerable secondary processing times are generated by the grinding wheel passing over the workpiece and by the reversing movement of the table.

16.3 Basics of Creep Grinding

16.3.1 Introduction

By means of CF grinding, or creep grinding as it is more concisely termed, considerable material removal rates were achieved with high surface qualities for the first time in the early 1950s. In order to harness, the potentials of this grinding process, machines, grinding tools, and grinding technologies were developed (Uhlmann 1994a). DIN 8589 part 11 defines that, in the case of CF grinding, the infeed depth must be relatively large and the feed rate accordingly low (DIN 85891). An extensive description of the process is given by Andrews et al. (1985).

16.3.2 Process Characterization

Peripheral longitudinal grinding (CF grinding) forms a negative profile of the grinding wheel in the workpiece, the total grinding stock being cut in one or only a few passes. In this method, the grinding spindle is parallel to the workpiece surface to be machined, the workpiece material being mainly cut with the grinding wheel circumferential side. The grinding wheel is fed orthogonally to the workpiece surface by the amount a_e relatively to the workpiece (Figure 16.5).

Creep grinding is usually characterized by infeeds a_e larger than 0.5 mm and feed rates v_{ft} smaller than 40 mm/s (Uhlmann 1994b). In metal cutting, infeeds a_e in the range of 0.1–30 mm and feed rates v_{ft} in the range of 25–45 m/s are normal. In contrast to reciprocating grinding, the sign of the speed relation q does not change.

$$q = \frac{v_c}{v_{ft}} \tag{16.3}$$

If grinding takes place in synchronous rotation, q is positive, in counter rotation, however, the q is negative. Through the finishing in one run, the tool–workpiece contact time approximately equals the total machining time. There are no downtimes during the grinding process, as in the case of reciprocating grinding. In the case of machining with diamond and boron nitride grinding wheels, grain protrusion, too, can be a difference between CF grinding and reciprocating grinding. If the infeed is larger than the grain protrusion from the bond, the process is called CF grinding.

All grinding processes, in which the wheel peripheral speeds are above the usual values of 35–45 m/s, can be classified as high-speed grinding. High-speed grinding can be used for the machining of materials such as steels, tungsten carbides, plastics, and ceramics with adapted grinding wheel specification. The advantages of high-speed grinding are higher surface qualities, shorter machining times, little tool wear, and low grinding forces (Minke and Tawakoli 1991).

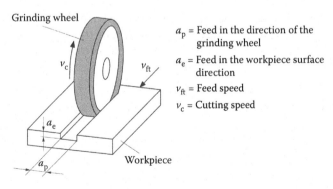

Grinding wheel

v_c
v_{ft}
a_e
a_p
Workpiece

a_p = Feed in the direction of the grinding wheel
a_e = Feed in the workpiece surface direction
v_{ft} = Feed speed
v_c = Cutting speed

FIGURE 16.5
Creep-feed grinding.

16.3.3 High-Efficiency Deep Grinding

High-performance deep grinding, with increased wheel peripheral speed and increased feed rate is a special case, requiring a cooling lubricant supply system, adapted in terms of pressure, and volume flow capability (Rowe and Jin 2001). Application limits for high-performance grinding result from wheel safety requirements, that is, the maximum admissible peripheral speeds for grinding wheels. These are approximately 125 m/s for grinding wheels with resin and vitrified bond. In the case of wheel peripheral speeds higher than 125 m/s, mainly CBN grinding wheels with sintered metal or plated metal bond in a metallic base body are used for steel materials. The advantages of high-speed grinding are higher surface quality, shorter machining times, lower wheel wear, and low grinding forces.

Table 16.2 compares typical process values for reciprocating, creep, and high-performance grinding (Minke and Tawakoli 1991). In the case of creep and high-performance grinding, infeeds in the millimeter range reach maximum values of approximately 30 mm. The specific material removal rates in the case of CF grinding differ within the range of the conventional values of $Q'_w = 3–5$ mm³/(mms) (Minke and Tawakoli 1991).

Due to the higher grinding forces in high-performance grinding processes, more demanding requirements are placed on the grinding machine and machine system:

- Rigid concrete or superior stiff structural design
- Robust guiding elements
- Effective spindle drives and mountings
- User-friendly cooling lubricant and cooling system
- Adapted grinding tools and dressing equipment
- Efficient control equipment

16.3.4 The Influence of the Set Parameters in CF Grinding

The most important setting parameters in creep grinding are cutting speed v_c, tangential feed rate v_{ft}, and infeed a_e. Either up- or downcut grinding can be set in CF grinding. The influence of the setting parameters on the work result will be explained below

16.3.4.1 The Influence of Cutting Speed v_c

Due to the large contact length and large infeed in CF grinding, grinding forces tend to be high, which, in turn, leads to high thermal stress in the working zone. High grinding forces lead to machine displacements and thus to dimensional inaccuracies on the

TABLE 16.2

Process Parameters of Reciprocating, Creep, and High-Performance Grinding Processes

Set Values Process	Infeed a_e	Feed rate v_{ft}	Cutting Speed v_c	Specific Material Removal Rate Q'_w
Reciprocating grinding	0.001–0.05 mm	1–30 m/min	20–60 m/s	0.1–10 mm²/mms
Creep-feed grinding	0.1–30 mm	0.05–0.5 m/min	20–60 m/s	0.1–15 mm²/mms
High-performance grinding	0.1–30 mm	0.5–10 m/min	80–200 m/s	50–2000 mm²/mms

workpiece. The grinding forces affecting the machine system can be reduced by minimizing the forces occurring at the single cutting edge during the grinding process. This can be realized by increasing the wheel speeds, which produces a smaller chip thickness and a lower number of active cutting edges. Furthermore, increasing the wheel speed generally results in a reduction of grinding wheel wear and in a lower surface roughness of the workpiece.

16.3.4.2 The Influence of Infeed a_e and Feed Rate v_{ft}

Machining time in the grinding process is crucially influenced by the infeed a_e and the feed rate v_{ft}. An increase of the specific material removal rate by increasing the infeed or the feed rate involves higher grinding forces and elevated thermal stress due to a higher chip thickness (Uhlmann 1994a).

16.3.4.3 The Influence of Dressing Conditions

In the case of CF grinding, continuous conditioning may be required during the grinding process. Continuous dressing takes place by separate grinding and profiling devices on the grinding machine. This procedure ensures that the grinding wheel has the necessary sharpness over the whole engagement time, has the desired profile, and sufficient chip spaces are available (Spur et al. 1993). A continuously dressed grinding wheel allows for low or constant grinding forces and process temperatures, and favorable wear behavior in the CF grinding process. The surface roughness and the condition of the subsurface of the workpiece are positively influenced with respect to the work result.

16.3.4.4 The Influence of Grinding Wheel Specification

Alongside the fully engaged cutting edges, there are additional grain tips engaged, situated deeper, which do not contribute to chip formation but only elastically and plastically deform the workpiece. In contrast to processes with small contact lengths, this occurrence must not be neglected in the case of CF grinding because of the large contact length. The friction in the working zone rises, causing higher thermal stress on the workpiece and the tool, as well as higher requirements on the tool drive.

Influencing parameters for the number of active cutting edges are

- Cutting speed v_c
- Tangential feed rate v_{ft}
- Infeed a_e
- Grinding agent concentration (grain number)
- Grinding agent geometry (grain size)

In order to avoid increasing the already high friction in CF grinding by additional friction and squeezing processes of the removed chips on workpiece and tool, the safe evacuation of the chips requires sufficient porosity of the bond structure (pores/chip spaces). This additionally counteracts the clogging of the grinding wheel. High thermal stress requires the transportation of sufficient cooling lubricant to the working zone, which is also aided by larger porosity in the grinding wheel bond. Grinding wheels for creep grinding should be softer than for reciprocating grinding for the same task.

16.3.4.5 The Influence of Up- and Downcut Grinding

The process variants up and down grinding can be set in the case of CF grinding, using the favorable alternative for the particular machining conditions. If the vectors of the cutting and feed motion have the same direction at the contact point of abrasive grain and material, the process is defined as down grinding. In the case of up grinding, the vectors of the cutting and feed motion show in different directions (Schleich 1980). The effect of up- or downcut grinding is discussed further below.

Because of the processes at the cutting edge engagement and the consequent chip formation, down grinding usually provides smaller surface roughnesses and thus higher-quality surfaces in the case of ductile materials. Due to better surface qualities, lower wear, and smaller grinding forces, down grinding is recommended, and can be traced back to more efficient chip formation of the single cutting edges approaching the workpiece surface with nearly maximum chip thickness. In the case of machining with small material removal rates, this process variant has only little positive effect on the maximum contact zone temperature. In the case of up grinding, the contact zone temperature drops faster because of better cooling lubricant conditions. There are definitively lower contact zone temperatures during machining with counter rotation with large infeed, high feed rates and, at the same time, high cutting performances, since the efficiency of cooling lubrication has significant effect here (Uhlmann 1994a).

The machining of metallic materials with CF grinding offers a number of advantages in contrast to reciprocating grinding (Spur 1989):

- Reduced grinding time by 50%–80% through higher cutting performance
- Lower edge wear leads to good profile stability
- Superior surface qualities
- Smaller single grain forces
- Lower temperatures in the contact zone

16.3.4.6 Process

The chip thickness and thus the cutting force at the single grain are smaller during CF grinding than during reciprocating grinding. Additionally, there are no repeat impact loads on the grinding wheel, so that the abrasive grains can be held longer by the bond matrix before the grains are broken out. On the other hand, thermal stress and the total cutting forces are higher. See also Figure 16.6.

In the case of CF grinding, the higher cutting forces require significantly higher static and dynamic rigidity and higher drive performance of the grinding machine and its ancillary units than necessary in the case of reciprocating grinding. The high thermal stress during CF grinding does not only represent a high load for tool and grinding machine, but also damages the workpiece material causing cracks and structural changes. Such structural changes do not occur or only to a small extent in the case of reciprocating grinding, being removed with the subsequent pass.

In the case of reciprocating grinding, the contact zone of tool and workpiece can be easily supplied with cooling lubricant owing to the short length of the contact zone. In the case of CF grinding, a much longer contact zone must be wetted, and, because of higher friction, more heat must be discharged. Therefore, the cooling lubricant must be fed with high pressure and volume flow through nozzles and conducting equipment with defined shapes (Table 16.3; Schleich 1980).

	Creep-feed grinding	Reciprocating grinding
	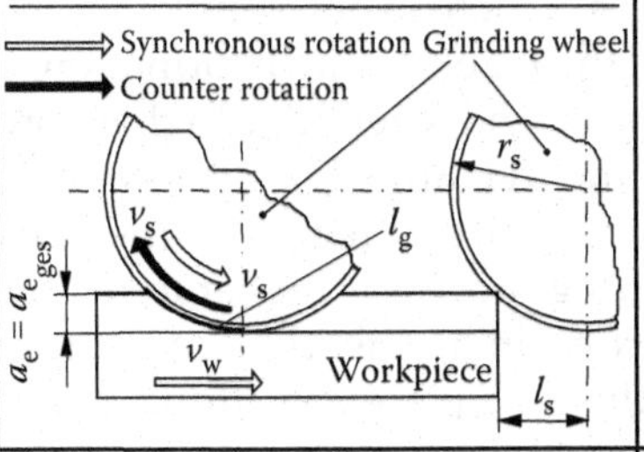 ⇒ Synchronous rotation Grinding wheel → Counter rotation	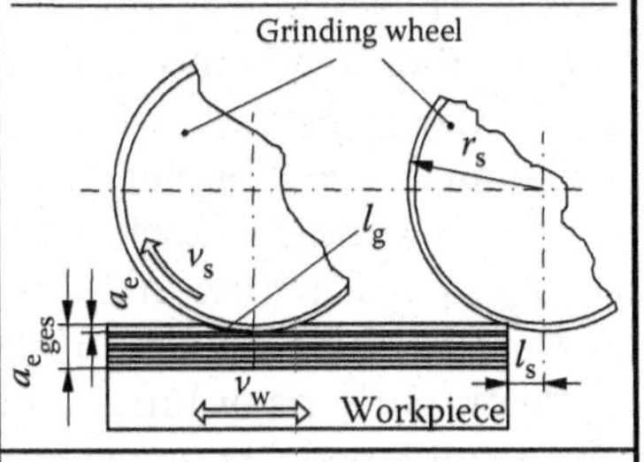 Grinding wheel
Infeed a_e (mm)	0.5–30	0.001–0.05
Feed rate v_{ft} (mm/s)	0.1–40	100–500
Speed ratio q (—)	3000–300,000	40–400
Geometrical contact l_g (mm)*	14–110	1.4–4.5
Number of grinding passes	Usually 1	Function of the overall infeed

*Radius of the grinding wheel r_s = 200 mm

FIGURE 16.6
Differences between reciprocating and creep-feed grinding.

TABLE 16.3

Differences between Reciprocating Grinding and Creep-Feed Grinding

	Reciprocating Grinding	Creep-Feed Grinding
Feed rate	High	Low
Infeed	Low	High
Number of tool passes	High	1 (or a few)
Up and down Grinding	Periodically changing	Adjustable
Secondary processing times	High	Low
Average chip thickness	Larger	Smaller
Average chip length	Smaller	Larger
Average grinding temperature/thermal stress	Lower	Higher
Grinding forces/cutting forces	Lower	Higher
Vertical shape deviation	Smaller	Larger
Surface roughness	Larger	Smaller
Radial wear	Larger	Smaller
Total wear on the grinding wheel	Smaller	Larger
Inclination to chatter	Larger	Smaller

16.3.4.7 Work Results

In CF grinding, surface roughnesses are much smaller due to the engagement of a higher number of cutting edges and due to smaller feed rates than in the case of reciprocating grinding. Since the functional surface properties of a workpiece are often important,

CF grinding has a clear advantage here. This can be traced back to the kinematics of creep grinding. Those grain cutting edges, which do not completely chip off due to their position in the bond or to advanced wear, through plastic deformation, contribute to the smoothing of the workpiece. On the other hand, there are higher forces and thermal stresses, which require much higher static and dynamic rigidity and effective driving and ancillary units of the grinding machine. In the case of reciprocating grinding, the number of cutting edges participating in the cutting process is higher in relation to the material volume. Based on the short contact time of grain and workpiece, a different surface is generated. The plastic curl-ups or burrs on the workpiece are, unlike in CF grinding, not smoothed by the simultaneously engaged neighboring cutting edges but pushed away into neighboring grooves by subsequent chipping processes. This results in temporary coverage of cut grooves by plastically deformed material. These curl-ups additionally increase surface roughness. Due to the stress during the process, these curl-ups do not correspond to the basic material, and there is no material cohesion as in the original state any more. Thus, there is an increased risk of particles detaching from the surface during subsequent use, for example, in the case of sliding bearing surfaces often machined by grinding, resulting in component breakdown due to friction and squeezing (Schleich 1980).

16.3.4.8 Grinding Wheels

The selection of the abrasive is mainly based on the properties of the workpiece material and on the secondary conditions during the grinding process, for example, the use of cooling lubricant. The larger CF grinding forces and thermal stresses in the working zone, however, require an adjusted bond of the grinding wheel.

In the case of reciprocating grinding, the wheel must absorb the impact stress due to the changing engagement of synchronous and counter rotation and to the high grinding forces at the single grain. In the case of CF grinding, impact and single grain forces are lower. In this case, high thermal stresses must be absorbed in the working zone.

A further aspect influencing wheel selection, is the chip shape, which is crucially affected by the workpiece material. In the case of reciprocating grinding, there are usually short, thick chips. In contrast, chips are relatively thin and long in the case of CF grinding (Uhlmann 1994a). The combination of abrasive grain/bond must be selected in a way that chips can be easily evacuated from the working zone without additional friction and squeezing on the workpiece and/or clogging of the grinding wheel. Therefore, there must be sufficient chip space available. An open structure of the grinding wheel allows for a significant increase of flow rate in the working zone through the cooling lubricant transport in the pores. This results in an enhanced heat transport, which is of special significance in the case of CF grinding.

16.3.4.9 Grinding Wheel Wear

Grinding wheel wear parameters are radial and edge radial wear, the latter usually being much higher during reciprocating and CF grinding than radial wear. Edge radial wear is usually higher in the case of reciprocating grinding than during creep grinding (Saljé and Damlos 1983).

The high wear during reciprocating grinding can be partially traced back to the short contact length. The shorter the tool—workpiece contact length at a constant material removal rate, the higher the stress for the single abrasive grain, since there is less time for cutting the material volume than in CF grinding. The material removal rate per grain increases, and thus the sum of cutting forces per grain, result in an increased wear. A further factor is the high impact stresses upon the repeated engagement in the workpiece after reversal. When, for instance, a high-speed steel was ground with an aluminum oxide grinding wheel, the tenfold increase of passes led to a 28% wear growth under otherwise identical conditions (Schleich 1980).

The total grinding forces, however, are usually lower during reciprocating grinding than during creep grinding. While a permanent effective self-sharpening takes place during reciprocating grinding, the large contact length during creep grinding leads to an increasing number of engaged grain cutting edges, the forces rising linearly as a function of the grinding time.

The two processes are further differentiated by the length of grinding path per wheel rotation. Due to the longer grinding path during reciprocating grinding, this process is clearly more sensitive to the grinding wheel out of roundness. Radial deviations are characterized by long waves on the workpiece. When certain deviations occur, they usually grow faster in the case of reciprocating grinding (Uhlmann 1994a).

16.3.5 Requirements for CF Grinding Machines

The technology of CF grinding places special requirements on the design of CF grinding machines. This relates to the machine frame and guidances, feed, and grinding spindle drives, and cooling lubricant and dressing devices. Due to higher forces, machine elements such as machine bed, support, table, grinding head, and guidances are to be designed much more rigid for creep grinding than for reciprocating grinding machines (VDI 3390). Furthermore, due to higher thermal stresses, measures must be taken in order to avoid the axial shift of the grinding spindle in the process. Alongside the process-friendly dimensioning of the machine elements, their relative position to each other is also significant. Thus, the force flow through the system should be short and directly counteract geometrical defects on the workpiece. Slideways, feed drives, and transmission elements must be designed in a way that impact-free motions and positioning accuracies are independent of the selected feed rates and occurring forces.

Grinding spindle drives must be infinitely variable and realize a wide range of cutting speed. Optimized grinding data must be kept constant during the process. In the case of conventional grinding wheels, the rotational speed must be adjusted automatically, in order to work with a constant cutting speed in the case of wear-related grinding wheel diameter loss. In favor of minimizing heat propagation in the workpiece, cooling lubricant should be fed synchronously as well as in contra-rotation to the grinding wheel.

16.3.6 Typical Applications

The advantages of CF grinding are shorter machining times, lower surface roughness, improved profile, and dimensional accuracy (Minke and Tawakoli 1991). In industrial mass production, there are two fields of application of CF grinding, in which high cutting performance and high component quality are required. In mass production, CF grinding is used to grind deep grooves with mostly parallel side walls and profiles in tight and deep profile contours and/or for difficult-to-grind materials. In the case of rotor manufacture, up to 30 mm deep slots are ground into the solid, mainly hardened material. A second field of application for CF grinding is turbine blade manufacture from nickel-based super-alloys. In the case of these workpieces with different profiles, high requirements are placed on geometrical accuracy and surface quality. Moreover, highest demands are made for the absence of heat-influenced workpiece subsurface layers. CF grinding satisfies these work conditions concerning high-removal rates and surface quality within the limitations of quickly rising contact zone temperatures (Werner and Minke 1981).

16.3.7 Economics of CF Grinding

Lower grinding times and high achievable surface quality with a high-removal rate argue for CF grinding in contrast to reciprocating grinding. In the range of medium to high batches, CF grinding is therefore more cost-efficient than reciprocating grinding. Grinding machines and ancillary units tend to be expensive resulting from higher thermal, static and dynamic stresses, and consequent requirements for special machine designs.

The choice of adequate grinding machines is a crucial prerequisite to completely utilize the potential of CF grinding. Requirements for economic CF grinding of ceramics are as follows (Uhlmann 1994a):

- High process forces require static and dynamic rigidity, high spindle drive performance, and load-independent rotational speed.
- High process temperatures and large contact lengths require cooling lubricant pumps with increased volume flow and pressure, adapted cooling lubricant nozzles.
- High dimensional accuracy and high forces require high-precision and rigid bearings and guidances, shock-free, and precise drives.
- High concentricity and adjustable grinding wheel topography requires integrated auxiliary conditioning equipment for diamond grinding wheels.

16.4 Basics of Speed-Stroke Grinding

16.4.1 Introduction

An innovation of surface grinding driven by the Japanese in the 1980s is a process called speed-stroke grinding that involves very high table speeds of 50–100 m/min and shallow depth of cuts of the order of 1 µm or less. The interest was initially centered around improved die manufacture and achieving high stock removal rates grinding ceramics, while keeping the depth of cut in the ductile grind regime. The earliest work was carried out on Elb grinders using toothed belt drives (Yuji 1990) and Okamoto (Akinori 1992). High-helix pitch ballscrews and more recently the introduction of linear motors has greatly expanded the scope in terms of speed and acceleration and deceleration rates.

The benefits and drawbacks of the process can be understood by going back to the basic equations relating uncut chip thickness h_{cu} to specific grinding energy e_c, force/grit f_g, surface finish R_t, and mean surface temperature (θ):

$$h_{cu} = \sqrt{\frac{v_w}{v_s} \cdot \frac{1}{C \cdot r} \sqrt{\frac{a_e}{d_e}}} \qquad (16.4)$$

$$e_c \propto \frac{1}{h_{cu}^n} \propto \sqrt{\frac{v_s}{v_w} \cdot C \cdot r \cdot \sqrt{\frac{d_e}{a_e}}} \tag{16.5}$$

$$R_t \propto \frac{h_{cu}^{4/3}}{a_e^{1/3}} \tag{16.6}$$

$$f_g \propto h_{cu}^{1.7} \tag{16.7}$$

$$T_{max} \propto \sqrt{a_e \cdot v_s \cdot C \cdot r} \tag{16.8}$$

For a constant stock removal rate chip thickness increases with work speed. Consequently specific energy decreases as table speed increases for constant removal rate, depth of cut, and surface temperature fall while surface roughness increases and force/grit increases.

Speed-stroke grinding therefore offers the ability to remove stock faster and with lower forces and less risk of burn, but roughness will be higher and compensation must be made in bond/grit strength for increased force/grit. Compensation for higher roughness can be made by controlling the process; that is, switching to a slow table speed for finish grinding. These predictions were demonstrated by Inasaki (1988) in the grinding of alumina and zirconia with resin diamond wheels on a lead screw driven machine at 36 m/min (Figure 16.7).

For the machine tool builder, the primary area of concern for machine design is being able to control the acceleration and deceleration rates without excessive vibration and without overshooting the part so much that the benefits of cycle time are lost.

The big advantage of the linear motor is that the pattern of the acceleration and deceleration can be readily adjusted so that forced vibrations can be minimized. Inasaki (1999), for example, demonstrated that a sinusoidal pattern for acceleration and deceleration can successfully suppress wheel head vibration (Figures 16.8 and 16.9).

Wheel grade selection is severely impacted by the effective high level of interrupted cut and high force/grit. Tönshoff et al. (1996) reported that wheel wear during speed-stroke grinding of alumina was very dependent on bond type. Electroplated and resin bonds showed little wear, but metal-bonded wheel wear was significant and vitrified bond wheel wear was dramatic. The differences in wheel wear were postulated to be due to the relative brittleness of the bond in the presence of vibration.

The Jung S320 is one of the first grinders to be offered commercially for speed-stroke grinding targeted at nonstandard punch grinding, slot grinding and tooling prone to vibration (Motion 2001). The grinder has a linear motor drive offering 50 m/min at 600 strokes/min and an X axis position accuracy of 3 μm on 15 mm stroke length.

Danobat has designed a machine capable of 200 m/min and a max accel/decel rate of 3 g. Speed-stroke grinding is expected to grow in numerous industries as original equipment manufacturers (OEMs) become more experienced in the use of linear magnetic motors.

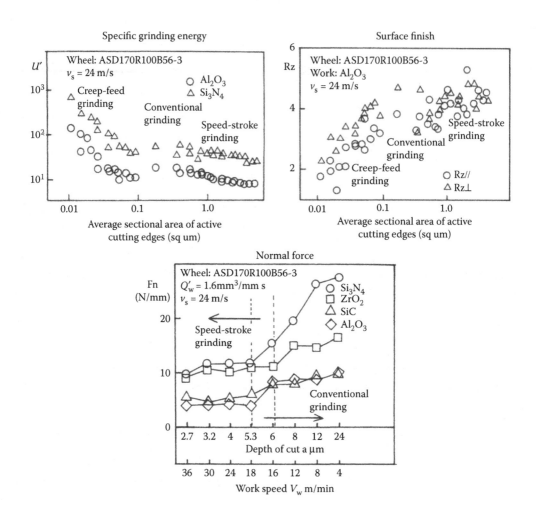

FIGURE 16.7
Process characteristics in speed-stroke grinding.

16.5 Successful Application of CF Grinding

16.5.1 CF Grinding with Vitrified Wheels Containing Alox and Silicon Carbide

In general, as depths of cut increase so do grinding forces and burn while uncut chip thickness and therefore roughness and force/grit decrease. However, when the depth of cut becomes extreme, that is, greater than 1–3 mm maximum grinding temperatures can actually fall. Grinding using a combination of slow table speeds and deep depths of cut defines the CF process. Interest started in this field in the 1950s but reached its zenith from a research viewpoint in the 1970s at the University of Bristol, United Kingdom driven by the aerospace industry and the need to grind highly burn sensitive nickel- and cobalt-based high-temperature alloys.

The CF process developed using soft (E or F), highly porous conventional wheels at relatively low wheels speeds of 3000–6000 sfpm to keep frictional heat to a minimum, limit required coolant pressure, and due to limits on wheel structural strength. The wheels

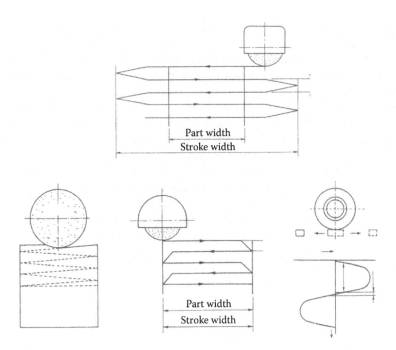

FIGURE 16.8
Strategies for velocity profiles for speed-stroke grinding.

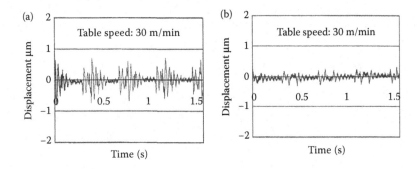

FIGURE 16.9
Impact of acceleration strategy on machine vibration for speed-stroke grinding. (a) Constant acceleration and (b) sinusoidal acceleration. (From Inasaki, I., 1999, Surface grinding machine with a linear motor driven table system. Development and performance test. *Annals of the CIRP.* 48(1), 243–246. With permission.)

were dressed using formed diamond rolls. Dressing was at first intermittent during the cycle but it was subsequently found that continuously dressing at infeed levels of 0.2–2.0 μm/rev of the wheel not only maintained the profile but kept the wheel sharp thus allowing much higher stock removal rates than had previously been seen. This latter process is the familiar continuous dress CF (CDCF).

16.5.2 Coolant Application in CF Grinding

Coolant application is absolutely critical to the process. CDCF is usually carried out using a water-based coolant that fills the highly permeable wheel structure when supplied correctly and under sufficient pressure. The coolant then becomes the primary source of heat removal and maintains the part surface at a temperature at or below 130°C the boiling point of the coolant under the hydrodynamic pressure conditions in the grind (Howes 1990). The coolant is excellent at maintaining surface temperatures by the efficient removal of heat until the heat flux exceeds the heat capacity of the fluid. At this point, the fluid boils effectively eliminating all benefits of the coolant and temperatures rapidly climb to those experienced in dry grinding (Figure 16.10). The phenomenon is known as "film boiling" (Howes 1990, 1991).

The heat capacity of the coolant is inversely proportional to the bulk temperature of the incoming fluid. Reducing the incoming coolant temperature by 40°C to 20°C raises the critical power flux significantly. Similarly increasing the coolant pressure to an optimum value where the coolant and wheel velocities are matched, maximizes the critical power flux.

It is interesting to note that straight oil, due to its lower heat capacity, is less capable of cooling the workpiece, and will cause the part surface to be hotter and much more likely to burn (Ye and Pearce 1984). However, the higher film boiling temperature of the oil (ca. 300°C) is less likely to cause rapid and catastrophic failure of the coolant. Water is the better coolant, unless finish or wheel wear is the overriding issue, because of the large capacity of the grinding wheel pores to hold coolant.

16.5.2.1 Film Boiling

The impact of film boiling can be very marked. As the water turns to steam, there is a rapid rise in temperature of the workpiece causing it to thermally expand. This lead to a sudden increase in depth of cut and additional heat generation followed by massive wheel breakdown. The workpiece then cools to leave a series of deep, usually blackened troughs in the surface (Figure 16.11). The effect can often be detected in process by surges in the wheel head spindle power.

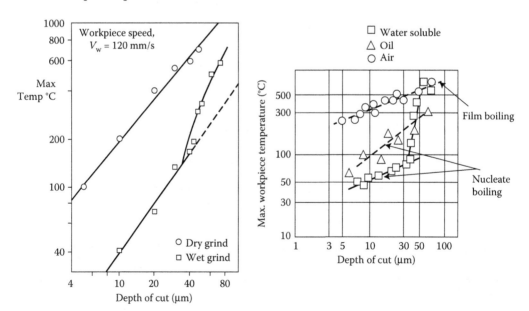

FIGURE 16.10
Illustrations of the impact that film boiling has on workpiece surface temperatures. (Adapted from Howes, T., 1990, *Annals of the CIRP.*, 39(1), 313–316; Howes, T., 1991, Avoiding thermal damage in grinding. *AES Conference.* http://www.nauticom. net/www/grind/therm.htm.)

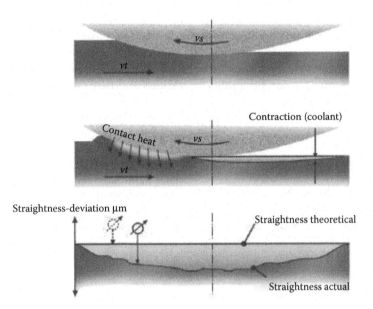

FIGURE 16.11
Effect of excessive heat generation in the grind contact zone. (From Noichl, H., 2000, CBN grinding of nickel alloys in the aerospace industry. *IDA Conference Proceedings, Intertech 2000*, Vancouver, Canada. July 21. With permission.)

16.5.2.2 Coolant Delivery System

Since coolant is such a key factor in CD grinding, close attention must be paid to all aspects of the coolant delivery system. Most CF applications either have interference with fixtures or have other issues that prevent an efficient shoe nozzle from being used. Therefore, coolant delivery must depend on coherent flow nozzle design with sufficient coolant volume and pressure. This discussion is valid for any grind operation where shoe style nozzles cannot be used. Extensive research and consolidation of available information on this subject has been carried out. Webster is included in the summary below (Webster 2000; Webster et al. 1995, 2002).

16.5.2.3 Coolant System Capacity

The standard coolant requirement is 1.5–2 gpm/hp of grinding power. A good estimation for *minimum* flow rate is therefore to take the spindle horsepower × 1.5 gpm. For a water-based coolant, at least 10 min is required for settling to allow the release of entrapped air and minimize foaming. This sets the tank capacity.

16.5.2.4 Coolant Pressure

The optimized coolant pressure at the nozzle is such that the coolant velocity matches the wheel velocity. The coolant velocity can be readily calculated from the coolant pressure using $V = (2\Delta P/\rho)^{1/2}$, where V is coolant velocity, ΔP is the pressure, and ρ is the coolant density. Note calculating the velocity based on the flow rate and the cross-sectional area of the nozzle aperture will always underestimate the velocity as the actual coolant jet cross-sectional area may constrict upon exit depending on the nozzle design. It is always best to measure the pressure as close to the nozzle as possible.

With that caveat, Tables 16.4 and 16.5 give the velocity as a function of pressure together with the associated maximum flow rates for various aperture cross-sectional areas.

TABLE 16.4

Metric Flow Rate Chart for a Nozzle with a Coefficient of Discharge of 0.95

Jet Speed (m/s)	Coolant Nozzle Pressure (bar)			Flow Rate (l/min) for Listed Nozzle Diameters (mm) and Areas (mm²)								
	Water SG = 1.0	Mineral Oil SG = 0.87	Ester Oil SG = 0.93	0.8 1	3.1 2	7.1 3	13 4	28 6	50 8	79 10	113 12	Area Diameter
20	2	2	2	0.9	3.5	8.1	15	33	57	90	129	
30	5	4	4	1.4	5.3	12	22	49	86	134	193	
40	8	7	7	1.8	7.1	16	29	64	115	179	258	
50	13	11	12	2.2	9.0	20	36	80	144	224	322	
60	18	16	17	2.6	11	24	43	97	172	268	386	
80	32	28	30	3.6	14	32	57	129	229	358	516	
100	50	44	47	4.4	18	40	72	162	287	448	645	
120	72	63	67	5.3	21	49	86	193	344	537	774	
140	98	85	91	6.2	25	56	100	226	401	627	903	
160	128	111	119	7.1	28	64	115	259	458	716	1031	
180	162	141	151	8.0	33	73	129	290	516	805	1160	
200	200	174	186	8.9	35	81	144	323	573	895	1289	

TABLE 16.5

English Flow Rate Chart for a Nozzle with a Coefficient of Discharge of 0.95

Jet Speed (fpm)	Coolant Nozzle Pressure (psi)			Flow Rate (gpm) for Listed Nozzle Diameters (in.) and Areas (in.²)								
	Water SG = 1.0	Mineral Oil SG = 0.87	Ester Oil SG = 0.93	0.003 1/16	0.012 1/8	0.028 3/16	0.049 1/4	0.077 5/16	0.11 3/8	0.15 7/17	0.196 1/2	Area Diameter
4000	30	26	28	0.6	2.4	5.5	9.7	15	22	30	39	
6000	67	58	62	0.9	3.6	8.2	15	23	33	45	58	
8000	119	104	111	1.2	4.8	11	19	30	44	59	78	
10,000	187	163	174	1.5	6.1	14	24	38	55	74	97	
12,000	269	234	250	1.8	7.3	16	29	45	65	89	116	
14,000	366	318	340	2.1	8.5	19	34	53	76	104	136	
16,000	478	416	445	2.4	9.7	22	39	61	87	119	155	
18,000	605	526	563	2.7	11	25	44	68	98	134	174	
20,000	747	650	695	3.0	12	27	48	76	109	148	194	
25,000	1166	1014	1084	3.8	15	34	61	95	136	185	242	
30,000	1680	1462	1562	4.5	18	41	73	114	164	223	291	
35,000	2286	1989	2126	5.3	21	48	85	132	191	260	339	
40,000	2986	2598	2777	6.1	24	55	97	151	218	297	388	

16.5.2.5 Piping

Values are given in Table 16.6 of pipe diameter against pressure and flowrate. Pressure is to be specified at the nozzle not at the pump. Piping should be kept as short as possible with minimal bends to avoid head losses. The pipe diameter should also be made as large as possible both to lower head losses but also to keep the flow laminar. Webster recommends a velocity of 6 m/s max to keep the Reynolds number of a water-based fluid below the level causing turbulence. This gives the following maximum flow rates for given pipe diameters (Table 16.7).

16.5.2.6 Nozzle Design

Coolant must be delivered from the pump via the piping, valves, and nozzle at the required pressure to match wheel velocity and in a laminar flow. Any turbulence or entrained air will create dispersion causing a rapid loss of momentum and preventing the coolant from overcoming the momentum of air generated by wheel drag; the so-called "air barrier."

Turbulence can be an inherent problem of the system governed by its Reynolds number (Re) which is defined as:

$$Re = \frac{v \cdot d}{v} \tag{16.9}$$

TABLE 16.6

Pipe Diameter for Various Pressures and Flow Rates

		Flow Rate (gpm)						
		Area of Nozzle (in.²) and Diameter Equivalent (in.)						
Pressure (psi)	Velocity (sfpm)	0.00307" 1/16"	0.01227" 1/8"	0.02761" 3/16"	0.04909" 1/4"	0.11045" 3/8"	0.19635" 1/2"	Area Diameter
25	3660	0.58	2.34	5.25	9.34	21	37.3	
50	5175	0.83	3.3	7.41	13.2	29.7	52.8	
75	6342	1.01	4.05	9.08	16.2	36.4	64.7	
100	7320	1.17	4.67	10.5	18.2	42.1	74.7	
150	8970	1.43	5.72	12.9	22.9	51.6	91.5	
200	10,350	1.65	6.61	14.8	26.4	59.5	106	
300	12,672	2.02	8.08	18.2	32.4	72.8	129	
400	14,620	2.33	9.33	21	37.4	84.8	154	
600	17,945	2.86	11.4	25.7	45.8	103	189	
800	20,706	3.3	13.2	29.7	52.9	120	218	

TABLE 16.7

Maximum Flow Rates for 6 m/s Velocity

Pipe ID (in)	1/2	3/4	1	1.25	1.5	2	2.5
Flow Rate (gpm)	12	27	48	75	109	193	302
Flow Rate (l/min)	45	103	183	285	410	730	1140

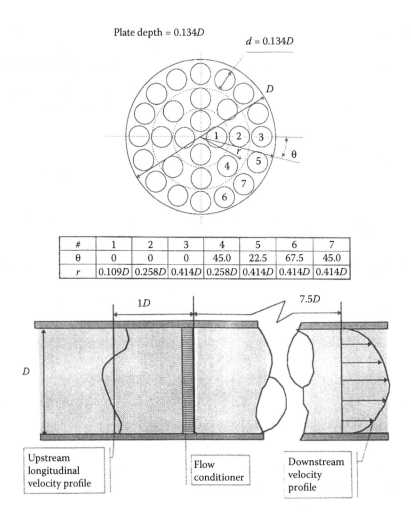

FIGURE 16.12
Mitsubishi flow conditioner (Adapted from Cui, C., 1995, Experimental investigation of thermo-fluids in the grinding zone. PhD dissertation, University of Connecticut.)

where v is the fluid velocity, d is the diameter of round jet or pipe, and v is the kinematic viscosity of fluid (10^{-6} m^2/s for water).

The Re is a measure of flow quality where smaller is less turbulent. Coolant jets are always turbulent. Laminar coolant supply nozzles are rare. Re values in excess of about 2000 will have turbulence; a value that encompasses virtually all grinding situations. The critical Re for turbulence is even lower for short orifices. Nevertheless values up to 100,000 can still give a reasonably constrained flow. The Re value should be improved where possible by chilling the coolant to increase viscosity or by reducing the exit diameter by using several small nozzles. Turbulence and pressure losses can also be generated within the pipes by bends, changes in pipe diameter or rough surfaces. This can be eliminated by the use of a flow conditioner just prior to the nozzle system. One very simple style proposed by Webster consists of a Mitsubishi in-line flow conditioner, originally designed for creating laminar flow in systems to measure city water supplies as shown in Figure 16.12. Its

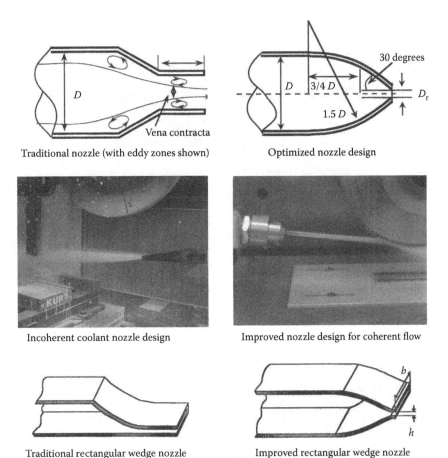

Traditional nozzle (with eddy zones shown)

Optimized nozzle design

Incoherent coolant nozzle design

Improved nozzle design for coherent flow

Traditional rectangular wedge nozzle

Improved rectangular wedge nozzle

FIGURE 16.13
Improvements in nozzle design to achieve more coherent flow.

most important feature is the ability to present a laminar flow to the nozzle entrance with virtually zero drop in pressure.

An optimized nozzle design is as simple as it is effective (Figure 16.13). The design developed by Webster, Cui, and Mindek at the University of Connecticut to generate coherent jets was based on 1950 fire hose technology. The beauty of the design is the way it balances all the forces from the coolant such that at the exit all flow is perpendicular to the exit. Ideally this balance is best maintained by using a round opening. However, the design is almost as efficient with a rectangular opening except for some small dispersion effects at the corners as long as the internal profile is balanced.

Complex shaped exit designs to follow the form of the wheel profile should be avoided as they are generally highly dispersing. Instead, several smaller round nozzles can be used far more effectively. When building the nozzles care should be taken to make sure the inner surface is as smooth as possible and that a sharp edge is maintained at the exit (no nicks).

The concept of using small round nozzle exits has been taken one stage further by a patented "card key" system whereby quick-change nozzle plates carrying several small round, internally contoured exits for each required wheel profile can be placed inside a plenum chamber. The system is offered as part of the field services program at HGTC,

FIGURE 16.14
Plenum chamber coolant delivery with quick-change "card key" nozzle plates. (Courtesy of Saint-Gobain Abrasives.)

Single nozzle arrangement
for creep-feed grinding of
jet engine blade dovetail

Pressure gauge

Single fire hose
round nozzle

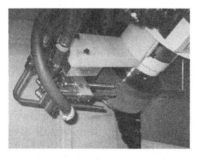

Multiple fire hose round
nozzle arrangement for
wide form application

FIGURE 16.15
Examples of fire hose round nozzle applications

Worcester, United States (Saint-Gobain Abrasives). (Figure 16.14). Other round nozzles are shown in Figure 16.15. A typical large coolant system is shown in Figure 16.16.

Oil coolant has a higher viscosity than water making jet coherency a lot easier. It also offers great benefits in terms of abrasive life (up to 100× with plated CBN). However, oil has a range of problems of its own. It is potentially a hazard both environmentally from

FIGURE 16.16
Oil coolant system for use with a 40-hp grinder operating with plated CBN wheels, including pumps, chiller, filter, settling tank, and mist collection. (Courtesy of Campbell Grinders.)

misting and as a fire risk. It also has a half to a quarter of the thermal conductivity and heat capacity, and is more prone to air entrapment. Gaulin (2002) infers the following coolant system requirements for oil:

Flow rate	1 gpm/wheel spindle hp
Settling time	25 min
Chiller capacity	4000 BTU/wheel spindle hp

Although the flow rates are actually lower than those for water-based coolant, the settling time and chiller capacity are both approximately triple. The risk of fire is real, although somewhat exaggerated. Every oil coolant has a lower and upper explosion limit based on the percentage of oil to air. The limits are typically 0.6%–7% by volume of oil, when the oil can actually be ignited. Flushing the grind area well or throwing down a curtain of oil to flood and knock down misting greatly reduces the fire risk. In the event of a fire, there is an overpressure created in the grinder of about 4.5 bar max. So grinders equipped for handling oil should have

1. Good encapsulation to prevent oil or mist from exiting
2. An efficient mist extraction system
3. A CO_2 fire suppression system
4. A spring loaded flap to release the initial pressure generated from an explosion and then automatically seal as the CO_2 system kicks in

It is interesting to note that a survey carried out in Germany between 1987 and 1994 found the leading causes of fire were (Ott and Storr 2001):

1. Workpiece jammed
2. Loss of coolant supply (sensor failure)

3. CNC operator error

4. Grinding wheel with steel core (e.g., electroplated) lost its coating and generated excess heat from rubbing on the part

The impact of oil on the physical size of a coolant system can be seen from Figure 16.16. This illustrates a self-contained, compact coolant system designed for a single grinder with a 40-hp spindle and plated CBN wheels. It consists of a 280-psi, 60-gpm high-volume pump, a 1000-psi, 5-gpm high-pressure scrubber pump, vacuum, paper filtration unit, chiller, air extraction unit, and a coolant tank large enough to allow a 10-min settling time. The system is actually larger than many of the grinders it is designed to supply. Coolant delivery to CF grinding is also a consideration when examining the importance of up versus down grinding.

To reemphasize comments made before, in upcut grinding, each individual grit experiences a high proportion of its time in contact initially just rubbing or ploughing the workpiece material (Figure 16.17). This consumes a large amount of energy in friction and elastic and plastic deformation. Only when the grit reaches its exit point, it makes a sufficiently deep cutting depth to initiate a chip. In downcut grinding, the chip is almost immediately created suppressing many of the friction losses. In up grinding, the coolant introduced at the point of entry of the wheel acts directly on the newly formed surface. In down grinding, coolant again added at the point of entry of the wheel must be carried through the grinding zone, during which it is heated up before it can react with the newly produced surface. In actual production situations, the ease of effectively introducing coolant into the grinding zone for a downcut grinding configuration appears to far outweigh any advantages of the kinematics of up grinding.

In summary, there are lower grinding forces and power and a lower risk of burning in down grinding. Up grinding, although able on occasion to give longer life, should be reserved for just the last pass in a finish grind mode at <50 µm d.o.c. and relatively high table speed to help remove artifacts from the rough grind such as burrs and edge faceting produced by coolant hydrodynamic force changes.

16.5.3 Continuous Dress CF

Continuous dress CF is characterized by high stock removal rates. For deep slots with stock levels of the order 10 mm values and Q' of over 100 mm^3/mm/s have been reported for both steels and nickel-based materials. However, in actual production due to either limited stock levels, access of coolant, or dimensional or thermal stability of the part, the removal rates are generally much lower. For example, Tables 16.8 and 16.9 give some recent case histories of optimized wheel performances from the Radiac Abrasives web site (2002). As can be seen, high-removal rates are achieved on relatively burn-insensitive materials and/or large depths of cut. But where stock amounts are limited especially with burn sensitive materials, weak clamping or lack of a heat sink (thin-walled surfaces) removal rates drop precipitously.

CDCF is a higher consumer of grinding wheels and diamond rolls. Remember, it is consuming the wheel by constantly dressing at a rate of 0.5–2 µm per wheel revolution. It also requires a high stiffness machine and dressing system. The large amount of abrasive swarf in the system is an additional problem that creates disposal cost and maintenance issues. Nevertheless, to achieve good form accuracy with high stock removal rates and deep depths of cut, it is still a process that is hard to beat.

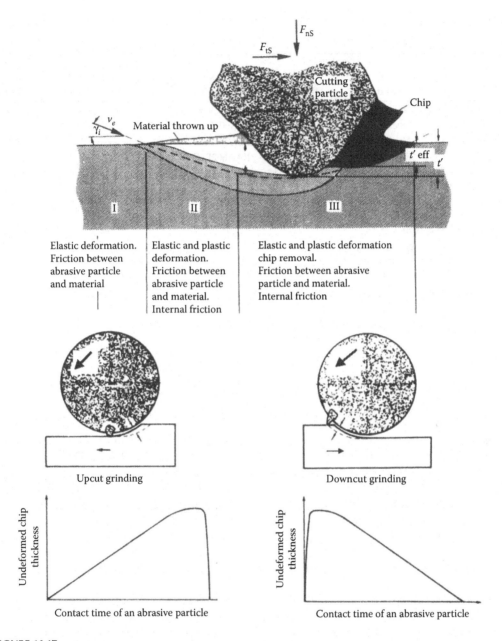

FIGURE 16.17
Schematic representation of chip formation and chip formation curves for upcut and downcut grinding. (After Köenig W., Schleich, H., 1982, *Ultrahard Materials. Application Technology, Volume 1.* De Beers Industrial Review. With permission.)

16.5.3.1 The Viper Process

Several alternatives to the CDCF process have been developed using conventional abrasive wheels. The first is the VIPER (Vitrified Improved PERformance) process patented by Rolls Royce (1999) with licenses to various builders of machining centers including Bridgeport and Makino. The VIPER process was developed around multisurface grinding

TABLE 16.8

Removal Rates in CDCF

Component	Material	Dress Mode	Wheel Specification	Stock Removal d.o.c. × Feed Rate (mm) × (mm/s)	Q^* (mm³/ mm/s)
Connecting rod	Steel 45HrC	CDCF + 0.80	RAA542FBOOVOS	5.7 × 15	85
Pomp slot	Steel 58HrC	CDCF + 0.80	9RA602F802FVOS	21 × 2.8	60
Blade dovetail	Ni alloy 37HrC	CDCF + 0.80	9RA461F85CVOS	8.6 × 1.5	13
Vane profile	Rene 37HrC	CDCF + 0.80	RAA602E8QOFVOS	2 × 3.4	6.8
Engine segment	Flame spray Ni	CDCF + 0.85	RAA602E800FVQS	1.65 × 1 7	2.8

TABLE 16.9

Removal Rates in the Viper Process for Grinding Turbine Blade Root Forms

Pass	Wheel Speed (m/s)	Table Speed (mm/min)	Depth of Cut (mm)	Dressing Depth (mm)	Q'_w (m³/mm/s)
1	35	1000	1.0	0	16.7
2	35	1000	0.5	0	8.3
3	35	1000	0.2	0.2	3.3
4	35	1000	0.1	0	1.7
5	35	1000	0.05	0	0.8
6	35	1000	0.02	0	0.3

of aerospace components in one clamping, where the wheel had to index to several positions. Standard positioning of the nozzle in each case was difficult for clearance reasons and CDCF impossible. The VIPER process uses an indexable nozzle to force coolant at 50–70 bar into the wheel structure ahead of its entry into the grind zone (Mohr 2000). The coolant is then thrown out tangentially just at the point of grind. Injecting coolant into the wheel ahead of the grind is a common practice and indexable nozzles have been used on grinding/machining centers since the 1980s. However, the novelty of this process is the precise optimized angling of the nozzle to direct the coolant for each operation. An example of the grind process is illustrated in Table 16.9 for grinding aircraft engine blade root forms.

Although a higher stock removal rate can be achieved in the initial rough cut, the wheel breakdown is such that feed rates have to be backed off to maintain form. Part deflection can also be an issue. The process requires small diameter (<250 mm), ultraporous wheels which limits life.

With the move to grinding and multitasking, several OEMs of machining centers now offer grinding including alternatives to VIPER such as CDCF complete with quick-change wheels and diamond rolls, Makino (2002) is illustrated in Figure 16.18.

An alternative to this process is the use of extruded ceramic grain wheels, trade name Altos from Saint-Gobain Abrasives. The latest generations of seeded-gel SG-based grains (TG and TG2) have increasingly greater length/diameter aspect ratios. As shown in Figure 16.19, the natural packing density decreases steadily as this aspect ratio increases. TG2, for example, has a natural porosity of almost 70%. Even allowing for up to 10% of vitrified bond, there is 60% porosity available with high permeability for coolant retention.

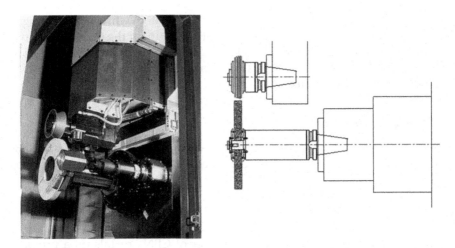

FIGURE 16.18
Makino machining center equipped with CDCF capability on quick-change HSK adaptors.

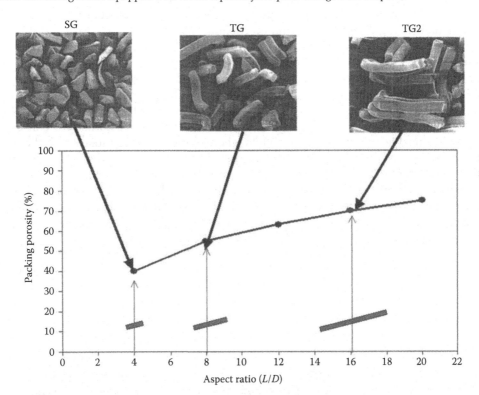

FIGURE 16.19
Large grain aspect ratios of SG and TG wheels.

Furthermore, the bond is concentrated at the cross-over points of grains like weld points giving a surprisingly strong structure.

In tests by the author, grinding Inconel 989, typical of land-based gas turbine materials, at 30 m/s and constant 1.25 mm depth of cut in water-based coolant (synthetic) Targa and Altos abrasives generated excessive heat at Q' values below 8 mm³/mm/s. However, for Q'

values between 10 and >50 mm/mm/s, burn was eliminated, while maintaining a finish <40 micro inches and a constant G ratio of 3–5. Grinding in oil the G ratio increased to 50 with no burn at any feed rate up to $Q' = 50$ mm^3/mm/s. A similar trend in performance was observed grinding Inconel 718, typical of aero-engine parts. In this case, however, the G ratio values increased by a factor of 3.

16.5.4 CF Grinding with CBN

CF grinding with CBN can be divided quite definitively between plated CBN and vitrified CBN in terms of distinctly different machine conditions, coolant, and operating parameters.

16.5.4.1 Electroplated CBN

Electroplated CBN wheels consist of a single layer of CBN crystals on a profiled metal hub. As discussed in Chapter 4, this leads to limits on finish and profile tolerances unless the surface is conditioned after plating which adds significantly to the cost. Therefore, the wheels are used more typically in roughing operations and under conditions that maximize life, namely high wheel speed and oil coolant.

The impact of oil on wheel life cannot be understated. Gaulin (2002) reports a typical life difference between grinding in oil and grinding in water-based coolant of 100:1. Certainly while grinding nickel-based alloys the author has observed very repeatable data showing 25–75 times more life with a good mineral oil compared to the latest synthetic or semisynthetic water-based coolant. The impact on life is so extreme that the economic use of plated CBN in water-based coolants, certainly as far as the aircraft engine industry is concerned, is very questionable except for small job lots where manufacturing must weigh the wheel cost against the cost of buying expensive formed diamond rolls to dress a bonded wheel and the equipment to utilize them.

The impact of high wheel speed is very similar to that discussed under external cylindrical grinding especially with operation under HEDG conditions. The biggest benefits of plated CBN in HEDG grinding comes about by achieving high stock removal rates with deep depths of cut and low U' vales, while maintaining a modest overall grinding power to keep forces within the range of normal machine stiffness. Not surprisingly, the typical examples seen in production are narrow slots in, for example, collets, spanners, and pump rotors (Figure 16.20).

FIGURE 16.20
Examples of HEDG grinding applications.

There is one particular application for electroplated CBN wheel that has been the driver for machine design to do HEDG grinding and that is radial groove grinding of aero-engine components. Many engine components, for example, vanes, nozzles, and casings, require narrow radial slots to be machined in them. Most engines are 500–1000 mm in diameter which sets the typical range of the slot diameters. Traditionally, these would either be milled with a very small diameter cutter which is very expensive or plunge ground with a 2A2-plated wheel of the same diameter as the groove. A 1000 mm diameter plated CBN wheel is also both expensive and requires a massive machine to utilize it. An alternative approach is to grind the slot with a small cup wheel of a very precise diameter and dish angle. Figure 16.21 shows the configuration. The groove of depth d has inner and outer faces defined by radii R_1 and R_2, respectively. The cup wheel or radius r is tilted $\theta°$ from the vertical. The critical part tolerance is usually the outer groove face taper which must be held to a tolerance of ≤ 0.0002 in. By comparison, the overall groove width has a tolerance of typically 0.002–0.004 in. Simple geometry calculation shows that for a part outer radius R_2, the maximum wheel radius to give a theoretical vertical face is given by $r = R_2 \sin\theta$. Applying this value of θ to the inner wall radius R_1 give a radial difference (taper) between top and bottom of the groove of:

$$\Delta R_1 = \frac{d^2(1+\sin^2\theta)}{2R_1\cos^2\theta} \qquad \begin{array}{c} R > \Delta R \quad r = R_2\sin\theta \\ R_2 > R_1 \end{array}$$

If the component drawing does not define an outer groove face taper but merely the overall taper from top to bottom of the groove, it is possible to further improve the overall taper by reducing θ about $1/4°$. It should be noted that ΔR_1 increases as d^2 but for small changes in angle reducing θ produces only a linear relationship of ΔR_2 with d. The final adjustment of θ is usually done using 3D CAD.

The θ usually lies in the range of $10°$–$20°$ which defines the wheel diameter in the range of 125–250 mm. This in turn defines the wheel capacity of the machine. Furthermore, the interpolation of the radius requires the coordinated movement of several axes. This has led to purpose-designed grinders such as the Edgetek HEDG grinder (Holroyd Machines, UK; Figure 16.22) but also the application of advanced tool and cutter grinders capable of

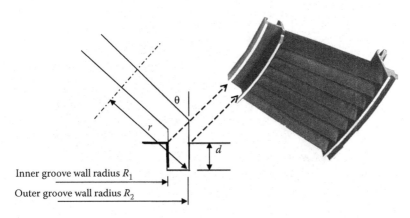

FIGURE 16.21
Arrangement for radial groove grinding using plated CBN wheels.

FIGURE 16.22
Edgetek HEDG grinder.

complex, accurate CNC path movements. An example of which from Huffman (Clover, SC, United States) is shown in Figure 16.23.

Under ideal conditions with deep depths of cut, oil coolant, and high wheel speeds, plated wheels can achieve $Q' = 100$–2000 mm^2/s depending on the material grindability. But as with CDCF, conditions are rarely ideal and stock removal rates may be only 10% of those indicated above. For example, grinding Inconel 718 or similar the maximum stock removal rate in the majority of applications based on burn and wheel life considerations is <8 mm^2/s.

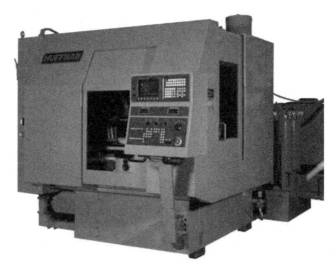

FIGURE 16.23
Huffman high-speed grinder.

16.5.4.2 Vitrified CBN

Vitrified CBN has seen a dramatic growth from 1995 in certain key areas of CF grinding especially aerospace and some tool steel. The impact on the grinding of Inconel-based engine components has been particularly dramatic since about 2000 with the introduction of near net shape parts, 6 Sigma (6σ) and just in time (JIT) manufacturing methods, resistance to oil coolant, and purpose-designed grinders (see below).

Vitrified CBN grinding of Inconel is incapable of the removal rates possible with CDCF being limited to $Q' \leq 8$ mm³/mms in water-based coolant for reasons of economics (abrasive cost) but this removal limit is increasing as bond technology improves. Certainly at removal rates of $Q' < 5$, the abrasive cost is often less than half that of conventional abrasives. The abrasive cost against plated CBN grinding in water-based coolant is even more extreme as illustrated in Figure 16.24. This compares the wear of various production specifications of plated CBN with a standard vitrified CBN production wheel. When the limit of wheel wear is less than 0.002 in., typical of most engine component tolerances, plated wheels wear four to seven times faster than vitrified CBN unless the grains are preconditioned as discussed in Chapter 5. At the end of this wear period, the plated wheel must be recoated by the wheel manufacturer whereas the vitrified CBN wheel can be redressed up to a 100 times dependent on layer depth.

Not surprisingly the abrasive cost can be 20 times higher with plated than with vitrified CBN in water-based coolant. However, in oil the life of the plated wheels increases by a far greater factor compared to the increase of vitrified CBN wheel life such that the abrasive costs become almost comparable.

16.5.4.3 Process Selection

The various new and competing technologies, each requiring somewhat different process conditions, requires careful consideration of the end users true needs especially when considering capital equipment expenditures. This can be readily understood by considering as an example the recent changes in the processing of nickel alloy-based aero-engine components, for example, high-temperature blades and vanes (Figures 16.25 and 16.26).

The choices before a process planner setting up a new production line include:

1. *Oil or water for coolant.* As discussed above oil allows justification of plated CBN and is compatible with TG-based SG abrasives. Water-based coolants justify vitrified CBN or CDCF with alox wheels and may permit TG-based abrasives in heavy stock applications.

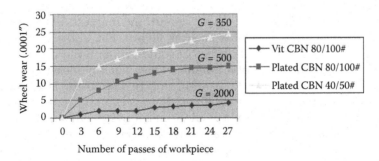

FIGURE 16.24

Comparison of the wear of vitrified and plated CBN wheels grinding IN 718 in water-soluble oil.

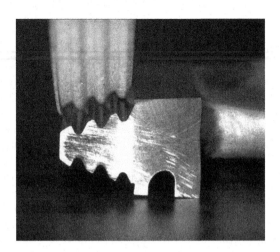

FIGURE 16.25
Rootform ("dovetail" or "fir-tree").

FIGURE 16.26
Nozzle guide vane grinding on aircraft engine blade.

2. *High-volume batch production or JIT cell manufacturing concepts.* Traditional manufacturing methods based on CDCF achieve very high production rates in
   both the aero-engine and land-based gas turbine markets. Extremely complex
   multispindle machines by OEMs such as Blohm, Elb, and Excello (n.d.) were
   designed in the late 1970s and 1980s that could grind all the surfaces of a blade
   in a cycle time of the order of a minute. Novel machine design concepts such
   as opposing twin wheel spindles could grind two opposing faces such as both
   dovetail forms at the same time to balance the forces at the higher metal removal
   rates (Salmon 1984). The output of these machines is still unrivalled but the
   process was found to have several drawbacks. First, the machines were very
   complex, some with up to five spindles and over 1000 control feedback inputs,
   making them difficult to maintain and requiring a very skilled workforce to
   maintain them. Second, the processes were geared to high production where

change over times could be lengthy. Consequently, large quantities of expensive components would be produced generating a high "work in progress" inventory and total manufacturing times in terms of months. In the 1990s, manufacturing moved toward a JIT strategy in conjunction with a policy for 6σ quality control. This in turn led to cell approach for manufacturing where all the equipment to make a part, including coordinate measuring machine (CMM) inspection and other manufacturing processes, such as electrical discharge machining (EDM), were contained within a single cell with a minimal number of operators. Grinders were now much simpler, lower priced, and with tooling designed for quick change over times where necessary. An operator under these conditions could only efficiently function with cycle times of the order of 4 min. The need for CDCF under these conditions was eliminated.

3. *Stock level.* Older casting technology produced casings with very heavy levels of stock removed up to 10 mm deep making them particularly suited to CDCF. Modern near net shape casting methods have now reduced these stock levels to as little as 1–2 mm on many aero-engine components, eliminating the need for a high Q' operation and making processing much more suited to a CBN solution.

4. *Part holding: Encapsulation or hard-point mounting.* Many aerospace components are complex but relatively weak structurally or have reference surfaces which can easily distort. Traditionally, these parts are held by encapsulating them in a low melting point metal alloy. This provides a solid robust structure with clearly defined and rigid location surfaces. However, the mounting and dismounting processes are expensive and add no processing value to the part. With the move toward lower stock levels and the lower Q' for grinding and its associated forces, end users are implementing hard-point mounting methods where the part is supported on its reference datum points with hardened steel or carbide tipped pneumatic or hydraulic clamps.

Hard-point mounting increases the processing options available to the end user. The following includes several that are not practical with encapsulation. For example

- The components can be processed in a series of inexpensive three axis grinders with a single grind operation per machine. This can be done either one component at a time, Figure 16.27 or ganged up in series, Figure 16.28.

- The components can be ground on several surfaces in a single clamping on four- or five-axis grinders using a combination of tilt and/or rotational axes to present the various grind surfaces in turn to the wheel. The wheel or wheel "pack" (multiple wheel assembly) is designed with a range of diameters and profiles to achieve the clearance to reach and generate the various forms on the part. The design of the fixture is pivotal to the whole process and often extremely complex (Figures 16.29 and 16.30). It can be readily seen that the size of the machine, wheel design, dresser design, and coolant delivery can only be defined once the fixturing is designed. Several companies have specialized in this type of fixturing, working closely with OEMs and wheel makers for turnkey operations.

- Several components can be mounted in a single fixture to present a series of different grind stages. The components are moved progressively from stage to stage taking off one finished blade each time at the final stage and adding one

FIGURE 16.27
Single encapsulated part.

at the initial stage. This approach is most suited to small, easy-to-handle, blades on four- or five-axis grinders (Figure 16.31). Fixture design is again critical. For these types of operations in combination with near net shape components, the need for high stock removal rates becomes less critical, as the majority of the cycle time is used for machine indexing. This makes the approach particularly suited to processing using small CBN wheels.

5. *Selection of metal removal process.* One of the most fundamental decisions for the process engineer is to decide which is the most suitable metal removal process. For example, EDM is a relatively slow method for cutting complex forms but EDM machines are inexpensive relative to a grinder. Therefore, it may be less expensive and more flexible to buy several EDM machines than a single grinder. The decision must be based on capital equipment costs, cycle times, wire/abrasive costs, and tolerances. Milling of nickel-based alloys is very expensive in perishable tooling costs, and with advances in CBN and ceramic wheel technology manufacturers of land-based engine components are beginning to convert to grinding (Figure 16.32).

Grinding is also being considered as a replacement for broaching of nickel alloys again due to the high capital and tooling costs, long tool change times, high broaching forces, and floor space required. For example, reported on a study for a consortium including

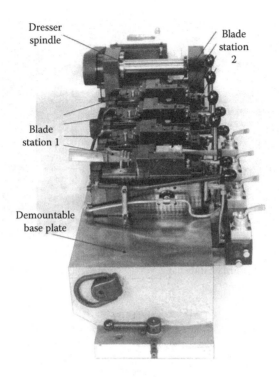

FIGURE 16.28
Parts ganged up in series.

FIGURE 16.29
Engine blade assembly.

Rolls Royce and SNECMA to replace the broaching of turbine compressor disk root forms with grinding; the root form being first rough ground with Altos wheels then finished ground using small profiled plated CBN pins (Figure 16.33).

It is interesting to note this study was carried out using a vertical machining center (VMC) and made significant reference to "multitasking." The move to multisurface processing of parts in one fixture has demonstrated the benefits in terms of improved tolerances from

FIGURE 16.30
Hard-point mounting fixture for nozzle guide vane.

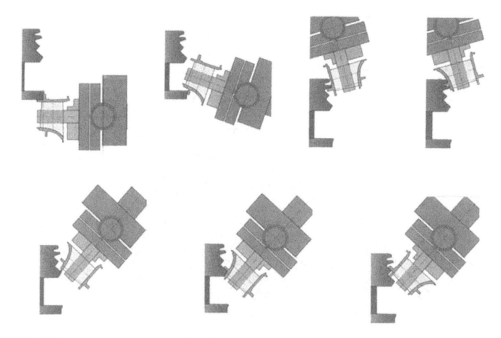

FIGURE 16.31
Example of multisurface processing of engine component in one fixturing using a profiled wheel pack. (From Blohm, n.d., *Super Abrasive Machining Center*. United Grinding Technology, Miamisburg, Ohio, Trade Catalog. With permission.)

processing components in a single clamping. However, there is no reason why each metal removal operation must be one of grinding. In multitasking a machine may be required to do several operations including grind, mill, turn, probe, burnish, and deburr. Machine tools now become much more generic with companies traditionally involved with machining centers offering grinding capability while manufacturers of grinding machining offer

FIGURE 16.32
Multiaxis machining center for grinding land-based gas turbine buckets with plated CBN wheels to replace milling and HSK automatic wheel changing.

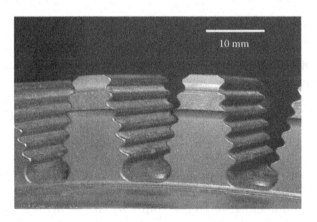

FIGURE 16.33
Turbine compressor disk root form and sketch showing blade in location. (Adapted from Burrows, J.M. et al., 2002. Point Grinding of Superalloys. Industrial Diamond Review, June.)

options common to machining centers such as quick-change HSK tooling, milling and turning capabilities, and a radically different working volume in terms of machine axes movement and reach.

Several of the large manufacturers of high-performance machining centers such as Makino and Bridgeport mentioned previously and Hitachi Seiki now promote "grinding centers"—machines with structural, drive and control components, coolant delivery

FIGURE 16.34
Makino A77 grinding center and components.

systems, and size control ability comparable or superior to many grinders. Figure 16.34 shows a Makino A77 grinding center. On the other hand, manufacturers of high-speed grinders such as Edgetek now refer to their processing as "abrasive machining." The distinction between a grinding machine and a machining center has become blurred in these instances.

The natural conclusion to this trend is for some end users to attempt to further reduce capital equipment costs by the use of low cost, standard produced, VMCs, and making modifications to them in-house typically to run plated CBN wheels. This approach has met with some success in application with noncritical tolerances but the end user should be aware that the stiffness, thermal stability, and repeatability will not be in the same class as a modern grinder and savings in capital equipment may be lost in process development with no guarantee of success as very few VMC manufacturers provide turnkey processes. Attention must also be paid to differences in guarding codes.

With so many options and new technologies to choose from the selection of the best is difficult. At the current time, the following approaches are being taken by manufacturing:

1. High volume dedicated manufacturers with near net shape parts are converting primarily to vitrified CBN wheels in water-based coolant in combination with.

2. Hard-point mounting for workholding.

3. High volume dedicated manufacturers with near net shape parts willing to grind in oil are converting to plated CBN in combination with hard-point mounting for workholding.

4. High volume dedicated manufacturers with high stock castings either remain with conventional CF wheels in conjunction with traditional CDCF or recent VIPER technology in water-based coolant, or advanced ceramic grain wheels in water or oil coolant. The shift to hard-point mounting is slower in these cases because of the higher forces imparted by the higher removal rates that in turn cause part deflection. This is especially a problem when the reference mount points are on thin aerofoil sections.

5. Job shop and small batch manufacturers willing to grind in oil coolant are using predominantly plated CBN with some advanced ceramic grain wheels on simpler form or flat surfaces.

6. Job shop and small batch manufacturers: Those unwilling to grind in oil coolant use a combination of conventional alox wheels with CDCF, plated CBN for radial grooves, and occasional vitrified CBN usage as volume justifies.

7. Heavy stock castings for land-based engines: Manufacturers are moving away from milling to grinding in oil with a combination of advanced ceramic grain (Altos) wheels for roughing and plated CBN for finishing. Existing grind applications with heavy stock remain as CDCF.

Recent developments in machine design have progressed hand in glove with the changes occurring in part processing. Dedicated machines for vitrified CBN use small wheels in the range 8–12 in diameter (200–300 mm) to minimize wheel cost while still maintaining an acceptable range for the dressing crush ratio across a typical 1" depth of form profile.

The working area within the machine is now more a cube to accept the bulky work holding fixture but give free access in all directions. In many cases, this allows a complete downsizing of the machine size, especially width, but with increased stiffness. Several machine tool builders of small mini CF grinders report machine loop stiffness values in the range of 750,000–1,250,000 lbf/in (130–220 N/μm). Examples of compact CF grinders are shown in Figure 16.35 while Figures 16.36 and 16.37 show Campbell 5-axis horizontal spindle grinders for multisurface grinding.

Wheel spindle motors must be variable speed with power and torque curve characteristics closely matched to the expected wheel diameter range and types of wheels to be run. This ensures sufficient power over the full life and expected operating speeds of each wheel type. Most AC servo motors have a peak power at a particular rpm and drop off progressively with rpm on either side of this value. Trying to combine different wheel types and diameters therefore becomes a compromise. The required wheel speed ranges and diameters for each wheel type are normally as shown in Table 16.10. Figure 16.38 shows typical power speed combinations for different abrasives.

It is possible to provide flexibility in order to use a combination of vitrified CBN and regular alox wheels or to use vitrified and plated CBN with a single machine configuration for all three compromises using the power available.

Some OEMs, such as Blohm (n.d.), have developed their spindle technology, Figure 16.39, to extend the range of RPMs over which they can maintain a constant high power level. This makes their grinders particularly flexible for the job shop environment by now allowing the use of all the three abrasive types without significant compromise on wheel diameters.

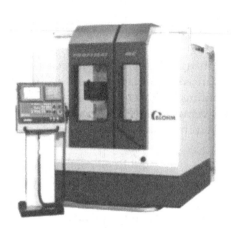

Blohm profimat MC

Campbell 1950

Jones & Shipman dominator 624

Micron machine

FIGURE 16.35
Examples of compact creep feed grinders.

In conclusion, the machine tool industry has undergone major changes over the last 5 years driven by changes in part processing philosophy and abrasive wheel technology. This has led at one extreme to dedicated grinders designed in many cases to the processing of a specific part. At the other extreme machine tools have become ever more flexible both in the use of a greater range of wheel technologies and to encompass multiple metal removal processes.

TABLE 16.10

Wheel Diameters Against Operating Speed

Bond System	Normal Operating Wheel Speed (sfpm)	Wheel Diameter Range (in.)	Wheel rpm Range
Vitrified Alox	4000–6500	14–18	850–1775
Vitrified CBN	6000–10000	10–12	1990–3820
Plated CBN	9000–14000	6–10	3440–8910

FIGURE 16.36
Example of five-axis horizontal spindle grinder for aircraft engine blade multisurface grinding. (Courtesy of Campbell Grinder.)

16.6 Face Grinding

16.6.1 Introduction

Face grinding is surface grinding using the face of a grinding wheel. Face grinding may be contrasted with the previous examples in this chapter in which grinding was performed predominantly with the peripheral surface of the grinding wheel. Face grinding

FIGURE 16.37
Example of five-axis horizontal spindle grinder for aircraft engine blade multisurface grinding.

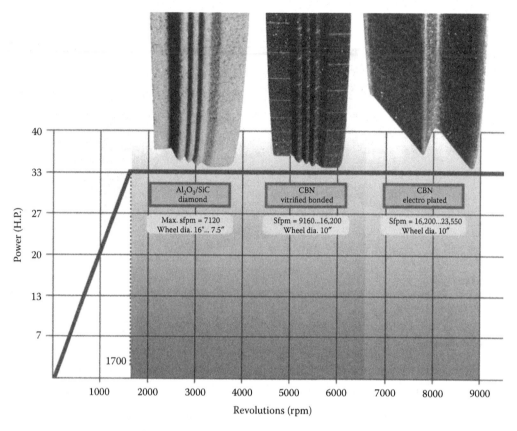

FIGURE 16.38
Power speed combinations for different abrasives.

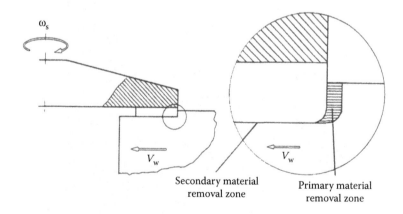

FIGURE 16.39
Material removal in vertical spindle face grinding with horizontal feed.

encompasses a range of processes characterized by a total conformity between the wheel and workpiece. These include segment grinding, double- and single-disk grinding, and fine grinding. For such a configuration, the standard equation for uncut chip thickness does not apply since $d_e = \infty$.

There are two fundamentally different versions of face grinding illustrated in Figures 16.39 and 16.40. In Figure 16.40a, the work is plunged at constant feed rate v_f into the face of the wheel. In Figure 16.40b, the work is fed tangentially at constant depth of cut and constant feed rate v_f across the face of the wheel. While this is commonly termed face grinding, with a newly dressed grinding wheel, material removal is primarily effective on the periphery of the wheel at the corner. As the corner breaks down due to wear, material removal from the work takes place due to the abrasive grains on the wheel face.

Malkin analyzed vertical spindle face grinding with linear horizontal feed and determined the maximum uncut chip thickness that occurs at the middle of the vertical step (Figure 16.40b). For a triangular chip cross section, the chip thickness is given as

$$h_{cu} = \sqrt{\frac{\sqrt{3}}{Cr} \cdot \frac{v_w}{v_s}} \tag{16.10}$$

where v_w is the table speed, v_s is the wheel speed, C is the number of active cutting edges per unit area of the grinding wheel surface, and r is the ratio of width to depth of the triangular chip cross section.

Similarly Malkin (1989) also analyzed face grinding with infeed only (Figure 16.40a). In this case, the uncut chip thickness is considered constant and given by

$$h_{cu} = \sqrt{\frac{2}{Cr} \cdot \frac{v_f}{v_s}} \tag{16.11}$$

where v_f is the infeed and v_s is the wheel speed.

For tangential feed, it is possible to predict factors affecting grinding energy and power for the process and by implication burn based as in Chapter 2 on:

$$e_c \propto \frac{1}{h_{cu}^n} \propto \sqrt{\frac{v_s}{v_f} \cdot C \cdot r} \tag{16.12}$$

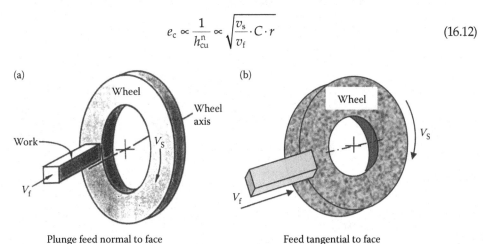

(a) Plunge feed normal to face

(b) Feed tangential to face

FIGURE 16.40
Face grinding (a) with plunge feed and (b) with tangential feed.

$$T_{\max} \propto \sqrt{a_e \cdot v_s \cdot C \cdot r} \tag{16.13}$$

For plunge feed normal to the wheel face as in Figure 16.40a, the maximum temperature and subsurface temperatures can be found from one-dimensional heat flow as shown by Rowe (2014, Chapter 18). It is found that the surface temperature constantly rises with time from commencement of plunge feed:

$$T_{\max} = \frac{q \cdot R_w \cdot t^{1/2}}{\sqrt{\pi \cdot k \cdot \rho \cdot c}} \tag{16.14}$$

where $q = P/A_w$ is the grinding power per unit area. R_w is the proportion of grinding heat that flows into the workpiece and $t = z/v_f$ is the time to remove a depth of material stock z. With $q = e_c \cdot v_f$ and assuming that specific energy varies as above,

$$T_{\max} \propto \sqrt{v_s \cdot C \cdot r \cdot z} \tag{16.15}$$

It follows that plunge feeding directly into the wheel face is a process that easily gives rise to burn and thermal damage. Both processes involve a long contact length which increases the tendency for wheel glazing.

Not surprisingly conventional wheels for face grinding are open-structure, coarse grit, and friable grain: Superabrasive wheels are low concentration to keep Cr values low.

Wheel speeds are also low or in the case of fine grinding very low at 1–3 m/s, in part because of the risk of burn but also because the wheels are in shear which leads to lower burst speeds than in cylindrical grinding.

Heat generation and the associated temperature profiles have been modeled for both linear feed and straight infeed grinding. Lin and Zhang (2001) modeled the temperature field for a cup wheel grinding with linear infeed (Figure 16.41).

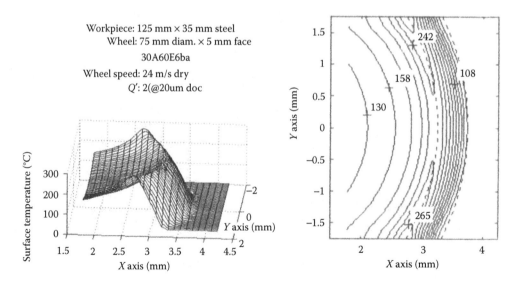

FIGURE 16.41
Temperature distribution in face grinding. (From Lin, B., Zhang, H. L., 2001, Theoretical analysis of temperature field in surface grinding with cup wheel. *Key Engineering Materials.*, 202–203, 93–98. With permission.)

They found that the temperature rise is extremely abrupt at the leading edge of the wheel and does not fall away quickly behind. The rise in temperature peaks within 2 mm of the edge of the wheel. The model is based on a simple arc-line heat source. In reality, there will be frictional rubbing from the flat secondary cutting area leading to additional heating. For this reason where the grind process allows, such as when using highly wear resistant superabrasive wheels, the width of contact is kept to less than 5 mm and ideally 3 mm.

Spur et al. (1995) calculated the temperature distribution in solid circular components during double-sided fine grinding (Figure 16.42). Even at low wheel speeds encountered in fine grinding (<4 m/s) quite significant increases in temperature could arise and be focused at specific points.

Such temperature distributions cause differential thermal expansion such that more material is ground off the hottest points on the part and flatness is compromised as the part cools. At higher speeds such as in double-disk grinding, especially of thin parts, coolant starvation can create more catastrophic effects akin to film boiling and resulting in severe burning in often quite localized spots. In high precision grinding or at high speeds, low heat generation and/or effective coolant delivery is critical.

16.6.2 Rough Grinding with Segmented Wheels

Grinding with segments represents the rough, high stock removal end of the flat grinding spectrum. The machine can be very large with powerful motors. The largest recently had a 200″ diameter chuck, 117″ diameter grinding wheel, and a 600 hp wheel drive custom built for a company to manufacture saws (Anon 1995).

The majority of segment grinders have rotary tables such as those manufactured in the past by Cone Blanchard and Mattison that offered machines with standard chuck diameters from 20″ to 136″, spindle motors from 15 to 450 hp, and wheel diameters from 11″ to 72″. There are still a huge number of these machines in production worldwide with processes that date back to well before 1940 and machine designs that have changed little in the last 20 years (Blanchard 2000; Mattison n.d.). Wheel speeds are 4200 sfpm or less. Coolant is delivered to both the outside of the wheel and through the spindle. Table speeds are such

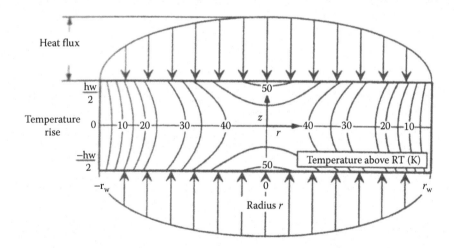

FIGURE 16.42
Temperature rise distribution in double-sided face grinding. (From Spur, G., Funck. A., Engel, H., 1995, Problems of flatness in plane surface grinding. *Trans NAMRI/SME*, XXIII, 97–102. With permission.)

that wheel/table rpm will be in the standard range 25:1 to 100:1 to allow adjustment for chatter and burn.

With wheel sizes so large, segments are the only practical way of keeping the weight manageable for mounting and demounting. They can also provide additional clearance for coolant access and flushing of swarf. Segments are available in a vast variety of shapes. Some are like pie segments, some form a tongue and groove to lock together, while others are for specific clamping mechanisms. The commonest shapes in the United States are for Blanchard, Cortland, Mattison, and Ferro chucks (Figures 16.43 and 16.44).

The commonest European grinders seen in the United States are Reform and Goeckel. These machines have linear tables and use smaller segments (Figure 16.45). They are

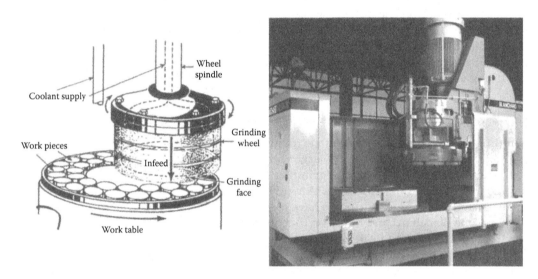

FIGURE 16.43
Vertical spindle face grinding with segmented grinding wheel and rotary work-table. (From Reform, n.d., Schleifmaschinen. Reform Maschinenfabrik Adolf Rabenseifner, Fulda, Germany, Trade Brochure. With permission.)

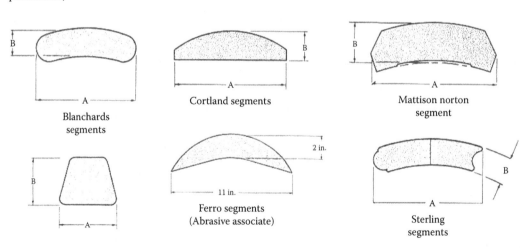

FIGURE 16.44
Grinding wheel segment shapes.

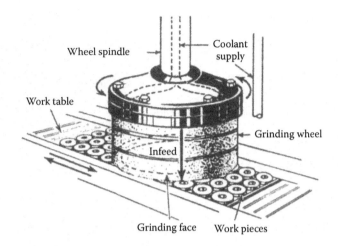

FIGURE 16.45
Vertical spindle face grinding with linear work-table.

popular to grind knives for the paper, ceramics, food, and cutlery industries (Reform n.d.; Goekel 1997).

Segments for rough grinding are usually vitrified, highly porous (E—G grades), and contain coarse grit. For the highest metal removal rates, 24#–30# grit is used, which with an appropriate spark out will give a good visual finish (63 RMS). A 36#–46# grit is used only for very fine finishes or for grinding difficult materials such as hardened tool steels or some stainless steels. The finest grit sizes available are 180# for the most difficult materials.

Abrasive types are almost exclusively conventional with the exception of some diamond for certain ceramics grinding. CBN is occasionally used, but only for the smallest wheel sizes as the wheel cost is prohibitive. Also where it is used, it is only for fine finish applications as the grit sizes required for roughing (>80#) are not available or cannot be held effectively in the bond (for the same reason grit sizes coarser than 24# are not used in conventional abrasive segments). The largest wheel noted in the literature is a 22" diameter resin wheel using metal bond segments and resin bond continuous rim CBN at 80# for grinding tool steels (Deming and Carius 1991).

SG is generally offered as an alternative to CBN to improve life and removal rates. Downfeed rates of 0.025 in./min are reported on D2 steel using SG abrasive in the hardened or unhardened state (Narbut et al. 1997). Stock removal rates are usually determined by monitoring power and burn and by observing wheel wear. Flatness tolerances are in the range of 0.0005"–0.005" for general roughing.

With proper grade selection, a wheel should be virtually self-dressing. Diamond impregnated tools or steel "star cutter" dressers can be used if required for initial truing or conditioning purposes.

A lot of rough grinding operations has seen competition from other processes especially milling and turning. This has led to some interesting shifts back and forth from one process to another. A case in point is the sealing surfaces of engine blocks. This has swung between grinding and lapping, to machining and lapping, then back to grinding or machining only as technology or block design has changed. Currently, grinding tooling cost is less than machining and can generate better sealing surfaces but machining centers have less capital equipment cost and can have much flexibility for other applications. The industry has therefore numerous solutions depending on the tolerances, sealing requirements, and capacity requirements.

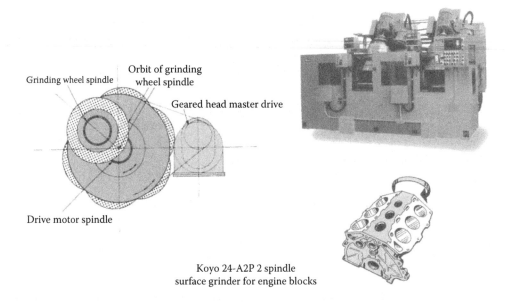

Grinding wheel spindle
Orbit of grinding wheel spindle
Geared head master drive
Drive motor spindle

Koyo 24-A2P 2 spindle
surface grinder for engine blocks

FIGURE 16.46
Koyo inclined head orbital face grinder.

Koyo 4 offers an inclined head orbital surface grinder to rough and finish grind engine blocks in a production line environment on a single machine. The orbital motion allows the use of a relatively small 355 mm wheel to minimize grinding pressure and part distortion (Figure 16.46).

16.6.3 Rough Machining/Finish Grinding

For manufacturers that choose to rough machine, the part (Thielenhaus n.d.) provides a grinder/superfinisher to Microfinish® the faces to create a good sealing surface (Figure 16.47).

FIGURE 16.47
Examples of engine block face-grinding setups. (From Sess, M., 1999, Machine tools are doing more. *Manufacturing Engineering*, November, 46–53; and Thielenhaus, n.d., *Microfinish® the Economical Precision-Machining Method in the Automotive and Motor Industry.* Thielenhaus Microfinish Corp, Novi, Michigan. Trade Brochure. With permission.)

This is particularly suited for applications such as aluminum cylinder heads with hardened liners that creates step problems for machining. The process uses small conventional abrasive segments in a rotating grinding head. Makino (Sess 1999) reported a postmilling process using small high-speed electroplated diamond wheels. A similar combination of turning and electroplated CBN grinding was recently developed to replace double-disk grinding of connecting rods. Three solutions from machine tool builders with traditions in grinding, superfinishing, and machining, respectively.

16.6.4 Single-Sided Face Grinding on Small Surface Grinders

This particular type of grinding refers to a range of small inexpensive grinders that use wheels from 18" to as small as 6" for a host of small part applications in tool rooms and small production shops (Figure 16.48). Typical examples include bearing rings, aircraft parts, punches, dies, seals, inserts, pump components, EDM electrodes, and ejector pins. The wheels may be segments or more commonly continuous rim and divided between conventional and narrow rim superabrasive grain. In many ways, this is a continuation of the grinding described in the previous section except using smaller stiffer machines to

DCM tech
IG 280M
rotary surface
grinder with
segmented wheel
detail

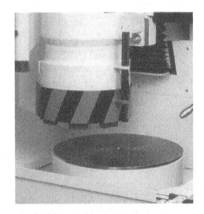

Delta
rotary surface
grinder with
manual single
point dresser
detail

FIGURE 16.48
Examples of small rotary surface grinders. (From Delta, 1997, *Surface Grinding Machines*. Delta s.p.a. Cura Carpignano, Italy, Trade Brochure; and DCM Tech, 1999, Industrial Rotary Surface Grinders IG 180M, IG 280M, IG 280 CNC, HB5400 Series. Series of Trade Brochures. With permission.)

obtain tighter tolerances, and with an increased wheel speed range on some models up to 10,500 sfpm. Common machine types in the United States include Delta (1997), DCM Tech (1999), and Swisher (2000).

16.6.5 High-Precision Single-Sided Disk Grinding

The grinders referred to in this section are an extension of those described above focused at flatness tolerances in the micron and submicron level often in competition with lapping (Table 16.11).

TABLE 16.11

Examples of Flat Surface Grinding on Small Rotary Grinders

Material	Hardness	Part Function	Rate Stock Removal	Tolerance		Surface Finish
				Thickness	Flatness	
A-2	60Rc	Punch	0.035"/min	0.0005*	±0.0002*	20 Ra
A-6	62Rc	Rachet ring	0.018"/min	±0.0001*	±0.0002*	16 Ra
D-2	60/62Rc	Form roll die	0.003"/min	±0.0003*	±0.0003*	24 Ra
M-50	55Rc	Aircraft brg	0.010"–0.020"/min (all 4 surfaces ground)	±0.0001*	±0.00015*	8–10 Ra
T-15	60/62Rc	Form tool	0.020"/min	N/A	±0.0002	8–12 Ra
4140	50/55Rc	Gear blank	0.070"/min	0.0005*	±0.0002	24 Ra
P/M	–55Rc	Hyo pump	Coarse feed/fine feed 0.050"–0.010"/min	±0.0001*	0.0001*	16 Ra
8620	60Rc	Spur gear	Coarse feed/fine feed 0.100"–0.020"/min	±0.0003*	±0.0003*	24–32 Ra
P/M	–50Rc	Hyo pump gear	Coarse feed/fine feed 0.040"–0.005"/min	±0.00015*	±0.00015*	32 Ra
P/M	–50Rc	Hyo pump gear	Coarse feed/fine feed 0.044"–0.005"/min	±0.00015*	±0.00015*	32 Ra
Carbide	—	Mechanical seal	0.012"/min	±0.0002*	±0.0002*	32 Ra
Carbide	—	Mechanical seal	0.002"/min	±0.0001*	±0.0001*	0–1 Ra
ALO$_2$-epoxy	—	Electronic comp	0.030"min	±0.0005*	±0.0005*	—
Carbon graphite	—	Seal	0.080"/min	±0.0005*	±0.0002*	—
Ceramic	—	Disk	0.020"/min (15 PCS/Chuckload)	±0.0002*	±0.0003*	8–12 Ra
1016	55/58Rc	Brg end plate	0.012"/min (25 PCS/Chuckload)	±0.0003*	±0.0005*	16–24 Ra
1018 Magnetic steel	Soft	Servo motor stator	0.020"/min (8 PCS/Chuckload)	±0.0002*	±0.0002*	32 Ra
430 Stainless steel	Soft	Antilock braking component	0.035"min	[2]±0.0005*	±0.0005*	32 Ra
Carbide	—	Insert	0.015"/min (45 PCS/Chuckload)	±0.0002*	±0.0002*	8–12 Ra
Carbide	—	Insert	0.001"/min (8 PCS/Chuckload)	±0.0001*	±0.0001*	0–1 Ra
Bakelite	—	Fuel control system component	0.020"/min (36 PCS/Chuckload)	±0.0003*	±0.0005*	32 Ra

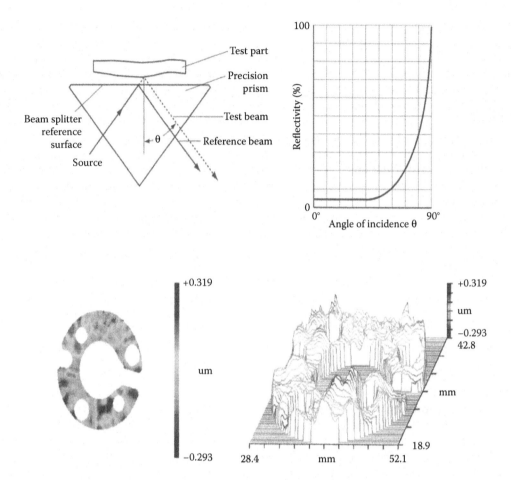

FIGURE 16.49

Flatness measurement by grazing angle interferometry. (From Tropel, 1996, *Tropel Build on Our Experience.* Tropel Corp, Fairport, NY. Trade Brochure. With permission.)

Measurement of flatness at the micron level is critical and care should be taken where flatness has been defined by a machine tool builder as to how that value was determined. Use of indicators on a gauge table, linear measurements on Talysurf style equipment, or counting fringes using an optical flat and monochromatic light source will generally give an over-optimistic value. The most accurate system currently available for general production is based on grazing angle reflection interferometry with equipment made by companies such as Tropel and Zygo (Figure 16.49).

The method relies on the interference of light, and the fact that the reflectivity of a rough surface increases dramatically when the light strikes it at a glancing angle allowing inspection of apparently non-reflective surface such as ceramics, plastics, metals, and composite. It is the same concept as when sunlight strikes water near sunset. The light is reflected much more than at noon. The system uses a laser beam to generate a well column of light which is divided into two beams by partial reflection at the prism surface. Part of the beam is then reflected off the part where it is then recombined after reentering the prism creating interference fringes. A camera system then captures data at about

60,000 coordinate points to compile a 3-D image of the test part. The interference pattern is phase shifted by slightly moving the prism and the sinusoidal variation in light intensity is then monitored for each data point. Software then analyze the phase shift to give the 3D height map. The system has a resolution of 0.01 µm, an accuracy of 0.1 µm, and a range of about 30 µm in height variation. Image processing time is about 30 s, which is quite sufficient for shop floor process monitoring. Systems are available to handle up to 500 mm parts.

Single-disk grinder technology, Figure 16.50, for the precision steel industries such as automotive components, hydraulics, seals etc. has been influenced significantly by developments from the semiconductor industry for wafer grinding. To achieve good flatness requires a very stiff machine combined with precision work handling equipment to hold and spin the part, and low grinding forces. Stiffness is achieved by using small wheels with a large spindle bearing to limit cantilever deflection at the wheel edge. Work-holders are belt or gear driven to spin the part and narrow faced wheels are used to limit grinding pressure. This naturally leads to the use of superabrasives for wear considerations.

Koyo, for example, has designed a range of single-disk grinders for targeting parts with flatness requirements of the order of 0.2–2.0 µm using several combinations of grind methods grinding either one or two parts at a time or roughing and finishing simultaneously with two-wheel heads (Figure 16.51). In one instance, the author ground hardened steel cylinders 25 mm in diameter with vitrified CBN wheels and achieved a flatness of 0.3 µm, finish of 0.15 µm Ra, stock removal of 100 µm in 4 s roughing, and a total production rate of 350/h dressing every 4000 parts. This was achieved in water-based coolant.

A similar approach, using narrow face wheels, has been taken by a number of machine tool builders that traditionally specialized in superfinishing. Their processes now often require stock removal placing them more into a grinding than a polishing regime. Supfina has reported a machine design for superfinishing of valve tappets. In this case, the tappet face requires a well-defined crown which is generated by the use again of a small face wheel but with its axis of rotation tilted slightly to that of the part (Figure 16.52). Research into the use of superabrasive wheels with narrow face widths has resulted in some surprising developments in production grinding in unexpected areas. If the contact width

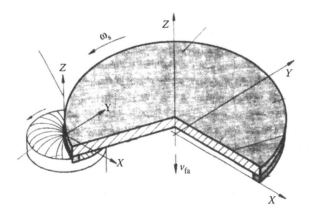

FIGURE 16.50
Single-disk grinder configuration.

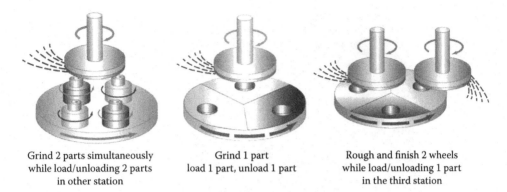

Grind 2 parts simultaneously while load/unloading 2 parts in other station

Grind 1 part load 1 part, unload 1 part

Rough and finish 2 wheels while load/unloading 1 part in the third station

FIGURE 16.51
Strategies for single-disk grinding. (From Koyo, 2000, *Koyo R Series Vertical Spindle Type, Rotary Table Type Surface Grinder*. Koyo Machine Industries, Osaka, Japan. Trade Brochure. With permission.)

is kept thin enough the wheel will maintain its flatness indefinitely resulting in grinders using vitrified CBN that do not require dressers. An example of this is "kiss" grinding of thrustwalls on camshaft and crankshafts as a replacement to turning to improve flatness (Figure 16.53; Hogan 2001).

In one example on crankshafts with Landis 3LB grinders the author achieved a flatness of <4 μm with a Q' of 40 mm^3/mms and a G ratio >1500. The CBN wheels ran at speeds up to 140 m/s in water-based coolant. The flatness could not be achieved by turning because the chip could not be broken without leaving a raised area.

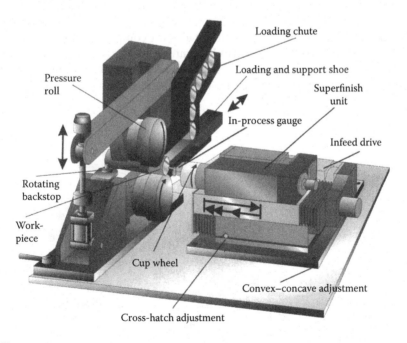

FIGURE 16.52
Superfinishing arrangement for grinding tappets. (From Supfina, 1996, *SUPERFINISH for the Engine Industry*. Superfina Grieshaber GmbH. With permission.)

FIGURE 16.53
Face grinding of crankshaft side walls.

16.6.6 Double-Disk Grinding

Double-disk grinding is a high production, high accuracy, processing for parts with flat and parallel sides. Two opposed grinding wheel disks mounted on separate spindles simultaneously grind opposite and parallel faces on components fed between them by a variety of techniques. Double-disk grinders can be configured in either a horizontal or vertical configuration referring to the axis of wheel rotation. Horizontal grinders are generally used for mid-range part sizes while vertical grinders are used for either very thin or very large parts (Figures 16.54 and 16.55).

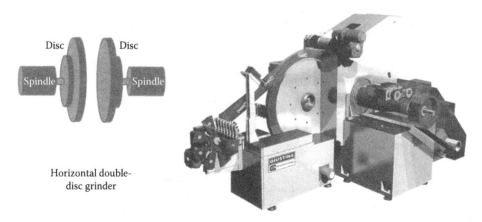

FIGURE 16.54
Horizontal spindle double-disk grinder. (Courtesy of Guistina. With permission.)

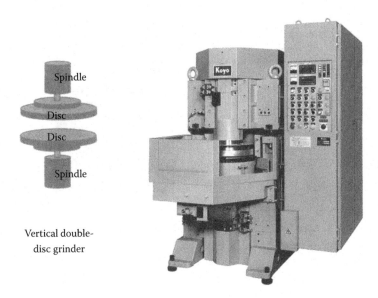

FIGURE 16.55
Vertical double-disk grinder. (Courtesy of Koyo, 1999, *Koyo Grinders*. Koyo Machine Industries, Osaka, Japan. Trade brochure. With permission.)

The grinders require extremely high radial and axial stiffness. Most are either designed around a combination of large axial thrust, angular contact, and roller bearing combinations (Daisho n.d.; Diskus 1996; Koyo n.d.), or full hydrostatic (Mattison 2 n.d.; Kubsh 1988). The motor drive is usually a synchronous or servo AC belt drive rather than a direct drive because of space limitation of the machine configuration. The spindle shaft and wheel-mounting surface are made from a one-piece steel construction with coolant passed through the center. The wheel is supplied either as an abrasive layer bonded to a steel back plate or as a layer with nuts inserted into the back face, which is in turn bolted to the spindle wheel-mounting face. The components of double-disk spindles (Landis Gardner 1989) and wheels are shown in Figure 16.54 through 16.57.

Coolant, passed through the center hole, is disseminated over the surface of the wheel with slots while holes relieve the pressure and provide chip clearance and coolant access.

Most wheels for double-disk grinding are resin or plastic bonded running at speeds of 35 m/s or less. Resin and plastic bonds can withstand more lateral strain and abuse than other more brittle bonds. Some oxy-chloride bonds are still used with conventional abrasives for the cutlery industry for dry grinding or where cool cutting is critical. Vitrified bond usage has been increasing especially with CBN for finish grinding on small machines specifically designed for using superabrasives. Vitrified bond usage has been increasing especially with CBN for finish grinding on small machines specifically designed for electroplated CBN. There has also been some interest in the spring grinding industry for electroplated CBN. Anker (1998) reported success in ends grinding of springs dry at 60 m/s on small purpose-designed machines using coarse grit ABN 600 although resin bonded CBN is more common especially in Japan.

Conventional abrasives still dominate virtually all general roughing processes and commercial type finishing and semifinishing operations. The 16#–36# grit is used for rough grinding, 36#–80# for commercial finishes and tolerances, and 60#–150# for fine finishes and tight tolerances. It is only in this latter category that CBN in particular has

FIGURE 16.56
Components of a double-disk grinder wheel spindle unit.

Components of a double-disc wheel

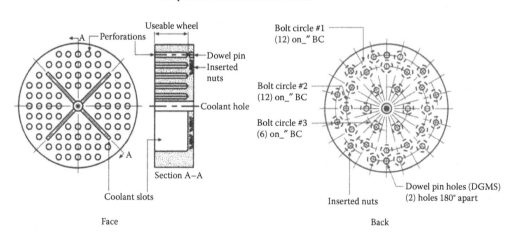

FIGURE 16.57
Components of a double-disk wheel.

been competitive with grit sizes in the range of 100#–600# (Deming and Carius 1991), for example, reported good success grinding hydraulic pump vanes on retrofitted Besly double-disk grinders using 30 in resin CBN wheels although they note on new dedicated machines the wheel size was reduced to 23 in.

Since thickness and flatness control is critical the grinder should be kept in a temperature controlled room with the coolant chilled to ambient. Some manufacturers even keep work-pieces in a coolant bath before grinding so as to stabilize the temperature. With coolant passing through the center of the spindle to remove heat, thermal growth can be held to a minimum. This is further aided by the machine configuration and spindle design inherent in double-disk grinding such that there is a tendency to compensate for thermal growth in the base with growth in the opposite direction for the spindle (Schlie and Rangarajan 1987). Nevertheless this is not perfect and compensation must still be added for wheel wear so postprocess gauging is common on precision operations. Timed compensation is more common for commercial tolerance work.

Conventional wheels are dressed using either stationary diamond or star cutter dress-ers mounted on powered radial arm dresser that swings between the two-wheels dressing both at the same time (Figures 16.58 and 16.59). The face of each wheel is dressed flat in

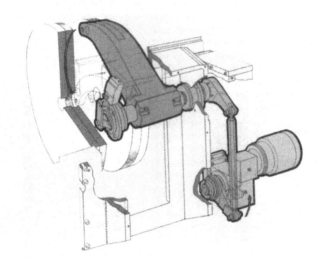

FIGURE 16.58
Dressing mechanism for horizontal double disk

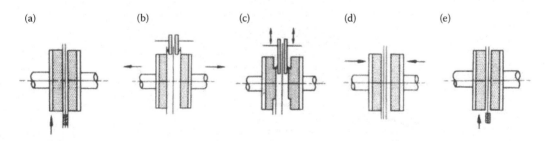

FIGURE 16.59
Automated dressing cycle for double-disk grinding. (a) Part feed, (b) wheels open, (c) dress, (d) wheels open, and (e) resume feed. (From Mattison, 2. n.d. *Double Disc Grinders*. Mattison Machine Works, Rockford, IL. Trade brochure. With permission.)

relation to the spindle alignment regardless of the head setting. For resin and CBN the radial arm has on occasion been retrofitted with a rotary dresser for truing in combination with either a grit blast type conditioning process or stick dressing. Diamond wheels are almost exclusively dressed using dressing sticks.

There are a variety of techniques for double-disk grinding depending on the stock removal and part tolerance requirements (Figure 16.60).

Rough grinding with the highest removal rates uses through-feed grinding in a shear mode. Through-feed grinding involved feeding the parts supported by rails top and bottom through the center of the wheel. The parts are driven at the entrance of the wheels by belts in contact with the side of the part. Passage through the wheels is driven by contact with the part behind. The wheels are angled with a "head setting" such that the gap between them is narrowest at the entrance where 75% of the stock is removed. Often 2 or even 3 grades are used across the face of the wheel to allow for changes in surface footage and prevent burning in the center portion where only rubbing occurs. The wheel wears over time causing a step to move progressively over the wheel face and toward the center.

Higher precision is achieved by progressively grinding using a rotary feed or carousel method (Figure 16.61). In this case the wheels are again angled but the maximum clearance is at the exit. The parts are fed in a rotary carrier with pockets that closely match the part

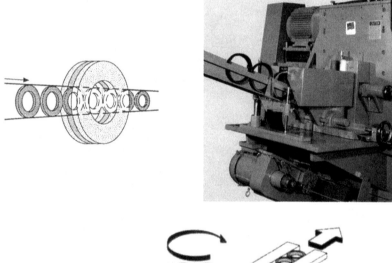

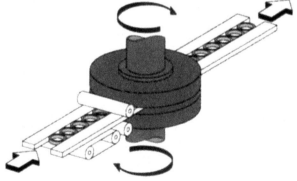

FIGURE 16.60
Double-disk grinding arrangements. (From Landis Gardner, 1989, *Horizontal Disc Grinding Machines and Systems*. Litton Industrial Automation, S Beloit, IL. Trade Brochure; and Diskus, 1996, *Diskus Double Disk Face Grinding Machines Series DDS*. Diskus Werke Schleiftechnik, Langen, Germany. Trade Brochure. With permission.)

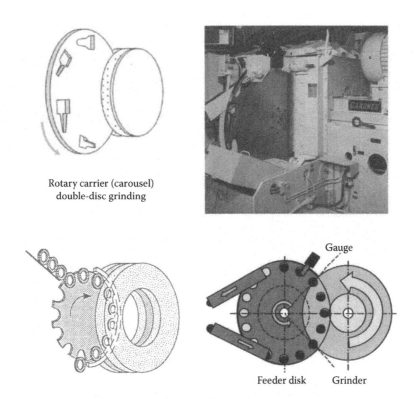

Rotary carrier (carousel)
double-disc grinding

Gauge

Feeder disk   Grinder

FIGURE 16.61
Rotary and carousel work feeding arrangements.

shape. Small parts especially those with a high aspect ratio are sometimes rotary fed by a geared tooth inner ring surrounded by a smooth outer ring.

The calculations for the feed geometry and head settings are relatively straightforward. The carousel should pass over the wheel such that the centerline of the part cuts the inner diameter of the abrasive face. Wheels for rotary grind double disk have a large center hole in order for limited variation in surface footage across the face. Most spindle heads have three or four bolts at pivot points that can be raised up and down monitored by a portable micrometer. The adjustment can be as simple as a single nut or a series of lockscrews and adjusting screws. There is some debate as to how the stock should be removed. For conventional wheels, the entrance gap should take 20%–40% of the stock by shear grind with the remainder removed as it passes through the wheels (Doubman and Cox 1997). For CBN, however, the cut should start about 20 mm into the wheel such that the outer surface of the wheel is kept wear free to just finish grind for flatness (Figure 16.62).

The wheel should initially be set at zero and then simple geometry allows a ready calculation for how much the entrance and rear pivot points need to be raised in order to equalize stock removal along the whole grind path. The entrance should then be double checked for clearance with an unground part with the wheels stationary.

Again, two-grade wheels are sometimes used. For example one design by the author uses a combination of coarse grade CBN pellets for rough grinding (for reducing contact area) with a fine grit CBN continuous rim. This increased stock removal capability on one application by a factor 3, Figure 16.63.

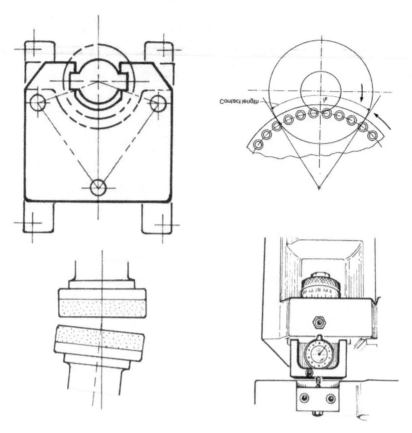

FIGURE 16.62
Three-point adjustment of upper spindle. (From Koyo, n.d. *Vertical Spindle Surface Grinders KVD Series*. Koyo Machine Industries, Osaka, Japan. Trade brochure. With permission.)

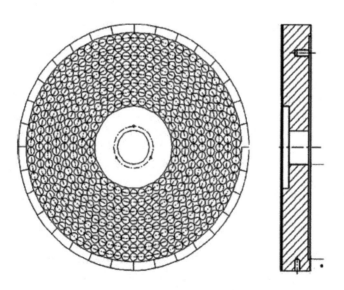

FIGURE 16.63
Two-grade combination pellet and continuous rim superabrasive disk wheel.

FIGURE 16.64
Test washer ground with low-lubricity coolant.

On a vertical disk grinder the bottom wheel should be set slightly higher than the exit plate. The parts should fit snugly in the carrier. Special attention should be paid in the case of round parts as flatness can be significantly improved if the part is allowed to spin in the holder (Figures 16.64 and 16.65). The spin is generated by the relative surface footage seen by one side of the part relative to the other. Obviously the spin can be greatly enhanced by rotating the wheels in opposition but this creates too much wheel wear and can damage the carrier. The wheels are therefore rotated in tandem with a small difference in rpm between the two to vary the spin. Excessive spin will cause wear of the carriers that are often made of only mild steel or fiber glass. In some cases carbide inserts are added to prolong life. The spin is such an issue that it can often be the primary factor governing the selection of coolant. Too high a lubricity can cause the part to lose traction and spin sporadically or even hydroplane. For this reason double-disk grinding often uses synthetic coolant even with superabrasives.

Coolant delivery is through the center of the spindle but must be dispersed by means of a baffle or other device to fling the coolant outward. This is made more difficult by the presence of the carrier. For thin parts the carrier can warp and may well end up being ground on occasion leading to an increased risk of burning.

FIGURE 16.65
Test washer ground with high-lubricity coolant.

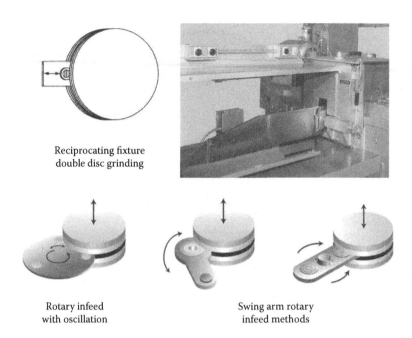

Reciprocating fixture
double disc grinding

Rotary infeed
with oscillation

Swing arm rotary
infeed methods

FIGURE 16.66
Examples of feed mechanisms for double-disk grinders.

For the highest precision, through-feed grinders are typically small, compact and use superabrasives. Examples include the Koyo models KVD 300 (305 mm wheels) and KVD 450 (455 mm wheels). Very few examples exist of larger wheels except for some older references to grinding pump vanes with 23 in. vitrified CBN (Navarre 1986; Landis Gardner 1989), and for finish grinding of piston rings with 600 mm resin CBN wheels taking 20 μm or less of stock (Buthe et al. 1992). The problem relates to the difficulty in achieving sufficient stiffness, especially from the cantilever effect that comes into play with the larger diameter wheels as the part exits.

Through-feed grinding in general has problems achieving precision tolerances in a single pass with an acceptable cycle time when heavy stock removal is involved (>200 μm). The alternative is to infeed the wheel with their faces parallel. Several variations on this method exist. One method is to reciprocate or gun feed a part between the wheels with linear oscillation. This is particularly effective for controlling wheel flatness when grinding large workpieces (Figure 16.66).

Alternatives include some form of rotary carrier or swing arm. The carrier may have its own drive mechanism to make the part spin to further improve flatness. Infeed grinding will give the very best accuracy but at the lowest production rate. It will also have problems maintaining flatness for prolonged periods of time because of differential surface footage between the outer and inner wheel radii.

The design of the component is important when considering double disk as a processing method. Well-qualified datum surfaces must be provided when trying to control thickness. The surfaces to be ground should be comparable in area, configuration and stock removal. Otherwise the stock removed will be uneven or a great deal of trial and error effort must be expended trying to vary grade and surface footage from one wheel to the other to equalize the removal rates. The face should also be presented to the wheel reasonably flat and parallel. For this reason the side faces locating in the carriers must also be reasonably square to the surfaces to be ground. There should be no functional surfaces projecting beyond the planes of the surface to be ground. For finish grind operations the part must be stress relieved, especially for thin parts, to avoid "potato-chip" or "saddle" distortion effects. When through-feed horizontal grinding the parts need to be in a stable balanced condition to prevent tipping as they are exiting the wheels.

Typical case history examples from the published literature and trade brochures are given in Table 16.12. Double-disk grinding is limited to tolerances of about 1–2 µm for

TABLE 16.12

Examples of Typical Accuracies in Double-Disk Grinding

Component	Major Dimension	Stock Allowance	Size	Flatness	Parallelism	Finish	Cycle Time	Reference	Year
Compressor cylinder	4.4" diameter	0.006"	0.001"	0.0002"	0.0002"	25RMS	18 s	Kibsh	1988
Compressor roller	1.5" diameter	0.003"	0.0002"	0.000060"	0.000060"	12RMS	19 s	Kibsh	1988
Compressor	0.94" × .16"	0.004"	0.0002"	0.0002"	0.0001"	12RMS	4 s	Kibsh	1988
vanes	0.657" × 16"	0.009"	0.0003"	0.0003"	0.0002"	12RMS	4 s	Kibsh	1988
Coil springs	1/8"–1.25"		0.001"		1 deg.		1.8 s	Litton	1989
Ferrite seals		0.012"	0.001"	0.0003"	0.0003"	20RMS	1.2 s	Litton	1989
Connecting		Rough 0.0625"					4.5 s	Litton	1989
rod		Finish −0.02"	0.0002"	0.0005"	0.0005"	20RMS	4.5 s	Litton	1989
Circlip		0.008"	0.001"	0.0005"	0.0005"	20RMS	0.072 s	Litton	1989
Aluminum Valve part		0.020"	0.01"	0.0015"	0.006"	63RMS	6 s	Litton	1989
Steel		Soft −0.025"	0.002"	0.001"	0.001"	63RMS	3.6 s	Litton	1989
Washer		Hard −0.005"	0.002"	0.001"	0.001"	63RMS	2 s	Litton	1989
pump rotor			2.5 µm	2.5 µm	2.5 µm	3.2Ry		Daisho	1998
cam-ring			2.5 µm	2.5 µm	3.0 µm	0.5Ra		Daisho	1998
cylinder			1.2 µm	1.5 µm	1.5 µm	0.5Ra		Daisho	1998
drive gear			2.0 µm	2.0 µm	2.0 µm	2.5Rz		Daisho	1998
ceramic seal		50 µm		3 µm	3 µm	1.2Ra	3.5 s	Koyo	c1995
Connecting rod		0.3 mm		15 µm	15 µm	3.2Rz	30 s	Koyo	c1995
Compressor roller		50 µm		1 µm	1 µm	1.2Rz	18 s	Koyo	c1995
								Koyo	c1995
Bearing ring		80 µm		0.5 µm	0.5 µm	1.0Rz	0.5 s	Koyo	c1995
								Koyo	c1995
Aluminum vane		100 µm		5 µm	5 µm	2.5Rz	8 s	Koyo	c1995
								Koyo	c1995
Gear		300 µm		2 µm	2 µm	3.2Rz	10 s	Koyo	c1995

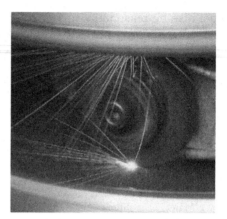

FIGURE 16.67
Comparison of block and electrodischarge truing of superabrasive wheels. (Courtesy of Koyo Machine USA, Novi, Michigan. With permission.)

flatness and size and stock levels under 100 μm for high production such as using rotary carrier feed. Slightly better control is possible by infeed grinding but at a considerable cost in cycle time. The best results are again on dedicated machines using small superabrasive wheels.

The limitation on flatness is due to the stiffness of the machine and wheels both in grinding and even in dressing. As stated above, the exit point between the wheels is where the most deflection will occur from being cantilevered furthest from the spindle axis. This is true both for grinding and dressing. Unfortunately, in most cases the pressure cannot be relieved as in single-disk grinding by reducing the contact width of the wheel.

Resin and many vitrified superabrasive wheels are usually dressed using some form of stick feeder, Figure 16.67.

For larger wheels the stick pressure becomes the limiting factor for wheel flatness. Koyo has therefore developed a contactless dressing process namely electrodischarge truing used in conjunction with metal bond wheels. This is reported to double flatness accuracy.

Double-disk grinding is an important process for the ceramics and semiconductor industry and there is interest here to increase accuracy at high production rates. One vertical spindle machine design reported by (Ueda et al. 1998) involves two novelties. The first is the use of live magnetic bearings to control thickness in conjunction with closed loop gauging. The second, and perhaps more interesting for its simplicity and general application potential, is to increase the grind length by positioning the parts behind the wheel head. This system grinds electronic filters and oscillators to +/− 0.5 μm (Figure 16.68).

Although spin and reduced wheel width techniques cannot be used on the majority of double-disk grinding applications because of fixturing, there is use of it where only a portion of each surface has to be ground. The most active example currently is brake rotor disk, see Figure 16.69. Grinding is carried out as a finishing operation to improve flatness after turning. The rotors are ground on the outer diameter areas which means they can be readily held and spun in a work holder. In the 1990s, these were ground using conventional wheels or segments (Anon 1996a; Thielenhaus n.d.) but more recently

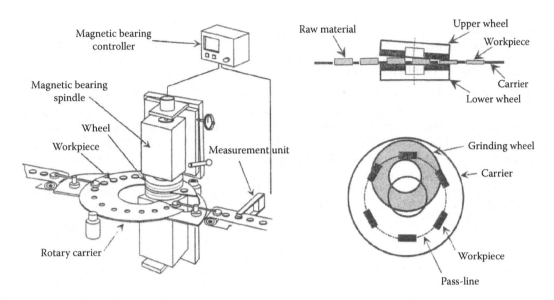

FIGURE 16.68
Double-disk grinder for grinding ceramics. (From Ueda, S. et al., 1998, *International Journal of the Japan Society for Precision Engineering*, 32(1), 9–12. With permission.)

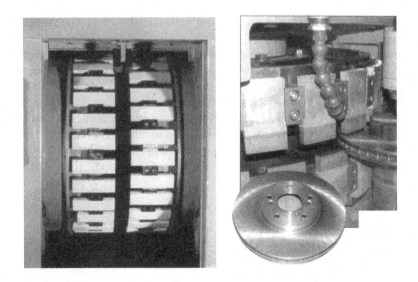

FIGURE 16.69
Examples of alox and superabrasive stone grinding of brake rotor disks. (From Daisho, 2001, Trade advertisement; and Thielenhaus, n.d., *Microfinish® the Economical Precision-Machining Method in the Automotive and Motor Industry*. Thielenhaus Microfinish Corp, Novi, Michigan. Trade Brochure. With permission.)

these have been replaced by metal-bonded CBN segments (Daisho 2001) or narrow faced metal-bonded CBN wheels by machine builders such as Thielenhaus and EMAG. The stock amount varies from 5 to 100 μm depending on the turning operation and the material (steel or cast iron). Wheel speeds are low at about 25 m/s or less with blocks but up to 60 m/s with thin-rimmed wheels.

16.7 Fine Grinding

16.7.1 Principles and Limitations of Lapping

Fine grinding, also known as flat honing, refers to a type of grinding being advanced by machine tool builders that previously made lapping machines. Lapping using free abrasive slurry has been the traditional method for achieving micron and submicron flatness (Figure 16.70). Stock removal occurs as a result of a rolling and sliding of abrasives grains between a lapping plate (single-sided lapping) or plates (double-sided lapping) and the workpiece, resulting in work material compaction, deformation, and finally failure as the material strength is exceeded (Hodge 1992; Schibisch 1997, 1998; von Mackensen et al. 1997). The process is slow, inefficient in abrasive use, and leaves a dull, pitted surface. The lapping debris must be continually flushed away during processing, as the abrasive cannot usually be recycled, leading to high disposal costs. The abrasive slurry must maintain a consistent layer on the plates. Therefore, lapping plate speeds are limited to surface speeds of 1 m/s to resist centrifugal force and prevent heat generation. Perhaps the most problematic issue of all is that the slurry is very dirty and clings to the parts. This creates the need for an additional expensive process of cleaning the finished parts while also producing an unpleasant work environment.

Nevertheless, lapping is a mature and well-developed process in terms of the understanding and control of the kinematics of producing flat surfaces. The more sophisticated lappers use geared carriers to hold the parts which are driven to describe a planetary motion covering the full surface of the lapping plates as shown in Figure 16.71. The drive mechanism consists of an outer and inner tooth or pin ring; the outer ring is usually fixed, while the inner ring rotates to create a series of epicyclic or hypocyclic motions relative to the motion of the bottom plate. The process is discussed in more detail by Ardelt in Appendix 1 (Hitchiner et al. 2001).

Less sophisticated, single-sided lappers have simple rings to hold them in place and maintain plate flatness, Figure 16.72. The rings rotate by friction created in the process and are moved in and out using simple adjustable arms to prevent the generation of a concave or convex surface. Pressure is applied to the parts by either hand weights, pneumatic pressure, or even, for larger components, just the weight of the part itself.

Lapping

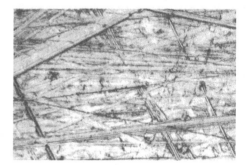

Grinding

FIGURE 16.70
Surface textures produced by lapping and fine grinding.

FIGURE 16.71
Geared fixturing arrangement for a double-sided lapper. (From Melchiorre, n.d., *Melchiorre Fine Grinding, Lapping, Honing, Polishing.* Melchiorre S R L, Milano, Italy. Trade Brochure. With permission.)

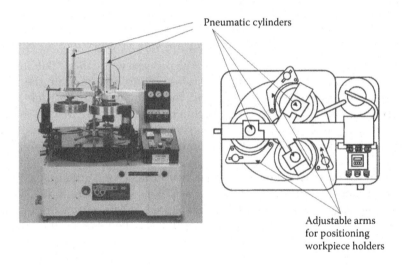

FIGURE 16.72
Lapping machine.

16.7.2 Double-Sided Fine Grinding

As discussed above, the problem with double-disk grinding is generating and maintaining flatness. This is either from the inherent limitation of trying to generate a flat surface while through-feeding into a wedge grind zone, or in the case of the slower infeed methods, maintaining a flat wheel with large differentials in surface footage from the outer to inner diameter.

Lapping kinematics using parallel plates is much better at maintaining wheel flatness and several lapping machine tool builders have taken this concept and made grinders using lapping kinematics together with superabrasive wheels. Although the removal rates are still much lower than a through-feed double-disk grinder, the grind is done in batches

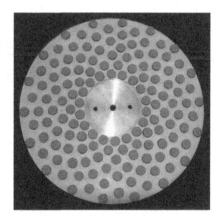

FIGURE 16.73
Wheel segments for fine grinding.

with coverage of up to 15%–20% of the plate area which can be 100 or more components. Consequently, cycle time/part becomes competitive with other grinding processes while grinding is done under controlled pressure and at low wheel speeds to minimize the risk of thermal damage or distortion.

The earliest efforts were initiated in the late 1970s by OEMs such as Hahn and Kolb using alumina and silicon carbide abrasive. The process was called "flat honing" (Stahli 1998) or "fine grinding" (Anon 1996b). This was able to demonstrate the viability of a fixed abrasive process but wheel wear, and the associated dress time and operator intervention, was a limiting factor. Although a number of these machines are running very successfully, grinding with superabrasive wheels was subsequently pioneered with OEMs such as Wolters 2000 (1 n.d), Stahli (1998), Melchiorre (n.d.), and Modler (n.d.).

Wheel design for fine grinding can take several forms based either on a 100% covered, segmented face with coolant slots or holes, or as pellets (see Figure 16.73). Wheels requiring low coverages use round pellets while hexagonal pellets are used for packing >90% or to keep gap distances low while grinding small parts. Pellets offer the potential to infinitely vary abrasive coverage across the face of the wheel in order to fine tune wear. They also provide good coolant access and swarf clearance. However, they are still problematic for small parts and more expensive to fabricate than segmented wheels especially when a high coverage is required.

The wheels are bolted to flanges that have a sophisticated internal labyrinth of coolant channels to control temperature. The coolant itself is fed through holes in the upper wheel. Low viscosity oil is the coolant of choice, especially in Europe, although some water based is in use in the United States albeit with reduced wheel life. Wheel speeds have remained for the most part comparable to those used for lapping 1–3 m/s. This is in part historic but also in part to keep heat generation to a minimum with only limited coolant access. More recently, however, speeds have increased significantly with machine builders pushing the envelope up to 10 or 15 m/s. In doing so coolant delivery and regulation has becomes much more critical as has spindle power, chiller systems, and stiffness.

Spindle horsepower is very dependent on table size and is typically double the power required for lapping because of the higher metal removal rates (Table 16.13).

TABLE 16.13

Wheel Spindle Power Requirements for Fine Grinding Machines

	Wheel Diameter Range (mm)							
	450–500	560	650–700	840	900–1000	1100–1200	1400–1500	1800–2000
Spindle power (kW) [Stahli]	3.0		9.5		19	26	32	52
Spindle power (kW) [Melchiorre]	3.0		9.2		18.5	22	30	
Spindle power (kW) [Modler]			7.5					
Spindle power (kW) [Wolters (2),3]		(4)	12	(5.5)	16	24		

Source: OEM data from 2000 and earlier. With permission.

Wheel sizes are large with 700–1000 mm being common (Figures 16.74 and 16.75). With the additional power available for grinding, the machine and tooling must be more rigid and rugged to withstand the extra forces. Downfeed systems are predominantly pneumatically controlled (Lapmaster, Stahli, and Wolters) but hydraulic (Modler) and mechanical servo feed (Melchiorre) are also used. The upper wheel head assembly is so heavy that the system is generally providing upward force to keep from allowing the full weight of the head to be applied to the grind. All grinding is carried out under controlled pressure. Pressures are expressed within the industry in "deca Newtons" or daN, where 1 daN = 10 N or 2.25 lb (Table 16.14).

Taking the machine design values for pressures and wheel dimensions, together with the fact that the table have a 15%–20% surface area coverage of components, maximum working pressures for grinding are in the range 7–21 N/cm^2 or 10–30 lb/in^2 which is about double the pressures used for lapping.

FIGURE 16.74
Peter Wolters Microline double-sided fine grinder.

FIGURE 16.75
Stahli DLM 700 double-sided fine grinder.

TABLE 16.14

Maximum Grind Pressures against Wheel Diameter

	Wheel Diameter Range (mm)							
	450–500	560	650–700	800–900	900–1020	1100–1230	1390–1500	1800–2000
Maximum grind pressure (daN) [Stahli]	300		1300		2400	3200	4400	7000
Maximum grind pressure (daN) [Melchiorre]	300		1200	1500	2500	3000	3500	
Maximum grind pressure (daN) [Modler]			2000					
Maximum grind pressure (daN) [Wolters (2),3]	250	700	700		1000	1000	2000	3000

Size is monitored in-process by a probe in the center of the spindle assembly and relies on the wheels wearing at submicron levels per load. The ability to monitor size is the limiting factor on cycle time in many cases as grinding removal rates can be up to 20 times faster than lapping. With the latest developments in bond technology and machine design, wear rates per load are often 0.2 μm or less.

16.7.3 Comparison of Fine Grinding with Double-Disk Grinding

The selection of the appropriate technology for generating parts in the flatness range of 0.5–2 μm is becoming quite subtle (Figure 16.76). Technological advances in both processes are accelerating. The view just 5 years ago was that fine grinding held the better flatness but double disk held the better size control. However, with Koyo's development of EDT dressing of metal bond wheels, flatness can now often be held to less than 0.5 μm.

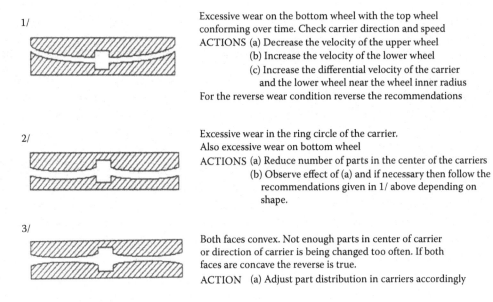

1/

Excessive wear on the bottom wheel with the top wheel conforming over time. Check carrier direction and speed
ACTIONS (a) Decrease the velocity of the upper wheel
(b) Increase the velocity of the lower wheel
(c) Increase the differential velocity of the carrier and the lower wheel near the wheel inner radius
For the reverse wear condition reverse the recommendations

2/

Excessive wear in the ring circle of the carrier.
Also excessive wear on bottom wheel
ACTIONS (a) Reduce number of parts in the center of the carriers
(b) Observe effect of (a) and if necessary then follow the recommendations given in 1/ above depending on shape.

3/

Both faces convex. Not enough parts in center of carrier or direction of carrier is being changed too often. If both faces are concave the reverse is true.
ACTION (a) Adjust part distribution in carriers accordingly

FIGURE 16.76
Effect of grinding wheel and ring carrier wear.

On the other hand, recent developments in magnetic probes for size control combined with new wheel technology for reduced wheel wear per load has dramatically improved size-holding capability for fine grinding. The drawbacks to fine grinding are that the process is a batch process, which is difficult to fully automate, and the process requires coolant. Disk grinding is a continuous feed process, which can be easily automated and runs well in water-based coolants. Fine grinding due to the number of parts that can be ground simultaneously and the recent increases in wheel speeds and head pressures, probably has a higher output. Disk grinding requires constant postprocess gauging to monitor size; fine grinding being a batch process, size is checked just once or twice per load, which can be done manually with minimal labor cost. On the other hand, if there is a size error, the entire batch must be scrapped for fine grinding, while it may only be a few parts for disk grinding.

Fine grinding is a technology driven by European machine tool builders, while disk grinding is driven by Japanese machine tool builders, and it is likely that this is influenced by the particular markets that they serve (Table 16.15).

16A.1 Appendix: Lapping Kinematics

16A.1.1 Introduction

Considerable research work has been carried out on grinding with lapping kinematics at the University of Berlin. Dr. Thomas Ardelt has provided detailed information on this research below.

During face grinding on lapping machines, several workpieces are moved simultaneously between two horizontally positioned grinding wheels (Figure 16A.1). The parts are

TABLE 16.15

Application Examples for Fine Grinding

Component	Material	Size (mm)	Stock Amount (µm)	Removal Rate (µm/min)	Rough Grind Time/Part (sec)	Flatness (µm)	Parallelism (µm)	Finish (µm)	Size Control (µm)	Wheel Spec
Bearing plate	Bronze	119	330	45	18	<1.5	<2	<0.5Ra	5	B
Bearing race pair	Steel 560HrC	125/70	30	10	9	<1.5	<1.5	0.25Ra	1.5	B
Cam-ring	Steel 63HrC	65	65	10	7	<1.5	<1.5	1.5Rz	1.5	B
Cam-ring	PM 95 HrB	35	150	16	2.5	<1	<1	<0.8Ra	2.5	B76
Cutting tool	It/12 62HrC	75	200	16	14	2.5	2.5	0.15Ra	2	B46
Gear	Steel 62HrC	20	25	7.5	2	<1	<1	<24Rz	2	B
Gear	PM45HrC	30.5	43	11	5	<1.2	<1.2	0.18Ra	1.25	B46
Injector plate	Steel 61HrC	20	60	20	2.3	<1	<1	0.15Ra	1.5	B18/B30
Oil pump gear	PM 43HrC	57	157	20	3.2		<1	0.<35Ra	1.25	B76
Oil pump gear	PM 95 HrB	84	280	35	4.9	<5	<3	0.55Ra	3	B76
Piston ring	90CRMoV18	91-5	30	60	0.8	<2	<2	1.8Rz/0.18Ra	2	B
Piston rod	Aluminum	90	560	135	8	<1.5	<1.5	<3.7Rz	10	B
Plate	Steel 62HrC	95 x 75	35	35	33	<3	<2	<2Rz	2	B
Pump plate	PM 70HrB	89	28	6.6	43	<2.5	<2.5	<0.2Ra	1.25	B46
Pump ring	Steel 58HrC	35	125	15	12	<2	<1	<2Ra	2	B46
Pump rotor	PM D39	52	220	25	29	<1.5 <2	2Rz	2.5	B	
Pump rotor	PM 52HrC	44	100	25	3	<1.5	<2	3Rz	1.5	B
Pump stator	PM C11	60	200	40	25	<1.5	<2	2Rz	2.5	B
Slide disk	31CrMo12	136/55	400	28	86	<5	<2	<2.4Rz	6	B
Tappet shim	100Cr6	25	150	25	1-1.5	<2	<2	2Rz	3	B
Thrust bearing disk	ss1819	200/192	71	10	70	<7	<3	<25Ra	10	B
Valve plate	ss1819	100	355	30	3	<40	<2	<38Ra	15	B
Valve plate	Steel47HrC	119	460	45	25	<2.5	<2	<0.6Ra	2.5	B
Seal disc	Alumina	42	405	250	0.5	<2	<2	<0.4Ra	5	D
Seal disc	Alumina	33/19	405	110	1.1	<2	<2	<0.4Ra	5	D
Seal disc	Alumina	37	610	250	1	<2	<2	<0.4Ra	5	D
Seal washer	Piezo ceramic	24	18	430	0.03	<1.2	<1.5	0.5Ra	5	D
Slide disc	Alumina	23	140	19	6.3	<1	<2	<.38Ra	6	D
Tile	SiC	63.5	1650	80	68	<5	<5	0.225Ra	5	D
Tool insert	Alumina	32	250	16	19	<1	<1	0.252Ra	2	D

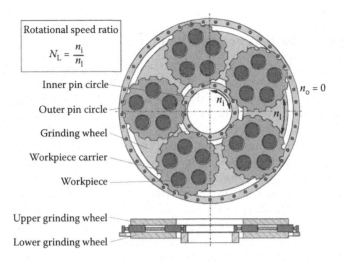

FIGURE 16A.1
Design of a double-wheel lapping or grinding machine. (From Uhlmann, E., Ardelt, T., 1999, *Annals of the CIRP*, 48(1), 281–284. With permission.)

fixed in workpiece carriers that are led between two-pin circles. This way, characteristic cycloidal path curves are generated between parts and grinding wheels that are similar to the movements in planetary gears. It was found that a variation of the types of planetary movement directly influences material removal rates and driving power and particularly the resulting quality of the produced parts (Ardelt 1999). As a basis of understanding these effects, kinematic possibilities of double-wheel lapping and grinding machines have to be analyzed.

16A.2 Kinematic Fundamentals

Since the mid-eighties the Institute for Machine Tools and Factory Management of the Technical University Berlin worked on an analytical description of the relative motions in cycloidal restricted guidance on lapping machines. A model was developed that calculates the profile wear of the lapping wheel as a function of the path curves of the workpieces on the wheels (Spur and Eichhorn 1997; Simpfendörfer 1988; Uhlmann and Ardelt 1999).

For all machines with fixed external pin circle, selectable kinematic parameters are the rotational speeds of the lower grinding wheel n_l, the upper grinding wheel n_u, and the internal pin circle n_i. Thus, a description of all path types and velocities related to the lower grinding wheel with the rotational speed ratio

$$N_L = \frac{n_i}{n_l}$$

is possible. N_L describes unequivocally for a single machine, the path type on which the workpieces move relative to the lower grinding wheel. All path types can be covered with different velocities within the scope of the possible rotational speeds of a machine (Ardelt 1999).

16A.3 Analysis of Path Types and Velocities

In the following, the connection between rotational speeds and emerging path curves is explained (Ardelt 2000). This kinematic analysis depends on the diameters of the grinding machine design. In this example, it is done for a machine type Duomat ZL 700 by Stähli Läpp-Technik GmbH, Germany. The outer pin circle of this machine is fixed, whereas the two grinding wheels and the inner pin circle can be driven individually.

Figure 16A.2 shows the characteristic path types that a workpiece center point covers as a function of the rotational speed ratio N_L. The path types designated with letters are classes which occur in certain ranges of N_L. The curves marked with numbers are special cases that arise only at a definite ratio of the two rotational speeds.

At high negative rotational speed ratios, all workpiece points cover stretched hypocycloids (a) on the grinding wheel. These path curves turn into stretched epicycloids (b) at −2.39. At the transition between these two path types the parts move on eccentric circular paths around the grinding wheel center (1). Whereas the cyclically recurring shape elements of all path types with $N_L < 0$ wind around the center of the grinding wheel, tight loops with strong bends emerge at $N_L > 0$, repeating themselves on one half of the wheel coating. At a rotational speed ratio of 2.15, the stretched epicycloids become interlaced epicycloids (c). At the transition point, common epicycloids emerge which show a reversing point in the form of a buckling (3). With increasing rotational speed ratio, circles around the holder center (4) occur which represent the transition point to interlaced hypocycloids (d). At $N_L = 11.09$, common hypocycloids (5) emerge which turn into stretched hypocycloids (e).

The occurring relative velocities are shown in Figure 16A.3. The graph gives the mean path velocity by 100% and the occurring maxima and minima related to this mean value for every rotational speed ratio. The path velocities vary more or less strongly depending on the shape of the curves. The related path velocity $v'(t)$ given in the diagram is defined

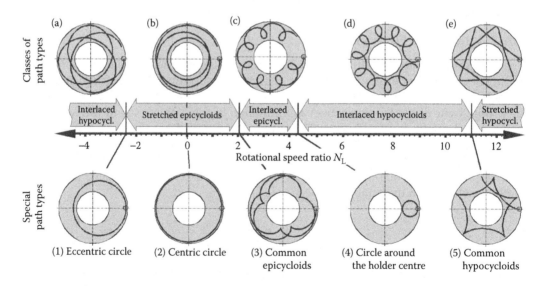

FIGURE 16A.2
Path types of the Duomat ZL 700. (From Ardelt, T., 2000, Einfluss der relativ bewegung auf den prozess und das Arbeitsergebnis beim Planschleifen mit Planatenkinematik, PhD thesis, TU, Berlin. With permission.)

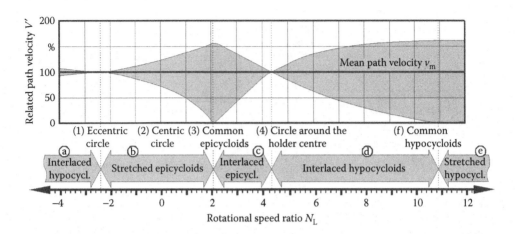

FIGURE 16A.3
Related path velocities of the Duomat ZL 700. (From Ardelt, T., 2000, Einfluss der relativ bewegung auf den prozess und das Arbeitsergebnis beim Planschleifen mit Planatenkinematik, PhD thesis, TU, Berlin. With permission.)

as the ratio of the path velocity $v(t)$ to the mean path velocity v_m. If the lines for minimum, mean, and maximum velocity intersect, the occurring path curves are covered with constant velocity. This effect occurs at $N_L = -2.39$ and 4.35. As could be seen in the previous graph, circular path types emerge for these two rotational speed ratios. Covering common epi- and hypocycloids at $N_L = 2.15$ and 11.09, the workpiece points minimum velocity is zero. On these path types, the workpieces are exposed to extreme differences in velocity which alternately reach a momentary standstill and an acceleration to over 150% of the mean path velocity.

Figures 16A.2 and 16A.3 demonstrate that the kinematic conditions at equal mean path velocity depend heavily on the rotational speed ratio and thus, on the shape of the path curve. To explain this effect, Figure 16A.4 shows two different workpieces moving along a stretched epicycloid path. Along this path curve, the relative velocity increases from the inner to the outer grinding wheel diameter. In a certain grinding time interval Δt_c, workpiece A moves a longer track length than workpiece B does due to their different velocities.

The upper grinding wheel descends simultaneously on all the workpieces. Thus, the achieved workpiece height difference after Δt_c has to be the same for all workpieces. This

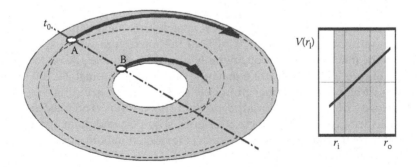

FIGURE 16A.4
Path curve sections of two workpieces after a certain grinding time interval. (From Ardelt, T., 2000, Einfluss der relativ bewegung auf den prozess und das Arbeitsergebnis beim Planschleifen mit Planatenkinematik, PhD thesis, TU, Berlin. With permission.)

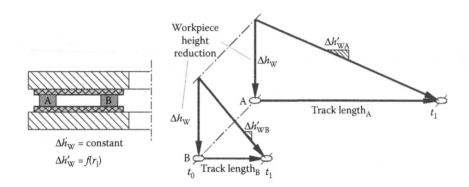

FIGURE 16A.5
Workpiece height reduction. (From Ardelt, T., 2000, Einfluss der relativ bewegung auf den prozess und das Arbeitsergebnis beim Planschleifen mit Planatenkinematik, PhD thesis, TU, Berlin. With permission.)

means for the mentioned example that in spite of the different relative track length both workpieces A and B lose the same height. It follows that the angle of grinding grit engagement has to be different for both workpieces, as the velocity vectors in Figure 16A.5 show. This effect strongly depends on the type of relative movement and on the distribution of relative velocities as a function of the lower grinding wheel radius.

Different angles of grit engagement may lead to different types of grit wear. It is possible to have one area of the grinding wheel surface working in the self-sharpening range, while another area shows grit flattening and looses its sharpness and grinding ability in short time. To set up stationary grinding processes, it is necessary to have equal microscopic wear conditions on the complete contact area, which can be achieved by analysis of the process kinematics.

16A.4 Kinematic Possibilities of Machines

The previously shown description of path types during machining on double-wheel lapping and fine grinding machines reveals a very broad range of the rotational speed ratio. This theoretical analysis was carried out, irrespective of the rotational speeds that can be actually set on real machines. Most standard designs permit rotational speeds at the inner pin circle of 50%–80% of the wheel speed. This means that a rotational speed ratio of 10 can only be reached with 5%–8% of the maximum wheel speed. This minimum usage of the machine power considerably limits the settable path velocities. To illustrate this fact, Figure 16A.6 shows for a machine of the type Duomat ZL 700 the rotational speeds that must be set at the inner pin circle and the lower grinding wheel to reach a mean path velocity of 50 m/min for all rotational speed ratios. The maximum rotational speeds possible are given by the horizontal lines. The perpendicular bar represents the range of settable rotational speed ratios, which is restricted in both directions by the inner pin circle (Uhlmann and Ardelt 1999).

Due to the technological progress in driving technology on the one hand and knowledge of lapping kinematics on the other, kinematic possibilities of machine systems are constantly expanded and improved. To illustrate this development, Figure 16A.7 compares the kinematic possibilities of various generations of the machine family ZL by Stähli GmbH.

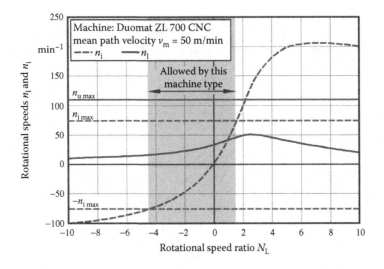

FIGURE 16A.6
Necessary rotational speeds to achieve a mean path velocity of 50 m/min at different rotational speed ratios NL. (From Ardelt, T., 1999, On the effect of path curves on the process and wheel wear in grinding on lapping machines, *Proceedings 3rd International Machining and Grinding Conference*, Cincinnati, Ohio.)

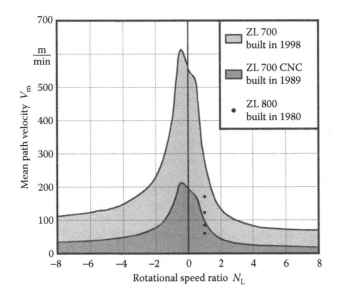

FIGURE 16A.7
Comparison of kinematic possibilities of different double-wheel machine designs. (From Ardelt, T., 1999, On the effect of path curves on the process and wheel wear in grinding on lapping machines, *Proceedings 3rd International Machining and Grinding Conference*, Cincinnati, Ohio. With permission.)

For the ZL 800, built in 1980, the rotational speed ratio between inner pin circle and lower lapping wheel is set firmly by a gearing. Changing the poles at the driving motor serves to realize four different velocities. Freely selectable rotational speeds permit any rotational speed ratios for the design ZL 700 CNC (1989). For the path velocities needed for grinding, the range of possible path types is strictly limited. Possible rotational speeds are considerably higher for the ZL 700 built in 1998. Here, both grinding wheels can be driven with up to $n_{u,l} = \pm 435/\text{min}$, the inner pin circle with up to $n_i = \pm 268/\text{min}$. These rotational speeds allow for mean path velocities v_m of more than 600 m/min or 10 m/s.

References

Akinori, Y., 1992, The press industry in Japan. *Pureu Gijutsu*, 30(1), 74–81.

Andrews, C., Howes, T. D., Pearce. T. R. A. P., 1985. *Creep Feed Grinding, Holt*, Rinehart and Wilson, New York.

Anker, A., 1998, Machining the ends of springs with ABN *IDR*, 4, 124–125.

Anon, 1995, Huge rotary grinder improves productivity by 20%. Manual, 62.

Anon, 1996a, Finish grinding improves quality of disc brake rotors. *Tooling & Production*, December, 42–46.

Anon, 1996b, Fine grinding can be the answer. *Tooling & Production*, June, 45–46.

Ardelt, T., 1999, On the effect of path curves on the process and wheel wear in grinding on lapping machines, *Proceedings 3rd International Machining and Grinding Conference*, Cincinnati, Ohio.

Ardelt, T., 2000, Einfluss der relativ bewegung auf den prozess und das Arbeitsergebnis beim Planschleifen mit Planatenkinematik, PhD thesis, TU, Berlin.

Blanchard, 2000, *Series of Trade Brochures on Models 11A-20–42HD/54HD/60HD Grinders*. Cone-Blanchard Corp, Windsor, VT.

Blohm, n.d., *Super Abrasive Machining Center*. United Grinding Technology, Miamisburg, Ohio. Trade Catalog.

Burrows, J.M., Soo, S.L., Ng, E.G., Dewes, R.C., Aspinwall, D.K., 2002, Point grinding of superalloys. *Industrial Diamond Review*, June.

Buthe, B., Wilson, E. M., Boemke, M., 1992, Double disc CBN grinding of piston rings. *IDR*, 52(3), 120–124.

CIRP, 2005, *Dictionary of Production Engineering II, Material Removal Processes*. Springer-Verlag, Germany.

Cui, C., 1995, Experimental investigation of thermo-fluids in the grinding zone. Ph.D. dissertation, Univ. Connecticut.

Daisho, n.d. *Daisho Double Disc Grinders DDG Series*. Daisho Seiki Corp, Osaka, Japan. Trade brochure.

Daisho, 1998, *Grind Master V Series*. Daisho Seiki Corp, Osaka, Japan. Trade brochure.

Daisho, 2001, Trade advertisement.

DCM Tech, 1999, Industrial Rotary Surface Grinders IG 180M, IG 280M, IG 280 CNC, HB5400 Series. Series of trade brochures.

Delta, 1997, *Surface Grinding Machines*. Delta s.p.a. Cura Carpignano, Italy trade brochure.

Deming, M., Carius, A. C., 1991, Introduction to Superabrasives. Course presented at Superabrasives '91, Chicago, IL. SME.

DIN 8589, Teil 11, Entwurf (05.2002): Fertigungsverfahren Spanen; Schleifen mit rotierendem Werkzeug; Einordnung, Unterteilung, Begriffe. Berlin, Beuth.

Diskus, 1996, *Diskus Double Disk Face Grinding Machines Series DDS*. Diskus Werke Schleiftechnik, Langen, Germany. Trade brochure.

Doubman, J. R., Cox, C., 1997, Double disc grinding in automotive parts manufacturing. *Automotive Manufacturing & Production*, October, 60–62.

Ex-Cell-O. n.d. *Grinding Centers XT820/XT285 for Guide Vanes And Turbine Blades*. Ex-Cell-O, Eislingen Fils, Germany. Trade brochure.

Gaulin, D., 2002, High speed grinding of difficult to grind materials. *IMTS SME Conference Proceedings*, Chicago.

Goekel, 1997, *Schleifmaschinen*. Goekel Maschinienfabrik GmbH, Darmstadt, Germany. Trade brochure.

Hitchiner, M., Willey, B, Ardelt, T., 2001, Developments in flat grinding with superabrasives. Precision grinding & finishing in the global economy. 2001 *Conference Proceedings*, Gorham, Oak Brook, IL.

Hodge, J. H., 1992, Lapping, honing and polishing. Engineered Materials Handbook, vol. 4. *Ceramics and Glasses*.

Hogan, B. J., 2001, Precision makes grinding vital. *Manufacturing Engineering*, February. 42–54.

Howes, T., 1990, Assessment of the cooling and lubrication properties of grinding fluids. *Annals of the CIRP*, 39(1), 313–316.

Howes, T., 1991, Avoiding thermal damage in grinding. *AES Conference*. Sourced from Internet: http://www.nauticom. net/www/grind/therm.htm.

Hughes, F., Dean, A., n.d. *Interrupted Cutting of Tungsten Carbide with Diamond Abrasive Grinding Wheels*. de Beers IDR Publication L16.

Inasaki, I., 1988, Speed stroke grinding of advanced ceramics. *Annals of the CIRP*, 37(1), 299–302.

Inasaki, I., 1999, Surface grinding machine with a linear motor driven table system. Development and performance test. *Annals of the CIRP* 48(1), 243–246.

Konig, W., Schleich, H., 1982, Deep grinding of high speed tool steel with CBN. In P. Daniel (Ed.), *Ultrahard Materials. Application Technology, Volume 1*. De Beers Industrial Review.

Koyo, n.d. *Vertical Spindle Surface Grinders KVD Series*. Koyo Machine Industries, Osaka, Japan. Trade brochure.

Koyo, 1999, *Koyo Grinders*. Koyo Machine Industries, Osaka, Japan. Trade brochure.

Koyo, 2000, *Koyo R Series Vertical Spindle Type, Rotary Table Type Surface Grinder*. Koyo Machine Industries, Osaka, Japan. Trade brochure.

Kubsh, L. M., 1988, Disc grinding: Yesterday, today and tomorrow. *MR88-593, 3rd International Grinding Conference*, Fontana, WI.

Landis Gardner, 1989, *Horizontal Disc Grinding Machines and Systems*. Litton Industrial Automation, S Beloit, IL. Trade brochure.

Lapmaster, 2000, *Model LFG 12 Dual Face Fine Grinding Machine*. Lapmaster International, Morton Grove.

Lin, B., Zhang, H. L., 2001, Theoretical analysis of temperature field in surface grinding with cup wheel. *Key Engineering Materials*, 202–203, 93–98.

Litton, 1989, *2V18 ® Vertical Disc Grinders*. Litton Industrial Automation, S Beloit, IL. Trade brochure.

Malkin, S., 1989, *Grinding Technology*. Ellis Horwood Ltd., New York.

Mattison, n.d. *Horizontal Surface Grinders*. Mattison Machine Works, Rockford, IL. Trade brochure.

Mattison, 2. n.d. *Double Disc Grinders*. Mattison Machine Works, Rockford, IL. Trade brochure.

Melchiorre, n.d. *Melchiorre Fine Grinding, Lapping, Honing, Polishing*. Melchiorre S R L, Milano, Italy. Trade brochure.

Minke, E., Tawakoli, T., 1991, Hochleistungsschleifen mit CBN-Werkzeugen. In *Wissenschaftliche Zeitschrift der Technischen Universität Otto von Guericke*. Band, Magdeburg, 35(4).

Modler, n.d. *Finimat 2000 Automatic Universal Finegrinder*. Johann Modler GmbH, Aschaffenburg, Germany. Trade brochure.

Mohr, H., 2000, Abgrenzung verschiedener flachschleifverfahren mit ergebnissen beim vollschnitt schleifen. Moderne Schleiftechnologie. Seminar April 13. Schwenningen, Germany.

Motion, 2001, Speed Stroke Grinding Machine Jung S320. August.

Narbut, N., Stafford, T., Tartaglione, J., 1997, Grinding with segments. *Cutting Tool Engineering*, December, 20–30.

Navarre, N. P., 1986, *CBN and Vitrified Bond: A New Focus*. Machine and Tool Blue Book, October.

Noichl, H., 2000, CBN grinding of nickel alloys in the aerospace industry. *IDA Conference Proceedings, Intertech 2000*, Vancouver, Canada. July 21.

Ott, H. W., Storr, M., 2001, Grinding fluids for the future. Oel-Held GmbH Stuttgart, Germany. Trade technical booklet. See also Ott and Storr. IMTS 2002 paper of same title. September 4, 2002. *SME Conference Proceedings*. Chicago.

Radiac, 2002, Radiac Abrasives Creep Feed Grinding. [Technical data sheets]. Salem, IL [cited 29 March 2002]. Available at http://www.radiac.com.

Reform, n.d., Schleifmaschinen. Reform Maschinenfabrik Adolf Rabenseifner, Fulda, Germany. Trade brochure

Rolls, Royce, 1999, Patent EP 0 924 026 A2.

Rowe, W. B., 2014, *Principles of Modern Grinding Technology*, 2nd edition. Elsevier, Oxford, UK and worldwide.

Rowe, W. B., in, T., 2001, Temperatures in high-efficiency deep grinding. *Annals of the CIRP*, 50(1), 205–208.

Saljé, E., Damlos, H.-H., 1983, Schleifscheibenverschleiß beim Profilplanschleifen—Vergleichende Untersuchungen für das Tief- und das Pendelschleifen. *VDI Zeitschrift* 125, 10.

Schibisch, D., 1997, Fine grinding with superabrasives. *Ceramics Industry*, December.

Schibisch, D., 1998, Fine grinding with CBN and diamond. An economic alternative to free abrasive lapping. *Abrasives Mag.*, February/March.

Schleifring, K., Jung GmbH G ppingen, Trade publication, Germany.

Schleich, H., 1980, Flachschleifen mit hohen Leistungen unter Verwendung thermischer Randzonenschädigung. *Technische Mitteilungen, <Haus der Technik e.V.>* 73, 11/12.

Schlie, D. R., Rangarajan, R. S., 1987, Solving manufacturing problems of today & tomorrow through disc grinding technology. *25th AES Conference*, October 19, Canton, Ohio.

Sess, M., 1999, Machine tools are doing more. *Manufacturing Engineering*, November, 46–53.

Simpfendörfer, D., 1988, Entwicklung und Verifizierung eines Prozeßmodells beim Planläppen. PhD dissertation, Technical University Berlin.

Spur, G., 1989, *Keramikbearbeitung – Schleifen, Honen, Läppen, Abtragen*. Carl Hanser, Verlag München, Wien.

Spur, G., Eichhorn, H., 1997, Kinematisches Simulationsmodell des Läppscheibenverschleißes. *IDR* 31(2), 169–178.

Spur, G., Funck. A., Engel, H., 1995, Problems of flatness in plane surface grinding. *Trans NAMRI/SME*, XXIII, 97–102.

Spur, G., Stöferle, T., 1980, *Handbuch der Fertigungstechnik*. Spanen, Band 3/2. Hanser-Verlag, München.

Spur, G., Uhlmann, E., Brücher, T. 1993, Werkstoffspezifische Schleiftechnologie—Schlüssel für erhöhte Prozesslauffähigkeit in der Keramikbearbeitung. In *Jahrbuch Schleifen, Honen, Läppen und Polieren*, Verfahren und Maschinen; 57. Ausgabe, Hrsg.: E. Saljé. Essen, Vulkan-Verlag.

Stahli, 1998, *Stahli Flat honing—The Range of Two-Wheel Flat honing and Lapping machines for Top Quality and High Output*. A. W. Stahli Ltd., Biel, Switzerland.

Stahli, n.d., *Lapping and Flat honing with Two-Wheel Machines*. A. W. Stahli Ltd, Biel, Switzerland.

Stahli, A. W., 2000, Flat honing with diamond and CBN grinding discs. *IDR*, I, 9–13.

Supfina, 1996, *SUPERFINISH for the Engine Industry*. Superfina Grieshaber GmbH, Germany.

Swisher, 2000, *Grinding Technology for the 21st Century*. Swisher Finishing Systems, Div. Crankshaft Machine Group, Jackson, MI. Trade brochure.

Tawakoli, T., 1993, Anforderungen an Kuhlschmierstoffanlagen bein Hochleistungsschleifen. In. Diamantanten, Rundschau, 1, 34.

Thielenhaus, n.d., *Microfinish® the Economical Precision-Machining Method in the Automotive and Motor Industry*. Thielenhaus Microfinish Corp, Novi, MI. Trade brochure.

Tönshoff, H. K., Meyer, T., Wobker, H. G., 1996, Machining advanced ceramics with speed stroke grinding. *Ceramics Industry*. July, 17–21.

Tropel, 1996, *Tropel Build on Our Experience*. Tropel Corp, Fairport, NY. Trade brochure.

Ueda, S., Takahashi, M., Nakagawa, T., Inagaki N., Yamada, T., 1998, Development of a high precision through-feed grinder. *International Journal of the Japan Society for Precision Engineering*, 32(1), 9–12.

Uhlmann, E. G., 1994a, *Tiefschleifen hochfester keramischer Werkstoffe*, Produktionstechnik. Berlin, Forschungsberichte für die Praxis, Band 129, Carl Hanser Verlag München, Wien.

Uhlmann, E., 1994b, Erhöhung der Bauteilqualität und der Wirtschaftlichkeit durch modifiziertes Tiefschleifen. In *Schleifen von Hochleistungskeramik*. Ergebnispräsentation des BMFT-Verbundprojektes, Univ. Kaiserslautern, Kaiserslautern.

Uhlmann, E., Ardelt, T., 1999, Influence of Kinematics on the face grinding process on lapping machines. *Annals of the CIRP*, 48(1), 281–284.

VDI 3390 10.1991. *Tiefschleifen von metallischen Werkstoffen*. Beuth, Berlin.

von Mackensen, V. et al. 1997, Fine grinding with diamond and CBN. *IDR*, 2, 40–43.

Webster, J., 2000, Effective coolant application in grinding. *IMTS 2000 Conference Proceedings*, Chicago.

Webster, J., Brinksmeier, E., Heinzel, C., Wittmann, M., Thoens, K., 2002, Assessment of Coolant Supply Effectiveness in Continuous-Dress Creep Feed Grinding. Presented to CIRP August.

Webster, J., Cui, C., Mindek, R. B. 1995, Grinding fluid application system design. *Annals of the CIRP*, 44(1), 333–338.

Werner, P. G., Minke, E., 1981, Technologische Merkmale des Tiefschleifens. Erhöhte Schnittkräfte und reduzierte Werkstücktemperaturen –Teil 1." *TZ für praktische Metallbearbeitung*, 75(3).

Wolters, 2000, *Peter Wolters Solutions in Precision Machining*. Peter Wolters Werkzeugmaschinen GmbH, Rendsburg, Germany. Trade brochure.

Wolters 1., n.d, *Microline Perfect Fine Finishing with Electronic-Pneumatic Control*. Peter Wolters Werkzeugmaschinen GmbH, Rendsburg, Germany. Trade brochure.

Wolters 2, n.d, *Peter Wolters Fine Grinding Systems*. Peter Wolters of America, Plainville, MA. Trade brochure.

Ye, N. E., Pearce, T. R. A. 1984, A comparison of oil and water as grinding fluids in the creep feed grinding process. *Proceedings of the Institution of Mechanical Engineers*, 195B, 229–237.

Yuji, N., 1990, High Speed Stroke Grinding. *Tsuru Enjinia*, 31, 96–101.

17

External Cylindrical Grinding

17.1 The Basic Process

17.1.1 Introduction

External cylindrical grinding refers to the grinding of an outer surface of a workpiece around an axis of rotation with the part held between centers. The surface may be a simple straight diameter or stepped, tapered, threaded, or profiled (Marinescu et al. 2004). The grinder may be configured for

1. Simple plunge operation for straight diameters and complex profiles.
2. Angle approach (typically 30° or 45°) for grinding diameters and shoulders without burn.
3. Universal grinders with several spindles on a rotating turret arrangement or drop down auxiliary spindle. Universal grinders are designed to allow use in combination with internal and face-grinding operations.
4. Combination of plunge and traverse operations. On modern computer numerical control (CNC) grinders, these are often canned operations and may be performed with straight, angle approach, or special profile wheels. The process has gained special appeal for high-speed grinding with cubic boron nitride (CBN) abrasives, although it is not limited to these conditions (Figure 17.1).

17.1.2 Work Drives

In all cases, the part is driven by a work drive. Short parts may be clamped at one end by a chuck or collet. More typically, the parts are held between live or fixed centers and driven by a dog, keyway, or face plate with a locator pin. Longer parts are supported along their lengths by steady rests. Centers are premachined in the part, and their location and quality are critical since runout causes chatter and leads to lobing. Burrs in keyways, used commonly in camshafts, create major problems, where orientation is required (Figure 17.2).

The workhead design varies greatly from one machine tool builder to another depending on load and sophistication of the equipment. On most standard machines, however, they are AC servo driven at speeds from 10 rpm to 400–1000 rpm and use hydrodynamic or roller bearing designs. Runout is always held to better than 1.2 μm and in ultraprecision grinders can be held to as tight as 0.1 μm. AC servo drives allow orientation to be monitored and controlled to seconds which in turn has allowed original equipment manufacturers (OEMs) over the last 25 years to develop CNC grinding of round parts such as punches and camshafts. (This will be discussed in more detail below.)

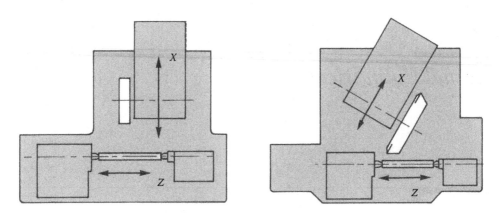

FIGURE 17.1
Machine configurations. (From Shigiya, 1996, CNC cylindrical grinders GPS-30. Shigiya Machinery Works, Hiroshima, Trade brochure. With permission.)

FIGURE 17.2
External grinding between centers with dog drive and diameter gauging. (From Shigiya, 1996, CNC cylindrical grinders GPS-30. Shigiya Machinery Works, Hiroshima, Trade brochure. With permission.)

17.1.3 The Tailstock

The tailstock is typically a plain sleeve bearing, although linear ball bearings are also used. Most have taper adjustment capability and hydraulic actuation for automatic load/unload capability. Some tailstocks have spring loading to compensate for temperature changes, while others have coolant ported through them (and the headstocks) for temperature stability.

17.1.4 Wheel Speeds

The vast majority of standard grinders still use conventional abrasives at wheel speeds of 43 m/s or less. The wheel life is relatively long, so any benefits from upgrading to CBN have to come from improved cycle time or quality. However, an analysis of about 30 different cylindrical grinders currently on the market reveals the following (Table 17.1) power capability per unit width of wheel as a function of wheel size and speed.

Some of the power is required to overcome the inertia of the wheel and flange assembly. So assuming an average power available given by median value for each speed and

TABLE 17.1

Wheel Speed, Wheel Diameter, and Power Required per Unit Width

Wheel Speed	350 mm	450 mm	600 mm	Q' max
33 m/s	30 W/mm	60 W/mm	90 W/mm	1.5–4
43 m/s	50 W/mm	100 W/mm	150 W/mm	3–9
60 m/s	100 W/mm	150 W/mm	200 W/mm	4–12

assuming typical wheel widths between one-third and the maximum allowed, the maximum stock removal rates Q' (mm³/mms) are also estimated based on hardened steel. These numbers are limited by the system stiffness and cannot be increased to justify CBN. The only exceptions therefore are grinding very difficult materials such as tool steels and Inconel, where there is a loss of cycle time from dressing, or for very narrow profiles where higher removal rates still remain within the machine stiffness parameters. One area vitrified CBN is used successfully, for example, at speeds of 60 m/s or less, is groove grinding in transmission shafts.

Since the system is weak, work/wheel speed ratios are comparable to those for internal grinding; 1:50 for the weakest grinding processes and up to 1:150 for stiffer grinding processes. However, the equivalent wheel diameter d_e always remains under 250 mm and more typically is under 75 mm. Coolant access is therefore relatively easy and wheel grades are more a function of material and finish requirement and readily available in wheel makers specification manuals.

Nevertheless d_e does have an impact once standard speeds are exceeded. A standard speed for alox wheels is 43 m/s and this is not only for safety reasons. At speeds up to this value, standard wheel grades can grind parts with the full range of d_e values using the spindle power provided. However, once 43 m/s is exceeded, the burn level is rapidly exceeded for all but the smallest d_e value such that by 60 m/s, the wheels are limited to <25 mm and burn insensitive workpieces, unless the wheel grade is reduced drastically.

Similar effects are seen with vitrified CBN as will be discussed below. The high-thermal diffusivity of CBN shifts the wheel speed limit. For example, grinding hardened steel in water, the value for CBN is shifted from 43 to 80 m/s. Essentially, machines designed specifically for Alox or for CBN abrasives require two completely different wheel speed ranges.

17.1.5 Stock Removal

Maximum stock removal rates require wheel speed, workspeed, and infeed speed to lie within an optimum range. This is illustrated by Figure 17.3 showing power and process limits as found experimentally by Rowe et al. (1986).

Part tolerance capability in terms of incoming stock is better than for internal grinding. The stock removed on diameter should be at least three times the roundness specification, six times the straightness specification, and overall twice as much stock on diameter as the total geometry tolerances.

Flat form traverse dressing of alox wheels is usually with a single point diamond or blade tool mounted on the foot or headstock. Form dressing is again either by interpolation with a stationary or rotary diamond or by plunge roll dressing. The former offers increased flexibility but adds significantly to cycle time. The most accurate forms,

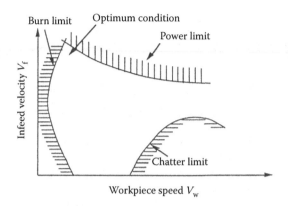

FIGURE 17.3
Power and process limits in grinding. (Adapted from Rowe, W. B., Bell, B., Brough, D., 1986, *Annals of the CIRP*, 35(1), 235–238.)

particularly in high-production applications such as the bearing industry, for example, twin raceway grinding where truth of radii tolerances are <1 μm, usually require formed diamond rolls.

17.1.6 Angle-Approach Grinding

Angle-approach grinding adds complexity to the grinding process. Its benefits are that it allows grinding with the periphery of the wheel not the side, and it grinds two surfaces at once. Its drawbacks are nonuniform wheel wear and temperature, and the fact that dressing is now necessary in two planes. If acoustic emission dress detection is used, it must make two independent contacts on orthogonal axes.

The effect of an angle approach is to reduce the equivalent wheel diameter on the face from total conformance to a finite number while making only a small impact on the d_e for the cylindrical surface. Malkin (1989) presents the following formulae for calculating d_e for face and outer diameter, where α = angle of approach

$$d_e(\text{face}) = (\text{Wheel diameter})/\sin\alpha \qquad (17.1)$$

$$d_e(\text{o.d.}) = (\text{Wheel diameter})/[(\text{Wheel diameter}/\text{Part diameter}) + \cos\alpha] \qquad (17.2)$$

Example: Wheel diameter = 450 mm. Part diameter = 25 mm.

	$\alpha = 0°$	$\alpha = 30°$	$\alpha = 45°$
d_e (face)	∞	900 mm	636 mm
d_e (o.d.)	23.7 mm	23.9 mm	24.1 mm

Angle approach requires conventional wheels to be run at no more than 45 m/s and Q' values <1 when removing stock from the face. Much more attention must also be paid to coolant delivery.

17.1.7 Combined Infeed with Traverse

The final method of grinding is a combination of infeed with traverse. This is a typical operation for long shafts and cylinders. An extreme form is roll grinding for the paper and steel industries. The nature of the wheel wear and the actual Q' values become less well defined, as the wheel is cutting on its leading edge. Since breakdown will be rapid with an alox wheel, infeed amounts are kept at <15 μm on diameter and the wheel cross fed up to one-third (cross feed one direction only) or one-fourth (cross feed both directions) of its width per revolution of the workpiece. The cutting action is distributed over the feed width with two overlaps at size for finish.

With modern CNC equipment, all the operations described above can be incorporated into grinding many surfaces in a single chucking as illustrated in Figure 17.4 (Okuma n.d.).

17.2 High-Speed Grinding

17.2.1 Introduction

The last 25 years have seen a dramatic growth in the use of vitrified CBN at wheel speeds of 80 m/s up to as high as 200 m/s in certain key cylindrical applications. Three such applications, camshaft grinding, crankshaft grinding, and peel grinding, are discussed to illustrate what the technology is and where it is likely to expand. In light of the limitations apparent when grinding with alox wheels at high speed, it is necessary to first review the current theories on heat generation and burn in grinding.

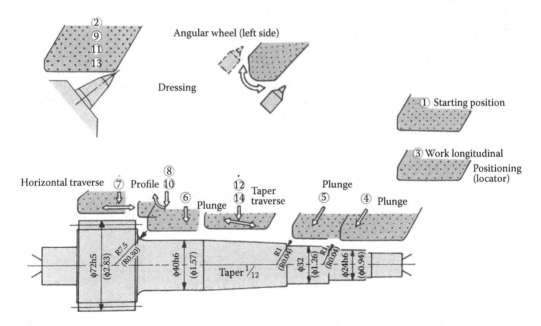

FIGURE 17.4
Combined infeed and traverse for face and diameter grinding. (From Okuma, n.d., GP-24N CNC plain cylindrical grinder, GA-24N CNC Angle Head Cylindrical Grinder. Okuma Machinery Works, Japan. Trade brochure. With permission.)

17.2.2 Energy and Temperature in High-Speed Grinding

A starting point is the common definition of uncut chip thickness:

$$h_{cu} = \sqrt{\frac{v_w}{v_s} \cdot \frac{1}{C \cdot r} \sqrt{\frac{a_e}{d_e}}} \quad \text{where } h_{cu} \ll a_e \tag{17.3}$$

where v_s is the wheel speed, v_w is the workspeed, a_e is the depth of cut, d_e is the equivalent wheel diameter, C is the active grit density, and r is the grit cutting point shape factor.

C and r are characteristics of the wheel and how it was dressed. They are usually treated as a single (Cr) factor. Uncut chip thickness is a reasonably reliable factor for predicting several grinding characteristics especially specific grinding energy e_c.

$$e_c \propto \frac{1}{h_{cu}^n} \propto \sqrt{\frac{v_s}{v_w} \cdot C \cdot r \cdot \sqrt{\frac{d_e}{a_e}}} \tag{17.4}$$

Since $C \cdot r$ is normally a constant for a given wheel/dress set of conditions, some OEMs will use this factor to characterize a given wheel or compare performance from one wheel grade or type to another. $C \cdot r$ for a plated CBN, for example, is only 20% that of a typical vitrified CBN wheel. Similar predictions can be made for surface roughness Rz and force/grit f_g'.

$$R_t \propto \frac{h_{cu}^{4/3}}{a_e^{1/3}} \quad \text{(Surface roughness)} \tag{17.5}$$

$$f_g \propto h_{cu}^{1.7} \quad \text{(Force per grit)} \tag{17.6}$$

Predicting power requirement as a function of grinding parameters especially wheel speed is the first step. Power can be predicted if typical values of specific energy for the particular wheel/material/setup are known. The second and more important step is to determine where the heat goes and what temperature will be created.

There is an enormous amount of experimental work on heat and temperature models in grinding as introduced in Chapter 2. Much of this work has been refined and the important factors isolated (Rowe et al. 1996; Rowe and Jin 2001; Marinescu et al. 2004; Rowe 2014). The maximum surface temperature depends on the grinding power per unit width ($F_t' \cdot v_s$), the grinding speeds, and material parameters. A maximum grinding temperature when grinding is given by:

$$T_{max} = C_{max} \cdot R_w \cdot \frac{F_t' \cdot v_s}{\beta_w} \cdot \sqrt{\frac{1}{v_w \cdot l_c}} \tag{17.7}$$

where the thermal parameters in the equation that affect grinding temperature are described as follows.

The C_{max} factor. The value is approximately equal to 1 for conventional grinding. The value is reduced for deep grinding and at high workspeeds, that is, high values of Peclet number,

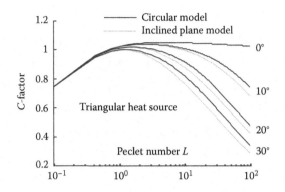

FIGURE 17.5
Factor C for max temperature. (Adapted from Rowe, W. B., Jin, T. J., 2001, *Annals of the CIRP*, 50(1), 205–208.)

L. Rowe and Jin (2001) give charts of C values for maximum temperature in Figure 17.5 and for finish surface temperature in Figure 17.6.

Peclet number L and workspeed. The effect of workspeed is defined by Peclet number L. Values greater than 5 represent reasonably high workspeed, although much higher values can be achieved and allow cool grinding. Peclet number L is given by:

$$L = \frac{v_w \cdot l_c \rho \cdot c}{4 \cdot k} \tag{17.8}$$

where k is the thermal conductivity, ρ is the density, and c is the specific heat capacity of the work material.

Contact angle ϕ. A large depth of cut and a small wheel diameter lead to a large contact angle $\phi = l_c/d_e$. The contact angle, the contact surface, and the finish surface are illustrated in Figure 17.7. The finish surface is the surface remaining after the material has been removed by grinding. At large depths of cut, temperature rise on the finish surface is lower than the maximum temperature rise on the contact surface. This feature is used to advantage in creep feed grinding and in high-efficiency deep grinding (HEDG).

The transient thermal property β_w of the workpiece material is given by $\beta_w = \sqrt{k \cdot \rho \cdot c}$.

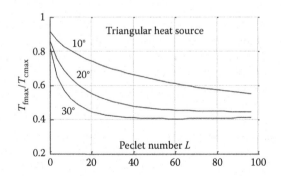

FIGURE 17.6
Maximum temperatures on the finish surface T_{fmax}, as a fraction of the contact temperature T_{cmax}. (Adapted from Rowe, W. B., Jin, T. J., 2001, *Annals of the CIRP*, 50(1), 205–208.)

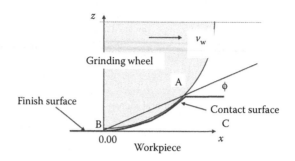

FIGURE 17.7
Contact angle ϕ, contact surface and finish surface.

Workpiece heat partition ratio. R_w is the proportion of the grinding energy that is conducted into the workpiece. R_w is a function of the wheel grain conductivity and sharpness and of the transient thermal property. Ignoring for the present, coolant convection and convection by the grinding chips, R_w approximates to R_{ws}. Hahn (1962) showed that

$$R_{ws} = \left(1 + \frac{k_g}{\beta_w \cdot \sqrt{r_0 \cdot v_s}}\right)^{-1}$$ (17.9)

where k_g is the thermal conductivity of the abrasive grain. Grain sharpness is related to r_0, the contact radius of the grain. R_{ws} is relatively insensitive to variations of r_0. Typically, R_{ws} for conventional grinding varies between 0.7 and 0.9 for vitrified wheels and between 0.4 and 0.6 for CBN wheels. After allowing for heat convected by the coolant and by the chips, R_w can be greatly reduced below R_{ws}.

The temperature equation for conventional shallow grinding ignoring heat taken by the coolant and chips can therefore be approximated for a given wheel/work/machine configuration to:

$$T_{max} \propto \sqrt{a_e \cdot v_s \cdot C \cdot r} \quad \text{Ignoring heat taken by coolant and chips}$$ (17.10)

In this case, it follows that increasing wheel speed, increasing depth of cut, or increasing the number of active cutting edges (by, e.g., dull dressing) increases the surface temperature. However, taking account of heat convected by the coolant and by the chips as follows shows that much lower temperature can be achieved than would be expected.

Heat convection by coolant and chips. In deep grinding, the long contact length allows substantial convective cooling from the grinding coolant. Also in high-rate grinding with low specific energy, the heat taken away by the grinding chips reduces maximum temperature very substantially (Rowe and Jin 2001). One of the advantages of high-rate grinding is that the specific energy is reduced.

Allowance can be made for convective cooling by subtracting the heat taken away by the coolant and chips as described by Rowe and Jin (2001). Allowance for convective cooling is essential for creep grinding as shown by Andrew et al. (1985). It has also been found important for other HEDG processes as employed for drill flute grinding, crankshaft grinding, and cutoff grinding. If allowance is not made for convective cooling the temperatures are greatly overestimated.

The maximum temperature equation modified to allow for convective cooling in the simplest form (Rowe 2014) is:

$$T_{max} = C_{max} \cdot R_{ws} \cdot \frac{F_t' v_s - \rho \cdot c \cdot T_{mp} \cdot a_e \cdot v_w}{\beta_w \sqrt{v_w l_c} + \frac{2}{3} h_f \cdot l_c} \tag{17.11}$$

where T_{mp} is a temperature approaching the melting point of the workpiece material. For steels, the material is very soft at 1400°C, and this temperature gives a reasonable estimate for the chip convection term.

The coolant convection coefficient is h_f, where traditionally coolant convection is defined proportional to mean surface temperature. In grinding, interest is focused on maximum temperature introducing a factor of approximately two-thirds. Fluid convection applies as long as the maximum temperature does not cause the fluid to burn out in the grinding zone. If burnout occurs, the convection coefficient is assumed to be zero. Burnout is a common condition in grinding but should be avoided in creep grinding and for low-stress grinding. Values estimated for convection coefficient when grinding with efficient fluid delivery were 290,000 W/m²K for emulsions and 23,000 W/m²K for oil (Rowe and Jin 2001). Applying a factor of two-thirds for maximum temperature rather than mean temperature, yields 174,000 W/mK for emulsions and 13,800 for oil (Rowe 2014). Simple models for fluid convection are given in Chapter 10 Coolants and further experimental data.

The contact conditions between the abrasive grain and the workpiece are very different from each other. When an abrasive grain slides against the workpiece, a particular point on the workpiece only contacts the abrasive grain for an extremely short period of time of the order of 1 µs. However, the grain is sliding on the workpiece for a very much longer time, typically 1000 times longer and of the order of a millisecond.

A key factor for the workpiece, therefore, is transient thermal property β. This term involves both thermal conductivity and heat capacity. Transient thermal property is defined as $\beta = \sqrt{k \cdot \rho \cdot c}$ and is particularly important for transient heat conduction into the workpiece with moving heat sources.

However, the abrasive grain very rapidly achieves its maximum surface temperature, so that the transient thermal property is unimportant. For the abrasive grain, it is thermal conductivity that is important. Pure CBN is known to conduct heat up to 40 times faster than alox. Typically, the value of thermal conductivity for CBN is 5–20 times higher than for alox depending on the purity of the CBN. The thermal conductivity of CBN abrasive relative to alox actually lowers temperature by almost 50% on steels which is highly significant.

Specific grinding energy. e_c is a measure of the input of energy. One factor that can significantly affect its value is coolant. Oil having much better lubrication will tend to lower this value.

The approximate Equation 17.10 for grinding temperature in shallow grinding allows a number of predictions based on change in depth of cut and wheel speed. First, temperature increases with depth of cut. The increase is proportional to the (depth of cut)$^{1/2}$. This is a good predictor of temperature at shallow cuts but it was found about 30 years ago that it failed at very deep depths of cut (Werner 1983). Tawakoli (1993) gives a more accurate trend for temperature in Figure 17.8. Power increases as expected and the approximate shallow grinding prediction of temperature from this is the dotted line.

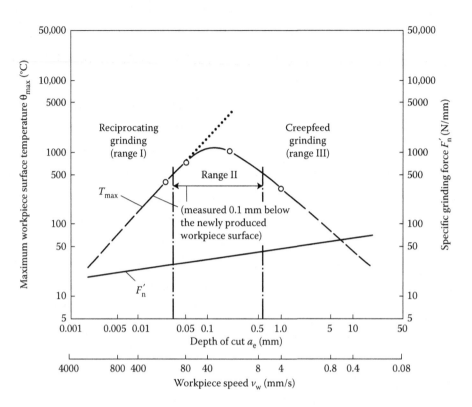

FIGURE 17.8
Predicted maximum temperatures at constant specific removal rate Q'_w. (Adapted from Werner, G., 1983, *Trens-Kompendium*, 2, 448–468.)

However, in zone III—creep feed grinding—it is believed that the contact length becomes so great that the coolant can dissipate over a large enough area for the temperature to fall. It should be noted that cylindrical grinding operations very rarely if ever, get into zone III. A much more interesting fact is that at very high workspeeds, the temperature should drop dramatically.

The second prediction is that temperature rises with wheel speed. This seems very reasonable in the same way that rubbing your hands faster makes them get hotter from frictional heating. This was also the observation in the grinding industry for most of its history. However, starting in the 1970s some curious results were obtained from German research suggesting that under certain circumstances if the wheel speed was increased enough the temperature would actually fall again (Tawakoli 1993). The explanation given by Tawakoli for this relates to the extremely brief length of time the abrasive grits are in contact with the workpiece. The surface is not in thermal equilibrium. The heat pulse initially spreads out over the surface before penetrating into the workpiece. The heated surface facilitates the removal of the next chip so reducing the grinding forces. However, before heat can spread down into the surface, the next chip is removed taking the heat with it. The critical wheel speed where these effects start to be apparent is about 100 m/s (Figure 17.9). When the speed exceeds the critical value, rapidly succeeding chip removals reduce the temperature.

It is still not clear whether these observations are for specific operating parameters or a more general effect. The issue gets clouded by problems such as nonoptimized coolant delivery and increased machine complexity.

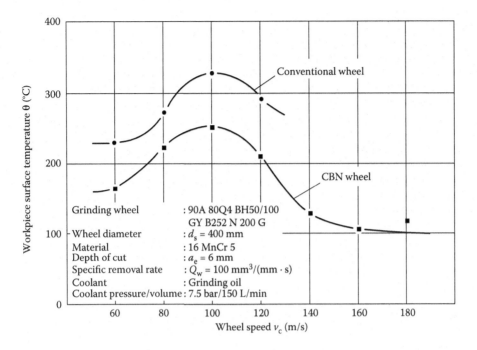

FIGURE 17.9
Effect of wheel speed on maximum temperatures at constant high removal rate and large depth of cut. (Adapted from Tawakoli, T., 1993, *High Efficiency Deep Grinding.* Mechanical Engineering Publications Ltd., London.)

It was shown by Rowe et al. (1986) that most research papers confuse wheel speed effects with chip thickness effects. It is possible to distinguish between chip thickness effects and wheel speed effects, if experiments are conducted in a special way. Most experimental results show the effect of increasing wheel speed with constant workspeed and feed rate. However, this leads to reduced chip thickness so that the two effects are confused. As wheel speed is increased, fracture forces on the wheel grits are reduced, while the amount of rubbing contact is increased. As is very well known, the result is that the grinding wheel is more likely to become dull. This is primarily a chip thickness effect but there is also a wheel speed effect, since each chip is removed in a shorter time leading to a faster chip temperature rise. Conversely, if wheel speed is reduced, the force on the wheel grits is increased due to increased chip thickness and there is a greater tendency for grit fracture. However, since the chip sees a slower temperature rise, there is a speed effect. In summary, there are two separate problems here. The first is to find the optimum chip thickness for a particular abrasive, work material and wheel speed condition. The second problem is to resolve the effect of speed on the removal process. The first problem can be solved holding wheel speed constant while varying v_w and v_f. The second problem can be solved holding chip thickness constant while varying v_s. From Equation 17.3, it can be seen that constant chip thickness requires v_s/v_w to be held constant and also v_w/v_f. Experimental results presented in this way are shown in Figure 17.10. The figure shows that there is an optimum speed for different depths of cut each corresponding to a particular value of v_w/v_f. It becomes clear that specific energy is reduced as depth of cut is increased but also that the optimum wheel speed increases as the depth of cut is increased.

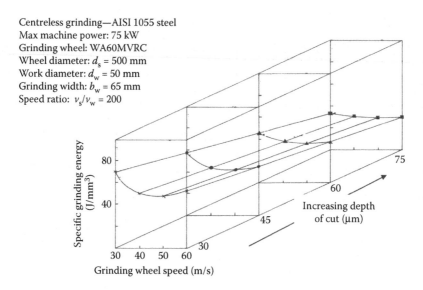

Centreless grinding—AISI 1055 steel
Max machine power: 75 kW
Grinding wheel: WA60MVRC
Wheel diameter: d_s = 500 mm
Work diameter: d_w = 50 mm
Grinding width: b_w = 65 mm
Speed ratio: v_s/v_w = 200

FIGURE 17.10
Wheel speed effect on specific energy at different depths of cut. (From Rowe, W. B., Bell, B., Brough, D., 1986, Optimization studies in high removal-rate centreless grinding. *Annals of the CIRP*, 35(1), 235–238. With permission.)

High productivity requires removal rate to be increased, while quality is maintained and burn is avoided. This is best achieved through low specific energy.

It has been found from experience over the last 10 years as more high-speed applications have been investigated that if an application creates burn at <100 m/s, then increasing wheel speed does not relieve the problem. However, if a process runs at 100 m/s without burn issues then usually it will also run at 160 m/s. Although most of the data are still anecdotal, it does suggest that the general trend in Figure 17.9 may be correct but that the temperature drop may be a lot less pronounced or even just level off after 100 m/s.

It is also possible to predict what type of applications or conditions are likely to be most successful at very high wheel speeds. The following practical recommendations can be made.

1. *Grind-burn insensitive material.* Wheel speed limits when cylindrical grinding with water-based coolant with vitrified CBN wheels are

Chilled cast iron	>160 m/s
Carbon steel 30HrC	>120 m/s
Carbon steel 60HrC	<100 m/s
D2 Steel 62HrC	<90 m/s
Inconel	<70 m/s

2. Use oil-based coolant instead of water.
3. Use electroplated CBN wheels instead of bonded wheels (smaller $C \cdot r$). For life as well as for lubrication, these wheels are best used with oil.
4. Use very high workspeeds (or very deep depths of cut).
5. Use CBN instead of alox (Figure 17.11).

Grind parameters

Machine:	Guhring
Wheel speed:	123 m/s
Component:	Crankshaft
Material:	Gray cast iron
Wheel:	Direct plated CBN
	Twin-wheel gring
Coolant:	Oil

Stock removal rate Q' = 135–166	
Roundness	= 4.5 um
Roundness Ra	= 0.75 um

FIGURE 17.11
High-speed crankshaft grinding. (Data courtesy of SGA Winters. Photo courtesy of GE Superabrasives.)

It was perhaps not surprising that one of the very first cylindrical applications to get the industry attention, grinding compressor cranks, met at least four of five of the optimum conditions. Gray cast iron was very easy to grind, it only proved economic to grind in oil, and used electroplated CBN wheels (Woodside 1988).

The next question that has to be answered when applying high speeds is to define what the benefits would be to justify the added expense. There has to be higher stock removal rates, better quality, or lower capitalization/operating costs.

There is a considerable amount of research that has demonstrated that very high-stock removal rates (Q' > 2000 mm³/mms) can be achieved with high speed. At these removal rates, the rubbing and ploughing fraction of the specific grinding energy is miniscule and the specific energy decays to an asymptotic level of about 10–15 J/mm³ at speeds of the order of 120 m/s and even as low as 7 J/mm³ at a speed of 180 m/s grinding steel (Rowe and Jin 2001). For chilled cast iron, the value is 30 J/mm³ (Figure 17.12; Wakuda et al. 1998). Furthermore, both the tangential and normal forces are reduced with increasing wheel speed (Toenshoff and Falkenberg 1996). Potential benefits are, therefore, higher stocks

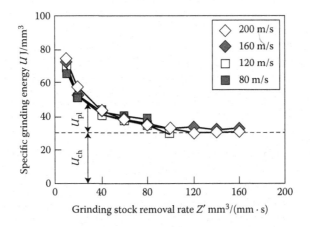

FIGURE 17.12
Effect of removal rate on specific energy. (From Wakuda, M. et al., 1998, *B JSPE*, 64(4), 593–597. With permission.)

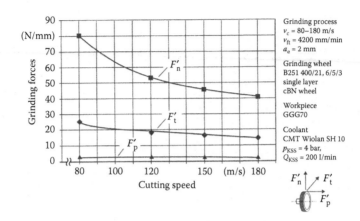

FIGURE 17.13
Effect of wheel speed on grinding forces. (From Toenshoff, H. K., Falkenberg, Y., 1996, High-speed grinding of cast iron crankshafts with CBN tools. *IDR*, 4, 115–119. With permission.)

removal rates and under the correct conditions or grinding configuration, higher energy efficiency, and less pressure on the part (Figures 17.12 and 17.13).

17.2.3 Coolant Drag and Nozzle Design in High-Speed Grinding

A major drawback with higher wheel speeds is the increased demand for nongrind power. This relates in part to the increased frictional drag on the spindle. However, a far more significant factor is coolant drag on the wheel.

The whole issue of coolant at high speed is problematic. Direct matching of coolant flow to wheel speed may require supply pressures over 1000 psi to achieve about 100 m/s. By 160 m/s, the coolant pressure required would be >2500 psi. Pressures over 1000 psi erode vitrified bonds, cannot maintain laminar flow, and create major misting problems. Coolant is therefore often applied at low pressure (60–150 psi typical) using a shoe-type nozzle such that the wheel accelerates the coolant into the grind zone. This creates large drag forces on the grinding wheel. Once in the grind zone the coolant creates enormous hydrodynamic pressure just like a hydrodynamic bearing. These, in turn, create high normal forces that manifest themselves as profile errors when grinding wide parts at speeds as low as 60–80 m/s using vitrified CBN wheels. At 160 m/s, the force exceeds 20 N/mm wheel width using oil (Brinksmeier and Minke 1993). The coolant also creates a great deal of resistance for wheel rotation. The combined effects of windage, spindle bearing, and coolant drag at 160 m/s can be as much as 2 kW/mm wheel width (Koenig et al. 1997). This is also confirmed by field experience (Figures 17.14 and 17.15).

17.2.4 Maximum Removal Rates

Taking both the beneficial and negative impacts of high speed, Table 17.2 gives the maximum stock removal rate as a function of wheel width for a 10 kW spindle.

At 160 m/s, to reap benefits from the speed, the spindle power requirement is a minimum 4 kW/mm of which 50% is actually available for grinding. This has led industry to two options: either use very narrow wheels and get the benefits of low forces (peel grinding) or build high stiffness cylindrical grinders with very large motors (camlobe grinding and crankpin rough grinding).

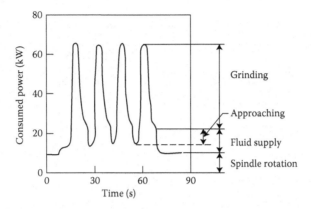

FIGURE 17.14
Power consumption in grinding cast iron camlobes at 200 m/s vitrified CBN wheel. (From Ota, M. et al., 1997, A cam grinding machine using an ultra-high speed and high power grinding wheel spindle. *1st French and German High Speed Machining Conference.* With permission.)

17.2.5 Peel Grinding

Peel grinding is an excellent solution to the above analysis and a testament to German research and application in this field.

The peel grinding process is sometimes viewed as turning with a grinding wheel (Figures 17.16 and 17.17). Figures 17.16 illustrates the geometry for the simplest version of the process with the wheel axis parallel to the part axis. The entire wheel width is no

FIGURE 17.15
Shoe nozzle on Schaudt camlobe grinder for diesel cams. (From UGT, 2000, Dana's long camshafts are no challenge for schaudt grinders. *Grind Journal.* Trade Brochure. With permission.)

TABLE 17.2

Maximum Stock Removal Rates as a Function of Wheel Width

Wheel Width (mm)	Idle/Coolant (kW)	Grind Power (kW)	Q' max Steel Electroplate CBN (mm³/mms)	Q' max Cast Iron Vitrified CBN (mm³/mms)
1	2	8	800	270
2	4	6	500	200
3	6	4	300	130
4	8	2	100	60
4.5	9	1	40	20
4.8	9.6	0.4	10	7
5	10	0	0	0

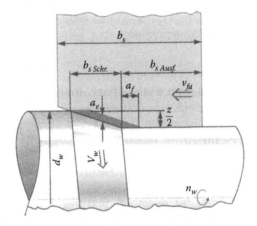

FIGURE 17.16
Wheel profile for parallel contour grinding. (From Lutjens, P., Mushardt, H., 2000, Hard turning or grinding. *European Production Engineering.* July. With permission.)

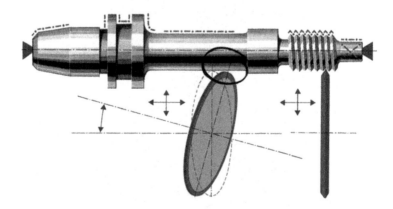

FIGURE 17.17
Principle of the Junker Quickpoint contouring cylindrical grind process. (From Junker, 1992, Junker Quickpoint CNC o.d. grinding with CBN or Diamond. Erwin Junker Maschinenfabrik GmbH, Trade Brochure. With permission.)

more than 7 mm. The operation removes the entire stock amount $z/2$ in a single pass. Roughing occurs over an angled length $b_{s\ Schr.}$ of 2–5 mm with a spark-out region $b_{s\ Ausf.}$. This length can be as little as 2 mm and still hold finish under 2 Rz. This is achieved by using a very high workspeed, up to 10,000 rpm, to obtain a sufficient overlap ratio (Koenig and Treffert n.d.). The high workspeed also meets one of the criteria above for low burn. For the process to be effective, the wheel profile must hold up under high-stock removal rates, $Q' > 80$ mm^3/mms which effectively demands oil coolant and high wheel speeds. Initially metal bond CBN wheels were used but these proved difficult to dress and OEMs have sacrificed some life by going to vitrified CBN for automatic dressing capability.

Numerous OEMs such as Landis, Okuma, Overbeck, Rouland Gendron, Schaudt, Tecchella TMW, and Weldon offer peel grinding in this configuration. Perhaps, the biggest proponent, however, has been Junker using a patented Quickpoint process (Junker 1993). This method greatly reduces the contact length by being able to tilt the wheel at an angle of 1.

Most applications in the field are currently running at wheel speeds of 90–120 m/s. Q' values are of the order of 50–100 mm^3/mms, although accurately defining a value can be difficult without knowing the exact wheel profile. Wheel spindle power ranges from 15 to 30 kW depending on the OEM. In general for a given grinding power, contour grinding will remove material three times faster with less than a sixth the normal force of conventional plunge o.d. grinding.

As reports are now being circulated from end users regards the performance of the process in actual production, the prognosis is very good. It appears to be highly flexible, allowing fast change over times, and generating improved quality from a consistent grind with low forces. Schultz (1999) from DaimlerChrysler reported that in grinding transmission shafts, peel grinding reduced manufacturing costs by 50% with better and more consistent part quality compared to external plunge grinding with alox wheels. Setup time was also reduced by 80% because the same wheel could grind a whole family of parts, while machining time was reduced by 45%. Even abrasive costs were reduced by 5%.

There has been resistance in the U.S. market because of the requirement for oil coolant. However, at least one application has been developed successfully on Weldon grinders in soluble oil grinding Inconel at 120 m/s and achieving $Q' = 120$ mm^3/mms. The wheels have to be made wider, reducing some of the low normal force benefits, to allow more wear back of the roughing step, while still maintaining a spark-out zone. But finish, size and roundness appear very stable and independent on spark-out length until a critical minimum value is reached.

With globalization and commonality of processes by larger end users, it is becoming apparent that oil coolant is starting to gain acceptance again in the United States. Improved coolant handling knowledge from Europe combined with increasing use of powdered metal for automotive components prone to rusting are additional drivers. Additionally, as mentioned above, improved vitrified CBN bond technology is permitting economic viability even in water-based coolants.

17.3 Automotive Camlobe Grinding

17.3.1 Incentives for Change

Camlobe grinding has been the leading driver of high-speed grinding over the last 25 years. It is a high-cost process, making up more than 40% of total camshaft

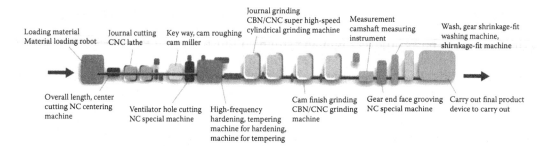

FIGURE 17.18
Flow chart for a modern automotive camshaft manufacturing line. (From UMS, 2002, United Manufacturing Solutions. UMS Ltd., Japan. Trade brochure.)

manufacturing costs. Wedeniwski (1989) and capital equipment intensive resulting in pressures from end users to reduce cycle times, while offering machine tool builders sufficient end sales volume to justify intensive research into innovative processing (Figure 17.18).

17.3.2 Camlobe Profile Features

The finish grinding of the camlobe is one of the most difficult operations in the manufacture of an automotive camshaft. Profiles must be held at micron levels with equally tight timing angle, straightness and finish requirements. The profiles are defined per degree of part rotation and supplied as a lift table with either a length dimensions from the center point or as a deviation from the base circle diameter value. Figure 17.19 provides the terminology for the various features of the camlobe. There are huge changes in equivalent wheel diameters going from the base circle through the ramp into the flank. On occasions the flank may even have a reentry or negative radius leading to near total conformance with the wheel, $d_e \to \infty$. Heat generation and burn are constant concerns at points along the flanks where conformance is greatest and/or the coolant is most masked from the grind zone. Thermal modeling of the process presented by Pflager (2002) illustrates the effect (Figure 17.20).

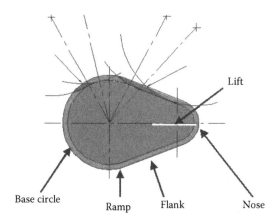

FIGURE 17.19
Features of a camlobe.

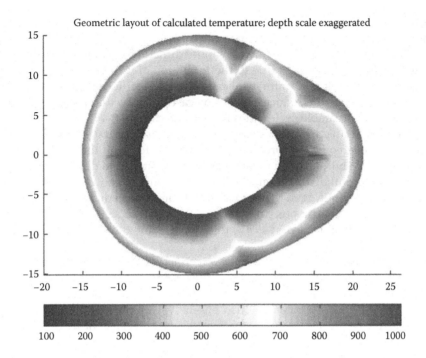

FIGURE 17.20

Temperature distribution with varying contact conformity (depth greatly exaggerated). (From Pflager 2000. With permission.)

17.3.3 Camshaft Grinders Using CBN

Camlobe grinding has been transformed in the last 20 years. Prior to 1980, all cams were ground on mastercam grinders, where the wheel slide movement was controlled by following a rotating template or mastercam. In the late 1970s, AC servo closed-loop systems with CNC controls to the workhead were able to control the rotation while maintaining orientation of the camlobe in synchronization with movement of the wheelhead to CNC generate the profile. It was not long after this that machine tool builders realized that the accuracy required for generating cam profiles were the same as those required to implement CBN wheels. The first camlobe grinder reported in the literature designed specifically for CBN, was manufactured by Toyoda Machine Works (TMW) and used resin-bonded CBN wheels. One such machine, a TMW GCB7-63 was supplied to Caterpillar c1980 to grind hardened steel diesel camshafts (Hanard 1985). The GCB7 had hydrostatic ways and wheel spindle, CNC profile generation, variable workspeed drive to control stock removal around the camlobe, and an acoustic touch sensor to determine the relative position of the wheel and diamond truer. The wheels were 600 mm diameter and, being resin bonded, required a post-true conditioning process using a dressing roll and free abrasive grains. This conditioning process was difficult to control and proved, together with the low resilience of the resin bond, to be the process-limiting factor. The wheel speed was only 50 m/s.

Following the fuel crisis of the 1970s U.S. automotive manufacturers strived to improve gas mileage by improvements in engine efficiency. One area that was targeted was reduction in engine friction by using a roller rather than a flat tappet to transfer the lift from the camlobes to the valves. Rollers reduced friction but generated higher normal forces that were found to exceed the strength of the cast iron materials in use at the time for camshaft fabrication. This in turn forced engine manufacturers to switch to hardened forged steels.

Hardened steel proved to be considerably more difficult to grind than cast iron because it was sensitive to grinding burn due to the formation of untempered martensite, a brittle layer causing spalling and premature cam failure. Steel forgings or assembled cams, despite having only a third the stock of cast iron casting, proved impossible to grind in the hardened state in comparable cycle times to cast iron camshafts.

Two different manufacturing strategies were, therefore, pursued by different engine manufacturers to overcome this problem. Some elected to grind the steel in the soft state, where burn was not an issue, then to heat treat. This route led to diminished dimensional accuracy due to thermal distortion in the heat treatment. By contrast, Ford Motor Co., for example, demanded grinding in the hardened state and proceeded to develop the technology, first by assessing continuous dressing and then by adopting plastic bonded conventional abrasive wheels. By the mid-1980s, this had become the standard processing route for much of the industry.

In the 1980s, increasing Environmental Protection Agency demands for fuel economy combined with improved emissions led engine designers to make radical changes to the camlobe. "Reentry" profiles were designed with concave radii of less than 350 mm in the ramp areas to speed up the rate at which engine valves open and closed. This further increased the risk of thermal damage during grinding because of the increased contact length. It also created a far more fundamental problem: production grinders of the time all used conventional wheels with a usable diameter range of either 750 to 600 mm or 600 to 450 mm; this was totally impractical to meet production for the new camlobe designs.

In 1984, Suzuki described the first results from a CBN camlobe grinder dedicated to grinding with vitrified CBN wheels. In line with camshafts still made in Japan, the data reported were grinding cast iron. The grinder, a TMW GCH 32, is described in detail by Tsujichi (1988). It had a fixed wheel speed of 80 m/s, this being considered the "state of the art" in high speed at that time. Wheels for this machine had been developed by TVMK, a joint venture between TMW and Unicorn Industries (Hitchiner 1991), and were only 350 mm to allow the machine to be designed with a small footprint.

In 1985, Ford completed the design for its 3.8L V6 engine with incorporation of a steel camshaft with the first reentry (concave) ramp profile. Ford Manufacturing faced with the capacity issue of a production line making 750,000 cams/year elected to purchase 16 of the GCH 32 machines and committed themselves to "making them work." This line was the first major installation of its kind in the world and represented the seminal technological and engineering R&D experiment to determine the viability of CBN for cylindrical grinding (Renaud and Hitchiner 1991).

There was no doubt as to the improvement in quality or ability to make capacity. The main problem was striking a balance between wheel cost/part by using hard grades and eliminating burn by going softer. This was a foretaste of the question of the optimum wheel speed for a given abrasive and application. The solution was to lower the wheel speed from 80 to 60 m/s which tripled wheel life.

Figure 17.21 gives the abrasive cost/part for the first 9 years of the machines installation. After process optimization, the abrasive costs were well below those of conventional wheels grinding a nonreentry profile cam. The biggest improvements though were in quality: Some of the achievements listed include

- 1992 zero defect award for zero defects reaching assembly lines.
- Several perfect performance audits.

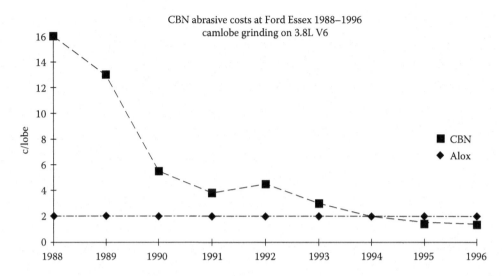

FIGURE 17.21
Reduction of CBN abrasive costs at Ford, Essex, 1988–1996.

- Machine uptime of greater than 95%.
- Only 1 camshaft in over 4 million returned from an engine in the field. This was traced to loss of grinder machine power due to lightning and was subsequently eliminated as a potential process problem.
- One operator for 18 machines (two additional grinders installed in 1992).
- One wheelwright for all wheel changes on a two-shift basis; changes scheduled several days in advance (average 1 wheel/week for the complete line).

Ten years after 60 m/s was established as the optimum speed for grinding hardened steel in water-based coolant, machine tool and wheel technology have improved this to about 80 m/s. But they have also tripled stock removal rates from $Q' = 10$ to >30 mm³/mms.

In the rest of the world, with the exception of diesel truck engines, most of the camshafts have remained cast iron based. Consequently, they saw far fewer problems using CBN. Also Europe uses oil coolant far more frequently. Wheel speeds of 80–100 m/s worked well at Q' values of >50 mm³/mms. In recent years with machine tool improvements, these values have now reached as high as $Q' > 100$ mm³/mms at 160 m/s.

Today there are probably more grinders in India running at 160 m/s grinding scooter, motor cycle, and small car cams than that are in the United States for all grinding; such is the importance of camshaft grinding and material type.

17.3.4 Raising Dynamic Thresholds and Work Removal Rates

Grinding camshafts is in large part a geometry problem or to quote Pflager (2000), "The problem with cams is that they are not round." High-speed grinding with high removal rates requires high workspeeds. The challenge of the machine tool builder is to generate slide movements that can produce these accurately at the required speed.

Machine movements are limited by the dynamics of the machine namely velocity, acceleration, and jerk (rate of change of acceleration) which must be calculated at all points

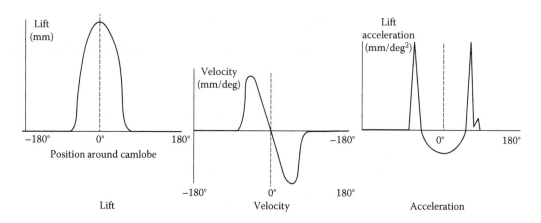

FIGURE 17.22

Lift, velocity, and acceleration profiles for one rotation of the camlobe. (From Landis 1996, Landis windows-based automatic workspeed generation system. Trade Brochure. With permission.)

around the lobe. This must be combined with the ability to take the lift data table at each degree and produce a smooth transition-free surface. The angular resolution has therefore to be much finer and rotational speeds in older machines were limited by control computational speeds (Figure 17.22).

The workspeed rpm on the base circle is constant and defines the stock removal rate for general definition of process capability. As the wheel moves from the base circle through the ramp to the flank area the workspeed must reduce rapidly. Since this time is finite due to the inertia of the system, there is a brief period where the contact length and stock removal rate both jump dramatically leading to a significantly higher Q' value. Although near instantaneous, this can be the life or cycle time limiting factor for the wheel (Figures 17.23 and 17.24).

The earliest machines were limited to about 100 rpm for workspeed based on computer speed. More recently workspeeds have increased to over 300 rpm both by control advances and also by the use of linear motors to reduce the inertia of the wheel slide.

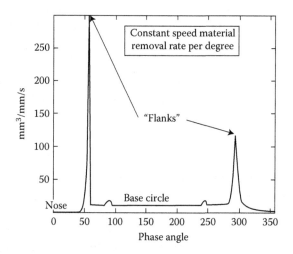

FIGURE 17.23

Q' against phase angle around camlobe at constant work speed.

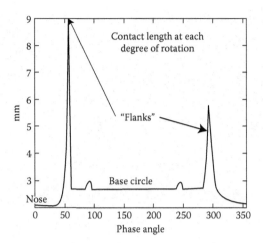

FIGURE 17.24
Contact length versus phase angle around camlobe. (From Pflager, W., 2000, High speed cam grinding. *IMTS 2000 Manufacturing Conf.* SME Chicago, IL. With permission.)

This has allowed a reduction in peak Q' values and thus pushed up base circle stock removal rate capability. Current Q' values are typically 20–40 mm^3/mms for hardened steel and 70–100 mm^3/mms for cast iron.

17.3.5 Improving Productivity within Burn Thresholds

During roughing the workspeeds are usually set at about 20% over the theoretical limit for maintaining the profile within tolerance. This is to maximize stock removal without thermal damage. Finish grind workspeeds are held at 20% below the limit for maintaining profile. The stock amount in finish should be greater than depth at which thermal damage may have been generated during roughing and is typically 50–125 μm.

The sensitivity to burn of the grinding process for hardened steel cams in particular is such that nondestructive inspection methods based on Barkhausen noise have generally been adopted on many production lines (Fix et al. 1990). Barkhausen noise is an inductive method that measures the noise generated by the abrupt movement of magnetic domain walls under the application of an alternating magnetic field. When a coil is placed near the sample, the change in magnetization created by the shift in the domain wall induces an electrical pulse. The sum of all pulses from all domain movements within the sample area provides the final signal or Barkhausen noise amplitude (BNA).

The BNA value is sensitive to several factors including the microstructure of the steel, hardness, and surface finish. However, for a given grade of steel kept within the standard limits of the process, the biggest factor affecting BNA is residual stress. In particular, the relaxation of compressive stress cause by retempering can be detected by an increase in BNA signal. This increase is directly correlated to the severity of the thermal damage so long as the transformation temperature is not exceeded. If the severity of the burn is such that untempered martensite is formed then the signal actually drops. Therefore, the method is primarily for ensuring a process stays in control at a level well below the point that significant softening of the steel occurs.

Figure 17.25 is a photograph of a Rollscan® multihead sensor system from American Stress Technology. The system simultaneously measures all 12 ground lobes of an AISI

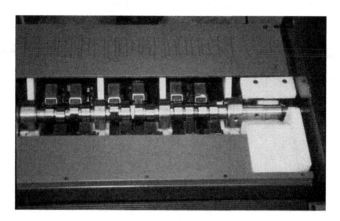

FIGURE 17.25
Rollscan multihead sensor system from American Stress Technology.

1050 steel camshaft (Figure 17.26). The system must be calibrated by using mastercams checked for various levels of burn by Nital etching. The value 50 is nominally set at burn, 45 is the upper process limit, while the process is under control for values less than 40. Figure 17.27 illustrates the change in BNA value on camshafts ground with a vitrified CBN wheel as a function of parts after dress. In this case, at 80 m/s, the chip load was too low and the wheel was glazing leading to increased grinding power. Thermal damage increases was such that the wheel had to be dressed after just 30 parts. The signal can also be plotted as a function of angle around the lobe. Figure 17.28 plots the BNA values around a lobe during optimization studies. The initial cycle 1988 with high levels of BNA shows peaks in the two ramps as predicted from the temperature analysis in the figure. This also correlated with Nital etch checks.

17.3.6 Camshaft Work-Rests and Force Equalizing

Camshafts are inherently weak. They have hollow centers for weight reduction that are preground on the journal surfaces prior to the lobe grind operation. During lobe grinding, the journals are supported by steady rests with flat PCD wear shoes (Figure 17.29).

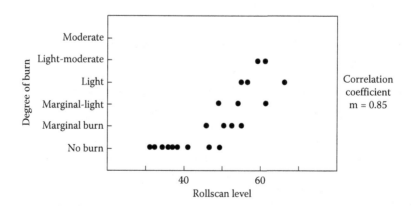

FIGURE 17.26
Calibration of a Rollscan system for detecting onset of burn.

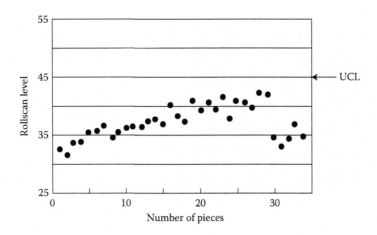

FIGURE 17.27
BNA variation with parts per dress (dress every 30 pieces).

Any chatter that may have been generated grinding the journals will translate directly to the lobe grind. Furthermore, steady rests are relatively expensive and therefore limited to the minimum number to achieve acceptable part quality. Too hard a wheel grade can readily cause chatter and profile errors from push off of the part. Tailstock pressure is applied hydraulically: too little pressure allows the part to bow due to the grinding force while too much will induce bow in the part and create taper. Similar taper problems can be created by incorrect setup of the steady rests. Identification of the root cause is usually achieved by evaluating the change in taper from one lobe to the next (Figure 17.30).

Camshafts can have up to 16 lobes with several in the same rotational phase. This offers the opportunity to grind two lobes at the same time. This concept has been taken further by building twin-wheelhead machines with two independently moving wheel slides. This

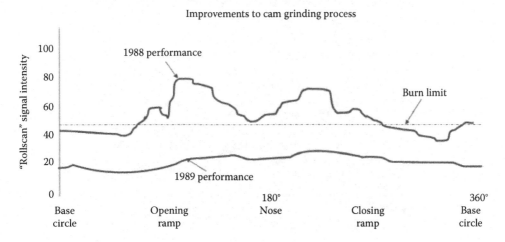

FIGURE 17.28
Rollscan multisensor system and readings.

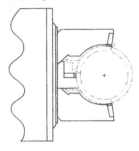

The part centerline is controlled regardless of part diameter

FIGURE 17.29
Aerobotech™ self-centering steady rest system for camshaft grinding. (From Aerobotech 1996, *Parts and Tooling*, June 1996; Anon, 1998b, Steady rest technology hold form in parts. *Tooling & Production*, June. With permission.)

FIGURE 17.30
Landis 3L CNC single-wheel CBN camlobe grinder.

allows two twin-wheel sets with different spacings between the abrasive sections to be used for added flexibility. Up to four lobes with two different spacings can be ground simultaneously (Figure 17.31; Landis 1996).

Reentry profiles for most engines for the U.S. market could be ground using wheels of >300 mm. However, smaller engines, especially for the European market used smaller cams with therefore smaller reentry radii. These required wheels as small as 50 mm (Figure 17.32).

For this type of application, an alternative approach has been required: to rough the cam-lobe using a large plated CBN wheel and to finish and generate the reentry profile using a small vitrified CBN wheel. The process is carried out in oil coolant to achieve the economics using plated CBN. The wheel speed for the vitrified wheel is limited due to size to <100 m/s. The smaller wheel spindle is on either a vertical slide (Pflager 2002) or a swing down arm (Schaudt n.d.) to prevent interference with the larger wheel (Figures 17.33 and 17.34).

These machine configurations created two additional benefits. The first was that since the small wheel was finishing only, it could be specified to produce a lower surface finish than a wheel required to rough and to finish grind. In some circumstances, this allowed the elimination of polishing—a process using conventional abrasive film and as expensive in terms of consumables as the grinding process.

The second benefit relates to grinding diesel camshafts. Gasoline camshafts have intake and exhaust lobes, a diesel camshaft has an additional injector lobe which is typically a

FIGURE 17.31
Landis twin-wheelhead camlobe grinder.

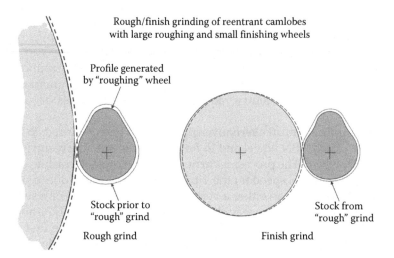

FIGURE 17.32
Large and small grinding wheels compared for reentrant camlobe.

FIGURE 17.33
Landis twin-wheel camlobe grinder for roughing and finishing reentry cam profiles. (From Flager 2002. With permission.)

different width to the other two. Diesel camlobes also have extremely stringent straightness specification of <50 μ″, which is comparable to the break-in depth of a vitrified CBN wheel. When trying to grind, all three lobes with one wheel, a step is rapidly formed and the process becomes uneconomic. However, a twin-wheel machine with the small auxiliary spindle configured with two vitrified CBN wheels can contend with the different widths.

17.4 Punch Grinding

A related but somewhat simpler process to camshaft grinding is the grinding of punches for the die and pharmaceutical industries (Figure 17.35).

FIGURE 17.34
Schaudt CF41 CBN twin-wheel camlobe grinder for roughing and finishing camlobes with reentry profiles.

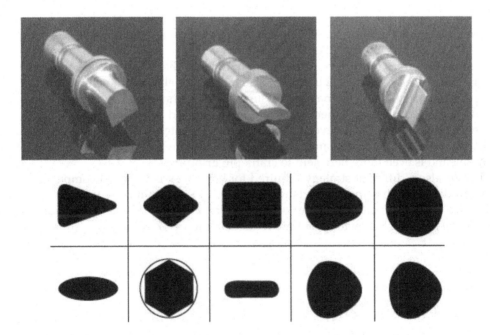

FIGURE 17.35
Punch profiles. (From Studer, Motion Schleifring August 2001, p. 29, Trade Brochure. With permission.)

Profiles for punches are simple radii and straight faces and machine controls are set up with canned programs to generate a given shape with the operator entering a few basic dimensions. Batch sizes are relatively small so flexibility and rapid change over times are critical. Parts are held by the shank in a simple three or four jaw chuck. Computational demands and workspeeds while still demanding are lower than in camlobe grinders as is the cost of the machine. OEMs include ANCA, Studer, and Weldon.

Most punches are made of tool steel making CBN particularly attractive where punch designs allow a constant wheel corner radius for the blend at the holder end. Cycle times

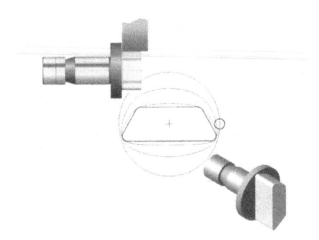

FIGURE 17.36
Punch grinding strategy. (From Studer, Motion Schleifring, August 2001, p. 29, Trade Brochure. With permission.)

are reduced by 30% (Kampf 2000) or greater over conventional abrasives using wheel speeds of 60–80 m/s in water-based coolant.

The problem areas for punch grinding are in holding sharp corners on square punches and holding taper values of 1–2 µm parallel and orthogonal to the punch axis as the result of wheel and coolant hydrodynamic pressure. Some errors in profile can be programmed out if they are consistent, others may require prolonged spark-out times if using alox wheels. For CBN wheels, however, the coolant can be switched off almost entirely during spark out to relieve hydrodynamic effects. The technique is called "trickle" or "dribble" grinding and is even more pertinent to crank pin grinding.

There is also a different strategy required for how stock is removed compared to cam-lobe grinding. Punch grinding starts with a round billet to limit inventory and maximize flexibility. There can be up to 30 mm of stock on larger parts. The optimized spiral wheel path to the final size is key to cycle time reduction (Figure 17.36).

17.5 Crankshaft Grinding

Whereas camshaft lobe grinding face difficulties associated with cylindrically grinding a nonround part at least all the lobes of the component are ground on a common centerline. Finish grinding crankshaft pins is albeit a simple round grind operation but each pin is on a different centerline governed by the "throw" of the crank. Additionally, the crankshaft is very weak and hence easily distorted if overclamped, is unbalanced at grind, and is prone to "unwind" if the heat treat is not carefully controlled. Consequently, the first step in designing a grind process for a crankshaft is an finite element analysis (FEA) analysis of the clamping forces and foot-stock clamping pressures and a subsequent determination of steady or workrest requirements (Figure 17.37; Pflager 2002).

Cranks are made of either nodular iron in the 150–250 BHN range for automotive engines or steels surface heat treated in the 25–40 HRC range for higher power and diesel applications. Most automotive crank pins are first turn broached to relieve the sidewalls and leave

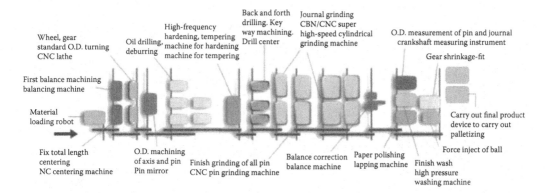

Wheel, gear standard O.D. turning CNC lathe

High-frequency hardening, tempering machine for hardening machine for tempering

Oil drilling, deburring

Back and forth way machining. Drill center

Journal grinding CBN/CNC super high-speed cylindrical grinding machine

O.D. measurement of pin and journal crankshaft measuring instrument

Gear shrinkage-fit

First balance machining balancing machine

Material loading robot

Carry out final product device to carry out palletizing

Fix total length centering NC centering machine

O.D. machining of axis and pin Pin mirror

Finish grinding of all pin CNC pin grinding machine

Balance correction balance machine

Paper polishing lapping machine

Finish wash high pressure washing machine

Force inject of ball

FIGURE 17.37
Flowchart for a modern automotive crankshaft manufacturing line. (From UMS, United Manufacturing Solutions. UMS Ltd., Nagoya, Japan. 2002, Trade Brochure. With permission.)

an undercut at each edge. This undercut is then fillet rolled, resulting in a finish flat grind with about 1 mm stock performed using vitrified CBN abrasive. For diesel cranks, there is no undercut for fear of generating stress raisers and the pin, edge radii and side walls are all ground. Wheels must be dressed to the exact width and profile. Most of the stock now occurs on plunging the depth of the sidewalls and can be up to 0.5″. Special sandwich wheels with split grades are used primarily with alox/SG combinations, although some vitrified CBN applications have been developed recently (Figures 17.38 through 17.40).

There are several machine design strategies for approaching the issue of grinding each individual pin on its centerline. The traditional method is to use a mechanical chuck to index each pin in turn onto its axis of rotation, and use automatic throw changers to compensate from one crank type to another. The chucks and indexing fixtures are complex, product specific, and being mechanical—potentially unreliable. The method loses a significant amount of cycle time for each indexing movement. An alternative approach developed in the early 1990s by Landis for high production, >200,000 cranks/annum, was to use simplified chucking on grinders dedicated to a specific pin (Anon 1994). The grinders were

FIGURE 17.38
Typical crankshaft showing oil lubrication holes in the pins.

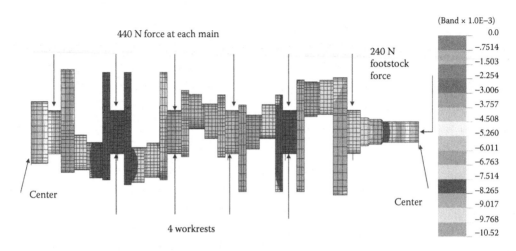

FIGURE 17.39
FEA analysis of the effect of forces on the crankshaft during grinding.

set up similar to a transfer line with a load/unload system that moved the crank from one single-pin grinder to the next until all the pins are ground. This reduced cycle time but at the risk that if a grinder had operational problems or required a wheel change then production was lost from all the grinders in the line. Fortunately, the reliability of modern grinders is such that, with proper preventative maintenance, down time from mechanical problems is minimal, while CBN technology has pushed wheel life to as great as 3 years.

The introduction of vitrified CBN to crankpin grinding in the late 1980s allowed machine tool builders to completely rethink how crankpins should be ground. The ideal way to grind the crankpins would be with the crankshaft being rotated on the same axis as seen by the engine. This requires the wheel to be moved by the throw amount using a CNC control akin to a camlobe grinder. Such a concept was impossible trying to hold micron size tolerances over throw lengths of the order of 100 mm with conventional wheels that varied in diameter during their life by 25% and wore microns/pin. However, it became practical with CBN wheels that varied in diameter by only 1% and wore nm/pin.

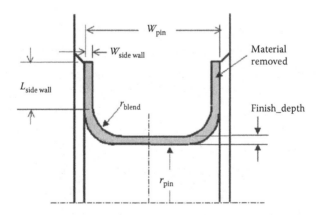

FIGURE 17.40
Stock removal in diesel crankpin grinding. (From Corallo, V., Gridley, T., Medici, M., 1993, Regrinding rolls for modern day requirements for size, shape, dimension and finish. *34th MWSP Conf Proc. ISS-ASME*, Vol 30. With permission.)

The process has been termed "CNC contour grinding" (Toenshoff and Falkenberg 1996), "pin chasing" by Naxos-Union Emag (2002), or "CNC orbital pin grinding" (Anon 1998a).

CNC contour grinding offers significant benefits in terms of flexibility, reduced tooling costs, for example, steady rest needs, and smaller machine footprint. Its greatest benefit, however, may be increased quality especially with regards to pin roundness. The standard automotive crankpin tolerance on roundness has historically been 5 µm but with more stringent 2 CpK capability requirements, this has now reduced to 2.5 µm. The major source of out of roundness in most crankpin grinding operations is created by the presence of oil lubrication holes drilled into the surface of the pins. These can be clearly seen in the photograph (Figure 17.39). These holes reduce the wheel/workpiece pressure both by physically reducing the contact length but more importantly by reducing the hydrodynamic coolant pressure. Depending on the stiffness of the crank, wheel speeds are typically limited to ≤80 m/s and finish grinding is carried out in "trickle" grind mode to reduce the out of roundness down to the order of 2 µm. With traditional grinding methods, this still consumes almost the entire available part tolerance and creates a very discrete and obvious dip in the roundness profile. CNC contour grinding with programmable crankpin angular position capability allows this 2 µm to be compensated for thereby improving overall roundness by 50%.

The drive for increased quality in crankpin grinding has therefore led to commonality in machine tool design with camlobe grinding to the point that machine tool builders are now offering a common machine for both applications and with the flexibility to grind mains too. For example, the Landis 3L twin-wheelhead nonindexing grinder illustrated in Figure 17.31 can be configured to grind camlobes and/or mains or crankpins and/or mains.

Interestingly, the grind process for a dual wheel pin grinder is more complex. For a constant workspeed, the instantaneous stock removal rate will vary from a maximum given by (rpm/60) · (stroke + pin radius) infeed/rev to a minimum given by (rpm/60) · π · (stroke − pin radius) · infeed/rev. For a single-wheel orbital pin grinder, it is possible to vary the instantaneous workspeed just like in cam grinding to even out the removal rate. When dual wheel pin grinding, the problem is all the pins are out of phase with each other. This makes optimization of the workspeed more complicated. For a dual wheel arrangement, it is also necessary to ensure the two wheel speeds are shifted by 1%–2% to avoid regenerative chatter.

The most recent developments in crankpin grinding have been to marry the CNC contour technology with modern plated CBN technology by increasing spindle motor power up to 80 kW and wheel speed up to 160 m/s. In conjunction with increased acceptance of oil coolant from improvements in its handling and containment, rough or "green" grinding has been gaining acceptance as a replacement to turn broaching. Laycock (1996) reported the following maximum stock removal rates for plated CBN wheels rough grinding cam and crank materials in oil:

Soft cast iron	360 mm³/mms
Soft steel	250 mm³/mms
Hard steel	200 mm³/mms

Pflager (2000) confirmed a more general value of >150 mm³/mms that is over 300% greater than hard turning and 50% greater than soft milling. Giese (1999) reported a plated CBN wheel life grinding pins including the undercuts of 40,000–60,000 crankshafts at comparable cycle times to turn broaching. Green grinding eliminated the need for the frequent replacement of carbide cutting tool inserts. Plated CBN has actually proved cost-effective grinding at half the cycle time of turn broaching. Since machine tools costs are comparable or less for high-speed grinders than for turn broaches (Anon 2000), green grinding offers significant capital equipment cost savings.

As speed and stock removal rates increase, limitations occur in the maximum power available for grinding. As discussed above, a narrow wheel reduces coolant drag losses making more power available to grind at higher Q' values and hence additional removal rate capability as a result of the lower e_c values associated with this. Several researchers and OEMs (TMW 1998; Toenshoff et al. 2001) have therefore proposed grinding pins and mains in multiple plunges to increase overall stock removal rates and increase flexibility to grind mains and pins of different widths.

17.6 Roll Grinding

The roll grinding market covers an incredibly broad range of applications from rolls for business machines of just an inch or less in diameter to rolls for the steel industry that can weigh up to 50 ton. Work piece materials include rubber, aluminum, steel, cast iron, ceramics, granite and exotic metals, and polymers. Often finish requirements must be held in a tight band of values, and surface integrity and cosmetic appearance—the elimination of feed lines—are paramount.

The largest areas of use, certainly in abrasive consumption, are the steel and paper industries. Rolls for the steel industry are made of either forged steel (48-52HrC) or chilled cast iron. Finish specifications and tolerances are driven by whether the rolls are used in hot strip mills or cold strip mills, with the demands for cold strip mills being considerably tighter as they impart the final finish on the rolled steel:

> *Hot strip:* 0.002 in shape deviation, 0.0020 in roundness, ≤40 μin. Ra finish
>
> *Cold strip:* 0.0005 in shape deviation, 0.0002 in roundness, ≤10 μin. Ra finish

A typical wheel for a steel roll hot strip application would be 900 mm × 75 mm and would last about 100 h grinding 100–150 rolls. Typical roll size (50 ton): 30 in diameter × 70 in long with 0.010 in stock on diameter. Most wheels are either conventional resin or occasionally vitrified bonded.

Grinding wheel selection for cold strip rolls is governed by the need to eliminate regenerative chatter. Consequently, very compliant bonds are used such as shellac, resin, epoxy, and even using cork abrasive.

Rolls for the paper industry are made from cast iron or chrome steel for general smoothing and kallandering while for some specialty applications, especially for fine paper production, granite, and polymer granite are used. Rolls can be up to 1 m in diameter and 12 m long. The shear weight of these in operation creates a sag or bow in the center of the roll that can be a significant proportion of the paper thickness to be smoothed. Consequently, an inverse profile to compensate must be ground in the roll. Surface finish requirements are ≤0.4 μm Ra after roughing and ≤0.1 μm Ra after finishing.

Most modern machine tools can maintain 1.25 μm taper, flatness, and profile on rolls up to 300 ton. Leading OEMs include Waldrich Seigen, Herkules, Pomini, Toshiba, Capco, and Schaudt. Other machines common in the industry include Voith, Farrel, Churchill, Craven,

Naxos Union, MSO, and INNSE. Small machines maintain the standard moving carriage design of a typical general-purpose cylindrical grinder, but the large machines for the steel and paper industries have a moving wheelhead, where the operator moves with the wheel while controlling the process.

The grinder consists of up to three independent slides on a massive casting. The whole machine is mounted on a massive isolation pad as in the example in Figure 17.41 of a Pomini machine. The machine requires an extensive laser alignment upon installation or repositioning to the required accuracy, while great care is taken in the design to minimize

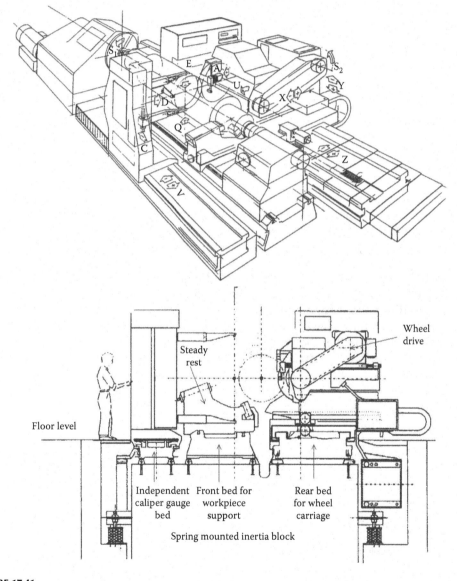

FIGURE 17.41
Pomini roll grinder general view and cross section. (From Oppenheimer and Gridley 1989; and Corallo, V., Gridley, T., Medici, M., 1993, Regrinding rolls for modern day requirements for size, shape, dimension and finish. *34th MWSP Conf Proc. ISS-ASME*, Vol 30.)

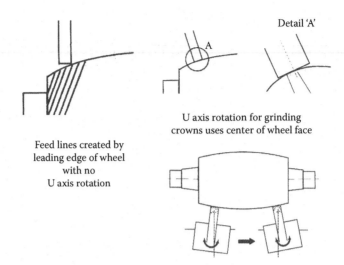

FIGURE 17.42
Roll grinding using a U-axis wheelhead rotation to generate crown and eliminate feedlines.

the effects of thermal movements. Residual repeatable slide errors can be compensated for in modern CNC controls.

The rear axis carries the wheelhead and its primary axes of motion X and Z. All slides are hydrostatic as, in general, is the wheel spindle. Grinders are also equipped with Y and U' tilt axes for microswiveling the wheelhead. These are required to maintain a normal position against the part when grinding crown profiles in order to minimize feed lines (Figure 17.42).

All machines are equipped with at least a load monitor for the wheel spindle. On older machines, the operator will constantly infeed the wheel to keep the load constant. On modern CNC machines, the control will either infeed after every stroke, or more typically, at a continuous preset rate during the grind, or adaptively under constant load. This is to compensate for the very significant breakdown of the wheel. The control will also carry out short stroke grinds at certain positions in order to remove excess stock on a new roll or when changing crown Figure 17.43. All the while the CNC is able to generate a taper, crown, or compound curve shape while compensating for bed error (Waldrich Siegen 1996).

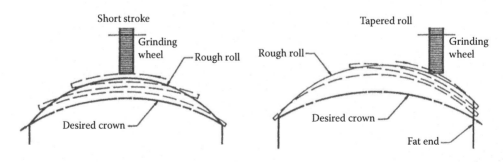

FIGURE 17.43
Short-stroke grinding strategies for crown shaping. (From Ehlers, J., 1991, Applications of computer controls and display to roll grinders *CH2973-6/91/0000 IEEE*. With permission.)

The front or center bed carries the workhead drive and tail stock. Being separate and rigid, it allows the isolation of any effect of loading a heavy workpiece. Usually, at least one axis of adjustment is available for initial roll alignment and taper correction.

The third carriage carries the gauging equipment—a two-point dimensional measuring caliper with optional Rollscan, eddy current, and finish measuring attachments. A longitudinal *V* axis moves the caliper independently of the wheel slide. If a separate carriage is not present, the gauging system rides on the rear carriage with the wheelhead.

Older grinders without automatic gauging relied on the operator stopping the grind operation to measure the roll using manual calipers, and then feeding with higher load in those area with higher stock; a process based on art and experience. Automatic gauging with roundness and shape fed directly into the CNC can now continuously measure and compensate.

References

Aerobotech, 1996, *Parts and Tooling*, Arobotech Systems Inc., Madison Heights, MI, June.

Andrew, C., Howes, T. D., Pearce, T. R. A., 1985, *Creep Feed Grinding*, Holt Rinehart and Winston, Eastbourne, UK.

Anon, 1994, Single-workhead grinding system for automotive crankshafts. Manual February.

Anon, 1998a, Single orbital crankpin grinder replaces six dedicated machines. Manual.

Anon, 1998b, Steady rest technology hold form in parts. *Tooling & Production*, June.

Anon, 2000, Dagenham takes a new turn. *Machinery and Production Engineering*, July 7.

Brinksmeier, E., Minke, E., 1993, High performance surface grinding—The influence of coolant on the abrasive process. *Annals of the CIRP* 42/1, 367–370.

Corallo, V., Gridley, T., Medici, M., 1993, Regrinding rolls for modern day requirements for size, shape, dimension and finish. *34th MWSP Conference Proceedings of the ISS-ASME*, Vol. 30, USA.

Ehlers, J., 1991, Applications of computer controls and display to roll grinders, *IEEE* CH2973-6/91/0000.

Fix, R. M., Tiitto, K., Tiitto, S., 1990, Automated control of camshaft grinding process by Barkhausen noise. *Materials Evaluation*, 48, 904–908.

Giese, T. L., 1999, Grinding it green *Tooling & Production*, December.

Hanard, M. R., 1985, Production grinding of cam lobes with CBN. *SME Conference Proceedings Superabrasives '85*.

Hitchiner, M. P., 1991, Systems approach to production grinding with vitrified CBN, *1991 SME Conference Proceedings, Superabrasives '91* MR91-148.

Junker, 1992, Junker quickpoint CNC o.d. grinding with CBN or diamond. Erwin Junker Maschinenfabrik GmbH, Trade brochure.

Junker, 1993, Automotive components being ground at high speed by CBN grinding wheels. Erwin Junker Maschinenfabrik GmbH, Trade brochure.

Kampf, E., 2000, Grinding of form punches with CBN. Swiss Quality Production, July.

Koenig, W., Klocke, F., Stuff, D., 1997, High speed grinding with CBN grinding wheels—Boundary conditions, applications and prospects of a future-orientated technology. *1st French and German Conference*.

Koenig, H. C., Treffert, C., n.d., High speed grinding of any contour using CBN grinding wheels. Laboratorium fur Werkzeugmaschinen und Betreibslehre. RWTH, Aachen.

Landis, 1996, Landis windows-based automatic workspeed generation system. Trade brochure.

Laycock, M., 1996, Recent developments in camshaft and crankshaft grinding. *IGT Annual Seminar*, Bristol, UK.

Lutjens, P., Mushardt, H., 2000, Hard turning or grinding. European Production Engineering, July.

Malkin, S., 1989, *Grinding Technology*. Ellis Horwood Ltd., Chichester, UK.

Marinescu, I., Rowe, W. B., Dimitrov, B., Inasaki, I., 2004, *Tribology of Abrasive Machining Processes*. William Andrew Publishing, Norwich, NY.

Naxos Union Emag, 2002, CBN Curbelwellen-schleifmaschinen. Naxos Union GmbH, Trade brochure.

Okuma, n.d., GP-24N CNC plain cylindrical grinder, GA-24N CNC Angle Head Cylindrical Grinder. Okuma Machinery Works, Japan. Trade brochure.

Oppenheimer, J. G., Gridley Jr., T. H., 1989, Automatic roll grinding in the nineties, *Proceedings of the Aluminum Association International Aluminum Sheet and Plate Conference*, Vol. 1, Nashville, TN 6/23/92, 185–205.

Ota, M., Ueda, H., Maeda, M., 1997, A cam grinding machine using an ultra-high speed and high power grinding wheel spindle. *1st French and German High Speed Machining Conference*.

Pflager, W., 2000, High speed cam grinding. *IMTS 2000 Manufacturing Conference*. SME Chicago, IL.

Pflager, W., 2002, High speed grinding in a mass production environment. *IMTS 2002 Manufacturing Conference*. SME Chicago, IL.

Renaud, W., Hitchiner, M. P., 1991, The development of camshaft lobe grinding with vitrified CBN. *SME Conference Proceedings. Superabrasives '91* MR95-163.

Rowe, W. B., Bell, B., Brough, D., 1986, Optimization studies in high removal-rate centreless grinding. *Annals of the CIRP*, 35(1), 235–238.

Rowe, W. B., Morgan, M. N., Black, S. C. E., Mills, B., 1996, A simplified approach to control of thermal damage in grinding. *Annals of the CIRP*, 45(1), 299–302.

Rowe, W. B., Jin, T. J., 2001, Temperatures in high efficiency deep grinding. *Annals of the CIRP*, 50(1), 205–208.

Rowe, W. B., 2014, *Principles of Modern Grinding Technology*, Elsevier, Oxford, UK and worldwide.

Schaudt, n.d., CF41 CBN Nockenform-schleifmaschine. Schaudt Maschinenbau GmbH, Stuttgart, Trade brochure.

Schultz, A., 1999, Precision grinding of transmission components using modern grit materials. *Precision Grinding and Grinding with Superabrasives Conference*. Gorham Advanced Materials, Chicago, IL.

Shaw, M. C., 1996, *Principles of Abrasive Processing*. Clarendon Press, Oxford, UK.

Shigiya, 1996, CNC cylindrical grinders GPS-30. Shigiya Machinery Works Hiroshima, Trade brochure.

Suzuki, 1984, Development of camshafts and crankshafts grinding technology using vitrified CBN wheels. SME MR84-526, Abrasives Conference, USA.

Tawakoli, T., 1993, *High Efficiency Deep Grinding*. Mechanical Engineering Publications Ltd., London.

TMW, 1998, GL63M and GL 100M. Toyoda Machine Works, Japan. Trade brochure (in Japanese).

Toenshoff, H. K., Falkenberg, Y., 1996, High-speed grinding of cast iron crankshafts with CBN tools. *IDR*, 4, 115–119.

Toenshoff, H. K., Friemuth, T., Becker, J. C., 2001, Next generation of crankshaft production. *Annals of the CIRP*, 2, 551–571.

Tsujichi, 1988, CNC cam grinder with small diameter CBN wheel. *SME Conference Proceedings, 3rd International Grinding Conference, Fontana, WI*, MR88-609.

UGT, 2000, Dana's long camshafts are no challenge for Schaudt grinders. *Grind Journal*. Trade brochure.

UMS, 2002, United Manufacturing Solutions. UMS Ltd., Japan. Trade brochure.

Wakuda, M., Ota, M., Ueda, H., Miyahara, K., 1998, Development of ultrahigh speed and high power cam grinding machine 1st report—Characteristics of ultrahigh speed grinding of chilled casting. *B JSPE*, 64(4), 593–597.

Waldrich Siegen, 1996, Worldwide partners for intelligent manufacturing sSolutions. Waldrich Siegen Werkzeugmaschinen, Germany. Trade brochure.

Wedeniwski, H. J., 1989, CBN-Nockenschleifen bei Hochgeschwindigkeit produktiver und ohne Rissbildung. *Werkstaff und Betrieb*, 122(9), 796–800.

Werner, G., 1983, Realisierung niedriger Werkstuckoberflachen temperaturen durch den Einsatz des Tiefschleifens. *Trens-Kompendium*, 2, 448–468.

Woodside, J., 1988, High-speed grinding proved in production. *Tooling & Production*, May.

18

Internal Grinding

18.1 Introduction

Internal grinding is the primary process for the precision finishing of internal surfaces or bores. The bores may be simple cylindrical surfaces or may be surfaces requiring the generation of complex and exact profiles for applications such as bearing and CV joint races, or fuel injection seats. Figure 18.1 shows a typical vitrified cubic boron nitride (CBN) grinding wheel for internal grinding together with the rotary tool used for dressing.

Most precision internal grinding operations require the capability to hold tolerances on size, roundness, straightness, taper, and cylindricity of the order of 0.5–10 µm, but in special applications such as fuel injection, the tolerances have become increasingly stringent in recent years to as tight as 0.25 µm or even lower. The majority of the highest volume applications are found in the bearing and automotive industries grinding hardened steel (carbon, alloy, PM, and M50) with relatively small wheels under 100 mm on horizontal spindle grinders with finish requirements under 0.5 Ra. There are also larger diameter applications in specialty bearing and aerospace engine assembly grinding bores up to 1000 mm more. The largest such parts are usually ground on specialist vertical spindle machines.

Grinding still dominates bore finishing for all the tightest precision tolerance work because of stock removal capability, accuracy, and cost. However, hard turning is making inroads for finishing larger bores with tolerances >2.5 µm. It is also becoming more common to see multipurpose machining centers for applications such as automotive gears that rough turn the bore, rough and finish the gear flanges, but then finish grind the bores in the same chucking when tighter accuracy demands. Diamond honing is also gaining favor and is reported to hold good straightness but stock removal rates are still limited. To be cost effective, stock removal must be kept to under 10 µm.

Modern computer numerical control (CNC) internal grinders provide a great deal of flexibility not only to plunge grind simple bores but also to face grind, profile grind, and even contour grind outer diameters. Multislide machines may even simultaneously grind inner (i.d.) and outer diameters (o.d.) and flange using two wheels (Figures 18.2 and 18.3).

18.2 The Internal Grinding Process

The basic internal grinding process is illustrated in Figure 18.4.

In the basic internal grinding process, the wheel is fed perpendicularly into the part usually accompanied by a short-stroke high oscillation along the axis of the wheel. The part

FIGURE 18.1
Internal grinding machine setup for power steering pump rings showing a vitrified CBN wheel in the dress position.

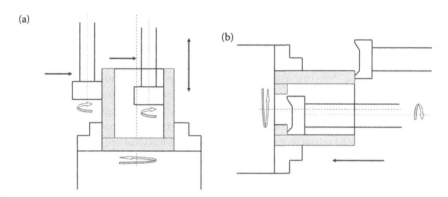

FIGURE 18.2
Diameter grinding and face grinding operations on an internal grinding machine. (a) Internal and external diameter grinding. (b) Face grinding.

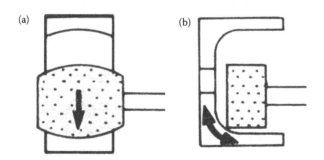

FIGURE 18.3
Profile grinding and contour grinding operations on an internal grinding machine. (a) Internal profile grinding. (b) Contour grinding.

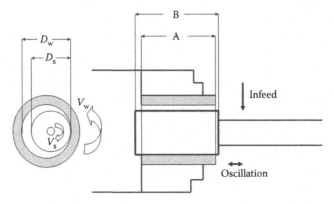

FIGURE 18.4
The basic internal grinding process.

is rotated in the opposite direction to the wheel (down grind) in virtually all applications except occasionally in finish grinding with large interrupted cuts, if roundness is an issue.

Internal grinding is a very weak system, where the primary weakness is the wheel mount or quill. This can readily deflect during the grind leading to problems of taper and shape. The grinding conditions are driven first and foremost by the system stiffness and the level of normal grinding force. The key factors to consider are discussed below.

18.3 Abrasive Type

18.3.1 Grain Selection

The first and most important decision in recent years is the choice of whether to use conventional, ceramic, or CBN abrasive. This will depend on the condition of the machine, dressing capability, and part size. In general, high volume, hardened steel parts under 15 mm with straight bores, for example, tappet rollers, lifter bodies, fuel injector bores, small bearings, and especially M50 steel aerospace, are usually most cost effective with CBN even on older machines. Between 15 mm and about 50 mm, CBN is becoming more and more common on new equipment. Over 50 mm, alox- and SG-based wheels dominate. There are several reasons for this. First is purely the abrasive cost in the wheel. Small wheel manufacture carries a high percentage of labor in its cost regardless of abrasive type. For CBN as the wheel size increases, the cost of abrasive soon dominates and wheel cost becomes proportional to size. Alox grain being much less expensive, the overall manufacturing cost does not increase as rapidly with diameter.

The second issue governing abrasive selection concerns normal force and its relationship to wheel/part conformance characterized by equivalent wheel diameter d_e, where

$$d_e = \frac{d_w \cdot d_s}{d_w - d_s}$$

The larger the equivalent wheel diameter, the greater the contact length and hence the higher the normal force. As d_w increases the ratio d_w/d_s must be reduced to reduce d_e.

TABLE 18.1

Recommendations for Vitrified CBN Internal Wheels

d_s (mm)	d_s/d_w	d_e (mm)	Q' (mm³/mm/s)	Ra (mm)	Specification	Application
5	0.9	50	<1	<0.15	B46 H200 VSS	Fuel injection
10	0.9	90	3	<0.25	B64 I200 VSS	Valve lifters
20	0.85	130	5	<0.40	B91 K200 VSS	UJ cups
30	0.8	150	5	<0.50	B91 K150 VSS	Pumps
50	0.7	165	5	<0.50	B126 K150 VSS	Gear bores
250	0.25	350	3	<0.80	B181 F200 VSS	Aero shrouds

Also, the grinding pressure must be reduced further by increasing the structure and grit size in the wheel specification. This also improves chip clearance and coolant access. Unfortunately, the range of grit sizes and bond structures available in vitrified CBN systems which will still allow the necessary high G ratios to be maintained is more limited than those for alox or ceramic wheels. Consequently, d_e for CBN wheels is limited to about 150 mm over which the process is compromised either by abrasive cost or by cycle time. This will vary by material type: for very difficult to grind materials such as M50 or tool steels, the maximum d_e value for CBN would be much higher due to gains in productivity by the elimination of dressing, while for simple AISI 52100 plain bearings, the maximum d_e could be nearer 100 mm. Some typical examples of bond specifications for vitrified CBN wheels grinding hardened steels are given in Table 18.1. The values for d_s are for wheels as new and will reduce by up to 20% during use. Bond specifications are based on the Universal (Saint-Gobain Abrasives) VSS bond system. Grades will vary from one wheel manufacturer to another.

For conventional wheels, the range of bond systems is well established by the wheel manufacturers. Table 18.2 provides typical grit size requirements based on d_e and Q' values. Again, wheel grade and structure number will vary depending on the manufacturer and grit type.

18.3.2 Impact on Grain Configuration

The selection of grain type has a significant impact on wheel length. The specifying of the length of an alox or ceramic wheel is based on the recognition that there is a significant

TABLE 18.2

Recommendations for Alox Grit Sizes as Function of Part Size, Roughness, and Stock Removal Requirements

d_e (mm)	$Q' = 1$ (mm³/mm/s)	Ra (μm)	$Q' = 3$ (mm³/mm/s)	Ra (μm)	$Q' = 5$ (mm³/mm/s)	Ra (μm)
125	150#	<0.15	120#	<0.25	100#	<0.5
250	120#	<0.15	100#	<0.25	80#	<0.5
375	100#	<0.15	80#	<0.25	70#	<0.5
500	100#	<0.15	70#	<0.25	70#	<0.5
750	80#	<0.15	70#	<0.25	60#	<0.5
1000	80#	<0.15	60#	<0.25	46#	<0.5

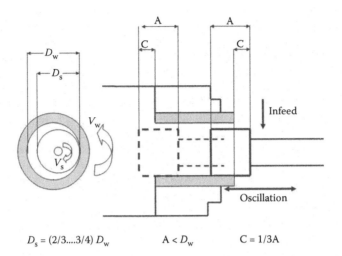

$$D_s = (2/3....3/4) D_w \qquad A < D_w \qquad C = 1/3A$$

FIGURE 18.5
Parameters for i.d. grinding with alox wheels.

level of wheel wear for each part ground. The length of the wheel is therefore kept short and the wheel oscillates a significant distance past the ends of the part to maintain straightness. The breakdown is sometimes so great that it is often dressed midcycle just prior to finish to reestablish size. Salmon (1984) makes the following general recommendations with reference to Figure 18.5, where the oscillation is a third of the wheel length past both ends of the part.

For CBN wheels, the wear is one-hundredth the wear with alox and the goal is to maintain the straightness of the wheel face and avoid excessive edge breakdown. Consequently, the wheel is made about 1–2 mm longer than the length of the part and a much smaller oscillation stroke applied such that the wheel normally remains in contact with the full length of the part for all or most of the time. Small adjustments of the end points can be made to adjust for taper and shape in the part. For example, if the back of the part is small, then the wheel oscillation end point is moved further back past the end of the part such that the wheel contact length is now slightly less than the full length of the part. This gives more time for the wheel to remove metal, where the part is small; in addition, the reduced contact length will reduce normal force and quill deflection further improving shape.

Using this method for taper compensation, any residual quill deflection, if constant, could be compensated for by adjustment to the wheelhead. Unfortunately, the normal forces change significantly both with wheel sharpness after dress and longer term with changing wheel diameter. The machine operator can attempt to adjust for these by either adjustment to the wheelhead alignment or the oscillation end points. This approach is usually unacceptable both for the degree of operator intervention and the lack of process control.

18.3.3 Quill Designs for CBN

The preferred approach is to ensure that the quill design has been optimized for use with vitrified CBN as illustrated in Figure 18.6. The quill length should be reduced to a minimum and the mounting diameter increased. The standard wheel screw used for

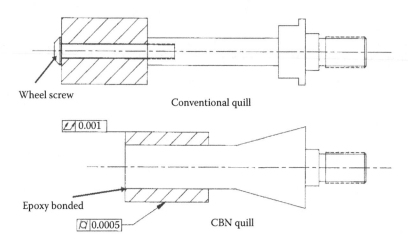

FIGURE 18.6
Quill design for CBN compared with conventional quills for alox.

conventional wheels should be eliminated and the wheel glued directly to the quill. This allows the wheel bore size to be maximized (note the lack of stiffness of the wheel bond itself). Stiffer quill materials may also be necessary and several materials are listed below. Carbide is up to four times stiffer than steel but is very brittle and is usually only used for the smallest wheels with very tight tolerances. Molybdenum is twice as stiff as steel and tough but expensive. Ferro-TiC is a less-expensive alternative. Tungsten alloys with trade names such as No-Chat and Wolfmet have the same stiffness as molybdenum but cost a third as much. However, they are almost twice as dense and any runout from mistreatment will reduce spindle life. Titanium has been used occasionally for extremely high rpm applications because its low density moves the resonant frequency of the quill above the rotational frequency.

For a cylindrical bar of radius r and length l,

$$\text{Deflection} = \frac{s = 4Wl^3}{3\pi r^4 E} \quad \begin{array}{l} W = \text{Load} \\ E = \text{Young modulus} \end{array}$$

Material	E 10^6 psi	Density Gm/cm^3
Steel	25–30	8
Titanium	17–27	4
Molybdenum	48–52	10
Tungsten	≈50	19
WC	75–100	13
Ferro-TiC	40–45	6
Vitrified bond	4–18	<3

Discussions with machine tool builders suggest that if the calculated deflection during the rough feed portion of the grind cycle exceeds 80% of part tolerances, then the quill should be redesigned as indicated above. A crude estimate can be made using the equation above by either direct measurement of power or by estimation, that is,

1. From spindle power (assume $F_n/F_t = 0.33$),

$$W \text{ (newtons)} = 3 \times \frac{\text{Power (Watts)}}{\text{Wheel speed (m/s)}}$$

2. From typical values $Q' = 5$ mm³/mm/s and specific energy $e_c = 60$ J/mm³,

$$W \text{ (newtons)} = 3 \times 5 \times 60 \times \frac{\text{Part length (mm)}}{\text{Wheel speed (m/s)}}$$

Once the CBN wheel length to diameter ratio exceeds about 1.5 then increased attention has to be paid to the level of quill deflection. Especially, for wheels under 10 mm, it may be necessary to resort to a shorter wheel and stroking the bore as with alox wheels.

18.4 Process Parameters

18.4.1 Wheelspeed

As discussed in Section I, wheel speed for internal wheels is limited to 35–42 m/s for most conventional wheels and machines, and 60 m/s for vitrified CBN in purpose designed grinders.

Some vitrified CBN wheels are used at up to 80 m/s with special high-strength bonds on steel cores and subject to the restraints in Figure 18.7. The practice is generally limited to narrow bearing race applications on specially guarded machines, and where local safety regulations allow. In general, the speed is limited both by bond strength considerations and by coolant hydrodynamic forces. The latter effect can be very deleterious on taper and shape as speed is increased especially for the denser vitrified CBN wheel structures. The pressure is sometimes relieved on long narrow CBN wheels by adding three or four straight narrow slots along the length of the wheel.

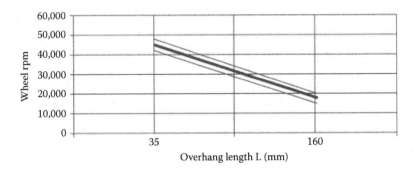

FIGURE 18.7
Maximum recommended wheel rpm as function of quill length.

18.4.2 Workspeed

Workspeed is usually expressed as a ratio of the wheel speed (not rpm). The value is influenced by dynamic stiffness, onset of burn, and roundness. A typical value would be 1:100 with values as high as 1:50 on stiff grinders for best roundness or 1:150 on weak grinders limited by chatter. As wheel speed increases, the ratio is usually dropped (i.e., the workspeed is increased) to reduce thermal damage. Grinding large bearings for applications such as steel mills work speeds are limited and the greater risk is chatter. Under these conditions the work speed velocity is shifted a few percent every revolution to avoid the onset of regenerative chatter.

18.4.3 Oscillation

Oscillation is provided to improve finish and smooth out wheel wear. Stroke length is limited to <3 mm typical but is high frequency (up to 20 Hz) and fast (up to 1.5 m/min). Oscillation can improve finish Ra and appearance somewhat and can also reduce spark-out times.

18.4.4 Incoming Part Quality

The incoming part quality is critical. The grind process is often used to "fix" problems earlier in the line and blamed for their deficiencies. The part must come in burr free from any premachining/heat treat process and free of any contamination such as tumbling media used for deburring. On a standard grinder, the out of roundness coming to the machine must be no more than one-eighth of the stock amount, while the total geometry error should be no more than about a quarter.

18.4.5 Dressing

Dressing has been discussed in Section I for both conventional and CBN wheels. The end user should be aware, however, of the importance of the diamond being on centerline with the wheel. This is especially critical if the wheel is dressed each part just prior to finishing for sizing purposes as illustrated in Figure 18.8 (Bryant 1995). If the diamond is off centerline, the wheel will actually extend further by a distance governed by its radius. Consequently, as the wheel gets smaller, this distance will increase leading to the part diameter getting bigger.

When traverse dressing vitrified CBN wheels attention must be paid to the limitations of the grinder and the dresser spindle. Very few grinders are equipped with AE dress sensor technology, so the dress infeed repeatability is limited by slide repeatability and thermal stability. It is necessary to determine by trial and error a safe total infeed amount to ensure repeatability and also determine the level of wheel wear if possible. (This may be possible from QC data on size or tracing a coupon of the wheel face.) If the wheel wear is a significant portion of the infeed repeatability, then it is best to make multiple passes at smaller infeed depths to ensure the last pass is at a near constant depth. For example, if the minimum total infeed amount were 3 μm and the wheel wear was 1.5 μm, then it would be best to make at least two passes at 1.5 or 2 μm. However, if the wheel wear was only 0.5 μm, then a single pass at 3 μm would be stable.

Internal grinding machines are very compact and most have evolved from designs based around single point diamonds, or the end user wishes to retrofit an existing machine for rotary dressing. Space for a motorized spindle is therefore very limited, and in the past,

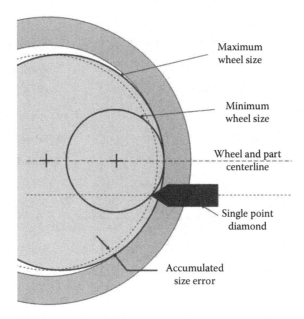

FIGURE 18.8
The effect on part size of diamond dresser being off centerline.

the grinders were equipped with air or small AC servo-driven electric dressing spindles with very limited torque. In turn, the dresser diameter is required to be very small, sometimes only 10–15 mm, and often could only be run counter directionally. The dresser configuration was usually that of a cup so that even though the dresser wear was high, the cup diameter remained constant to maintain a constant crush ratio. Hydraulic motors offered better torque but had problems of heat generation especially during the prolonged period of new wheel dressing, where additional heating would cause thermal errors. Also, industry is now eliminating hydraulics on new grinders where possible to reduce maintenance costs. Fortunately, many of these issues have been overcome in the last few years by the introduction of high power density DC servo motors to eliminate torque deficits, while the use of synthetic needle diamond in dressing cups and disks has reduced dresser wear.

Plunge roll dressing of alox wheels for internal grinding has been standard for alox wheel for 40 years with hydraulic and electric-motorized spindles usually running counter directionally. Limitations in available equipment and system stiffness have held back the conversion of such processes to CBN abrasives, although this is now changing rapidly.

18.4.6 Grinding Cycles

The grind cycle on a standard internal grinder consists of a series of fixed infeed rates and amounts, namely a rapid approach, rough grind, finish grind (10–30 µm), spark out, retract, and unload/load. Figure 18.9 graphs the programmed slide moves with time for one grind cycle.

Figure 18.10 illustrates the associated grinding power. There is a rapid climb during roughing leveling off during the finish grind, then going through an exponential decay during spark out. The spark-out time should ideally be three time constants after which no further material removal or improvement in finish will occur as the threshold force

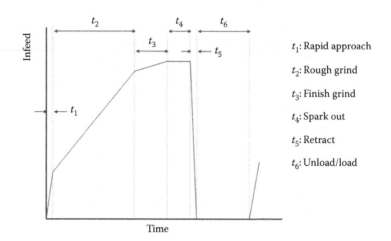

FIGURE 18.9
A typical grind cycle.

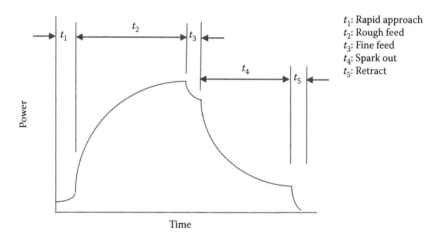

FIGURE 18.10
Variation of power in a typical grinding cycle with set feed rates.

is reached. The graph shows an ideal situation for a part with zero runout entering the grind cycle. In reality, the part will typically have 50–100 μm runout and the smooth climb in power during roughing will have superimposed on it rapid oscillations with the frequency of the work rpm. The amplitude of this will reduce but can still often be seen during spark out demonstrating the level of out of roundness still remaining in the part after grind.

In addition to a superimposed oscillation from runout, any excess stock will show up as a spike during rapid infeed. Any such power spike will be deleterious to wheel life.

18.4.7 Automatic Compensation of Process Variations

A fixed infeed system has major limitations from the inability to adapt to changes in incoming stock or wheel conditions. Much research has been carried out to develop systems to control the system based on force either directly using normal force or indirectly by monitoring power (tangential force).

One of the earliest versions was the Heald-controlled force (CF) system pioneered by Hahn and the Heald Corporation in the 1960s. This applied a controlled normal force from the wheel to the part via a hydraulic ram. The system was relatively crude but highly effective for the appropriate application. Hundreds of these machines are still in production for plain bore applications.

Their use can be problematic when internal plunge profile grinding unless the incoming part preform is well controlled. In bearing manufacture, the race is previously turned in the soft state and then heat treated during which the bearing profile may often distort slightly. To compensate for this, the bearing race is turned to a slightly smaller radius to ensure cleanup after grind. Unfortunately, this leaves excess stock at the two tangent points to the outside diameter. A CF-type process sees only the average force over the whole bearing surface to be ground and will plunge in a rapid approach rate through the excess stock leaving wear steps in the wheel.

In the 1980s with advances in closed loop AC servo drives, it became possible to closely monitor power. Initially this was used to detect part contact during rapid approach and to control spark-out times based on decay time constants. The efforts culminated in the 1990s with controlled power grinders such as the Bryant UL2 (Figures 18.11 and 18.12).

In this type of grinder, all the feed points are still fixed but all the infeed rate parameters are now at a fixed power level controlled by a fast response power monitoring system. The control inputs include a stock load sensor for first contact, rough grind, a rough grind power trip differential for alox wheels (power must drop below a given level at a fixed feed point i.d. dressing midcycle), finish grind, and a spark-out power differential (power decays below a given power level). The grinder also takes into account the initial problems of break in for a new CBN wheel. After the initial wheel dress, the first parts are fed at a set percent of full feed power setting and then ramped up at a given percent increment every given number of parts. This prevents loading of the wheel when dull and allows wheel several grades harder to be used increasing parts/wheel by up to five times.

As the wheel wears through its life its diameter can reduce by 20% significantly affecting the grinding conditions. First the physical amount of CBN becomes less increasing wear

FIGURE 18.11
First installation of Bryant grinders for universal-joint cup grinding.

FIGURE 18.12
Universal-joint cup grinding arrangement.

and reducing parts/dress. Second, d_e, and hence contact length, is reduced allowing a faster infeed rate for a given adaptive power setting. Fast feed rates will further accelerate the rate of wheel wear. The Bryant UL2 has an equivalent wheel skip decay function option that allows the end user to input a parts/dress algorithm as a function of wheel diameter. This is nonlinear and can vary from 200 parts/dress for a new wheel to as little as 10 parts/dress at minimum wheel size. In addition, to prevent the wheel getting too sharp and breaking down exponentially, there is CBN adaptive dress trip rate that is set to a given maximum allowed infeed rate. If this value is exceeded, the wheel is automatically dressed.

This system has allowed enormous improvements in CBN wheel life. Adaptive power grinding systems are now available from several machine tool builders either as software integrated into custom controls similar to that described above or as separate control systems that can be added to an existing grinder. In addition, some of the latest CNCs have options available to control feed to a given power. Its use though is limited to a range of wheel sizes from about 6 mm diameter up to 30 mm. Below about 6 mm, the power signal to noise ratio is too low to get a clean signal, while at diameters greater than about 30 mm, the power detection and reaction times are not yet fast enough to prevent loading with harder CBN wheel grades. Linear motor technology is expected to improve on this.

Systems based on normal force offer the ultimate in control as a direct measurement of system deflection. Hahn again pioneered work in the 1990s on direct measurement of normal forces and compensation for quill deflection. In a series of articles (Hahn and Labby 1995a,b; Hahn 1997, 2000), Hahn analyzed the taper deflection problem and proposed a method based on detection of normal force, calculation of quill deflection and microswiveling of the wheelhead to compensate.

The system deals with the one factor that cannot be influenced by adaptive power control namely threshold normal force. Figure 18.13 plots a typical slide feed cycle for a simple rough grind and spark-out operation together with the actual position of the wheel for a new and used condition. As shown in Figure 18.14, the threshold forces at the end of cycle can be less than half for a worn wheel than a new one. This will result in a significant change in part size and taper over the life of the wheel regardless of spark-out time or roughing infeed rate.

Hahn developed a microangle/force sensing subplate to both measure the force and make the compensation for deflection (Figure 18.15).

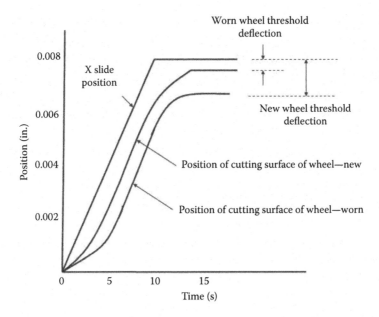

FIGURE 18.13
Actual position of the grinding wheel surface compared with the feed position due to spindle deflections and wheel wear.

The system can also measure work removal parameter Λ ($= Q'/F_n$) which is a key quantitative measure of wheel sharpness related to specific energy. Although Λ cannot be measured directly using power only, it is a key parameter in determining maximum feed rates in weak systems limited by burn. Λ is especially pertinent to alox wheels that become dull with time after dress and may cause burn, if threshold forces are too high.

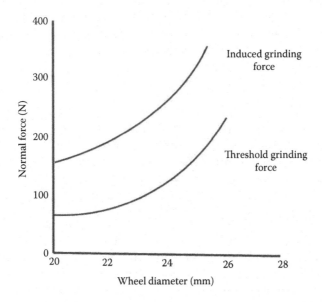

FIGURE 18.14
Variation of threshold force in a grinding cycle with wheel diameter.

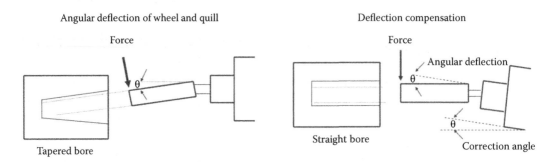

FIGURE 18.15
Compensation for deflection by adjustment of angle in response to force.

Having taper compensation based on normal forces also allows faster overall cycle times for weak systems susceptible to chatter. New wheels with relatively large d_e values can sustain relatively high-feed rates without the onset of regenerative chatter. As d_e is reduced, contact length and regenerative chatter can occur faster. Normally, the feed rates and wheelhead position are set to those that are stable for the smallest diameter. With taper compensation, faster feed rates may be used for larger wheels and algorithms applied to gradually reduced rates to those of the fixed infeed grinder as the wheel gets smaller.

Alternate systems based on measuring the physical deflection of a quill using opposed air gauges have reportedly been used in the bearing industry for some years. A system was recently presented by UVA (n.d.) called GPC-PSH which provides the combination of a sensor in the spindle to measure deflection and a compensating method for swiveling the head. The method is claimed to improve straightness and cylindricity by up to 40% grinding small fuel injection components.

Interest in force controlled grinding is probably much greater than current original equipment manufacturer (OEM) publications and literature would suggest. Competitive industries such as bearing manufacturers are very secretive with highly skilled research groups. Several build their own grinders. The potential for controlled force grinding was presented in 1988–1990 by Heald in combination with The Torrington Company (Bell et al. 1988; Matson et al. 1990; Vaillette et al. 1991). These papers focused on several of the aspects of controlled force grinding discussed above. Part of the interest was in how to optimize the use of CBN that was then still quite new, and process optimization by, for example, gap elimination. Another area of research, however, was in the use of Λ to help improve roundness. During rough grinding, once continuous contact has been made between the wheel and part, the level of eccentricity in the part is given by $(F_{max}-F_{min})/K_s$, where F_{max} and F_{min} are the maximum and minimum normal force, respectively, seen in the power consumption plot as a once per work revolution spike, and K_s is the system stiffness. The number of revolutions needed to round the workpiece up to an acceptable running truth is a function of system stiffness and Λ. Feed rates are automatically regulated to ensure sufficient revolutions of the workpiece before reaching finished size.

In-process gauging is a common, if somewhat delicate, method of controlling size. Longanbach and Kurfess (1998) reported using in-process gauging for the real-time measurement of out of roundness, waviness, or chatter during grinding which was incorporated into a production grinder (Figure 18.16).

Acoustic emission (AE)-based sensors have also been applied to internal grinding. Inasaki (1991) reported on an AE sensor water coupled to the wheel quill as part of an

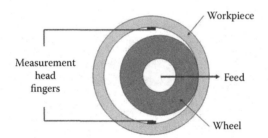

FIGURE 18.16
In-process diameter gauging.

integrated system including power monitoring for developing unattended grinding. AE was used for gap elimination, dressing, and chatter detection. Several machine tool builders offer AE systems as an option for gap elimination and/or touch-dressing CBN wheels, although use of AE for dressing is still limited. Drake (2000) has incorporated the AE sensor in the wheelhead. Studer (2001) has incorporated a ring sensor into the work chuck flange for small wheel applications. EMAG Reinecker (2001a,b) offer sensory analysis modules that include AE for gap elimination and dressing as well as power controlled adaptive grinding. Similar options are offered by Toyo (2000), while Okuma (2000) offers AE for touch-dressing CBN. Finally, the size of AE sensors has reduced to the point that they can now be fitted inside small DC servo-electric dressers.

18.5 Machine Tool Selection

18.5.1 Introduction

In 1938 in the first edition of "Grinding wheels and their uses" Heywood wrote "Internal grinding machines are being improved so rapidly that a machine only a few years old is practically certain to make the internal grinding operation cost more than it would with a later model." The most striking fact about this statement is that in the intervening 65 years, it is true as ever!

When selecting a grinder, the end user must be clear on the part dimensions and tolerances, required production rate, and required grinding operations. Some applications require high volume, dedicated machines while others may have a range of part sizes or require the grinder to do multiple grinding operation not limited to just a single internal grind operation or even just grinding. The following are some examples of common machine types and applications.

18.5.2 Fuel Injection

Grinding fuel injectors is characterized by very high-production requirements, small wheel diameters (2–6 mm), limited coolant access, and extremely tight finishes and tolerances (Figure 18.17). Bore roundness is especially stringent and must be held to <0.5 μm. Taking the requirement for CpK capability into account, the requirement is nearer 0.25 μm. There is usually a bore grind and an angled seat. For diesel injectors, there are often also faces and combination grinds.

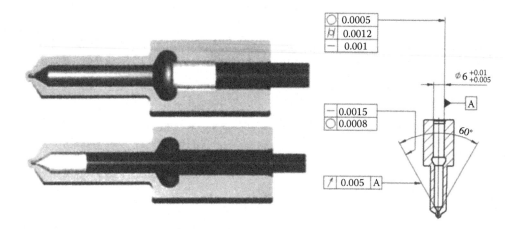

FIGURE 18.17
Example of a fuel injector body. (From UVA, n.d., UVA U80/U88 for internal grinding of small to medium size workpieces. UVA International AB, Bromma, Sweden (includes GPC-PSH system) Trade brochure. With permission.)

Machine tool builders focused on this industry such as UVA and Bahmuller build modular grinders, where the slides and spindle can be easily interchanged or modified thus maintaining flexibility and economics of scale in manufacture. Wheelheads run up to as high as 240,000 rpm (UVA) but speeds in the range of 140,000–180,000 rpm are more standard (Bahmuller 1997; Micron 2000; Okuma 2000; Studer 2001). Spindle power is low, <3 kW, at these speeds. CBN wheels dominate at around 325#–500# (B54–B36) grit size and quills are carbide.

There are numerous ways of mounting the wheels. Gluing direct to the carbide by the end user is problematic. Wheels are, therefore, often supplied on wheel screws. However, producing quills in carbide with high accuracy electrical discharge machining-cut internal threads for these to screw into is expensive. Only a few companies such as Hämex Härdmetallverktyg, Linköping, Sweden can produce to the necessary tolerances. Wheels are therefore also supplied on straight carbide shanks and held in a collet wheel mount.

Getting coolant into the grind zone is difficult. Fortunately, even at the wheel rotational speeds mentioned above, the wheel velocity is still barely 20–30 m/s which limits the risk of thermal damage. Also, low-viscosity oil, chilled and filtered to <5 μm, is used exclusively. If spray holes exist in the injector tip, it may be fed with coolant from behind the wheel. Many wheel spindles can also handle coolant through the center, some as high as 120,000 rpm. Specialist wheel companies can produce precision wheels on hollow wheel screws with fine perforations in the wheel face for seat or end grinding to deliver coolant evenly right at the grind point. For new applications, especially grinding at the end of long bores, finite element analysis (FEA) analysis of the dynamic stiffness and resonant frequencies of the quill is required.

The work drive is the most critical component for its running truth governs the running truth of the finished part. Aerostatic and angular contact ball bearing drives have been used but the most common is hydrostatic. Industry standard for precision is currently 0.1 μm. Speed of rotation is up to 4000 rpm. Since the workhead is also one of the stiffest parts of the grinder, some OEMs take advantage of this and mount the dressing diamond as a ring around the chuck.

Diaphragm, collet, and jaw chucks are all used for part holding (Figure 18.18). The primary concern is to avoid distortion of the part and to maintain accurate running truth.

FIGURE 18.18
Twin wheelhead machine for seat and bore grinding of injectors. (From UVA, n.d., UVA U80/U88 for internal grinding of small to medium size workpieces. UVA International AB, Bromma, Sweden (includes GPC-PSH system) Trade brochure. With permission.)

The jaws should be regularly ground; the same wheel used as for the bore grinding can grind high-speed steel jaws. Otherwise, dressable vitrified diamond wheels are available for grinding carbide jaws in situ.

Most slideways are hydrostatic with ballscrew/AC servo drives accurate to 0.1 µm. The high oscillation/short-stroke length required has recently led numerous OEMs to provide linear magnetic motors. All state of the art machines have sophisticated CNCs with a variety of the AE dressing and gap elimination options described above.

Load unload is carried out with a pick and place arm or gantry, as the parts are too small or awkward in shape to load by a gravity fed method.

18.5.3 Automotive Components (Lifters, Tappets, UJ Cups, and Plain Bearings)

This refers to a range of products using wheels in the 5–25 mm diameter range to grind extremely high volumes of components on dedicated grinders (Figure 18.19). Tolerances now become somewhat relaxed compared with those for fuel injection and fall in the 1–5 µm range.

The key to these machines though is cycle time, much of which is won or lost in nongrind time. For example, a 20-mm diameter component with 0.2 mm stock radial can be rough ground at $Q' = 5$ mm³/mm s in 2.5 s. However, the total cycle time can vary from over 15 s on an old hydraulic/mechanical grinder to less than 7 s on a modern CNC grinder (Table 18.3). The savings are in improved load/unload mechanisms, faster slide movements, gap elimination, optimized spark-out times, and the ability to use CBN wheels (Table 18.3).

18.5.4 Machine Layout

The machine design starts with the overall layout of the components. Since only a limited amount of travel is required, the X and Z axes are on independent slide systems

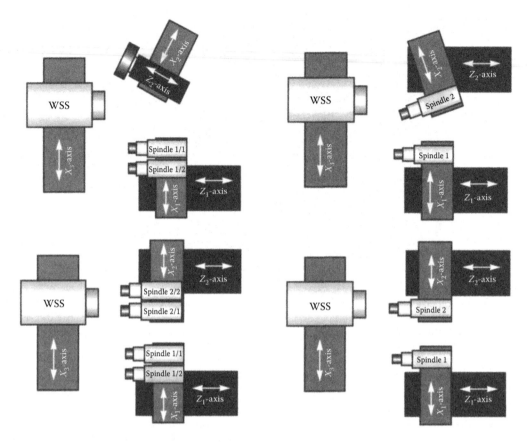

FIGURE 18.19
Modular machine designs allow for different configurations of multiple grind operations on seats, bores, and faces. (From Bahmuller, 1997, High precision internal grinding with the modular I-line. Wilhelm Bahmuller, Pluderhausen, Germany. Trade brochure. With permission.)

as in the example from Bryant (Figure 18.20; Koepfer 1995). This allows all the moving components, wheelhead and workhead close to the machine bed and ball-screws for improved stiffness and thermal stability. Most builders use hydrostatic slides for rigidity and speed (up to 40 m/min), and especially on the Z axis with the need for high-speed oscillation.

TABLE 18.3

Comparison of Cycle Times from Machine Design Changes in Last 20 Years

Grind Step	Pre-1980 m/c (s)	2000 m/c (s)	Advances
Load/unload	5.0	2.0	High-speed loaders
Rapid approach	3.0	0.5	AE/adaptive grind/CNC
Rough grind	2.5	2.5	
Dress	1.0	0	CBN replaces alox
Finish grind	1.0	1.0	
Spark out	3.0	1.0	Adaptive grind
Total cycle time	15.5	7	

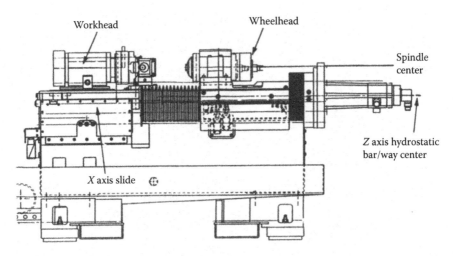

FIGURE 18.20
Bryant UL2 high-production grinder configuration. (From Koepfer, C., 1995, Production I d grinding—Change in the wind? *Modern Machine Shop*, December. With permission.)

18.5.5 Wheel Speeds

Wheel speeds with CBN of 30–60 m/s are now readily achievable with available spindles operating at 40,000–70,000 rpm with a good torque range. Higher speed spindles are available but these appear to lack the same rigidity, power, or durability. Most modern grinders will maintain a constant wheel velocity, as the wheel gets smaller. It is important to check there is available rpm/torque for the full life of the wheel size and abrasive type especially if planning to use both CBN and alox.

18.5.6 Work-Spindle Run-Out

Workhead spindle runout is held within the range 0.25–1.00 μm depending on part tolerances. Spindle speeds can vary within the range of 500–3500 rpm depending on the particular application.

18.5.7 Work-Loading Mechanisms

The selection of the appropriate load/unload mechanism is a major source of cycle time savings. Options include

1. *Through the work-spindle loading*: Small parts are fed through the workhead into a chuck. Each new part ejects the preceding part that falls or is held by a receiving arm. Grinding resumes as soon as the ground part and receiving arm are clear.

2. *Linear arm and two-arm loaders*: Combination swing and linear motion for front loading. Single-arm version loads only the part being ejected into an unloading chute. Two-arm loader: Latest designs use a combination of swing and linear motion with high-speed two-axis AC servo control (Figure 18.21).

3. *Flow systems to two-roll and shoe centerless support arrangements*: A shoe centerless work support arrangement grinds the i.d. concentric to the o.d. The latest flow or

FIGURE 18.21
Swing-arm part loader. (From Danobat, 1992, Gear-box—Grinding machines gear grinding. Danobat Corporation, Spain. Trade brochure. With permission.)

gravity fed load devices can load/unload in 0.5 s (Toyo 2000, n.d.). A gravity fed loader is used with a simple ejector arm to kick finished parts off the shoes. A variation on this is a two-roller and one-shoe arrangement (Figures 18.22 and 18.23).

18.5.8 Coolant

Coolant is more likely to be water-based than oil in the United States, although the reverse is true in Europe. The difference in performance can be as great as the difference between

(a)   (b)

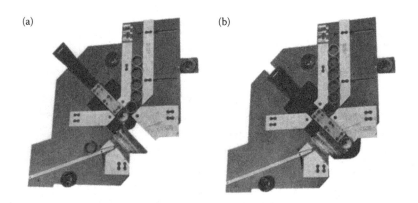

FIGURE 18.22
Bryant UL2 loader for shoe centerless mode grinding. (a) At load/unload position. (b) At grind position. (From Bryant, 1995, The ultraline UL2 high speed grinding machine. Bryant Grinder. Trade brochure. With permission.)

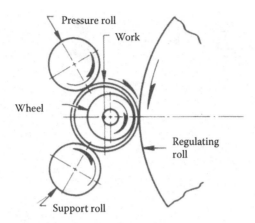

FIGURE 18.23
Two-roll internal centerless grinding machine.

fixed infeed and adaptive power grinding. For plain bores, coolant is supplied from both front and back. For blind end applications, through-the-spindle coolant is typical.

18.5.9 Gauging

Gauging is standard as either in process as discussed above or more commonly postprocess using air gauges to measure size, taper and even profile. Typical systems will plot a running performance SPC graph and make automatic adjustments on size and sometimes even taper. The systems average out the values of a given number of preceding parts (usually three).

18.5.10 Flexible Multipurpose Grinders

This refers to a class of grinders that must have the flexibility to perform a range of grinding operations that may also include face and o.d. grinding in a broad range of parts. Multispindle grinders also fall into this category. These can be assembled either on separate linear axes or on turret arrangements that may carry up to four spindles (Figures 18.24 through 18.26).

FIGURE 18.24
Linear and turret spindle arrangements. (From Okuma, 1996, GI 20N CNC internal grinder. Okuma Corp, Aichi, Japan. Trade brochure. With permission.)

FIGURE 18.25
Four-spindle turret on Tripet Grinder. (From Anon, 1989, Anything but the same old grind. *Machine and Tool Blue Book*. With permission.)

More recently, numerous machines have appeared on the market for simultaneously grinding bores and flanged o.d. (Figures 18.27 and 18.28).

Grinder designs will often shift from having independent X and Z axes slides to compound slide arrangements to increase the travel range and hence range of part sizes. Stacking the axes also allows contouring moves relative to the part or dresser. Cycle times are now longer as the slide movements are greater and loading systems have to be flexible or in many cases, loading is manual. However, they do offer the great advantage to the end user of performing several operations in one chucking (Figure 18.29).

The last category of machines is the natural result of the consolidation in machine tool manufacturing methods, tighter dimensional tolerancing, and JIT manufacturing

FIGURE 18.26
Three-spindle turret on Voumard grinder. (From Voumard, n.d., Grinding machines. Voumard Machines CO SA, Jardiniere, Switzerland. Trade brochure. With permission.)

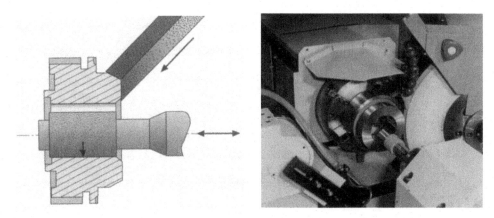

FIGURE 18.27
Simultaneous grinding of bore and flange. (From Bell, W. F., Brough, D., Rowe, W. B., 1988, Practical achievement and monitoring of high rate internal grinding—Part 1—Machine instrumentation and data acquisition. *3rd International Grinding Conference*, Wisconsin, USA, April 10, 1988. SME MR88-611. With permission.)

demands. This refers to the combination of machining and grinding in a single chucking. This has been receiving increased impetus since the mid-1990s and has produced some impressive improvements in productivity and capital equipment savings.

Several OEMs have designed vertical spindle machines, where the workhead is above the grinding wheel. An example of this from EMAG (2001b) is shown in Figure 18.30a and b. Having the workhead above the wheel allows the part to be loaded by gravity through the spindle into the chuck giving a load time of 1.5 s. Alternatively, the part is brought on a conveyor into the chuck from below. Load times are still as little as 4 s.

Another example from Campbell uses a conventionally configured vertical spindle grinder for processing gear bores and faces. Originally conceived as a purely grinding operation, the introduction of turning for the gear face reduced cycle time by 70% (Figures 18.31 and 18.32).

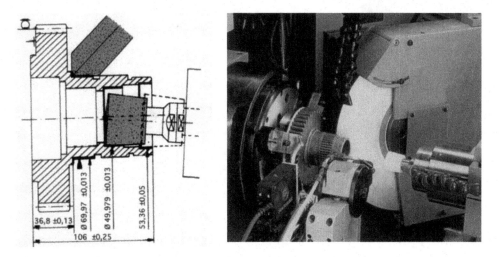

FIGURE 18.28
Simultaneous grinding of bore, external diameter, and face. (From Danobat, 1992, Gear-box—Grinding machines gear grinding. Danobat Corporation, Spain. Trade brochure. With permission.)

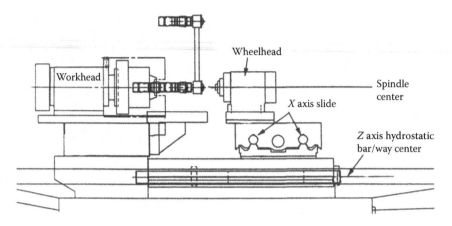

FIGURE 18.29
Compound slide grinder layout. (From Koepfer, C., 1995, Production I d grinding—Change in the wind? *Modern Machine Shop*, December. With permission.)

FIGURE 18.30
Combination grind and turn arrangements with part loading through the vertical workhead. (From EMAG Reinecker, 2001b, Vertical grinding center, vertical turning and grinding center. EMAG Maschinenfabrik GmbH, Salach, Germany. Trade brochure. With permission.)

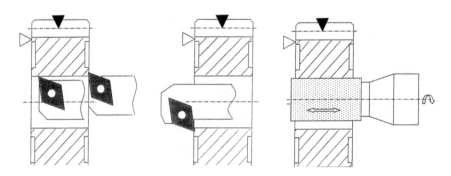

FIGURE 18.31
Strategy for turning/grinding of gear bore and faces.

FIGURE 18.32
Combined grinding and turning machine. (Courtesy of Campbell Grinder with permission.)

18.6 Trouble Shooting

When diagnosing problems with an internal grinding process, it is often necessary to go back to basics, as for the initial setup of the machine. For internal grinding, the key requirement is to ensure there is a straight line of contact between the grinding wheel and the part. There are four principal qualifications

1. X and Z axes must both produce a straight line of motion.
2. The workhead axis must be parallel to the Z axis.
3. The wheelhead axis must be parallel to the Z axis and lie in the same plane as the workhead axis.
4. The diamond must lie in the plane established by the wheelhead and workhead axes.

If the X or Z axes are not true or worn, the wheel path will not be straight. During dressing, the wheel will be profiled or tapered creating high points in the grind with excessive wear, part taper, and even feed lines. This cannot be corrected by adjusting the wheelhead. A similar problem will occur if the plane of the diamond is incorrect.

Bellmouth is another common problem which is produced either by worn slides or misalignment of the workhead axis with the wheel path axis. This cannot normally be corrected by reducing the stroke length.

Center height errors were discussed above in terms of the impact on dressing. Another effect is when misalignment of these axes results in the axis of the wheel crossing the centerline of the diamond. In this case, the wheel will be dressed with a hyperbolic shape again resulting in a high point with its attendant problems.

Once these issues are resolved then the more normal corrective actions can be taken. The commonest are

Feed line on workpiece	Dressing problems such as worn or poorly clamped diamond
	Dresser spindle runout
	Wheel grade too hard causing loading
Scratching of the part	Wheel too soft or coarse causing grit pullout
	Poor coolant filtration
	Wheel speed too slow for wheel grade causing breakdown
Burn	Wheel grade too hard
	Poor coolant delivery
	Work speed too slow
	Wheel speed too high
	Wheel infeed rates too high
Wheel glazed	Wheel too hard or too fine
	Dressing not aggressive enough or diamond is worn
	Too high an oil content in coolant—lower concentration, increase volume (alox wheels)
	Not working the wheel hard enough—increase work and infeed rates
Wheel loaded	Wheel grade too hard or fine
	Dressing not aggressive enough or diamond worn
	Poor coolant/delivery/filtration
	Not working the wheel hard enough—increase work and infeed rates
Bellmouth	Overtravel of wheel
Taper	Wheel breakdown
	Insufficient quill stiffness (CBN—back of part small)
	Wheel not dressed aggressively (CBN—back of part small)
	Incorrect stroke end points for oscillation
Out of round bore	Nonclean up
	Overclamping or balance problem with workhead/chuck
	Heat distortion during roughing of face grinding operation
Chatter	Wheel too hard
	Runout or balance issues with quill/wheel/spindle usually after a crash or as wheel bearing fails
	Machine vibrations
	Workspeed too high
	Dresser play, worn diamond
Chipping/cracking	Contamination in incoming parts (shot peen, tumbling media)
	Burrs from turning operation
	Expansion of arbor by temperature curing glue for mounting CBN wheel
	Incorrect rapid approach position for stock level or too high a trip level on gap elimination
	Poor coolant delivery

References

Anon, 1989, Anything but the same old grind. *Machine and Tool Blue Book*.
Bahmuller, 1997, High precision internal grinding with the modular I-Line. Wilhelm Bahmuller, Pluderhausen, Germany. Trade brochure.

Bell, W. F., Brough, D., Rowe, W. B., 1988, Practical achievement and monitoring of high rate internal grinding—Part 1—Machine instrumentation and data acquisition. *3rd International Grinding Conference*, Wisconsin, USA, April 10, 1988. SME MR88-611.

Bryant, 1995, The ultraline UL2 high speed grinding machine. Bryant Grinder. Trade brochure.

Danobat, 1992, Gear-box—Grinding machines gear grinding. Danobat Corporation, Spain. Trade brochure.

Drake, 2000, GS-I Internal grinding system specification. Drake Manufacturing, Warren OH. Trade brochure.

EMAG Reinecker, 2001a, Internal grinder ISA 103 CNC, ISA 102 CNC. Reinecker Karstens Corporation, Neu Ulm, Germany. Trade brochure.

EMAG Reinecker, 2001b, Vertical grinding center, vertical turning and grinding center. EMAG Maschinenfabrik GmbH, Salach, Germany. Trade brochure.

Hahn, R., 1997, Microangling/force—Sensing subplate raises internal grinding precision. *Abrasives Magazine*, December/January.

Hahn, R., 2000, PC-control. Sensors for production i d grinding. *Abrasives Magazine*, December/January 2000, 17–21.

Hahn, R. S., Labby, P., 1995a, Deflection compensation stops bore taper. *American Machinist*, June 1995, 42–46.

Hahn, R. S., Labby, P., 1995b, Normal force sensors raise internal grinding productivity. *1st International Machining and Grinding Conference*, Michigan, USA, September 12, 1995, SME MR95-180.

Inasaki, I., 1991, Monitoring and optimisation of internal grinding process. *Annals of the CIRP*, 40/1.

Koepfer, C., 1995, Production I d grinding—Change in the wind? *Modern Machine Shop*, December.

Longanbach, D. M., Kurfess, T. R., 1998, Real-time measurement for an internal grinding system. *NAMRC XXVI*, May 19, 1998, Georgia, USA, SME MS98-261.

Matson, C. B. et al., 1990, Practical achievement and monitoring of high rate internal grinding—Part 2—Evaluation of collected data. *4th International Grinding Conference*, Michigan, USA, September 10, 1990. SME MR90-08.

Micron, 2000, Super precision internal grinder MIG-101. Micron Machinery. Yamagata, Japan. Trade brochure.

Okuma, 1996, GI 20N CNC internal grinder. Okuma Corp, Aichi, Japan. Trade brochure.

Okuma, 2000, CNC internal grinder for mass production GI-10N. Okuma Corp, Aichi, Japan. Trade brochure.

Salmon, S. C., 1984, *Abrasive Machining Handbook*. Book Korber AG, Germany.

Studer, 2001, Process optimization for internal cylindrical grinding. *Motion—Trade Magazine for Schleifring*, September.

Toyo, 2000, T-11L series CNC internal grinding machines. Toyo Advanced Technologies, Hiroshima, Japan. Trade brochure.

Toyo, n.d., T-117 CNC T-157CNC CNC internal grinding machine. Toyo Advanced Technologies, Hiroshima, Japan. Trade brochure.

UVA, n.d., UVA U80/U88 for internal grinding of small to medium size workpieces. UVA International AB, Bromma, Sweden (includes GPC-PSH system). Trade brochure.

Vaillette, B. D. et al., 1991, CBN—Adaptive grinding. *Superabrasives 91 Conference Proceedings*, SME, Chicago.

Voumard, n.d., Grinding machines. Voumard Machines CO SA, Jardiniere, Switzerland. Trade brochure.

19

Centerless Grinding

19.1 The Importance of Centerless Grinding

Centerless grinding is used for fast production and high accuracy. The process is best for larger batches, where setup times are small compared to machining time. Small rollers and needles, for example, are machined to close tolerances in quantities of millions by through-feed centerless grinding.

Batch quantities do not need to be large to be cost-effective. Setup times can be reduced with appropriate setup fixtures. An advantage of centerless grinding is that center holes are not required for the purpose of location. Location of the workpiece during machining is provided from the newly machined surface itself. This eliminates a production operation and saves cost. It also avoids shape errors associated with out of roundness of the center holes. Another advantage is that the workpiece does not need to be clamped to a drive device. This saves time, and accuracy is improved, since driving devices are a source of errors.

The range of workpieces and materials produced by centerless grinding is wide. Notable examples include

Steel bar stock

Needles

Ball and roller bearings

Bearing rings

Glass rods

Plastic rods

Bottle corks

Silicon bars for wafer production

Uranium rods

Formed valve spools

Axles

Valve stems

Centerless grinding is used for rapidly cleaning up, roughing, and finish grinding. Workpieces may be straight or formed cylindrical shapes. Workpiece surfaces can be either internal or external cylindrical shapes. The process is often automated with barrel vibration feeders or other devices using sequential pneumatics, hydraulics, or robots. This means that machines can continue production with minimal supervision.

The following table lists features characterizing the efficiency of the process and factors that make these features possible.

Feature	Associated Factor
Sometimes faster than turning	Wide grinding wheel
Long wheel life	Large diameter wheel
Fast loading and unloading	Clamping unnecessary
Accuracy	Center errors avoided
Hard materials machined	Very hard wheel grits

19.2 Basic Processes

19.2.1 External Centerless Grinding

Centerless grinding is compared with between-center grinding in Figure 19.1. In center grinding, the workpiece is located on centers by means of center holes drilled in the workpiece. In centerless grinding, the workpiece is located on the workpiece perimeter. This can be a significant advantage for achieving part roundness and also for achieving part straightness since barreling due to workpiece bending is avoided. However, if the setup geometry is wrong, it can lead to poor roundness.

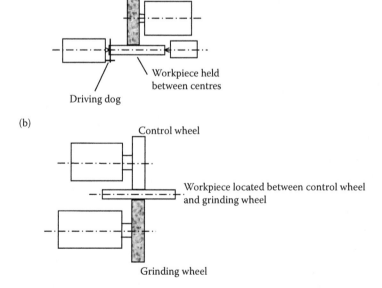

FIGURE 19.1
Work holding in (a) center grinding and (b) centerless grinding.

Figure 19.1a shows a plan view of external center grinding. The workpiece is mounted between centers and usually driven by a driving dog attached to the workpiece. The rotational speed is controlled by a workhead spindle drive.

This contrasts with the arrangement for centerless grinding shown in Figure 19.1b. In centerless grinding, the workpiece is pushed against the grinding wheel (GW) by a control wheel (CW), alternatively known as a regulating wheel. The workpiece is supported underneath by a workrest (WR) as in Figure 19.2a.

A CW controls workspeed by friction. In normal operation, the workpiece has the same surface speed as the CW. The advantage of this is simplicity. The workpiece is pushed into the grinding position with no requirement for clamping. Plunge grinding and through-feed grinding operations may be performed as described in later sections.

19.2.2 Approximate Guide to Work Height

The workpiece center in Figure 19.2a is usually higher than the GW and CW centers. This is because the height of the workpiece is set above center to achieve a rounding action as described later.

An approximate guide to work-center height is

$$h_w = \frac{1}{16} \cdot \frac{1}{[1/(d_s + d_w) + 1/(d_c + d_w)]} \tag{19.1}$$

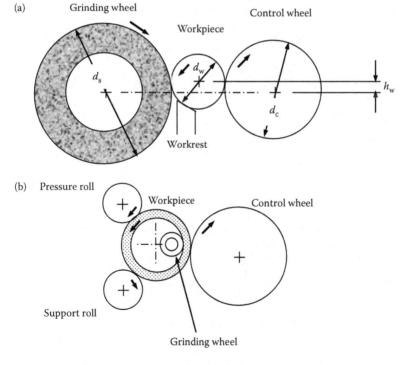

FIGURE 19.2
Basic centerless grinding processes: (a) external and (b) internal.

where d_s is the GW diameter, d_c is the CW diameter, and d_w is the workpiece diameter.

It should be emphasized that Equation 19.1 is only an approximate guide. Detailed guidance on setup values is given in sections below.

19.2.3 Internal Centerless Grinding

Internal centerless grinding is illustrated in Figure 19.2b. In the example shown, a CW and a support roll locate and support the workpiece. A pressure roll is used to firmly hold the workpiece against the support roll and the CW. In this setup, the external diameter needs to be ground prior to the internal diameter. The external surface becomes the reference for concentricity and roundness.

Using the process illustrated, an internal bore can be ground without the need to clamp the workpiece as required in conventional internal grinding.

19.2.4 Shoe Centerless Grinding

Shoe centerless grinding is used for workpieces of short length to diameter ratio. Shoe centerless grinding is illustrated in Figure 19.3. Figure 19.3a shows an arrangement for external shoe grinding, and Figure 19.3b shows an arrangement for internal shoe grinding. The workpiece is located axially by a magnetic drive plate. The workspeed is controlled by the speed of the magnetic drive plate.

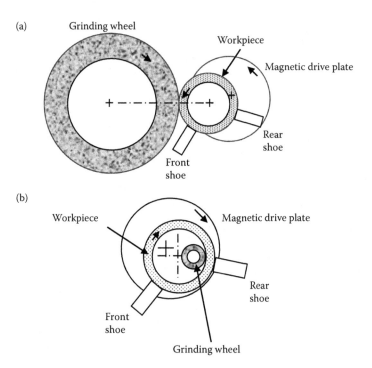

FIGURE 19.3
Shoe centerless grinding: (a) external and (b) internal.

19.2.5 Roundness and Rounding Geometry

The rounding action in centerless grinding depends on the setup geometry. The setup geometry is illustrated in Figure 19.4. Angles α and β have a strong effect on rounding action (Dall 1946). Angle β is termed the tangent angle. The angles depend on the work height h_w and the WR angle γ as given below. Good roundness in centerless grinding requires the use of appropriate values of these angles. It is therefore advisable to plan a procedure to ensure the required values can be quickly set up and checked. Typical values are

$$\gamma = 30°$$

$$\beta = 6° - 8°$$

For example, if $\beta = 7°$ and $\gamma = 30°$, $\alpha = 57.3°$, where $\alpha = 90 - \gamma - \upsilon \cdot \beta$ and $\upsilon = 0.38$ for a GW 300 mm diameter, CW 178 mm diameter, and workpiece 25 mm diameter. These calculations can be deduced from the following expressions.

Angles α and β are related to work height and WR angle as follows:

$$\alpha = \frac{\pi}{2} - \gamma - \beta_s \tag{19.2}$$

$$\beta = \beta_s + \beta_{cw} \tag{19.3}$$

where

$$\beta_s = \sin^{-1}\left(\frac{2h_w}{d_s + d_w}\right) \tag{19.4}$$

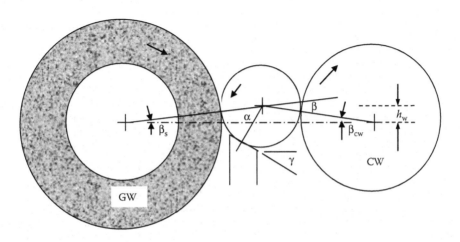

FIGURE 19.4
Centerless rounding geometry.

$$\beta_{cw} = \sin^{-1}\left(\frac{2h_w}{d_{cw}+d_w}\right) \tag{19.5}$$

The work height corresponding to a required tangent angle β less than $10°$ ($\equiv 0.1745$ rad) is, therefore, given by

$$h_w = \frac{\beta}{2[1/(d_s+d_w)+1/(d_{cw}+d_w)]} \tag{19.6}$$

where β is expressed in radians.

Centerless grinding conducted under best conditions is very accurate. Under nonoptimal conditions, involving vibrations and poor setup geometry, the process may produce lobed workpieces. The analysis of roundness in centerless grinding is discussed in Section 19.10. In summary, odd-order lobing—3, 5, 7, etc., lobes—is not quickly removed with values of tangent angle less than $4°$. With larger values of tangent angle, there is a danger of both higher order even lobing and higher order odd lobing.

Figure 19.5 illustrates experimental rounding results for tangent angles in the range, $\beta = 0°-10°$ with $20°$ and $30°$ WR (Rowe et al. 1965). The results were obtained by grinding workpieces on which a flat had been previously ground along the length to create a controlled initial error. The depth of the flat was $9.2\ \mu m$. The controlled initial error on the workpieces allowed rounding tendencies to be precisely determined for each setup. The optimum range for tangent angle was found to be $6°-8°$ with a $30°$ WR. Smaller and larger tangent angles reduced the rounding tendency. A $20°$ WR led to 16-lobe instability at a $10°$ tangent angle. The $30°$ WR was more stable than a $20°$ WR.

It has been suggested that low roundness errors can be obtained by varying the setup angles during the grinding operations (Harrison and Pearce 2004). For example, a larger tangent angle may be employed for rough grinding to eliminate odd-order lobing and a smaller tangent angle for finish grinding.

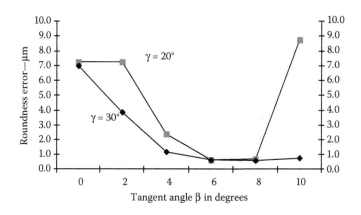

FIGURE 19.5
Roundness errors measured after grinding a workpiece with a flat using two different workrest angles: Initial workpiece roundness error $9.2\ \mu m$.

19.2.6 System Interactions

A machining process is an interaction of geometric, kinematic and dynamic, physical, chemical, and tribological phenomena. It is not necessary to fully understand all these interactions to achieve a satisfactory process, but it is necessary to be aware of the factors that can influence the outcome and seek to determine best practice for each part of the system. For more detail on system interactions, the reader is referred to "Tribology of Abrasive Machining Processes" by Marinescu et al. (2004).

Examples of the influences on centerless grinding accuracy are illustrated in Figure 19.6. The input to the process is viewed as the feed motion and the output as workpiece shape and quality. Quality is defined by size, roughness, roundness, straightness, surface integrity, etc. For high accuracy work, there are many factors that determine accuracy and quality. For example

- Accuracy of the machine setup affects parallelism and size accuracy.
- The shape of the GWs affects roundness.
- Wheel dressing affects surface roughness.
- The composition of the GW affects grinding forces, process stability, wheel wear, temperature rise, and surface integrity.
- Wheel wear affects consistency of size accuracy, shape accuracy, and surface roughness.
- Rounding geometry affects elimination of out of roundness.
- Friction affects speed control.

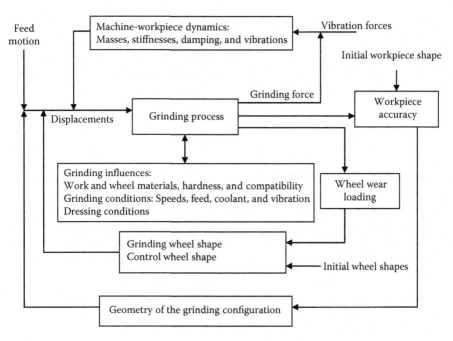

FIGURE 19.6
System interactions and accuracy in centerless grinding.

- Process fluid (coolant) affects cooling, flushing, process lubrication, and avoidance of wheel loading.
- Machine and workpiece dynamics affect process stability and avoidance of vibrations.
- Rotating masses, motors, drive belts, and feed drives affect forced vibrations.

So far, discussion has been limited to accuracy and quality. There will also be concern for productivity and costs. Productivity is the rate at which goods can be produced to specified quality levels. Productivity and costs are closely related.

An improvement in accuracy capability of a machine is usually accompanied by higher productivity. This is because the manufacturer works to specified tolerances and quality levels. If the machine is capable of better quality, the manufacturer can increase the rate of production and remain safely within tolerances. The manufacturer also seeks to reduce process costs for specified quality levels. This can often be achieved by increasing wheel-speed, changing to high-performance abrasives, improved machine structures, improved feed cycles, and improved control systems. The following sections discuss relevant aspects for optimizing system performance.

19.3 Basic Relationships

The basic relationships for parameters such as removal rate, power, and grinding conditions are not in every case the same as for grinding between centers. The basic relationships for centerless grinding are as follows.

19.3.1 Depth of Cut

The real depth of cut at the commencement of grinding is less than expected due to machine and system deflections. Depth of cut and deflections build up to a steady value with uniform infeed. The steady value of real depth of cut is

$$a_e = \frac{v_f}{2n_w} = \frac{1}{2} \cdot \pi \cdot d_w \cdot \frac{v_f}{v_w} \tag{19.7}$$

where v_f is the infeed rate, n_w is the work rotational speed, d_w is the work diameter, and v_w is the workspeed.

It is important to include the factor of one-half. This factor arises because feed is relative to the workpiece diameter. As the workpiece diameter reduces, the workpiece center retreats from the GW so that the depth of cut is halved. The depth of cut is necessary for determination of removal rate.

19.3.2 Removal Rate

The rate material is removed from the workpiece is given by

$$Q_w = a_e \cdot b \cdot v_w \tag{19.8}$$

It is more useful when assessing the overall efficiency of material removal to consider specific removal rate per unit width of contact of the GW.

$$Q'_w = \frac{Q_w}{b} \tag{19.9}$$

19.3.3 Power

In many cases, power can be measured directly from the machine. If there is provision for measurement of tangential grinding force on a machine, power can be derived from:

$$P = F_t \cdot v_s \tag{19.10}$$

where F_t is the tangential grinding force and v_s is the grinding speed.

19.3.4 Specific Energy

The grinding energy to remove a volume of material is a measure of the efficiency of material removal. The specific energy is given by

$$e_c = \frac{P}{Q_w} \tag{19.11}$$

Specific energy is a measure of the difficulty of grinding a work material under particular grinding conditions. Hard materials give rise to higher specific energy requirement than soft materials. Typical values of specific energy range from 10 to 200 J/mm³ of material removed. Lower and higher values than these are possible. A value of 10 J/mm³ represents an easy-to-grind material with very efficient grinding conditions. A value of 200 J/mm³ is typical of many finishing operations at lower process efficiency.

The process engineer having established a database for particular materials and GWs uses specific energy to determine machine power requirements and machining rates.

19.3.5 Contact Length

The total contact length is due to contact force and grinding geometry. Geometric contact length must be added to dynamic contact length (Rowe 2014) according to:

$$l_c^2 = l_g^2 + l_f^2 \tag{19.12}$$

Geometric contact length: The geometric contact length, based on the undeformed shape of the GW, depends on the depth of cut.

$$l_g = \sqrt{a_e \cdot d_e} \tag{19.13}$$

where the effective GW diameter is given in Section 19.3.6.

Dynamic contact length: Dynamic contact length depends on the force between the GW and the workpiece in the same way that contact area between an automobile tire and the road depends on the weight of the automobile. Dynamic contact length (Rowe et al. 1993) is given by

$$l_f = \sqrt{\frac{8R_r^2 \cdot F_n \cdot d_e}{\pi \cdot b \cdot E^*}} \tag{19.14}$$

where R_r is a roughness factor that depends on the roughness of the GW. Typically, R_r is of the order of 10 for a vitrified wheel. F_n is the normal grinding force, b is the width of contact and is usually equal to the width of the GW or the length of the workpiece. E^* is the effective modulus of elasticity of the GW and the workpiece materials. The value is given by

$$\frac{1}{E^*} = \frac{1 - v_1^2}{E_1} + \frac{1 - v_2^2}{E_2} \tag{19.15}$$

19.3.6 Equivalent GW Diameter

The wear of the GW is greatly affected by the equivalent GW diameter. The effective diameter in internal grinding is much greater than in external grinding, even though internal GWs are normally much smaller than external wheels. Equivalent diameter is defined as the wheel diameter that gives the same contact condition as in flat surface grinding. The equivalent diameter is based on the relative curvature of the two surfaces in contact.

$$\frac{1}{d_e} = \frac{1}{d_s} \pm \frac{1}{d_w} \tag{19.16}$$

The negative sign applies for internal grinding. Due to the conformity in internal grinding, a softer GW is required to withstand the increased length of contact and the consequent increase in rubbing contact. The longer geometric contact length increases rubbing wear and tends to cause the wheel to glaze.

19.3.7 Equivalent Chip Thickness

Equivalent chip thickness is a measure of the depth of penetration of the abrasive grains into the workpiece. Equivalent chip thickness is given by

$$h_{eq} = a_e \cdot \frac{v_w}{v_s} \tag{19.17}$$

Equivalent chip thickness is the thickness of the layer of workpiece material removed at wheelspeed. Since the speed ratio is typically 100, the thickness of the layer removed at wheelspeed is approximately one-hundredth of the thickness of the layer removed at workspeed. With 0.05 mm depth of cut, equivalent chip thickness is typically 0.5 µm.

In practice, depth of penetration of the abrasive grains into the workpiece varies over a much greater range due to variable spacing between grains, variable depth of grains below the wheel surface and surface roughness. A rough worn GW causes much greater grain depths than the same wheel freshly dressed. However, equivalent chip thickness is useful for comparing grain penetration for different grinding operations using a similar abrasive structure.

19.3.8 Grinding Ratio

Grinding ratio is a measure of the suitability of an abrasive for a particular grinding operation. Grinding ratio is defined as

$$G = \frac{\text{Volume of work material removed}}{\text{Volume of wheel material removed}} = \frac{V_w}{V_s} \tag{19.18}$$

Hard work materials and interrupted cutting operations tend to reduce the grinding ratio. In precision grinding with easy-to-grind materials, a G ratio of 5000 or more may be achieved. However, for a difficult-to-grind material operating under adverse conditions, the G ratio may drop to 1. In adverse conditions, it may be important to increase wheel-speed to increase G ratio.

19.4 Feed Processes

19.4.1 Plunge Feed

In plunge-feed grinding, the workpiece is fed in a radial direction toward the GW as shown for a stepped shaft in Figure 19.7. In the example shown, the operation simultaneously grinds two diameters. The plunge-feed rate v_f is the speed of radial approach.

Plunge feed may be implemented either through a grinding wheelhead slideway or through a control wheelhead slideway. Larger machines have slideways for both wheelheads for convenience of machine setting. Often, it is convenient to feed the CW with both the GW and the WR remaining fixed in position. When the feed slide is fully retracted, clearance is created between the CW and the WR. On retraction, the workpiece rolls down

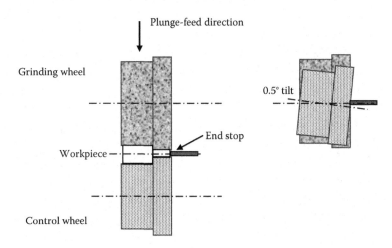

FIGURE 19.7
Plunge feed of a stepped shaft. The control wheel is tilted 0.5°–1° to locate the workpiece against the end-stop.

the WR to a chute. This allows ease of workpiece unloading from the machining position. The workpiece can be removed either manually or by an automated device.

A plunge-feed cycle may consist of several phases as illustrated in Figure 19.8. After loading the workpiece, the GW makes a rapid approach to take up the clearance between the GW and the workpiece. At a small distance from the workpiece, rapid approach is ceased and speed is reduced to a fast grinding feed rate. The fast feed rate may remove between half and 80% of the stock to be removed. Feed rate is then switched to a slow feed rate in order to improve roundness, reduce roughness, and bring the workpiece close to finish size.

As the feed movement reaches the position for finish size, the feed is stopped and there is a period of dwell. The purpose of the dwell is to allow "spark out." During dwell, the process produces sparks for a few seconds or longer due to the elasticity of the machine and GWs. The purpose of spark out is to allow the deflections of the machine and GWs to spring back and to allow the grinding force to decay to a low level.

Spark out is very important for size accuracy, roundness, and low roughness. After spark out, the wheel is smoothly retracted to allow unloading of the workpiece and reloading of the next workpiece.

Intermediate feed-cycle changes and final retraction may be signaled to a computer numerically controlled (CNC) controller using in-process size gauging of the workpieces. A size gauge indicates when a programmed size has been reached and the next phase of the feed cycle is triggered. When the final size has been reached, the system retracts the feed drive. For this system to work effectively, the system has to establish a relationship between feed axis position and workpiece size. This relationship is set up as part of the initial setting procedure. Usually, automatic compensation will be made for GW wear and GW dressing.

An alternative to spark out is the use of a very fine feed rate. The use of a very fine feed rate ensures that the workpiece reaches finish size without significant delay. The GW tends to act slightly softer when using a very fine feed rate instead of a dwell.

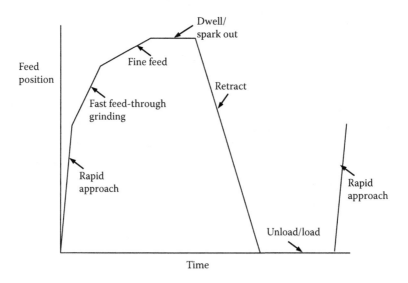

FIGURE 19.8
A plunge-feed cycle.

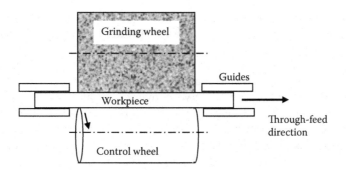

FIGURE 19.9
Through-feed grinding.

19.4.2 Through-Feed and Through-Feed Rate

In through-feed grinding, the GW is set at a fixed distance from the CW as in Figure 19.9. Through-feed is employed for longer workpieces and also for a continuous stream of short workpieces. In most cases, the length to diameter ratio of the workpieces is greater than one. There are relatively few investigations of through-feed grinding. Some examples are Meis (1980), König and Henn (1982), Pande and Lanke (1989), and Goodall (1990).

The workpiece is fed between guide plates into the space between the GW and the CW. The axis of the CW is inclined at a small angle Φ to the horizontal so that the axial component of the control wheelspeed drives the through-feed motion. The rate of through-feed depends on the angle of inclination of the CW.

The surface speed of the workpiece in stable grinding is normally equal to the surface speed of the CW. Differences in workpiece and CW surface speed are usually negligible. The resultant workspeed v_r can be resolved into two components. By convention, the slowest component of workspeed is termed the feed speed v_f and the fastest component is termed the workspeed v_w. The workspeed is therefore the circumferential component of the resultant workspeed and the through-feed speed is the axial component.

$$v_w = v_r \cdot \cos \Phi \tag{19.19}$$

The surface speed of the CW has a horizontal axial component that provides the axial feed rate of the workpiece otherwise known as the through-feed speed:

$$v_f = v_r \cdot \sin \Phi \tag{19.20}$$

where Φ is the angle of inclination and v_f is the feed rate.

19.5 Centerless Wheels and Dressing Geometry

19.5.1 The Grinding Wheel

Centerless GWs tend to be larger in diameter and wider than in center grinding. This means there are many more cutting edges available for grinding. Since the cutting action is shared between many more abrasive grains, wheel life between dressing operations tends to be greatly increased. This makes for economic production with little need to interrupt the process for redressing.

Most often, the type of abrasive used for the GW is vitrified aluminum oxide or silicon carbide. Other types of abrasive may used, but the vast majority of operations are undertaken with one or other of these two conventional abrasives. These abrasives are inexpensive compared to superabrasives and it is therefore reasonable to experiment with different wheels until an optimum wheel specification is achieved. There are many variations of wheel specification and abrasive types used in centerless grinding. The selection of a particular wheel will depend strongly on compatibility with the work-material, work-material hardness, the roughness limit, and factors such as size tolerance and risk of thermal damage.

Recent developments in conventional abrasives include monocrystalline grains for greater fracture resistance, sol gel grains that are designed to wear by microfracture while remaining sharp, and high aspect-ratio grains designed to remain sharp even with heavy wear.

For special operations where extremely high accuracy and low workpiece roughness are required, alternative types of wheel can be used in keeping with modern grinding technology. For example, it is possible to grind ceramics using metal-bond diamond wheels with electrolytic in-process dressing (ELID). For low temperature grinding of ferrous materials, vitrified cubic boron nitride abrasive is increasing in use for center and surface grinding. Vitrified CBN wheels can also be applied in centerless grinding. When using CBN, significant advantages can be obtained but it is important to pay careful attention to CBN grinding technology (Marinescu et al. 2004). CBN is notable for its natural hardness and high thermal conductivity. Both properties are ideal for grinding.

19.5.2 Grinding Wheel Dressing

"Dressing" is a term that commonly includes truing the GW to achieve a true form and dressing to achieve a suitable cutting edge distribution and sharpness. "Conditioning" is another term used for preparation of a GW particularly in the case of CBN. Conditioning, as a term, is used in different ways in different contexts. Conditioning a CBN wheel is often taken narrowly to mean a post-truing operation to open up the wheel surface by using a dressing stick or by machining an easy-to-grind material for a short period prior to grinding workpieces. Conditioning is sometimes used as a general term to include truing and dressing.

Examples of truing a GW are illustrated in Figure 19.10. The first example shows a truing movement to produce a stepped GW. The dressing tool is inclined to allow the diameter and the face of the GW to be trued.

The second example is for through-feed grinding. A small taper is trued on the lead-in area of the GW to spread the main stock removal over a greater area of the wheel. Without the taper, the main stock removal will be concentrated on the leading edge of the wheel causing rapid wear in this area. It is not absolutely essential to include a lead-in taper because a taper automatically develops as a consequence of rapid wheel wear in the region of rapid stock removal. As wear develops, the taper spreads across the surface of the wheel. However, it is sometimes considered good practice to dress a lead-in taper to provide a more controlled wheel wear process and prevent the possibility of scroll marks from the leading edge of the wheel.

Dressing is primarily important for the accuracy of the GW profile. Dressing is also important for the effectiveness and economics of the grinding process. Figure 19.11

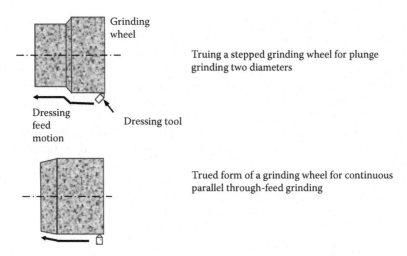

Truing a stepped grinding wheel for plunge grinding two diameters

Trued form of a grinding wheel for continuous parallel through-feed grinding

FIGURE 19.10
Examples of dressing operations for plunge and through-feed grinding.

compares roughness and process power for two different dressing tools. One was a single-point diamond tool. The other was a multipoint impregnated-diamond dressing tool. The two dressing tools give basically similar results although in this example, the sharp single-point diamond yielded lower workpiece roughness after grinding and lower grinding power. A single-point diamond that is allowed to become blunt leads to higher grinding power and poor surface roughness.

The dressing traverse rate is seen to be important. Dressing at a higher rate increases the roughness of the GW leading to higher workpiece roughness and lower grinding power. This allows faster grinding rates.

19.5.3 The Control Wheel

The CW is usually a rubber bond wheel. The rubber bond allows the regulating wheel to fulfill three important functions. These are to provide sufficient friction for speed control,

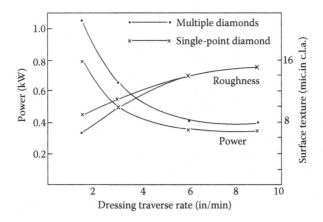

FIGURE 19.11
Effect of dressing traverse rate on power and workpiece roughness.

to provide sufficient flexibility for error averaging and to provide a surface that can be easily machined to provide accurate positioning. The rubber bond satisfies each of these requirements.

For multidiameter work, the CW must be stepped. The step height on the regulating wheel must match the step height on the workpieces to ensure support for the workpiece along an adequate proportion of the length. Normally, support will be provided along the entire length of the workpiece.

Angular adjustment of the control wheelhead may be provided in the horizontal plane. This adjustment is convenient for achieving parallel setup and elimination of small work-piece taper errors.

Angular adjustments should also be provided in the vertical plane for provision of an axial feed force. This introduces a further requirement for dressing feed angular adjustments to achieve the necessary CW profile as discussed in the next section.

19.5.4 Control Wheel Dressing

The axis of the CW is inclined at an angle Φ sufficient to provide traction in the axial direction. A small angle is employed for plunge grinding and a larger angle for through grinding as explained previously. This means the axis of the CW is not parallel with the axis of the workpiece. If the CW is dressed as a straight cylinder, the angle means there is point contact between the workpiece and the CW instead of line contact. Clearly, point contact will not ensure correct positioning of the workpiece parallel to the GW axis.

Line contact can be ensured between the CW and workpiece by dressing the CW along the required line of contact. A similar result can be achieved by dressing the CW parallel to the required line of contact but directly opposite in the same horizontal plane. This is illustrated in Figure 19.12.

The dressing tool must be set at a height h_d that is related to the work center height h_w. The dressing height is

$$h_d = h_w \cdot \frac{d_c}{d_c + d_w} \tag{19.21}$$

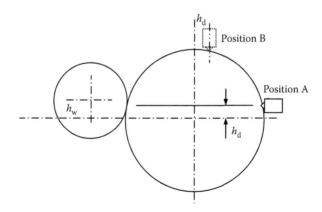

FIGURE 19.12
The control wheel is dressed at an offset height h_d to ensure line contact with the workpiece. Dressing may be carried out parallel to the line of contact at position A or at an angle to the line of contact at position B.

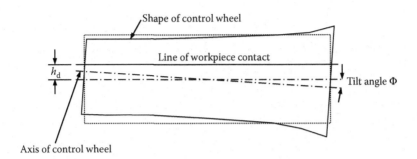

FIGURE 19.13
The control wheel dressed at an angle and above center.

The CW is no longer a true cylinder when tilted at an angle Φ and trued at a height h_d. It becomes a hyperboloid. The shape of the CW is shown schematically in Figure 19.13. The line of contact corresponds to the line of truing. Truing defines the diameter at each section of the CW. As the tool traverses across the CW, the height of the tool becomes higher relative to the center of the wheel. This increases the diameter of the wheel. The CW becomes larger at the low end of the wheel than at the other and satisfies the requirement that the workpiece is supported on a parallel line of contact with the GW.

If the dressing is carried out at position B in Figure 19.12, it is necessary to make additional adjustments for the angles involved. The dressing slide must be rotated relative to the plane of the line of contact by an angle approximately equal to the angle of tilt. The required rotation is made in a plane parallel to the tangent to the CW at the point of contact with the dressing tool. An adjustment is made to the angle for larger diameter workpieces. The dressing angle for position B as given by Jessup, in the book by King and Hahn (1986), is

$$\Phi_d = \frac{\Phi}{\sqrt{1 + d_w/d_c}} \tag{19.22}$$

19.5.5 Control Wheel Runout

Hashimoto et al. (1983) demonstrated a major source of errors in centerless grinding is due to run-out of the CW. To prove this point, the WR assembly was removed to allow the CW to be advanced up to the GW. By this means, the CW was dressed by the GW with much greater precision than possible by the normal method of dressing.

It was found that roundness errors were reduced from 1.7 to 0.2 µm with high-precision dressing and surface roughness was reduced from 0.32 to 0.12 µm Ra. It was also found that the new dressing method gave the CW increased wear resistance. This result has been confirmed by experiments conducted by the author.

19.6 The Workrest

The support surface of a WR should be harder than the workpiece. The WR also has to be tough and capable of withstanding the knocks encountered on the factory floor. The wear

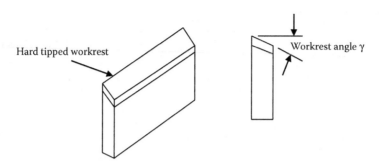

FIGURE 19.14
A typical workrest.

resistant surface of the WR is selected according to the types of work-material ground. For steels, tungsten carbide inserts provide a suitable surface. A typical tipped WR is illustrated in Figure 19.14.

Formed workpieces require the WR to be shaped to support the workpiece at more than one position along its length. For example, if the workpiece is stepped, the WR must also be stepped. The step height on the WR should correspond to the step height on the workpiece.

The top surface of the WR is almost invariably angled as shown. The WR angle γ may lie between 0° and 45°. For most work, the angle chosen is 30°. The angle 30° is generally found to give a strong rounding action. A WR angle of 0° is usually avoided, since no rounding action is achieved. The effect of WR angle is introduced in Section 19.2.5 on rounding geometry.

It is important that the WR is set horizontal and parallel to the GW surface to avoid problems of taper errors on the workpiece and forward ejection of workpieces from the machine.

19.7 Speed Control

For light workpieces, contact with the GW is sufficient to provide the driving torque to ensure workpiece rotation. A freshly dressed rubber CW gives good friction and prevents workpieces from spinning out of control. With sufficient friction, the WR and the CW both resist rotation of the workpiece against the driving torque provided by the GW. Under these conditions, the workpiece follows the speed of the CW.

19.7.1 Spinning Out of Control

With insufficient friction, the workpiece speeds up (Goodall 1990; Hashimoto et al. 1998; Gallego 2007).

This causes the friction coefficient between the workpiece and the CW to further reduce. The loss of friction causes the workpiece to spin out of control. To avoid this situation, the surfaces of the CW and WR must be clean. Also, the CW surface must provide adequate friction. The condition for spinning can be illustrated for the condition illustrated in Figure 19.15 for zero work height above center.

The GW exerts a normal force F_n and a tangential force F_t. The force ratio may be expressed as a GW friction coefficient μ_{gw}. The normal force N_{wr} exerted by the WR approximately balances F_t. The frictional force on the WR is given by the coefficient of friction, μ_{wr}

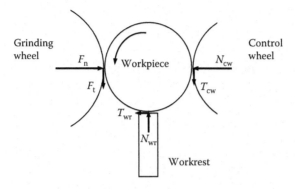

FIGURE 19.15
The forces on a workpiece.

for the two materials under the sliding conditions. Forces on the workpiece have to balance in the horizontal and in vertical directions and in rotation. These conditions lead to:

$$\text{Horizontal forces: } F_n = N_{cw} + T_{wr} \tag{19.23}$$

$$\text{Vertical forces: } F_t = \mu_{cw} \cdot F_n = N_{wr} - T_{cw} \tag{19.24}$$

$$\text{Rotational forces: } T_{wr} = \mu_{wr} \cdot N_{wr} = F_t - T_{cw} \tag{19.25}$$

Solving for the normal force on the WR and the friction force on the CW:

$$N_{wr} = \frac{2\mu_{gw} \cdot F_n}{1 + \mu_{wr}} \tag{19.26}$$

$$T_{cw} = \frac{1 - \mu_{wr}}{1 + \mu_{wr}} \cdot \mu_{gw} \cdot F_n \tag{19.27}$$

$$T_{wr} = \frac{2\mu_{wr}}{1 + \mu_{wr}} \cdot \mu_{gw} \cdot F_n \tag{19.28}$$

$$N_{cw} = \frac{1 + \mu_{wr}(1 - 2\mu_{gw})}{1 + \mu_{wr}} \cdot F_n \tag{19.29}$$

For a typical grinding operation, assuming $\mu_{gw} \approx 0.3$ and a coefficient of friction between WR and workpiece, $\mu_{wr} \approx 0.15$:

$$F_t = 0.3 \times F_n$$

$$N_{cw} = 0.92 \times F_n$$

$$N_{wr} = 0.52 \times F_n$$

$$T_{wr} = 0.078 \times F_n$$

$$T_{cw} = 0.22 \times F_n$$

The minimum value of the CW coefficient of friction for these conditions is

$$\mu_{cw} = \frac{T_{cw}}{N_{cw}} = \frac{0.22}{0.92} \approx 0.24$$

The usual condition is where the workpiece is above center, as in Figure 19.1. The friction on the CW becomes even more critical as work height is increased. A detailed analysis of force equilibrium is given by Jameson et al. (2008). As the height above center is increased, there is an increased risk of spinning.

19.7.2 Failure to Turn

A different problem can arise with heavy workpieces if there is insufficient CW friction to turn the workpiece on contact with the GW. Heavy workpieces are unlikely to suffer the problem of "spinning." In fact, heavy workpieces are more likely to suffer the problem of failing to turn at all under the action of the grinding force. If the workpiece fails to turn, the GW can machine a groove along one side of the workpiece. This situation is potentially dangerous since as the size of the groove increases, the tangential grinding force also increases to the point where the workpiece suddenly starts to rotate, leading to an excessive depth of cut and exceptionally heavy forces on the GW. It is important that this situation is never allowed to arise since it could lead to wheel breakage. When setting up a grinding process, it is important to check that the workpieces readily rotate and stop the machine if rotation fails to occur.

For heavy workpieces, additional provision must be made to ensure workpiece rotation before contact with the GW. An additional pressure roll may be employed to push the workpiece against the CW and this may be sufficient. For very large workpieces such as axle shaft housings, it will be necessary to provide an external driving device to grip and rotate the workpiece.

19.8 Machine Structure

19.8.1 The Basic Machine Elements

Main elements of a centerless machine include

- The machine base and table
- The grinding wheelhead, spindle bearings, and spindle drive
- The control wheelhead, spindle bearings, and spindle drive
- The infeed drive
- The GW dressing slide and drive
- The CW dressing slide and drive

- The workrest slide
- The fluid delivery and return system

The design of the machine must allow the machine:

- To accurately position the workpieces in relation to the GW
- To drive the GW, CW, and workpieces at precise speeds
- To accurately dress the wheels

In this section, aspects of machine design are considered in relation to performance. Results from a conventional machine of the type shown in Figure 19.16 are compared with results from a stiff research machine having the box structure layout illustrated in Figure 19.17. Further information is available in publications by Rowe (1974, 2013) and Rowe et al. (1987).

19.8.2 The Grinding Force Loop

Figure 19.16 shows the layout of a basic centerless machine. In this example, the grinding wheelhead forms part of an integral table and fluid containment tray. There is provision of a slideway on the table to allow feed of the control wheelhead. No provision is shown to feed the grinding wheelhead nor is provision for angular adjustments to the wheelhead. These features are incorporated into many machines for ease of operation and setup.

Forces in the process cause deflections of the machine leading to the possibility of size inaccuracy and lack of straightness and roundness. Machine design affects both accuracy and production rate, since a machine that is more compliant usually has to be operated more slowly to achieve the same accuracy. Often a more compliant machine is incapable of producing the same accuracy as a stiffer and better-designed machine. It is demonstrated below that a stiffer machine reaches the size specification faster than a compliant machine.

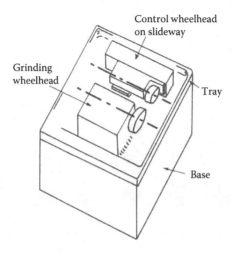

FIGURE 19.16
Main elements of a centerless machine.

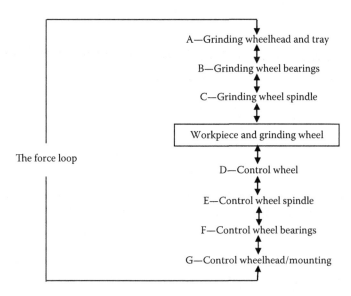

FIGURE 19.17
Constituents of the grinding force loop.

The forces in the process are reacted through a chain of machine elements. The overall compliance of the machine depends on the build up of compliance through the elements in the force loop. The main elements of the loop for the grinding force are shown in Figure 19.17. The compliances of various elements are shown in Figure 19.18. The elasticity of the GW is not shown, although this is of a similar order of magnitude to the compliance of the CW although rather stiffer (Rowe 1974). Other loops may be constructed for the dressing forces.

Each of the elements in the loop plays a critical role in resisting deflections. In this example, the low-speed plain CW bearings are a source of variability and low stiffness under finishing conditions. In addition, the workpiece also features in the force loop. If the workpiece is a thin-walled cylinder, the stiffness of the force loop will be greatly reduced.

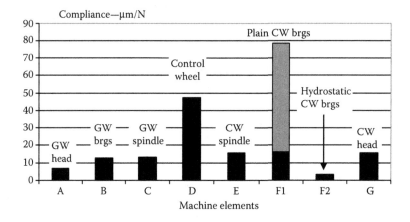

FIGURE 19.18
Compliance of the main elements in the grinding force loop.

19.8.3 Structural Layout

A conventional machine layout constrains the workpiece within a U-section, whereas ideally the workpiece should be constrained within a closed box layout. This point is illustrated in Figure 19.19. The U-section is inherently subject to deflections in the same manner as a tuning fork. Ideally, a machine should be constructed in a different way designed to resist deflections.

A very stiff ultraprecision research machine was designed and built to demonstrate this point. The layout is illustrated in Figure 19.19. The research machine allowed grinding results to be compared with results from a compliant conventional machine. The structural design was shaped by the following criteria:

1. The line of measurement to be coincident with the line of separating force.
2. The neutral axis for bending to be coincident with the line of force.
3. The dressing point to be opposite the grinding point on the GW and opposite the CW contact point on the CW, that is, along the line of force.
4. The dressing cam plates to be on the same line of force.
5. Hydrostatic bearings to be employed for high stiffness and accuracy.

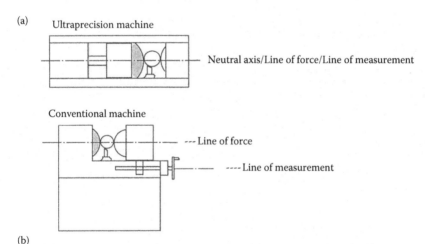

FIGURE 19.19
Design for ultraprecision: (a) an ideal layout compared with a conventional layout and (b) an ultraprecision machine. (From Rowe, W. B., 1979, Research into the mechanics of centreless grinding, *Precision Engineering*. IPC Press, 1, 75–84.)

6. A highly accurate and highly damped feed drive to be provided by a wedge-shaped cam driven between hydrostatic bearings.

7. The same principles to apply in the plan view. The wheels to be supported symmetrically between the spindle bearings.

The resulting wheel-workpiece stiffness was 3.5×10^7 N/m, five times greater than the stiffness of the conventional machine. The resonant frequency was increased from 78 to 500 Hz.

The stiff machine gave different grinding results from a conventional machine. Size control and roundness using a similar plunge grinding cycle were improved but the range of workspeeds over which accurate grinding could be achieved was greatly increased as shown in Figure 19.20.

At low workspeeds, workpiece roundness levels were similar for the two machines. High roundness errors at low workspeeds are partly because there are insufficient revolutions of the workpiece to reduce the roundness errors. Higher roundness errors can result from a stiff machine in some tests due to the rounding process. This aspect is discussed below and also in Sections 19.11 and 19.12 in relation to static compliance.

Results at high workspeeds are in distinct contrast to low speeds. Chatter was experienced with the compliant conventional machine at higher workspeeds, whereas with the stiff machine good accuracy was achieved at much higher speeds. This illustrates that roundness is not only a question of rounding geometry but also of the dynamic performance of the grinding machine. Since production rate is related to workspeed and wheelspeed, there is an implication that a dynamically stiff machine allows much higher levels of productivity.

The important point is that vibration levels should be low. Roundness is not simply dependent on static stiffness. In fact, high static stiffness with high vibration levels worsens roundness. Also a slow rounding process tends to reduce errors better than a fast

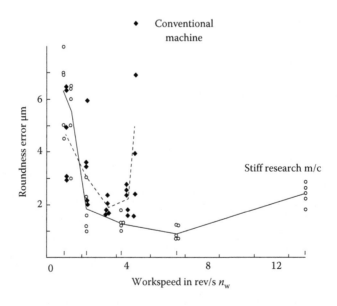

FIGURE 19.20
Effect of machine design on roundness accuracy and operating speed range.

process. A compliant machine slows down the rounding process and can under the right conditions allow better roundness to be achieved. This was demonstrated by replacing the soft rubber CW with a cast iron CW. It was found that roundness errors were increased with a stiff cast iron wheel.

19.8.4 Spindle Bearings

The CW spindle bearings were the most compliant element in the force loop compliances shown in Figure 19.19. The plain hydrodynamic bearings of the CW spindle are shown as element F1. Plain hydrodynamic bearings are reliable over a long period but at low speeds, they exhibit low and variable compliance with load. The compliance becomes very high under light loading as when sparking out. This is a serious disadvantage for consistency of grinding. Hydrodynamic bearings are much better at high speeds as when applied in GW bearings. High-precision rolling elements are better than plain bearings for low speeds, but need to be replaced regularly due to wear. Hydrostatic bearings give improved performance over the speed range having high stiffness, high accuracy, and low wear properties. The disadvantage of hydrostatic bearings is higher cost.

Hydrostatic bearings were substituted for the hydrodynamic CW bearings of the conventional machine. The hydrostatic bearings were provided with diaphragm control to improve the system stiffness, consistency, and rotational accuracy. The replacement bearings are shown as Element F2 in Figure 19.19. The GW bearings were also replaced by hydrostatic bearings giving marginally improved stiffness but much improved rotational accuracy. The result was much better roundness accuracy of the grinding process. The replacement of the CW bearings gave much improved size-holding capability. It was possible to achieve a specified size tolerance within a shorter dwell period for spark out.

Figure 19.21 shows how bearing stiffness affects deflection of the CW in plunge grinding for a workpiece 25.4 mm diameter by 50 mm long. The maximum deflections are two and half times larger with the plain bearings than with the hydrostatic bearings for the CW spindle.

Greater deflections of the grinding system require a longer spark out dwell to reduce the size, roundness, and surface texture variations. Controlled grinding experiments demonstrate this effect and the results are seen in Figure 19.22.

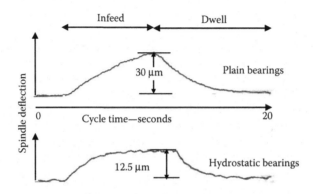

FIGURE 19.21
Control wheel deflections during a grinding cycle showing the effect of bearing stiffness.

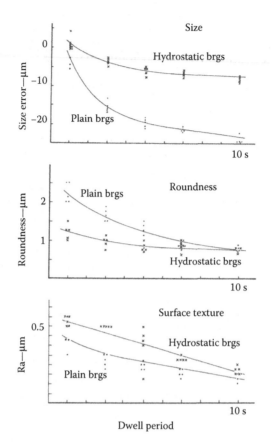

FIGURE 19.22
Workpiece accuracy with dwell period. Comparing stiff hydrostatic control wheel bearings with compliant plain hydrodynamic control wheel bearings.

Size variations are much smaller with the stiff hydrostatic bearings than with the compliant plain bearings. Size variations are reduced to a negligible level in half the dwell period required by the compliant bearings. Roundness errors also decrease during the dwell period. Stiff hydrostatic CW bearings reduce roundness errors faster. In later experiments, the GW bearings were replaced by hydrostatic bearings and the roundness errors were further reduced.

The results for surface texture may seem surprising. Surface roughness reduces during the dwell period as would be expected. However, surface roughness is greater for the stiffer hydrostatic bearings than for the compliant plain bearings. Experience with other machines, confirms that stiff systems produce slightly higher surface roughness in grinding. On reflection, this would be expected since a stiffer system imposes greater variations in grain depth than a soft system. A stiff system can produce low roughness but the requirement for low roughness is a system where grain depths are small and uniform. This can be achieved with high-quality small-grain wheels and a low level of vibrations.

Overall stiffness of a system in relation to grinding forces is expressed by the machining-elasticity parameter K. The parameter K is the true depth of cut divided by set depth of cut and is explained mathematically in Section 19.11.4. Evaluating K for the hydrostatic bearing machine and the plain bearing machine in a particular grinding setup gave the following results:

Plain bearing machine: $K = 0.23$

Hydrostatic bearing machine: $K = 0.44$

In summary, replacing compliant plain wheel bearings by stiff hydrostatic bearings almost doubled the effective system stiffness.

19.9 High Removal Rate Grinding

19.9.1 Introduction

Manufacturers wish to increase grinding rates for several reasons

- To increase output from each grinding machine
- To increase output by each machine operator
- To reduce cost per part

High-speed grinding techniques were developed in Germany (Opitz and Gühring 1968) and were also applied to centerless grinding (Schreitmuller 1971).

An experimental investigation of high removal-rate centerless grinding applied to steel and cast iron allowed process limit charts to be defined (Rowe et al. 1986). These materials are easy-to-grind materials. Limit charts show operating boundaries for infeed rate, workspeed, and wheelspeed. Examples are presented below (Figure 19.23). With large GWs as

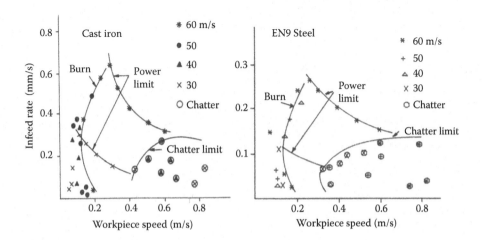

FIGURE 19.23
Process limits for typical material–machine combinations. 75 kW machine. Workpieces: Gray CI—40 mm diameter by 65 mm: EN9—50 mm diameter by 65 mm.

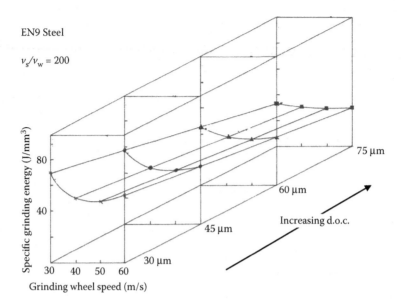

FIGURE 19.24

Specific energy with increasing depth of cut and wheel speed for constant v_s/v_w.

used in centerless grinding, the rate of wheel wear was low. It was not possible to achieve an excessive wheel wear condition, although this might have been possible if more power had been available. The machine had a 75-kW motor. The shape of the diagram means, there is an optimum point of operation for high removal rate within the region enclosed by the boundaries. This forms the basis of a control strategy.

Characterization of the process is developed by attention to the concepts of a "speed effect" and a "size effect." Results for the speed effect are distinguished and presented separately from results for the size effect to illustrate different physical effects.

A two-dimensional surface is presented showing the variation of grinding energy with variations in infeed rate, workspeed and grinding wheelspeed. Because speed and size effects are separated on the surface, it is possible to read directly the optimum grinding wheelspeed for minimum energy (Figure 19.24).

19.9.2 Routes to High Removal Rate

Some of the key features for achievement of high removal rate are as follows:

Increasing the number of active grits. The main route to increasing removal involves increasing the number of abrasive grains that remove material within the grinding period. The number of active grits can be increased by

- Increased GW diameter
- Increased GW width
- Increased grinding wheelspeed

A larger number of active grits means more material can be removed in a given time. This is because active grits can only remove so much material per engagement. Productivity is therefore directly related to the number of grits engaged per unit time.

Increasing the removal rate per grit. Another route to increasing production rate is to optimize the material removed per grit engagement. In practice, the material removed per grit cannot be easily determined. This means that overall efficiency of the process and removal rate must be investigated by grinding trials. The effect of grinding cycle parameters on process efficiency is considered further in Section 19.10 on process economics.

Power consumed increases with the number of grits removing material per unit time. Highest removal rate is achieved when removal rate per grit is optimized. If then, grinding wheelspeed is increased, removal rate per grit is reduced from the optimum value. This reduces efficiency. To optimize removal rate per grit, feed rate must be increased in the same proportion as wheelspeed. This restores removal rate per grit to the optimum value and increases overall removal rate. The technique of changing wheelspeed and feed rate in a constant proportion, maintains kinematic similarity to maintain an optimized process. Ideally workspeed should also be increased in the same proportion.

Longer redress life. It is important to achieve long wheel redress life. Material removal per grit should be maximum consistent with required quality and consistent with long wheel life between redressing. Large diameter and wide GWs mean that long redress life can usually be achieved more easily than in other grinding processes. Redress life is further explored under process economics. Redress life is strongly dependent on the combined nature of the abrasive, the workpiece material, and the severity of the grinding conditions.

Improved abrasives. Subject to achievement of quality levels, harder grits, and bonds allow longer redress life and larger stock removal. The bond must not be too hard for the grinding operation and the grits should be sufficiently friable. Otherwise grits eventually become blunt and are retained if grain forces rise to unacceptable levels. Blunt grits lead to high specific energy and glazed wheels.

Grits that are too weak and a bond that is too soft lead to rapid wheel wear and high surface roughness. An advantage of a soft wheel is that grinding forces tend to be lower than with a hard wheel.

For high wheelspeeds, it is necessary to consider the type of abrasive and wheel design for safe high-speed operation. Superabrasive wheels can be operated at higher speeds than conventional abrasives. Specialist conventional abrasives are available designed for speeds up to 140 m/s. These may prove to be the most economic choice for high-speed centerless grinding.

Grinding trials. For sustained and repeated batch manufacture, it is worth conducting grinding trials with different wheels and grinding conditions to improve process economics. In Section 19.10, effects of grinding wheelspeed on grinding rate are considered.

19.9.3 Improved Grinding Machines and Auxiliary Equipment

Consideration should be given to effects of machine design on grinding rate. Some features to be considered are

- *Machine power.* Many machines have insufficient power and this may frustrate attempts to increase removal rate. The full potential of the GW and high-grinding wheelspeed can only be realized if the machine has ample power. Speed and power available at the GW are particularly important.
- *Variable speed drives.* Variable speed drives allow wheel and workspeeds to be adjusted to avoid resonant machine conditions. The same requirement applies to

rotary dressing tool drives. Highest production rates and best quality are achieved when the machine speeds are adjusted to avoid resonant conditions. In this context, quality levels refer particularly to size, roundness, and roughness. The range of wheelspeeds must be sufficient to allow high removal rates.

- *High-speed GWs.* A 60 m/s is reasonable for conventional wheels rated up to this speed. Much higher speeds are possible with special-purpose wheel designs and superabrasives. High wheelspeeds greatly increase the potential for high removal rates while allowing quality levels to be maintained.

- *Machine stiffness.* A high-speed grinding machine needs to be stiffer than a low-speed grinding machine. This implies a sound structural layout and sufficient rigidity. Slideways should allow accurate movement and the feed drives must allow accurate control and adequate feed rates for grinding and for rapid loading and unloading. Other features to be considered include ease of setup and adjustment of machine settings.

- *Effective fluid delivery.* Fluid delivery assumes increased importance in high-speed grinding. Flowrates and supply pressures must be adequate to ensure transport of fluid into the grinding contact between the GW and the workpiece. It is only the "useful flowrate" that enters the grinding contact. Only useful flowrate acts to provide process lubrication and ensure cooling at the high temperature contact interface. The fluid should be delivered as a high-speed sheet of fluid directed as far as possible straight into the grinding contact (Gviniashvili et al. 2004).

- *An effective grinding fluid.* The grinding fluid must be chosen for tribological properties of lubrication, cleaning, cooling, material removal, corrosion protection, biocide action, fire risk, and friendliness to the operators and environment. This is a specialist field that requires careful consideration.

- *Dressing equipment.* Dressing slides should be sufficiently rugged and accurate for precise control of dressing operations. The type of dressing tool used is important. Multipoint diamond tools allow greater consistency over a longer period than single-point tools. Diamond rotary tools may be preferred for high-grinding wheelspeeds, particularly for superabrasive wheels.

- *Wheel guarding.* Wheel failures should be an extremely rare event. However, for high-grinding wheelspeeds, adequate guarding must contain burst wheels within the guarding enclosure. Wheel guarding must conform to safety regulations.

19.9.4 Process Limits

The process limits define the permissible range of speed conditions for stable grinding. Typical process limits are shown in Figure 19.23. Two examples are illustrated for a 75-kW centerless grinding machine. The gray cast iron is an easy-to-grind work material and specific material removal rates were achieved in excess of 40 mm²/s. Grinding EN9 steel, the specific removal rate achieved was 20 mm²/s. These removal rates were achieved using conventional abrasive wheels at a speed of 60 m/s and a grinding width of 65 mm.

The rate of grinding can be increased up to process limits of maximum machine power available, onset of thermal damage to the workpieces and chatter. Surface roughness depends primarily on the GW employed and the dressing process. Initial trials to ascertain the size of the operating area is essential as a first step to process optimization.

Results clearly show the benefits of increasing wheelspeed. The removal rates at 60 m/s are more than double the rates achieved at 30 m/s.

An interesting feature of the limit charts is the similarity of the shape for two different materials. This supports the conclusion that these diagrams have general validity.

The charts show that high workspeeds increase the probability of chatter. Low workspeeds increase the probability of burn. Low workspeeds concentrate the process energy in the contact zone for a longer period and so increase the susceptibility to thermal damage. Thermal damage or burn as it is generally termed can also give rise to a form of chatter that occurs at low workpiece speeds.

High-grinding wheelspeeds allow higher infeed rates to be employed for the same grinding forces and thus allow higher removal rates.

Effect of infeed rate. Material removal rate increases with infeed rate. However, as infeed rate is increased, material removed per grit also increases. This is apparent from the equivalent chip thickness expressed in terms of infeed rate and wheelspeed.

$$h_{eq} = \frac{1}{2} \cdot \pi \cdot d_w \cdot \frac{v_f}{v_s} \tag{19.30}$$

Increasing chip thickness, leads to higher stresses on the grinding grits causing greater wear and fracture. A consequence is "self-dressing," where the grits maintain or increase their sharpness due to the grinding stresses. This yields the benefit of reducing the energy required to remove a volume of material and produces lower specific energy. The disadvantage of the grit fracture process is higher surface roughness and faster wheel wear. The process where specific energy is reduced due to increased feed rate is termed the "size effect." The size effect can be explained in several different ways but basically the conclusion is that increased chip thickness or increased volume of material removed per grit reduces specific energy.

Increasing infeed rate, with other parameters constant, tends to increase grinding forces, increase roughness, reduce redress life, and reduce specific energy. The process tends to become more efficient until the optimum chip size is exceeded. Excessive infeed rate leads to high wheel wear, a low-grinding ratio, and rapid wheel breakdown.

It is particularly interesting to note in this large set of trials, that chatter was more frequently noted at low infeed rates than at high infeed rates. Explanations for this type of chatter have been found in centerless grinding where the chatter is particularly related to the setup geometry (Rowe 2014).

Effect of wheelspeed. Higher wheelspeeds allow higher infeed rates to be employed. Increasing wheelspeed without increasing infeed rate reduces chip thickness as can be seen from Equation 19.30. In this case, roughness is reduced and grinding forces are reduced, although specific energy is increased due to the size effect. The purpose of increasing wheelspeed is to allow infeed rate to be increased, thus increasing production rate, while maintaining quality levels and process efficiency.

Effect of workspeed. Within the stable range, increasing workspeed has a relatively small effect on the process. As workspeed is increased, specific energy is increased as may be deduced from Figures 19.23 and 19.24.

At high workspeeds, the probability of chatter is increased, so there is effectively a maximum workspeed due to the chatter. These results are for a particular workpiece diameter. It is therefore necessary to undertake trials for the particular workpieces to be produced. Figure 19.20 shows the chatter limit can be increased by improved machine design to raise the principal resonant frequency.

At low workspeeds, the probability of thermal damage to the workpiece increases. The burn boundary can be moved outward by using a sharper abrasive to reduce the specific energy.

19.9.5 Specific Energy as a Measure of Efficiency

Specific grinding energy can be considered as a measure of grinding efficiency. A low specific energy implies low energy consumed to remove a given volume of material. Low specific energy allows greater removal rate before meeting the power limit. Since specific energy is proportional to the grinding forces, it usually means that removal rate can be greater before meeting the burn limit, or the power, wheel wear, and roughness limits.

Specific energy reduces with increasing infeed rate due to the size effect as discussed previously. An example is shown in Figure 19.24.

From Equations 19.7 and 19.30, equivalent chip thickness can be expressed as $h_{eq} = (1/2) \cdot \pi \cdot d_w \cdot (v_f/v_s) = (1/2) \cdot \pi \cdot d_w \cdot (v_f/v_w) \cdot (v_w/v_s)$. This shows that maintaining constant speed ratio allows constant chip thickness and constant depth of cut to be maintained as wheelspeed is varied. This is demonstrated in Table 19.1, where the value of speed ratio v_f/v_w is given for each depth of cut and value of equivalent chip thickness. For these constant speed ratios, the chip removal conditions are kinematically similar. Figure 19.24, therefore, demonstrates at each depth of cut, the effect of varying wheelspeed on specific energy for kinematically similar conditions. It shows for each depth of cut and chip thickness the optimum wheelspeed.

If grinding wheelspeed was increased on its own, the size effect would reduce efficiency. However, the purpose of increasing grinding wheelspeed is to allow increased feed rates. By increasing wheelspeed and feed rate in direct proportion to each other, the size effect is eliminated. By this means, speed may be increased maintaining depth of cut and chip thickness constant.

It is found that specific energy reduces to a minimum as wheelspeed is increased at constant chip thickness and then starts to increase again. This is clearly illustrated in Figure 19.24. Specific energy starts to increase again at higher wheelspeeds possibly due to an increased rubbing contribution at higher wheelspeed. As feed rate is increased, chip thickness is also increased. This increases the value of optimum wheelspeed and suggests that even higher wheelspeeds would be an advantage.

Figure 19.24 is typical of grinding processes and suggests an optimum wheelspeed applies for particular depth of cut and feed rate conditions. For this example, it appears that the optimum wheelspeed is close to 45 m/s at the lower depth of cut and is close to 60 m/s at the higher depth of cut. The shape of Figure 19.24 suggests that much higher values of wheelspeed would be beneficial at very high removal rates. This is the explanation for the modern trend toward speeds exceeding 120 m/s.

TABLE 19.1

Depth of Cut, Feed Rate, and Equivalent Chip Thickness, $d_w = 50$ mm Diameter and $v_s/v_w = 200$

Depth of Cut, a_e (μm)	Equiv. Chip Thickness, h_{eq} (μm)	Infeed Rate/Workspeed
30	0.150	0.000382
45	0.225	0.000573
60	0.300	0.000764
75	0.375	0.000955

19.10 Economic Evaluation of Conventional and CBN Wheels

19.10.1 Introduction

Under particular through-feed centerless grinding conditions, results presented by König and Henn (1982) showed that the minimum cost per part was achieved at removal rates between 1 and 6 mm³/mm s. Increasing or reducing removal rate from this range increased costs. Such results are useful for demonstrating an approach to optimization. In this section, a general approach to optimization is demonstrated.

The following analysis by Rowe and Ebbrell (2004) demonstrates how a systematic approach can allow many different factors to be taken into account leading to large increases in productivity and reductions in costs while maintaining quality levels. The analysis applies for repeated batch manufacture. Costs in grinding can often be reduced by increasing removal rate. However, this is not the only factor and it is necessary to take account of mean cycle time including dwell time for spark out and costs related to labor, the grinding machine, GWs, dressing, and maintenance of required quality levels.

As demonstrated in Section 19.9, high wheelspeeds allow removal rates to be increased while maintaining quality levels. However, high wheelspeeds increase the cost of the grinding machine. Stiffer machines are required. More sophisticated equipment and techniques are required for power delivery, fluid delivery, dressing, wheel balancing, and spindle bearings. Special GWs are required for high speeds and better machine guarding. It is therefore necessary to give consideration to costs and benefits of these various aspects. Advantages may be obtained by introducing superabrasive wheels to allow much higher wheels speeds and long redress life.

Redress life results were obtained in grinding between centers using smaller wheels to minimize wheel costs. However, the principles are generally applicable. Repeated batch manufacture is a situation, where it is required to maximize production throughput and minimize process costs while achieving specified quality levels.

19.10.2 Cost Relationships

The following cost analysis takes account of labor cost, machine cost, and abrasive cost. Workpiece quality and redress life are related through the number of parts ground before roughness or roundness tolerance is exceeded.

19.10.3 Wheel Cost/Part

C_s is the wheel cost c_s divided by the number of parts produced per wheel N_w. The number of parts per wheel, is given by

$$N_w = \left[\frac{(d_{s\max} - d_{s\min})/2}{(r_s + a_d n_d)} \right] N_d \qquad (19.31)$$

where $d_{s\max}$ is the maximum wheel diameter, $d_{s\min}$ is the minimum wheel diameter, r_s is the radial GW wear per dress, a_d is the dressing depth, n_d is the number of dressing passes, and N_d is the number of parts per dress. The wheel cost per part is, therefore,

$$C_s = \frac{2c_s(r_s + a_d n_d)}{(d_{s\max} - d_{s\min})N_d} \qquad (19.32)$$

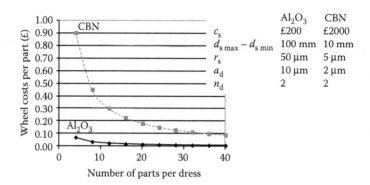

FIGURE 19.25
Dependence of wheel cost per part on number of parts per dress.

Figure 19.25 shows typical wheel cost/part varying with number of parts/dress for alumina and CBN wheels. Redress life has a strong effect using CBN due to initial cost but cost/part may be offset by long redress life. A high wheelspeed increases redress life by reducing chip thickness. Wheel cost/part becomes negligible with long redress life.

Initial wheel costs in centerless grinding are higher than in grinding between centers due to the larger diameter and width of centerless wheels. However, redress life is correspondingly increased so that wheel costs are spread over many more workpieces. The wheel costs in Figure 19.25 could be more than doubled in centerless grinding, although the wheel cost per part is likely to be reduced.

19.10.4 Labor Cost/Part

Labor cost per part C_l is the product of labor rate c_l and total cycle time t_t. Labor rate includes general overhead costs other than the cost of the grinding machine. The total cycle time is given by

$$t_t = t_s + t_d/N_d \tag{19.33}$$

where t_s is the basic cycle time and t_d is the dressing time. The basic cycle time allows for the diametric stock removal d_{ww}, the diametric standoff d_{ss}, and dwell time for spark out t_{so}, so that

$$t_s = \frac{(d_{ww} + d_{ss})\pi d_w}{Q'_w} + t_{so} \tag{19.34}$$

where Q'_w is specific removal rate and d_w is the workpiece diameter. Dressing time depends on the wheel width b_s the number of dressing passes n_d and the dressing feed rate v_d. The dressing time is

$$t_d = \frac{b_s n_d}{v_d} \tag{19.35}$$

The dressing feed rate depends on the effective width of the dressing tool b_d and the dressing overlap ratio U_d. Typically, the overlap ratio should be of the order of 3–10. The higher value gives lower roughness but higher grinding forces.

$$v_d = \frac{b_d}{U_d} \times \frac{v_s}{\pi d_s} \tag{19.36}$$

The total cycle time is, therefore

$$t_t = \frac{(d_{ww} + d_{ss})\pi d_w}{Q'_w} + t_{so} + \frac{b_s n_d}{v_d N_d} \tag{19.37}$$

Multiplying total cycle time by the labor rate gives the labor cost/part is

$$C_l = c_l \left[\frac{(d_{ww} + d_{ss})\pi d_w}{Q'_w} + t_{so} + \frac{b_s n_d}{v_d N_d} \right] \tag{19.38}$$

Labor cost/part is affected by removal rate, dwell time, number of dressing passes, dressing feed rate, and by number of parts/dress. With many parts/dress, the last term of Equation 19.38 becomes negligible. With several dressing operations per part, the last term becomes large.

The labor rate with overheads was based on £75/h. Figure 19.26 shows high removal rates reduce labor cost/part.

Figure 19.27 shows that redress life can strongly affect labor cost per part. Redress life is particularly important if redress life is short. The effect of dressing cost/part becomes negligible in this example beyond four parts/dress.

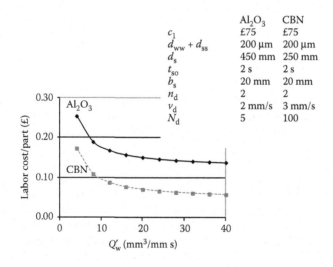

FIGURE 19.26
Dependence of labor cost per part on removal rate.

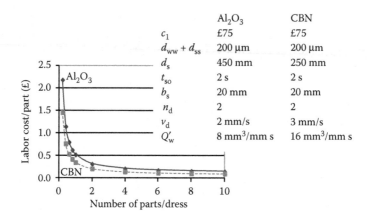

FIGURE 19.27
Dependence of labor cost per part on redress life.

19.10.5 Machine Cost/Part

Machine cost per part C_m is given by the cost of the machine C_{mc} divided by the number of parts N_{mc} produced within the payback time y_t. The number of parts produced is

$$N_{mc} = \frac{y_t}{t_t} = \frac{y_t}{(t_s + t_d/N_d)}$$

(19.39)

19.10.6 Total Variable Cost/Part

The total of the variable costs per part is the sum of wheel cost/part, labor cost/part including an overhead contribution, and machine cost/part. Constant workpiece material cost/part does not affect the selection of process conditions and can therefore be left out of this discussion on process costs. The total process cost/part for wheels, labor, and machine is, therefore,

$$C_t = \frac{2c_s(r_s + a_d n_d)}{(d_{s\,max} - d_{s\,min})N_d} + \left[\frac{(d_{ww} + d_{ss})\pi d_w}{Q'_w} + t_{so} + \frac{b_s n_d}{v_d N_d} \right]\left(c_1 + \frac{C_{mc}}{y_t} \right)$$

(19.40)

The number of parts/dress is a factor in all three costs contributing to the total process cost. This explains the importance of redress life. Inclusion of machine cost and redress life allows realistic evaluation of high-speed grinding and application of superabrasives.

19.10.7 Experiment Design

To evaluate costs, it is first necessary to determine the effect of grinding conditions on redress life. The effect of long redress life was studied grinding the bearing material AISI 52100 and the effect of low redress life with the nickel alloy material Inconel 718. Redress life was defined as the number of parts ground before either roughness of 0.25 μm Ra or roundness error of 1 μm was exceeded. By this means, the specified quality levels are built into the test procedure. The following procedure allows a comprehensive analysis to be performed with a minimum number of grinding and measuring trials.

TABLE 19.2

An $L_8 2^7$ Experimental Plan

Parameter	A	B	C	D	E	F	G
Trial				Level			
1	1	1	1	1	1	1	1
2	1	1	1	2	2	2	2
3	1	2	2	1	1	2	2
4	1	2	2	2	2	1	1
5	2	1	2	1	2	1	2
6	2	1	2	2	1	2	1
7	2	2	1	1	2	2	1
8	2	2	1	2	1	1	2

An example of an experimental design for a two-level investigation of seven variables using eight trials is given in Table 19.2. In the example shown four trials are conducted for each value of a parameter. The result for each parameter is the mean value for the combinations of the other parameters shown. In each case, level 1 represents the low value of a parameter and level 2 represents the high value. A large number of workpieces are ground for each trial, thus allowing redress life and G ratio to be assessed.

The investigation was conducted in three stages

Stage 1. Basic trials.

Basic trials were conducted to produce direct-effect charts for specific energy, roughness, roundness, size, and G ratio. Many factors were varied including wheelspeed, workspeed, dwell period, dresser speed, dressing direction, dressing overlap ratio, and dressing depth. Two levels were chosen for each variable. The levels were selected based on experience to give a reasonable span of the operational range. The two-level experiment leads to "direct-effects" charts.

Stage 2. Select best conditions and confirm.

Best conditions were selected for each GW by selecting a combination of values from consideration of the effects of each variable on production rate and quality from the direct-effect charts produced in the basic trials. A combination of values was selected to achieve the quality levels with shortest cycle time. The optimized conditions were validated by confirmation trials.

Stage 3. Cost comparisons.

Costs per part were evaluated for each GW using redress life from the trials. Cost comparisons were made based on conditions yielding best results from previous trials.

Figure 19.28 shows typical redress life results. The end of the redress life is indicated when the process no longer yields roughness, roundness, or any other quality parameter within tolerance.

19.10.8 Machine Conditions and Cost Factors

The cost factors used in the evaluation are given in Table 19.3.

A high-speed vitrified CBN wheel was tested on a special-purpose high-speed grinding machine for speeds up to 140 m/s. The results were compared with conventional alumina wheels and CBN wheels tested on a conventional machine at speeds at 45 m/s.

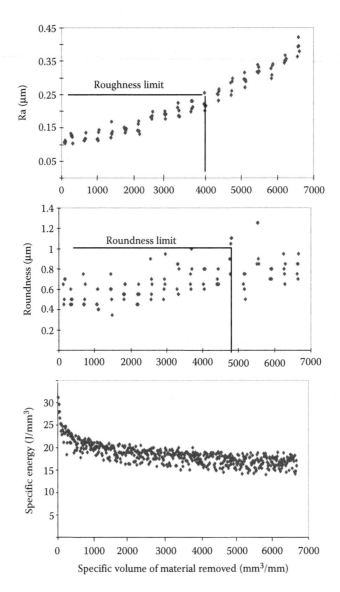

FIGURE 19.28
Redress life trials in high-speed CBN grinding with 2 s dwell and 20 mm³/mm s removal rate.

The special-purpose machine was designed to employ smaller diameter wheels so that wheel costs were lower for the highest wheelspeeds. The cost of the special-purpose machine was two and a half times the cost of the conventional machine.

For high-speed CBN grinding, it was necessary to employ a rotary-disk diamond dressing tool with a hydraulic drive. "Touch dressing" was employed to allow very small dressing cuts to be taken. The application of such techniques is essential for the successful application of vitrified CBN. The higher costs have to be justified by increased productivity and lower total cost per part.

A 6-month payback period was assumed consisting of 20 working days per month at 16 h/day.

TABLE 19.3

Wheel, Labor, and Machine Cost Factors

Wheel Details	All Al$_2$O$_3$	SG	CBN (45 m/s)	CBN (60–120 m/s)
c_s	£200	£320	£3000	£1700
d_{smax}	450 mm	450 mm	450 mm	250 mm
d_{smin}	350 mm	350 mm	438 mm	240 mm
c_l	£75	£75	£75	£75
$d_{ww} + d_{ss}$	0.2 mm	0.2 mm	0.2 mm	0.2 mm
d_w	40 mm	40 mm	40 mm	40 mm
b_s	25 mm	25 mm	25 mm	17 mm
C_{mc}	£100,000	£100,000	£100,000	£250,000
y_t	1920 h	1920 h	1920 h	1920 h

19.10.9 Materials, Grinding Wheels, and Grinding Variables

19.10.9.1 AISI 52100 Steel Trials

Five GWs were employed for easy-to-grind bearing steel AISI 52100. The wheels are listed in Table 19.4. The medium-speed wheels were larger than the high-speed wheels as indicated in Table 19.3.

A two-level experimental arrangement for eight factors based on an $L_{16}2^8$ array was used to determine the best conditions using high-speed CBN. Other trials were designed to determine best conditions for conventional speeds. Typical values of the parameters are given in Figure 19.29. Further details are available from Ebbrell (2003).

For best results, the following dressing and grinding conditions were selected from the basic trials and confirmed by confirmation trials for AISI 52100:

- *Conventional-speed Al$_2$O$_3$ and SG wheels.* These wheels were dressed using a single-point diamond, with $a_d = 10$ μm, $v_d = 3.18$ mm/s, and $n_d = 2$. Grinding conditions were $v_s = 45$ m/s, $v_w = 20$ m/min, and $t_{so} = 10$ s. For the basic aluminum oxide wheels, specific removal rate, $Q'_w = 1\,mm^3/mm\,s$. For the SG wheel, $Q'_w = 2.5\,mm^3/mm\,s$.

- *Conventional-speed CBN wheel.* The conventional-speed B91 CBN wheel was updressed with a rotary-disk dresser. The wheel was dressed at a speed, $v_r = -12.6$ m/s, $a_d = 2$ μm, $v_d = 3.9$ mm/s, and $n_d = 2$. Grinding conditions were $v_s = 45$ m/s, $v_w = 26$ m/min, $t_{so} = 10$ s, and $Q'_w = 4\,mm^3/mm\,s$.

- *High-speed B91 CBN wheel.* The high-speed B91 CBN wheel was updressed at $v_r = -42$ m/s, $a_d = 2$ μm, $v_d = 1.7$ mm/s, and $n_d = 2$. Grinding conditions were $v_s = 120$ m/s, $v_w = 26$ m/min, $t_{so} = 2$ s, $Q'_w = 20\,mm^3/mm\,s$.

TABLE 19.4

Grinding Wheels and Machines for Grinding AISI 52100

Abrasive Type	Wheel Specification	Wheel Speed (m/s)
Vitrified Al$_2$O$_3$	A46 K5V	45
Vitrified Al$_2$O$_3$	A80 J6V	45
Vitrified SG	A60 J8V	45
Vitrified CBN	B91 (medium speed)	45
Vitrified CBN	B91 (high speed)	60–120

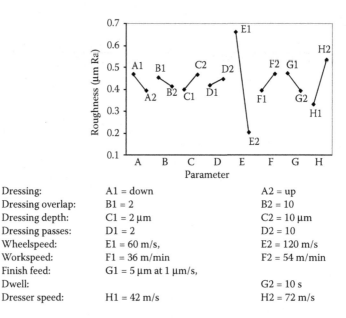

Dressing: A1 = down   A2 = up
Dressing overlap: B1 = 2   B2 = 10
Dressing depth: C1 = 2 μm   C2 = 10 μm
Dressing passes: D1 = 2   D2 = 10
Wheelspeed: E1 = 60 m/s,   E2 = 120 m/s
Workspeed: F1 = 36 m/min   F2 = 54 m/min
Finish feed: G1 = 5 μm at 1 μm/s,
Dwell: G2 = 10 s
Dresser speed: H1 = 42 m/s   H2 = 72 m/s

FIGURE 19.29
Direct-effects chart for workpiece roughness: High-speed CBN grinding of AISI 52100.

19.10.9.2 Inconel 718 Trials

Three GWs listed in Table 19.5 were each tested using an $L_8 2^7$ orthogonal array to determine the most favorable conditions.

The aluminum oxide wheel listed having 42% porosity gave better results than a number of other aluminum oxide wheels tested for grinding Inconel 718. The CBN wheels listed also had increased porosity, 40% compared with 35% and showed improved grinding performance for Inconel 718. Porosity was selected based on work by Cai (2002).

Best dressing and grinding conditions for the three wheels used for Inconel 718 were

- *Conventional-speed Al_2O_3 wheel.* The wheel was dressed with a single-point diamond, $a_d = 2$ μm, $v_d = 1.2$ mm/s, and $n_d = 2$. Grinding conditions were $v_s = 45$ m/s, $v_w = 20$ m/min, $t_{so} = 10$ s, and $Q'_w = 2\,mm^3/mm\,s$.
- *Conventional-speed vitrified B151 CBN wheel.* The wheel was dressed at $v_r = -12.6$ m/s, $a_d = 2$ μm, $v_d = 1.4$ mm/s, and $n_d = 2$. Grinding conditions were $v_s = 45$ m/s, $v_w = 20$ m/min, $t_{so} = 10$ s, and $Q'_w = 2\,mm^3/mm\,s$.
- *High-speed B151 CBN wheel.* The wheel was dressed at $v_r = -42$ m/s, $a_d = 2$ μm, $v_d = 8.5$ mm/s, and $n_d = 2$. Grinding conditions were $v_s = 120$ m/s, $v_w = 26$ m/min, $t_{so} = 2$ s, and $Q'_w = 2\,mm^3/mm\,s$.

TABLE 19.5

Grinding Wheel and Machine Combinations for Grinding Inconel 718

Abrasive Type	Wheel Details	Wheel Speed (m/s)
Vitrified Al_2O_3	A80 J6V	45
Vitrified CBN	B151 (medium speed)	45
Vitrified CBN	B151 (high speed)	60–120

19.10.10 Direct-Effect Charts

Direct-effect charts are the result of the experimental design described in Section 19.9.3. An example is shown in Figure 19.29. The example is for workpiece roughness in high-speed CBN grinding of AISI 52100. The results clearly show that high wheelspeed reduces roughness.

A chart for grinding ratio is shown in Figure 19.30 for the same conditions as in Figure 19.29. Figure 19.30 shows that wear of the GW is greatly reduced by increasing wheelspeed from 60 to 120 m/s. This is correspondingly reflected by longer redress life as indicated in the following cost comparisons.

19.10.11 Redress Life and Cost Comparisons

19.10.11.1 AISI 52100 Comparisons

Costs for grinding AISI 52100 were compared for the best conditions found for each GW and speed condition. The results are presented in Figure 19.31.

An A60-SG abrasive designed for high removal rates gave the same quality level and reduced costs compared with conventional alumina abrasives. Vitrified CBN allowed an

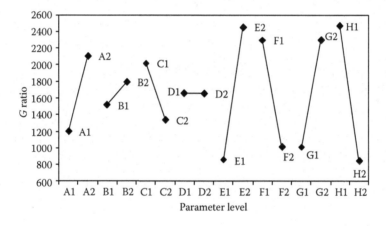

FIGURE 19.30
Direct-effects chart for *G* ratio. High-speed CBN grinding of AISI 52100. Parameters as in Figure 19.29.

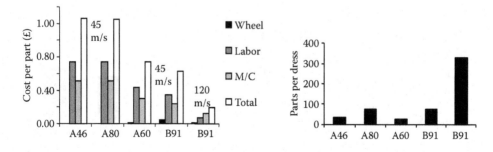

FIGURE 19.31
Comparisons of cost per part and redress life when grinding AISI 52100 with different grinding wheels and grinding conditions.

even higher removal rate at conventional speed further reducing costs and CBN at high speed allowed the highest removal rates and produced the lowest costs.

Costs per part were lower with high-speed CBN despite using a much more expensive machine for this speed. Redress life as indicated by the number of parts ground per dress was also increased using CBN and was further increased at high wheelspeed.

19.10.11.2 Inconel 718 Comparisons

Figure 19.32 compares costs/part when grinding Inconel 718 under the best conditions found with different GWs. As with easy-to-grind material, CBN allowed lower costs at conventional wheelspeed. However, this was at the same removal rate as used for the alumina wheel. Removal rate could not be increased due to the requirement to stay within tolerances over a reasonable redress life. One part/dress was achieved with alumina and 25 parts/dress with CBN. Using CBN at high wheelspeed further improved redress life to 30 parts/dress. This improvement was achieved by limiting removal rate to 2 mm^3/mm s for all three conditions.

19.10.12 Effects of Redress Life

With AISI 52100, redress life was high, so that abrasive cost, dressing cost and machine cost became negligible compared to removal rate. With high redress life, costs depend primarily on removal rate and dwell time. This was also found when grinding at high speed with CBN. Increasing wheelspeed allowed removal rates to be increased and dwell time to be reduced while maintaining redress life and quality levels.

However, redress life was short for Inconel 718. In this case, removal rates had to be kept low to allow redress life to be increased. Increasing redress life was the only way to reduce costs. Increasing wheelspeed was important to increase redress life.

Figures 19.31 and 19.32 show that high-speed CBN grinding reduces total cost/part for both materials. This is largely due to increased redress life. With AISI 52100, a high redress life was maintained with a high removal rate and short dwell time. With Inconel 718, removal rate had to be kept low to increase redress life. However, high wheelspeed also allowed spark out time to be reduced while maintaining workpiece quality and increased redress life.

19.10.13 Economic Conclusions

- Vitrified CBN offers advantages at conventional speeds and at high speeds even with a much more expensive machine. This is largely due to longer redress life.

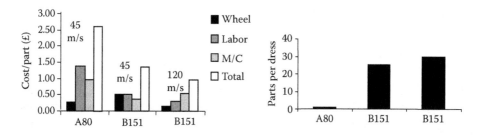

FIGURE 19.32
Comparison of costs per part and redress life when grinding Inconel 718 with different grinding wheels and grinding conditions.

- Increased removal rates are only economic if achieved with long redress life and good quality levels. Long redress life is particularly important in cost reduction for difficult-to-grind materials.

- With high removal rates, long redress life and short spark out periods, machine, and abrasive costs become negligible.

19.11 The Mechanics of Rounding

Rounding geometry was introduced in Section 19.2.5. A question of interest is whether centerless grinding is ultimately capable of the highest standards of roundness. In recent years, it has become apparent that achievable standards of roundness are approaching the limits of available measuring machines (Hashimoto et al. 1983).

The question of ultimate roundness is less likely to be posed in this way for most other processes such as cylindrical grinding between centers, where the potential achievement of high accuracy tends to be taken for granted and is often assumed to depend only on practical considerations. This may be partly justified, although high standards of roundness are often more easily obtained by centerless grinding rather than by grinding between centers, where center holes need to be lapped to minimize transmission of shape errors to the cylindrical surface being produced.

19.11.1 Avoiding Convenient Waviness

Three- and five-lobe shapes may sometimes persist after centerless grinding. Fortunately, three- and five-lobe shapes may easily be removed by grinding at the correct height above center. Odd-order lobing tends to persist with a zero or small value of the tangent angle. Odd-order lobing is a constant diameter shape sometimes known as a Gleichdicke shape. A constant diameter Gleichdicke shape can exist and rotate between parallel tangents as provided by the GW and the CW at zero center height. Gleichdicke deviations from circularity remain undetected by diameter measurement.

An example of a five-lobe shape is shown in Figure 19.33. The large roundness error resulted from grinding a workpiece at a zero tangent angle. The GW, CW, and WR shapes

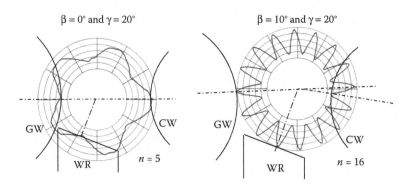

FIGURE 19.33
Examples of convenient waviness related to the setup geometry.

are superposed on the greatly magnified deviations from circularity of the workpiece. Normally, it is impossible to see roundness errors with the naked eye.

The angle of the WR has no effect on lobing at $\beta = 0°$. This is because any small movements toward or away from the WR simply cause the workpiece to move vertically up and down the vertical tangent at the CW contact point. As the workpiece rotates the odd-lobe shape is accommodated in the constant distance between the GW and the CW.

Even-lobe shapes are more likely at large values of tangent angle. This situation is also illustrated in Figure 19.33. In the example for $\beta = 10°$ and $\gamma = 20°$, 16 lobes are produced, that is, $n = 16$. An explanation for the susceptibility of this setup to 16 lobes can be found by considering the angular pitch of the waviness, $p = 360/n$. Each lobe in this example is separated by a pitch, $p = 360/16 = 22.5°$. We can consider in turn the corrective effects of the WR and the CW on this number of lobes:

1. Sixteen lobes fit neatly so that the waviness is almost unaffected by movement against the WR. This can be demonstrated as follows.

   For $\beta = 10°$ and $\gamma = 20°$, the value $\alpha \approx 66°$ so that $\alpha/p \approx 2.93 \approx 3$ is equal to an integer.

   Due to the integer relationship, 16 lobes is practically unaffected by the contact with the WR. A peak in contact with the WR coincides with a peak at the grinding contact. Corrective action due to contact with a WR is explained in more detail in Section 19.10.2.

2. Sixteen lobes are also unaffected by contact with the CW since:

$$\frac{180 - \beta}{p} = 7.55 \approx 7.5 \text{ is a half-interger}$$

   The implication for this condition is that when the trough of a lobe contacts the CW, the crest of another lobe contacts the GW and there is no corrective action. This is as illustrated in Figure 19.33.

It might be expected that $n = 18$ lobes with a pitch $p = 20°$ would also be a problem for $\beta = 10°$ since $180° - \beta = 8.5 \cdot p$. This provides the half-integer relationship required for insensitivity to the CW contact. However, in this case, $\alpha \approx 3.3 \cdot p$, which is well removed from an integer value and is closer to being a half-integer value. The WR, therefore, has a stronger corrective action on 18 lobes than on 16 lobes.

19.11.1.1 Rules for Convenient Waviness

Based on the above reasoning, three conditions can be established: If $180° - \beta$ gives a half-wavelength, if $360°$ gives a multiple of a whole wavelength and if α gives a multiple of a whole wavelength, waviness will persist. These three conditions may be summarized as

- The number of lobes, n is an integer.

$$180 - \beta \approx \left(n_1 - \frac{1}{2} \right) \cdot \frac{360}{n} \tag{19.41}$$

$$\alpha \approx n_2 \cdot \frac{360}{n} \tag{19.42}$$

where n_1 and n_2 are integers.

Miyashita (1965) gave two conditions for convenient waviness as

$$\frac{n \cdot \alpha}{180} = \text{Even number} \tag{19.43}$$

$$\frac{n(180 - \beta)}{180} = \text{Odd number} \tag{19.44}$$

Moriya et al. (1994) recommended that α/β should be equal to an odd number to avoid convenient waviness.

It is important to avoid geometry vulnerable to convenient waviness. In the above example, changing the WR angle to 30° and changing the tangent angle to 7° greatly reduces the risk of convenient waviness. This conclusion can be checked using a spreadsheet or a calculator to check values of n for near-integer values of n_1 and n_2.

From a practical viewpoint, it is important to avoid a workspeed that aligns a known source of excitation with a convenient waviness. For example, if $n = 22$ is a convenient waviness and wheelspeed is 20 rev/s, it would be important to avoid a workspeed of 1.1 rev/s since $20 \times 1.1 = 22$ and any change in wheel unbalance will directly excite the convenient waviness.

Another practical possibility for reducing roundness errors is to vary the setup geometry during the grinding operation (Harrison and Pearce 2004). Perhaps the simplest way to do this is to employ a larger tangent angle for roughing than for finishing. However, it is suggested that if two setup geometries are employed, both setups should provide a strong rounding tendency.

The analysis of rounding geometries is explored further in Sections 19.11 and 19.12 with reference to process stability and machine resonance.

19.11.2 Theory of Formation of the Workpiece Profile

Dall (1946) in the first analytical approach to predicting roundness used two basic parameters, the tangent angle β and the top WR angle γ shown in Figure 19.4. He calculated the magnitude of the error reproduced on the workpiece as a result of one already occurring at its point of contact either with the WR or with the CW. The calculations assumed a perfectly rigid machine.

Yonetsu (1959a,b) derived relationships between pregrinding and postgrinding amplitudes of harmonics of the workpiece profile. Theoretical relationships were obtained for the case of a sudden infeed, and these were compared with experimental results for infeed made over three revolutions of the workpiece. The theory showed qualitative agreement with the results. Three conclusions reached by Yonetsu were

- Odd lobes below the eleventh harmonic are better removed with a large tangent angle β.

- Even lobes below the tenth harmonic are best removed with a small angle β.

- Other errors are generated including even and odd harmonics that vary with β and it was suggested that these are related to the infeed motion.

Unfortunately the technique is unsuited to a conventional method of stock removal.

Rowe (1964) presented extensive experimental results for the influence of support geometry on roundness and supported the findings with the first computer simulation of centerless grinding. Computer simulation allowed a wide range of practical effects to be taken into account including the effect of contact length filtering, machine compliance, and spark out. Geometric and dynamic analyses of regenerative rounding instability were also presented.

Geometric stability of the process was also investigated by Gurney (1964). It was concluded that errors will be generated during the spark out period and that the stiffer the machine the greater will be the errors. It was also suggested that the GW may not always remain in contact with the workpiece during the spark out period. A recent analysis of process stability was offered by Pearce and Stone (2011) who largely follow the theory of rounding and stability given by Rowe (1964) but offer new insights into dynamic instability and application of the graphical method. A recent development of rounding analysis supported by experimental results provides a convenient method for setting optimal workspeed in relation to setup geometry and machine resonant frequencies (Rowe 2014).

Becker (1965) suggested an optimum geometrical configuration for rounding 2, 3, 5, and 7 lobes. This suggestion was reached by investigating a parameter $\delta s/\delta R$. The term δs is the difference in the apparent depths of cut when the particular shape is contained in its two extreme positions in the grinding configuration, and $\delta R = \Delta$ is the roundness error. It was proposed that the larger this parameter is, the greater will be the tendency to remove the shape. On the assumption that the optimum value of the tangent angle is $\beta = 6°$, it was suggested that the optimum WR angle $\gamma = 23°$. Decreasing the WR angle to $-10°$ improves the situation slightly for 3 and 5 lobes but worsens the situation for the ellipse and 7 lobes. Increasing the WR angle is detrimental for 2, 3, and 5 lobes and if over 35° for 7 lobes too. This method enables definite results to be obtained and is simpler than the method of Yonetsu. However, it does not allow an assessment of the final shape to be made nor does it take into account the elasticity of the machine.

It was realized in the early days of computers that it would be possible using a mainframe computer to carry out a realistic simulation of the plunge grinding process taking a number of important factors into account that would otherwise be ignored such as spark out, machine deflections, vibrations, loss of contact, GW curvature, and CW curvature (Rowe 1964). Importantly, by means of simulation, it was possible to show the rate of build up of regenerated roundness errors. In some cases, a geometric configuration might be theoretically unstable but relatively stable for practical purposes due to the high frequency of the instability and limitations of build up due to the large arc of contact with the GW.

The technique revealed several new aspects of the process including some contradictory indications for the effect of machine stiffness. It was, for example, found that work-regenerative vibrations due to geometric instability grew more rapidly with a very stiff machine. However, the elimination of roundness errors in a stable rounding process was usually improved with a stiff machine. It was also possible to explore the introduction of roundness errors due to the plunge-feed process. All these results were verified by experiments and dynamic analysis. The simulation technique and some experimental results are reported below.

Geometric work-regenerative instability was shown to be a special case of more general work-regenerative dynamic vibration instability (Rowe and Koenigsberger 1965). Other workers such as Rowe and Richards (1972) have since presented geometric stability charts for use in selection of rounding geometry for use in grinding. Extensive results for centerless grinding either below-center or above-center were presented by Johnson (1989).

Miyashita (1965) analyzed the role of machine vibrations in the generation of workpiece errors. Reeka (1967) following on from Becker, demonstrated the importance of the number of workpiece revolutions in reducing roundness errors. Roundness and effects of vibrations have been analyzed by various workers including the effects of machine design. Among these should be included Furukawa et al. (1971), Yoshioka et al. (1985), Rowe et al. (1989), Klocke et al. (2004), and Rowe (2014).

19.11.3 Workpiece Movements

This discussion deals only with the case of plunge feed, although the analysis has significance also for through-feed.

Figure 19.34 defines position on the workpiece.

A line of origin OX can be considered to rotate with the workpiece. Positions on the workpiece surface are defined by the angle from the line of origin. The position A is defined by

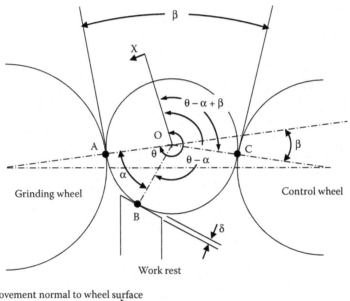

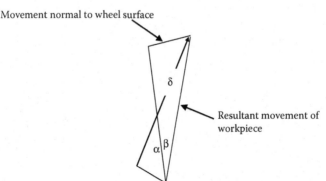

FIGURE 19.34
Workpiece position and movement due to an irregularity of magnitude δ on the workpiece at point B.

the angle θ so that as the workpiece rotates the angle θ increases. When $\theta = 0$ or $\theta = 2n\pi$ the line OX is coincident with OA.

In a similar way, point B at $\theta - \alpha$ defines the point of contact with the WR. Point C at $\theta - \pi + \beta$ defines the point of contact with the CW.

If an irregularity on the workpiece arrives at the WR as shown in Figure 19.34, the center of the workpiece will be displaced. Assuming the workpiece movement is constrained to slide along the tangent of workpiece contact with the CW, the movement can be calculated. The workpiece will be pushed away from the WR at point B. As a consequence of the movement the workpiece will move away from the GW at point A. In this way, an irregularity at point B gives rise to a further irregularity at point A. The irregularity can be considered as a local shape error.

The error at point A due to an irregularity of magnitude δ_1 on the workpiece at point B is

$$\frac{+\sin\beta}{\sin(\alpha+\beta)} \cdot \delta_1$$

The error at point A due to an irregularity of magnitude δ_2 on the workpiece at point C is

$$\frac{-\sin\alpha}{\sin(\alpha+\beta)} \cdot \delta_2$$

A reduction in radius from an initial reference circle is considered as an error. If the machine and GWs were absolutely rigid, the apparent reduction in radius $R(\theta)$ may be calculated in terms of the infeed movement $X(\theta)$ considered in a direction OA and the δ_1 and δ_2 errors. Defining $\delta_1 = -r(\theta - \alpha)$ and $\delta_2 = -r(\theta - \pi + \beta)$,

$$R(\theta) = X(\theta) + K_1 \cdot r(\theta - \alpha) - K_2 \cdot r(\theta - \pi + \beta) \qquad (19.45)$$

where

$$K_1 = \frac{\sin\beta}{\sin(\alpha+\beta)} \quad \text{and} \quad K_2 = \frac{\sin\alpha}{\sin(\alpha+\beta)}$$

The machine and GWs are not rigid and therefore Equation 19.45 is termed the apparent reduction in radius. The true reduction in radius depends on the deflection of the system. The true reduction in radius at the GW contact point A is $r(\theta) = R(\theta) - x(\theta)$, where $x(\theta)$ is the deflection at the grinding point. Equation 19.45, therefore, becomes

$$r(\theta) = X(\theta) - x(\theta) + K_1 \cdot r(\theta - \alpha) - K_2 \cdot r(\theta - \pi + \beta) \qquad (19.46)$$

For the case where $\beta = 10°$ and $\alpha = 55°$, the value of K_1 is approximately equal to 0.2 and the value of K_2 is approximately equal to 0.9. Clearly, an irregularity arriving at the CW contact point has a stronger effect than an irregularity arriving at the WR contact point.

19.11.4 The Machining-Elasticity Parameter

The machining-elasticity parameter K is a measure of the stiffness of a grinding system. The parameter provides a convenient way to account for elastic deflections of the system due to grinding force. The use of the machining-elasticity parameter works best at frequencies well below the dominant resonant frequency of the system.

$$K = \frac{\text{True depth of cut}}{\text{Set depth of cut}} = \frac{a_e}{A} \tag{19.47}$$

Values of K lie between 0 and 1. High-grinding forces and low machine stiffness lead to a low value of K. A low K might be less than 0.1. A high K might be greater than 0.4. The value of K is also dependent on grinding width. A large grinding width reduces K as described below.

It is possible to determine K, by measuring two stiffness values of the grinding system. This may be demonstrated as follows by considering the relationship between depth of cut and machine system deflections.

The set depth of cut is the sum at any instant of the deflection x and the true depth of cut a_e. The value K can therefore be written in the form

$$K = \frac{a_e}{x + a_e} \tag{19.48}$$

True depth of cut and deflection are both approximately proportional to normal grinding force so that

$$K = \frac{\lambda_0}{k_s + \lambda_0} \tag{19.49}$$

where k_s is the cutting stiffness and λ_0 is the static machine stiffness. It follows that $1/K = 1 + k_s/\lambda_0$, where k_s/λ_0 is the ratio of grinding force stiffness to machine static stiffness. Large grinding forces lead to a low value of K.

The cutting stiffness depends on workpiece hardness, GW sharpness, width of cut, and depth of cut. High cutting stiffness leads to high-grinding forces. Static machine stiffness includes the combined effect of the GW, workpiece, and machine structural elements.

A simple alternative method for measuring K depends on measuring any parameter proportional to depth of cut during "feeding in" or "sparking out." If the grinding machine is fitted with a suitable force transducer or a power-meter capable of showing instantaneous power, the following technique can be used.

In sparking out, set depth of cut is removed only after a number of revolutions. The rate of decrease of depth of cut depends on the value of K. If the apparent depth of cut has a value a_0 at the commencement of dwell, a_{e1} is the real depth of cut in the first half-revolution. The depth of cut diminishes each half-revolution.

$$a_{e1} = K \cdot a_0 \tag{19.50}$$

$$a_{e2} = K \cdot (a_0 - a_{e1}) = (1 - K) \cdot a_{e1} \tag{19.51}$$

After n half-revolutions

$$\frac{a_{em}}{a_{em-n}} = (1 - K)^n \tag{19.52}$$

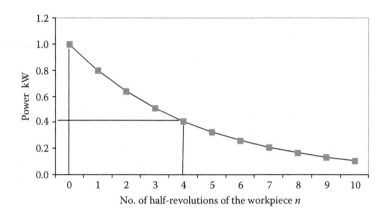

FIGURE 19.35
Power decay during spark out for $K=0.2$.

If true depth of cut is proportional to grinding power P

$$\frac{P_m}{P_{m-n}} = (1-K)^n \tag{19.53}$$

The combined effect of grinding force and deflections can be measured by taking the ratio of two values of grinding power separated by n half-revolutions of the workpiece during spark out as illustrated in Figure 19.35.

A typical value of K for a grinding system, taking a 50 mm wide cut, is $K = 0.2$. This represents a situation where the real depth of cut is one quarter of the total elastic deflections in the system. The power will decay to 0.41 of the starting value after four half-revolutions of the workpiece. The value of K can be found by plotting the log values of the power decay against n to obtain a straight line. The slope of the graph is $\log_e(1 - K)$.

19.11.5 The Basic Equation for Rounding

The true depth of cut at any position θ is the difference in workpiece radius at that position and the value of radius at the same position one revolution earlier.

$$a_e(\theta) = r(\theta) - r(\theta - 2\pi) \tag{19.54}$$

Similarly, the set depth of cut is the difference between the set reduction in radius at a particular position and the true reduction in radius one revolution earlier.

$$A(\theta) = R(\theta) - r(\theta - 2\pi) \tag{19.55}$$

Substituting Equations 19.45 into 19.55 and from Equations 19.47 and 19.54, gives the equation of Constraint.

$$r(\theta) = K[X(\theta) + K_1 \cdot r(\theta - \alpha) - K_2 \cdot r(\theta - \pi + \beta) - r(\theta - 2\pi)] + r(\theta - 2\pi) \tag{19.56}$$

Equation 19.56 with some restrictions may be used to simulate a wide variety of conditions that occur in centerless grinding.

19.11.6 Simulation

Simulation is a technique that became popular with the advent of computers. In 1961, the author realized that simulating the process on the Manchester University Mercury computer would make it possible to compare experimental workpiece shapes with predicted shapes and hence discover many important aspects of how the process works (Rowe and Barash 1964). The technique is reviewed here for its usefulness in explaining the rounding process. More recently, Cui et al. (2015) carried out similar work on a modern computer and were able additionally, to incorporate stiffness and damping factors and also mass elements at the workpiece contact points. Good agreement between theory and experiment was achieved by tuning the strength of these contact elements.

The workpiece circumference is divided into small steps with respect to a radial line of origin. Steps of 1° were sufficient. Initial workpiece radius errors were stored for positions around the workpiece. The error values represent an initial set from $r(-360)$ to $r(-1)$ when $\theta = 0°$. These values along with any suitable increments for the infeed motion can be fed into Equation 19.56. The angle θ is increased one step at a time and values calculated for $r(\theta)$. Calculated values are stored and represent the developing profile of the workpiece. It is thus possible to examine the effects of different infeed functions including the importance of the dwell period. Various initial shapes can be investigated, as these are a main source of variability in final shapes after grinding. It is even possible to consider the effects of machine vibrations on roundness.

It quickly becomes apparent that restrictions need to be applied to prevent radius increasing. This is the metal replacement restriction and is applied by incorporating a conditional statement. This and other restrictions required are as follows:

- *Metal replacement restriction.* The value of $r(\theta)$ is compared with the value of $r(\theta - 2\pi)$ and replaced by the previous value whenever metal replacement is implied.

- *GW interference.* With high frequencies and large radius variations from step to step, it is not possible for the shape to be accommodated. This is because the radius of the GW will interfere with workpiece positions adjacent to the point of contact. This is illustrated in Figure 19.36.

  Grinding wheel interference has a practical benefit in grinding in that it limits the development of higher frequency vibrations and has the effect of smoothing the profile. The restriction is applied at nearby positions by calculating the clearance

$$W_\psi = \left[\frac{1}{\cos \psi} - 1 \right] \cdot \frac{d_w}{2} \tag{19.57}$$

and the interference

$$I_\psi = r(\theta) + W_\psi - r(\theta - 2\pi) \tag{19.58}$$

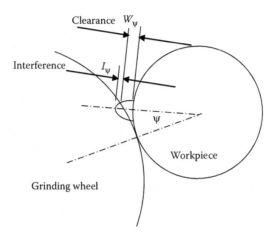

FIGURE 19.36
Grinding wheel interference from positions adjacent to the grinding point.

Modifications only need to be made for positive interference. An amount of material is removed at the adjacent position equal to the interference I_ψ multiplied by the machining-elasticity parameter K.

- *CW interference and WR interference restrictions.* Interference at the CW and WR are calculated similarly to the previous example. The largest interference is found and subtracted from the usual term at each contact.

Figure 19.37 shows typical simulation results for a uniform infeed. In both cases simulated, errors due to the infeed are less than 1% of the infeed per revolution after 45 revolutions.

With small values of β, 3 and 5 lobes predominate as explained previously. In the case, where $\beta = 10°$ and $\gamma = 20°$, the rounding geometry is unstable for 16 lobes. The resulting

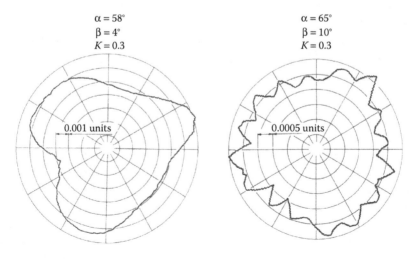

FIGURE 19.37
Simulated roundness for a uniform infeed of 1 unit/rev for 45 revolutions.

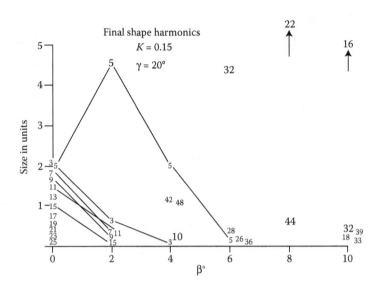

FIGURE 19.38
Simulated grinding of a workpiece with a flat 41 units deep with a 20° workrest.

roundness errors are small because there was very little excitation of the instability in the simulation. In practice, roundness errors are very much affected by level of excitation.

Poor roundness will result if the machine resonates at a frequency of instability. The workspeed should therefore be set to avoid excitation of an unstable frequency. For this geometry, and a resonant frequency of 72 Hz, workpiece speeds should be avoided close to $n_w = 72/16 = 4.5$ rev/s.

Initial shape errors vary greatly in practice depending on the previous machining process. Effects of initial shape on the final ground shape were investigated using a controlled error. A small flat was machined along the length of the workpiece as described in Section 19.11.7. The workpiece with a flat has a similar effect in grinding to an impulsive force and excites a wide range of frequencies each time the flat passes a contact point.

A series of simulations was performed for a workpiece with a flat. Amplitudes of the workpiece shape harmonics determined by Fourier analysis are illustrated in Figure 19.38. It is seen that 5 lobes predominate at a low tangent angle and even lobes at larger tangent angles. At 10°, 16 lobes are strongly evident. At 8°, 22 lobes appear and also 44 lobes. Lower numbers of lobes tend to present a greater problem than higher numbers of lobes. This is because higher frequencies are more likely to be filtered by interference with the GW, CW, and WR as explained in Section 19.11.6. Higher frequencies may also be smoothed by workpiece inertia.

Figure 19.39 shows simulation results for a 30° WR. Despite assuming a more compliant system, $K = 0.08$, the rounding action was much stronger than in the previous case. The build up of even-order waviness was less pronounced with the 30° WR than with a 20° WR.

19.11.7 Roundness Experiments and Comparison with Simulation

All the roundness problems predicted by simulation were experienced in practice when grinding a workpiece with a flat and sometimes when grinding an initially circular workpiece (Rowe 1964). The simulation technique was tested by comparison with an

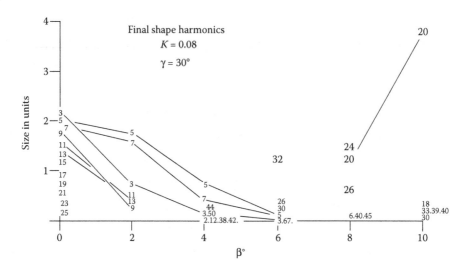

FIGURE 19.39
Simulated grinding of a workpiece with a flat using a 30° workrest.

experimental shape measured after grinding. The pregrinding shape of the workpiece with a flat is shown in Figure 19.40. The workpiece was then ground to the depth of the flat. That is, the radial stock removed was 57 μm. The postgrinding shape is shown in Figure 19.41.

The shape predicted by simulation is shown in Figure 19.41. Agreement with the experimental shape is excellent considering the duration of the grinding cycle over 50 workpiece revolutions. An interesting aspect of the simulation was that it confirmed that the particular grinding setup represented a very compliant system with $K = 0.08$. This means the depth of cut at any instant was less than 10% of the deflections.

Good correlation between theory and experiment is obtained when the initial workpiece irregularity is large, so that the effect of initial shape predominates over other effects. Usually, grinding results, when magnified to a high degree, exhibit an almost random spread of shapes and magnitudes reflecting the variety of pregrinding shapes and the build up of errors on the GW and CW. The advantage of working with a large initial roundness error is that it allows the systematic rounding effect to be clearly seen (Figures 19.5 and 19.42).

Starting with a round workpiece, it might be expected that a perfectly round workpiece should result. Simulation shows that this is not necessarily so. Even with a uniform infeed motion and careful sparking out, small errors result due to the geometry of the rounding

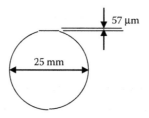

FIGURE 19.40
The initial workpiece shape used for experimental determination of rounding effect.

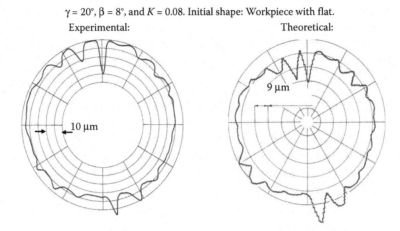

FIGURE 19.41
Experiment compared with simulation.

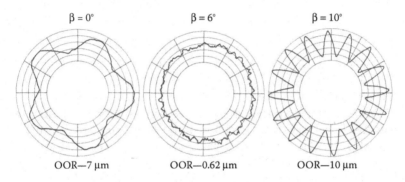

FIGURE 19.42
Experimental roundness resulting from grinding a workpiece with a flat using a 20° workrest.

action. With a compliant system and optimum setup of the rounding geometry, errors due to infeed motion are extremely small. Typically, the magnitude of the roundness errors due to the infeed motion was less than two-thousandths of the magnitude of the infeed per revolution. This means, for example, that if the depth of cut were 10 μm, the roundness error would typically be 0.01 μm. In other words, the effect is negligible compared with typical roundness errors ranging in precision grinding from 0.2 to 1.0 μm.

In practice, the best roundness depends on a number of factors including machine vibrations, bearing precision, dressing precision, and wheel wear as outlined in the discussion of the system in Section 19.2.6.

19.12 Vibration Stability Theory

The motivation to analyze stability is the need to design and operate systems for best roundness and to avoid the build up of roundness errors due to instability taking account of machine resonances. In principle, most grinding systems will be unstable due to wheel-regenerative instability. However, in most cases, wheel-regenerative instability builds up very slowly at high frequencies. When roundness or poor surface finish becomes apparent after a lengthy grinding period, it is simply necessary to redress the GW to restore good workpiece quality. In practice, it is possible to arrive at systems that yield very low roundness errors. The remainder of this chapter is mainly concerned with work-regenerative instability and achievement of satisfactory grinding conditions taking into account workspeed and machine resonances.

Section 19.12 introduces concepts of geometric stability and dynamic stability in the context of the rounding process and leads into techniques for comparing rounding stability for differing geometric setups and speed conditions. The subject is further developed in Section 19.13 with examples.

19.12.1 Definitions of Stability

The following are definitions relating to stable, unstable, and boundary conditions in grinding chatter. Forced vibration is distinguished from chatter vibration.

Marginal stability: Marginal stability is where waviness neither builds up nor decays as grinding proceeds without direct excitation of the marginally stable frequency. In practice, a linear system excited at a frequency of marginal stability will experience large vibrations only limited by system nonlinearity. Convenient waviness was described earlier in this chapter. Convenient waviness is waviness that can persist and is not corrected by the rounding geometry. Convenient waviness that neither builds up nor decays is termed marginal geometric stability.

A stable system: In a stable system, it is implied that roundness errors are reduced as grinding proceeds. In other words, waviness of the surface decays as grinding proceeds. However, it should be remembered that a stable system produces waviness when there is a source of sustained vibration excitation otherwise known as forced vibration.

An unstable system: An unstable situation is where waviness and vibration build up as grinding proceeds without a need for direct excitation. If the waviness builds up primarily on the workpiece, the instability is termed "work-regenerative chatter." Another type of instability is where waviness primarily builds up on the GW circumference or on the CW circumference. This second type of instability is termed "wheel-regenerative chatter." Wheel-regenerative chatter often builds up slowly with the grinding of successive workpieces so that instability may not be apparent until a number of workpieces have been ground. The waviness on each succeeding workpiece increases as the waviness builds up on the GW.

Chatter: In machining, an unstable system is said to chatter, since in metal turning and milling, unstable vibrations are characterized by audible vibrations or "chatter." In grinding, unstable regeneration of waviness and vibrations may occur, but due to a more gentle character the vibrations are not necessarily audible. The term chatter is used to describe regeneration of waviness and vibration due to an unstable system, whether it is audible or inaudible. Chatter should not be confused with forced vibration.

Forced vibration: Forced vibrations may occur in stable systems and do not imply chatter or instability. Forced vibrations occur due to the action of the normal machine forces. Examples of forced vibrations are the vibrations forced by GW unbalance, motor unbalance, uneven belt drives, out of roundness of the wheel spindles, and cyclic forces due to worn or inaccurate rolling elements in the bearings. Forced vibrations are mostly characterized by a repetitive and constant frequency that can be related to the speed of the offending machine element. It is particularly important to avoid coincidence between a forced vibration frequency and a marginally stable frequency. This situation can usually be avoided by judicious choice of workspeed.

19.12.2 A Model of the Dynamic System

The dynamic system is represented in Figure 19.43 (Rowe 1979).

In Section 19.11.3, the geometry relationships were defined in terms of angle. A dynamic system also depends on time and acceleration. Referring to Section 19.11.5, the main relationships defined in terms of angular position on the workpiece are

$$r(\theta) = X(\theta) - x(\theta) + K_1 \cdot r(\theta - \alpha) - K_2 \cdot r(\theta - \pi + \beta) \quad \text{waviness} \tag{19.59}$$

$$a_e(\theta) = r(\theta) - r(\theta - 2\pi) \quad \text{depth of cut} \tag{19.60}$$

$$x(\theta) = \frac{F_n(\theta)}{\lambda(\theta)} \quad \text{deflection} \tag{19.61}$$

$$F_n(\theta) = k_s \cdot a_e(\theta) \quad \text{normal force} \tag{19.62}$$

$$r_s(\theta) = C_{sw} \cdot F_n(\theta) + r_s\left(\theta - \frac{2\pi\Omega_w}{\Omega_s}\right) \quad \text{grinding wheel wear} \tag{19.63}$$

Ideally, CW wear and WR wear should be included in addition to the GW wear effect. However, these effects are usually much less significant.

Analysis of system stability starts from a general solution of waviness as a function that develops exponentially with time and has complex solutions of the complementary function of the form: $r(t) = c \cdot e^{pt}$. The roots of the complementary function have the

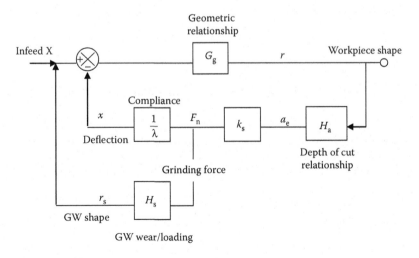

FIGURE 19.43
Representation of grinding as a dynamic system.

form $p = \sigma + j\omega$. The real part σ governs the growth rate of the system. A negative value implies a stable decay of the function. A positive value represents an unstable exponentially increasing value. The complex imaginary part ω represents frequency. A zero value, $\sigma = 0$, corresponds to solutions of the form: $r(t) = c \cdot \sin \omega t$. This is a waviness condition of marginal stability and constant amplitude.

If the system has an input function of the form $u_i = c \cdot \sin \omega t$, the output of a linear stable system will have the form $u_0 = D \cdot \sin(\omega t + \phi)$, where the amplitude is changed and there is a phase difference between the input and the output sine waves. It is important to realize that a sinusoidal forcing function at a frequency of marginal stability leads to an infinite response. A zero growth rate is therefore unstable.

Waviness expressed in terms of angle is: $r(\theta) = c \cdot \sin(2\pi\omega/\Omega) = c \cdot \sin 2\pi n$, where Ω is the workpiece speed in rad/s and n is the number of waves on the workpiece.

19.12.3 Transfer Functions

Transfer functions give the amplification and phase between the input and output for each block in the system. G_g gives the amplitude and phase of the wave output produced on the workpiece at the grinding point due to the setup geometry for a sine wave input at the grinding point. Thus, if the feedback produces a workpiece reduction in radius in phase with the existing waviness, the amplitude will tend to grow. If the feedback produces a small reduction in radius in anti-phase the amplitude will tend to decay. A reduction in radius leading or lagging the existing waviness will create a wave that is constantly shifting around the workpiece periphery.

In Laplace form, the transfer functions shown in Figure 19.43 are

$$G_g(s) = \frac{r(s)}{X(s)} = \frac{1}{[1 - K_1 e^{-\alpha s} + K_2 e^{-(\pi-\beta)s}]} \quad \text{geometric function} \tag{19.64}$$

$$H_a(s) = \frac{a_e(s)}{r(s)} = 1 - e^{-2\pi s} \quad \text{depth of cut function} \tag{19.65}$$

$$\frac{1}{\lambda(s)} = \frac{x(s)}{F_n(s)} \quad \text{compliance function} \tag{19.66}$$

$$H_s(s) = \frac{r_s(s)}{F_n(s)} = \frac{C_{sw}}{[1 - e^{-2\pi s/\Omega_s}]} \quad \text{wheel wear function} \tag{19.67}$$

Work-regenerative stability depends on the main feedback loop shown in Figure 19.43. The main open-loop transfer function (OLTF) for the main feedback loop relates the vibration deflections $x(s)$ fed back at the grinding point to the prescribed infeed motion $X(s)$.

$$\frac{x(s)}{X(s)} = \frac{G_g(s) \cdot H_a(s) \cdot k_s}{\lambda(s)} \quad \text{Open-loop transfer function} \tag{19.68}$$

For steady sinusoidal motion, we may write $s = jn$, where j is the complex operator. Frequency is expressed in terms of number of lobes around the workpiece, where

$n = \omega/\Omega = f/n_w$. The number of lobes n is equal to the vibration frequency ω in rad/s divided by the constant workspeed Ω expressed in rad/s. Alternatively, n is equal to the vibration frequency f expressed in Hz divided by the work rotational speed n_w expressed in rev/s. For practical work, it is more convenient to employ the familiar form $n = f/n_w$.

Expanding Equation 19.68 leads to the following form for the OLTF.

$$\frac{x(s)}{X(s)} = \left(\frac{k_s}{\lambda_0}\right) \cdot \left(\frac{\lambda_0}{\lambda(jn)}\right) \cdot \left(\frac{1 - e^{-j2\pi n}}{1 + K_2 e^{-jn(\pi-\beta)} - K_1 e^{-jn\alpha}}\right)$$

The first term k_s/λ_0 is the grinding force factor k_s divided by the static stiffness λ_0 of the dominant machine vibration mode. The second-term $\lambda_0/\lambda(jn)$ is the machine compliance function in the dominant vibration mode, as illustrated below for a second-order single degree of freedom system in Figure 19.44. The third term is the product of the depth of cut function $H_a(jn)$ and the geometric function $G_g(jn)$

At the limit of stability, the OLTF is equal to –1. Therefore, the condition for limiting stability becomes

$$\left(\frac{k_s}{\lambda_0}\right) \cdot \left(\frac{\lambda_0}{\lambda(jn)}\right) \cdot (1 - e^{-j2\pi n}) + 1 + K_2 e^{-jn(\pi-\beta)} - K_1 e^{-jn\alpha} = 0 \quad \text{Stability limit} \qquad (19.69)$$

System compliance can be approximated for very low frequencies by writing $\lambda(jn) = \lambda_0$. This is useful, when the lobing corresponds to a frequency well below the principal machine resonance.

In a similar manner, an OLTF may be developed for the reduction in GW radius in the wheel-regenerative loop.

$$\frac{r_s(s)}{X(s)} = k_s \cdot G_g(s) \cdot H_a(s) \cdot H_s(s) \qquad (19.70)$$

Wheel-regenerative vibration builds up lobing on the GW with increasing duration of grinding. When roundness errors exceed an acceptable level, the GW must be dressed to restore accuracy.

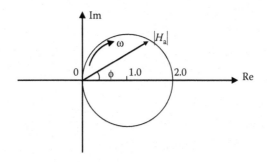

FIGURE 19.44
The frequency characteristics of the depth of cut function.

19.12.4 Nyquist Test for Stability

The Nyquist criterion can conveniently be employed to test for stability. The Nyquist criterion states that a system will be unstable, if the OLTF with increasing frequency encircles the point (–1, 0) on the negative real axis in a clockwise direction. In this plot, the coordinates are the real in-phase and imaginary quadrature parts of the function, (Re, Im) at each frequency. Applied to the stability limit given by Equation 19.69, instability is indicated by clockwise encirclement of the origin (0, 0).

It can be seen that the product $G_g(s)\,H_a(s)$ features in both the work-regenerative and the wheel-regenerative loop transfer functions. It is therefore important in avoiding instability to ensure that this term is stable. The frequency characteristics of the functions can be explored in the usual way by substituting $s = j\omega$ or as shown above by $s = jn$.

The Nyquist criterion is initially applied below to the geometric function shown in Figure 19.43 and later to the loop containing the geometric function and the machine vibration characteristics. The effect of added static compliance is also explored. The following paragraphs initially explore stability using Nyquist diagrams. Nyquist diagrams are used for simplicity compared to other methods and provide a rigorous test of stability margin. A further simplification is introduced in Section 19.13 leading to formulation of a new dynamic stability parameter for centerless grinding (Rowe 2014).

19.12.5 The Depth of Cut Function

The depth of cut function $H_a(j\omega)$ varies with frequency as illustrated in Figure 19.45. The function $H_a(j\omega) = 1 - \cos\omega T + j \cdot \sin\omega T$ is a circle of unit radius centered at the point (1, 0) on the real axis. The phase angle between the depth of cut and the existing wave is represented by the term ωT and varies from 0 to $2n\pi$. The period T is related to the workpiece rotational frequency Ω expressed in rad/s by $T = 2\pi/\Omega$. The number of lobes is given by $n = \omega/\Omega = f/n_w$, leading to the function $H_a(j \cdot n) = 1 - \cos2\pi n + j \cdot \sin2\pi n$.

The depth of cut function has real values at 0 and +2. The real axis represents depth of cut in phase with the lobing. The point where the locus crosses the real axis at 0 is an integer wave condition, where the new wave $r(\theta)$ is in phase with the old wave $r(\theta - 2\pi)$ passing the grinding point one revolution earlier with zero depth of cut. For a constant in-phase amplitude of lobing, excitation of the grinding force is zero. This condition occurs with convenient waviness. Excitation of the force is zero for a large waviness. For integer waves, at the limit of stability, grinding forces and compliances have no effect.

The point where the locus crosses the real axis at +2 corresponds to the half-integer wave condition, where the new wave is exactly 180° out of phase with the old wave. This condition causes a depth of cut amplitude that is double the amplitude of the wave. Instability at

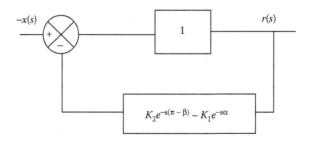

FIGURE 19.45
The geometric function $Gg(s)$ represented as an inner feedback loop.

frequencies close to this condition corresponds to the harsh nature of chatter experienced in cutting and milling operations. Excitation of the chatter force may be large compared to the case of convenient waviness.

19.12.6 The Geometric Function and Geometric Stability

In this section, we consider the nature of geometric instability and compare instabilities for different setup geometries and numbers of lobes.

The geometric function $G_g(s)$ as illustrated in Figure 19.45 represents an inner feedback loop in Figure 19.43.

The two feedback terms form circles as shown in Figure 19.46 when varying frequency. The larger circle in Figure 19.46a is the CW feedback and the smaller circle is the WR feedback. If the two terms are added, the feedback locus is a spiral of approximate circles varying in size depending on the number of lobes. Figure 19.46b shows a plot of the geometric function $1/G_g(j \cdot n)$ for 16–18 lobes for a geometric setup represented by particular angles α and β. Encirclement of the origin in a clockwise direction indicates geometric instability for a frequency close to 16 lobes but geometric stability for 18 lobes.

Near the origin, a wave at the CW position, for 16 lobes, is approximately in anti-phase with the wave at the grinding point. At the WR position, the wave is approximately in phase with the wave at the grinding point as described in the earlier discussion in Section 19.11 on convenient waviness.

If the geometric locus passes the origin on the negative side, waviness builds up because the feedback adds to the wave at the grinding point. This allows us to define a geometric stability parameter "A" defined as the real part of the geometric feedback loop. When the in-phase feedback A becomes negative and the quadrature part B is zero, the system is geometrically unstable. Plotting these two parameters forms the locus $1/G_g(j \cdot n)$

$$A = 1 + K_2 \cos[n(\pi - \beta)] - K_1 \cos[n\alpha] \tag{19.71}$$

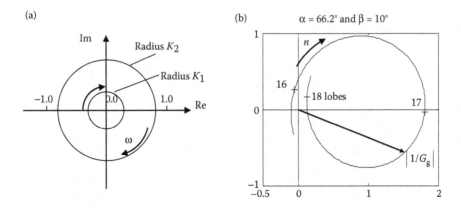

FIGURE 19.46
Stability test for the geometric function $1/G_g(jn)$ for frequencies from 16 to 18 lobes revealing geometric instability close to $n = 16$ lobes employing a 20° workrest and 10° tangent angle.

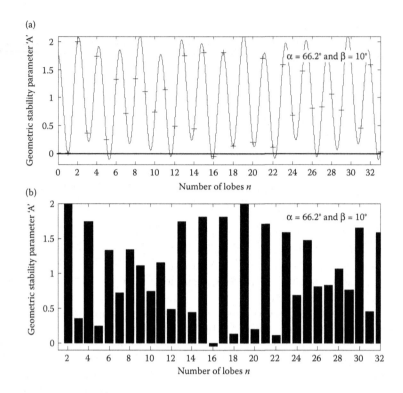

FIGURE 19.47
Geometric stability parameter: $\gamma = 20°$ workrest and $\beta = 10°$ (a) noninteger and (b) integer lobes.

$$B = K_2 \sin[n(\pi - \beta)] - K_1 \sin[n\alpha] \tag{19.72}$$

The function A defined by Equation 19.71 is termed *the geometric stability parameter*.

The plot in Figure 19.47b is negative for the unstable 16 lobes and for slightly less than 16 lobes, where the real component achieves a maximum negative value. The distance of the negative real component from zero is indicative of growth rate in the closed loop response for a rigid machine. The plot is slightly positive for the marginally stable 18 lobes indicating a small decay rate of the waviness in the closed loop response. The plot indicates a strong rounding tendency for $n = 17$ lobes.

Figure 19.47a is a plot of the geometric stability parameter A for 0–33 lobes including nonintegers. Values are shown for a tangent angle $\beta = 10°$ and a WR angle $\gamma = 20°$ and integer positions are marked with a " + " sign. This plot allows geometric stability at all the low frequencies to be investigated. It appears there are several instabilities over the range. The most serious is at 16 lobes because the instability coincides with an integer number of lobes, where the quadrature component is almost zero. The implication of integer lobes is that the feedback is in phase with the wave already existing on the surface. It does not matter if the feedback is small because the depth of cut continually adds to the existing waviness instead of reducing it. The system is always unstable in this case even if the machine is rigid. In fact, the build up of waviness will be faster with a rigid machine since the positive feedback is more strongly imposed through the depth of cut.

Instability is also indicated for approximately five and a half lobes. However, in this case the number of lobes is no longer an integer. This means, the existing waviness must be continually removed and replaced by a new waviness of larger amplitude produced at a new position around the workpiece circumference. In-phase deflections due to static compliance of the machine or wheel surfaces will reduce the depth of cut and make it difficult for the waviness to grow. A rigid machine increases the growth rate. Instability for noninteger lobing must be considered together with the force and compliance function as part of a dynamic analysis. The geometry of Figure 19.47 is discussed further in Section 19.13.

A local minimum or trough for integer lobes is of concern, even where the geometric stability parameter A is positive. For example, it can be seen that 18 and 20 lobes in Figure 19.47a have troughs on the stability chart. While these numbers of lobes are stable, large roundness errors will result if a forcing frequency is allowed to directly excite lobing. There is also a negative trough close to 22 lobes, and again it would be essential to avoid forcing at the corresponding frequency.

Geometric stability tested for integer lobes is shown in Figure 19.47b. The only integer instability up to $n = 32$ is at $n = 16$ lobes. However, figures such as Figure 19.47a reveal potential geometric instability for integer and noninteger lobes. For example, there is instability close to $n = 22$ lobes. This means that an initial 22-lobe shape will not be rounded up but will increase with time. However, geometric instability grows very slowly compared with dynamic instability. Often geometric instability remains undetected until a workpiece is ground with a large initial roundness error or when there is proximity to a resonance or a forcing vibration at the corresponding frequency. As shown in the following sections, proximity of a geometric instability to a resonance is likely to create dynamic instability.

Negative values of A define growth rate as shown in Figure 19.48 for a range of β and γ values and integer lobes. At each combination of β and γ, only the most unstable number

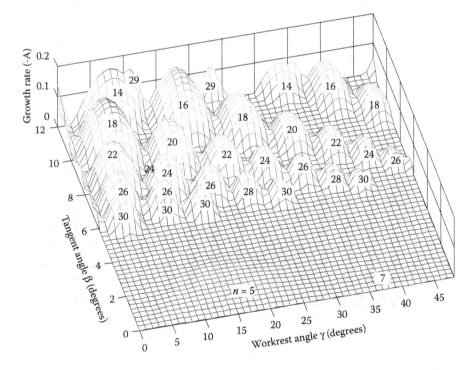

FIGURE 19.48
Maximum growth rate for integer lobes against tangent angle β and workrest angle γ.

of lobes is shown. This figure has been constructed for a machine having a CW diameter that is roughly 60% of the GW diameter. It was assumed that with small workpieces this yields a value $v = 0.38$, as in the example described in Section 19.2.5.

Geometrically unstable zones are shown in Figure 19.48 for $n = 3$–30 lobes. For a tangent angle less than 5°, the areas of slight geometric instability are for 5 lobes having low growth rate and 7 lobes having very low growth rate. Most of the region is stable. However, a tangent angle less than 5° fails to provide rapid rounding action for 5 lobes. Therefore, the tangent angle is usually made larger. A roughly periodic pattern can be seen for the number of lobes as β and γ are varied. There are small zones that appear more favorable than others, such as $β = 11°$ and $γ = 30°$. Another combination, $β = 7.5°$ and $γ = 30°$, appears to be stable for integer lobes up to $n = 32$. However, $n = 24$ and 26 integer lobes are only just stable as shown by the example in Figure 19.49a. Plotting the geometric stability parameter for integer and noninteger lobes as shown in Figure 19.49b reveals the geometric instability nearby to 26 lobes but not exactly at 26 lobes.

Figure 19.49 for a 30° WR may be compared with Figure 19.47 for a 20°WR, where there is an instability at $n = 16$ lobes. Stability appears to be more favorable in Figure 19.49 on two counts. There is no geometric instability in Figure 19.49 at an integer number of lobes. Also, the first marginal instability is at a higher frequency. This is an advantage because

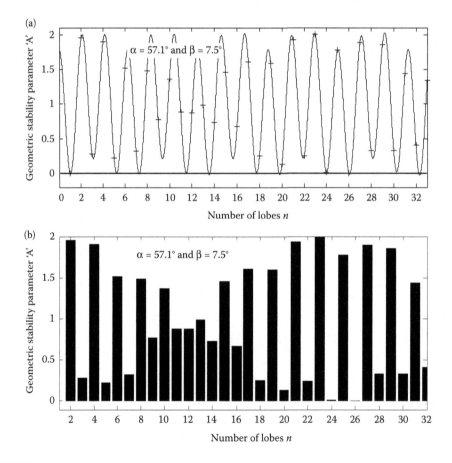

FIGURE 19.49
Geometric stability parameter: 30° workrest and 7.5° tangent angle for (a) integer and (b) noninteger lobes.

higher frequencies reduce the maximum amplitude of waviness as explained in Section 19.10.6. As explained later, there is also a need to consider workspeed in relation to resonant frequency and potential dynamic instability.

19.12.7 Static and Dynamic Compliance Functions

In this section, we consider the effect of static deflections and then consider the effect of dynamic deflections. Consideration is also given to where static deflections reduce geometrically unstable growth rate.

19.12.7.1 Static Compliance

Static compliance depends primarily on the elasticity of the wheels, bearings, slideways, and machine structure. System behavior analyzed below, including the effect of static compliance, confirms the results given by simulations in Section 19.10. The static compliance function at zero frequency is

$$\frac{x}{F_n} = \frac{1}{\lambda_0}$$ (19.73)

The static stiffness is λ_0. The OLTF for purely static compliance expressed in terms of n by writing as previously $s = jn$ becomes

$$\frac{x(jn)}{X(jn)} = \left(\frac{k_s}{\lambda_0}\right) \cdot \left(\frac{1}{1 - K_1 e^{-\alpha jn} + K_2 e^{-(\pi-\beta)jn}}\right) \cdot \left(1 - e^{-2\pi jn}\right)$$ (19.74)

The stability test with static compliance is given as explained above by equating (19.74) to –1, leading to

$$\left(\frac{k_s}{\lambda_0}\right) \cdot (1 - e^{-2\pi jn}) + 1 - K_1 e^{-\alpha jn} + K_2 e^{-(\pi-\beta)jn} = 0$$

Applying the stability test, Figure 19.50 shows how grinding force, static compliance, and depth of cut modify Figure 19.46 which was purely for the geometric function. A high value of the grinding force factor k_s increases deflections whereas a high value of grinding machine static stiffness λ_0 reduces them. Sizes and positions of the circular loops are modified. The extent of the modification depends on the magnitude of k_s/λ_0.

It is found that static compliance slightly reduces growth rate at frequencies close to the integer 16 lobes but cannot make the unstable situation stable. This is because the positions of the locus for integer lobes are independent of the static deflections due to the depth of cut function being zero at the stability limit as explained in Section 19.12.4. However, when the geometric feedback is negative, depth of cut will not be zero and in-phase deflections reduce feedback for an integer number of waves.

19.12.7.2 Dynamic Compliances

Dynamic compliances typically depend on the effective mass and stiffness in tuning-fork modes of vibration between the GW and the CW. The effect of a simple frequency

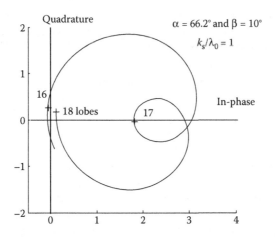

FIGURE 19.50
Stability test including static compliance for comparison with Figure 19.46 for a rigid machine. Geometric instability at 16 integer lobes remains unstable.

response function for machine system compliances can be demonstrated using a second-order single degree of freedom mass-spring-damper system defined by

$$\frac{\lambda_0}{\lambda(j\omega)} = \frac{\lambda_0 x(j\omega)}{F_n(j\omega)} = \frac{1}{(1-(\omega^2/\omega_0^2)+(j\omega/Q\omega_0))} = \frac{1}{(1-(n^2/n_0^2)+(jn/n_0 Q))} \qquad (19.75)$$

where
$\omega_0 = \sqrt{\lambda_0/m}$ is the undamped natural frequency
$Q = \lambda_0/c\omega_0$ is the dynamic magnifier; an inverse measure of damping c. Equation 19.75 expressed either in terms of ω or n can be employed in Equation 19.68

An ideal frequency response function of a machine is shown in Figure 19.51 for a second-order single degree-of-freedom mode of vibration. The frequency is expressed as number

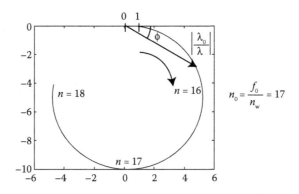

FIGURE 19.51
Ideal machine response with frequency expressed in number of lobes. Natural frequency n_0 at $n = 17$ lobes and dynamic magnifier $Q = 10$.

of lobes for a particular workspeed by writing $n = \omega/\Omega$ as previously. The natural frequency of the machine for the example shown in Figure 19.51 coincides with 17 lobes. This corresponds to a natural frequency of 68 Hz in the tuning-fork mode for a work rotational speed of 4 rev/s. The machine is lightly damped as indicated by the dynamic magnifier, $Q = 10$, a value that corresponds to a critical damping ratio of 0.05. This amplifies the static deflection by a factor of 10 at the resonant frequency.

The stability limit for the grinding process including the dynamic response of the principal vibration mode is given by the procedure explained above:

$$1 + K_2 e^{-jn(\pi-\beta)} - K_1 e^{-jn\alpha} + \left(\frac{k_s}{\lambda_0}\right) \cdot \frac{(1 - e^{-j \cdot 2\pi n})}{(1 - (n^2/n_0^2) + (jn/Qn_0))} = 0 \qquad (19.76)$$

A plot of the stability limit for frequencies in the vicinity of the natural frequency for the unstable 16-lobe geometry yields Figure 19.52, where the natural frequency and workspeed combination is set so that n_0 is at 16 lobes rather than at 17 lobes. Static compliance in Figure 19.52 is set much lower than in Figure 19.50 so that static deflections are much smaller. However, dynamic deflections are large due to the nearby resonance. This result demonstrates the effect of the interaction between the large deflections of the machine at frequencies close to 16 lobes with the geometric instability close to 16 lobes. It is seen that 16 lobes remain unstable as in Figures 19.46 and 19.50 but the main frequency of the instability is moved even lower below the frequency corresponding to the integer 16 lobes. It is also found that the loop between 17.5 and 18 lobes previously stable is now unstable due to the nearby resonance. This demonstrates the clear importance of setting workspeed to avoid resonance.

The situation of marginal geometric instability is often made worse by a resonance close to a frequency of marginal instability since resonance increases the probability of forced vibration amplitudes already in existence independent of the grinding process. These vibrations force roundness errors even if the geometry is stable. The user should therefore

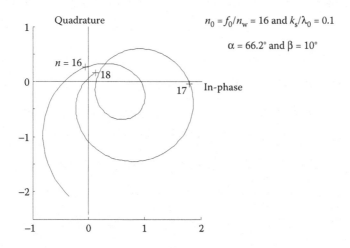

FIGURE 19.52

Stability test of dynamic system. Compare with Figures 19.46 and 19.50. Natural frequency n_0 at $n = 16$ lobes. Geometric instability at $n = 16$ is unchanged but instability is increased at nearby frequencies due to resonance.

treat a resonance in the same way as a frequency of forced vibration and avoid proximity to a convenient waviness. This may be achieved by adjusting the workspeed.

The effect of increasing compliance in the system model may be demonstrated by increasing the term k_s/λ_0 as shown by Rowe (2013). Increasing the compliance term increases dynamic deflections and increases the tendency to dynamic instability. However, additional static compliance can occur in a different and sometimes slightly beneficial way as follows.

19.12.8 Added Static Compliance

It has been shown above that the effect of resonance is to destabilize the process. It has also been argued that static compliance can reduce growth rate of regenerative vibration. In this context, we are referring to static compliance that does not contribute toward a phase shift between grinding force and depth of cut. In other words, added static compliance is taken to imply that the added compliance does not affect amplitudes in the resonant mode or reduce the corresponding natural frequency. A natural frequency of a machine is the combined effect of a substantial mass and the supporting stiffness of the structure. However, it is possible to incorporate compliance at the grinding point by making the abrasive structure of the GW compliant or by making the CW compliant. This does not significantly affect the stiffness in dominant tuning-fork modes of the machine. If the workpiece has low mass, compliance can be incorporated at the CW surface without significantly affecting the dominant resonances. The relatively soft rubber CW commonly fulfills this function.

An explanation for the stabilizing effect of added compliance is that the frequency response function of the machine is shifted to the right on the real axis. When the machine response function is added to the other terms in the characteristic equation, the shift to the right reduces the probability of enclosing the origin. However, as explained in Section 19.12.7, compliance does not affect stability of integer lobing but static compliance has benefit at nearby frequencies.

Additional static compliance can be demonstrated mathematically by introducing a secondary static compliance term $1/\lambda_2$. The stability test is given by Equation 19.77. The stability test is plotted in Figure 19.53 which may be compared with Figure 19.52.

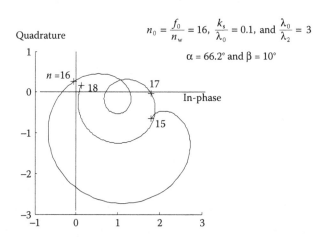

$$n_0 = \frac{f_0}{n_w} = 16, \ \frac{k_s}{\lambda_0} = 0.1, \text{ and } \frac{\lambda_0}{\lambda_2} = 3$$

$$\alpha = 66.2° \text{ and } \beta = 10°$$

FIGURE 19.53
Stability test of dynamic system with added static compliance for comparison with Figure 19.52.

$$\left(\frac{k_s}{\lambda_0}\right)(1-e^{-j2\pi n}).\left(\frac{\lambda_0}{\lambda_2}+\frac{1}{(1-(n^2/n_0^2)+(jn/Qn_0))}\right)+1+K_2e^{-jn(\pi-\beta)}-K_1e^{-jn\alpha}=0 \quad (19.77)$$

It can be seen that the additional compliance has stabilized the loop between 17.5 and 18 lobes that was previously unstable. The instability shown in Figure 19.52 just below 16 lobes is seen to be slightly softened in Figure 19.53 but not completely overcome. The next section considers more closely the combined effects of workspeed and natural frequency.

19.13 Dynamic Stability Limits

In this section, stability charts are presented by the classic method (Rowe and Koenigsberger 1965) and a number of features of instability in centerless grinding are explained. It is demonstrated that stability charts produced by the classic technique are difficult to interpret and onerous to plot. A new dynamic stability parameter A_{dyn} is presented that allows a broad spectrum of conditions to be simply presented in an easily interpreted form (Rowe 2014). The new technique copes with integer lobe values. The new technique also clearly shows how workspeed must be set to overcome practical rounding problems. The technique is validated for a typical grinding machine having several resonant frequencies.

19.13.1 Threshold Conditions

Substituting $s = j\omega$ in Equation 19.69, it is possible to solve the characteristic equation to find the chatter frequencies and the corresponding limits of grinding force stiffness at the threshold of instability. The following analysis was developed and first published by Rowe (1964). Noting that $f = n \cdot n_w$, the frequencies are given by

$$\left(\frac{n \cdot n_w}{f_0}\right)^* = \frac{1}{2Q}\left(\frac{B-A\cdot C}{A+B\cdot C}\right)+\sqrt{\left\{\left[\frac{1}{2Q}\left(\frac{B-A\cdot C}{A+B\cdot C}\right)\right]^2+1\right\}} \quad (19.78)$$

where A and B are the real and imaginary parts of the geometric function as previously defined in Equations 19.71 and 19.72 and

$$C = \frac{(1-\cos 2\pi n)}{\sin 2\pi n} = \frac{\sin \pi n}{\cos \pi n} \quad (19.79)$$

The threshold value of grinding force ratio is termed $(k_s/\lambda_0)^*$ and for various n is given by

$$\left(\frac{k_s}{\lambda_0}\right)^* = \frac{\{A\cdot[(n\Omega/\omega_0)^2-1]-B\cdot(n.\Omega/Q\omega_0)\}}{(1-\cos 2\pi n)} \quad (19.80)$$

Stability thresholds are independent of the grinding force for integer n and stability depends only on the geometric function. For integer n, the threshold value of $(k_s/\lambda_0)^*$ from

Equation 19.80 approaches infinity and becomes indeterminate. For integer n, Equations 19.78 and 19.80 become independent of both ω/ω_0 and (k_s/λ). Unstable integer values of n are particular examples of geometric instability. Noninteger values of n can also be geometrically unstable for an absolutely rigid structure as shown previously.

For half-integer values of n, the frequency is given by

$$\left(\frac{n \cdot n_w}{f_0}\right)^* = \frac{-A}{2Q \cdot B} + \sqrt{\left[\left(\frac{-A}{2Q \cdot B}\right)^2 + 1\right]} \tag{19.81}$$

It is essential to know whether the stable region is for values of k_s/λ_0 less than or greater than a boundary value $(k_s/\lambda_0)^*$. This can be discovered by checking phase angle χ between waviness vector $r(t)$ and deflection vector $x(t)$. The system is unstable for $k_s/\lambda_0 > (k_s/\lambda_0)^*$ if

$$-\frac{\pi}{2} < \chi < +\frac{\pi}{2} \tag{19.82}$$

where

$$\chi = \frac{3 \cdot \pi}{2} - \phi - \pi \cdot n \tag{19.83}$$

and

$$\phi = \arctan\left\{\frac{\left(n \cdot n_w / Q f_0\right)}{1 - \left(n \cdot n_w / f_0\right)^2}\right\} \tag{19.84}$$

Figure 19.54 illustrates phase relationships between normal grinding force, depth of cut, waviness, and deflection. The sign convention adopted is that a positive grinding force exerted by the GW on the workpiece gives a positive depth of cut. The workpiece applies an equal and opposite force on the GW that causes deflection. The deflection lags the grinding force by the phase angle ϕ. The sign convention for the deflection is based on the assumption that a positive deflection reduces depth of cut.

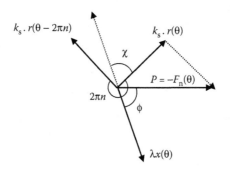

FIGURE 19.54
Vector diagram to determine phase angle χ.

When $|\chi|$ is less than 90°, deflection increases waviness. This means increasing grinding force increases waviness and the system becomes dynamically unstable.

The frequency ratio f/f_0 is the frequency of vibration divided by natural frequency. Since the vibration frequency equals number of lobes times workspeed, frequency ratio can be expressed in terms of number of lobes for particular workspeed and natural frequency. The frequency ratio is

$$\frac{f}{f_0} = \frac{n \cdot n_w}{f_0} \tag{19.85}$$

Figure 19.55 illustrates dynamic stability thresholds computed using Equations 19.78 through 19.85. Only the most marginal stability boundaries are shown. The approximate

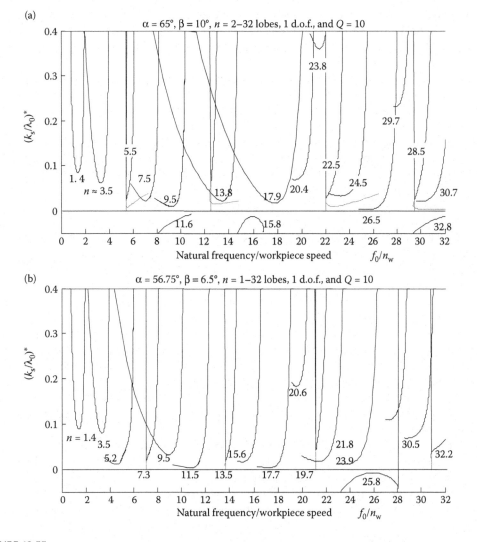

FIGURE 19.55
Dynamic stability limits: (a) $\gamma=20°$, $\beta=10°$ and (b) $\gamma=30°$, $\beta=6.5°$.

number of lobes is shown for each stability boundary. Some boundaries have been truncated for clarity. The chart is discussed in greater detail in Section 19.13.2.

19.13.2 Classic Dynamic Stability Charts

The classic technique for producing dynamic stability charts in a machining operation is developed here for centerless grinding and shown to produce a complex picture which makes interpretation of favorable stability conditions difficult. Following an initial presentation of dynamic stability charts, a simpler method is presented for evaluation of grinding conditions. The new method is based on geometric and dynamic stability parameters.

Dynamic stability charts following the classic approach are shown in Figure 19.55. The charts give the ratio of grinding force stiffness to static machine stiffness at the limit of stability. In center grinding, it is usually found that increasing this ratio through higher grinding force or lower machine stiffness makes the process unstable. However, in centerless grinding, it is not possible to reach this general conclusion. The following explanation provides examples of dynamic stability charts for centerless grinding and the complex nature of instability. Finally, methods are proposed to clarify the requirements for selection of workspeed.

Integer values of n are not included in dynamic stability charts due to the indeterminacy in Equations 19.80 and 19.81. Stability of integer lobes must be separately investigated employing the geometric function as described in Section 19.12.6 above.

Figure 19.55a shows examples of threshold values of k_s/λ_0 against f_0/n_w for the unstable geometry, $\alpha = 65°$ and $\beta = 10°$, for a 20° WR and a CW of the same size as the GW. Boundaries are plotted for different numbers of lobes. For example, at $n \approx 3.5$ lobes, threshold grinding force is given by $(k_s/\lambda_0)^* \approx 0.06$. This means that if workspeed is set so that $f_0/n_w \approx 3$, the threshold grinding force will be rather low. In other words, at this high workspeed, the process would be susceptible to regeneration of 3.5 lobes.

Stable zones appear below the solid black boundary lines and unstable zones appear above the lines. The stable zones indicate permissible workspeeds and grinding force levels. A permissible workspeed n_w is seen to depend on the natural frequency f_0. The value $f_0/n_w = 4.2$ appears to be suitable although such a low value of f_0/n_w indicates a high workspeed. Practical experience tends to indicate that low workspeeds are more stable than high workspeeds for a particular value of depth of cut. This is confirmed by the analysis that follows and the reasons are also discussed in Chapter 8 Dynamics of Grinding. Low workspeeds reduce the probability that marginally stable low frequency lobes will be brought into proximity with a resonance. Furthermore, low workspeeds require a large number of lobes around the workpiece. A large number of lobes is more likely to be attenuated by contact length filtering.

Most of the boundaries shown as solid black lines in Figure 19.55 indicate unstable conditions for a particular frequency if the grinding force is greater than the threshold value. This is the condition for "up" boundaries, where $k_s/\lambda_0 > (k_s/\lambda_0)^*$. Upboundaries are shown as solid black lines.

It is notable that some threshold boundaries occur for negative force conditions at workspeeds, where the condition is predicted to be unstable for all grinding forces. Examples of negative upboundaries are shown in the range $n = 11$–12 lobes and in the range $n = 15$–16 lobes. There are also cases shown on the original author version as dotted lines in Figure 19.55a, where the unstable condition is found for grinding forces less than the threshold value. These are "down" boundaries. For example, a dotted red line was shown in the range $n = 5$–6 lobes. Comparison of these exceptional ranges with the corresponding ranges in Figure 19.47 shows that negative upboundaries and positive downboundaries are closely associated with negative values of geometric rounding parameter A. In other words, such conditions correspond to geometric instability.

It should be remembered that a dynamic stability limit requires the particular work-speed to fulfill the relationship with natural frequency expressed by Equation 19.85. This has the consequence that the critical frequency ratio f/f_0 falls between 0.1 and 2 for most values of n.

In practice, dynamic instability is experienced where the product $n \cdot n_w$ is close to the natural frequency f_0. This means that dynamic stability problems will not usually be experienced if workspeed is selected to avoid resonances and potentially unstable lobing.

Figure 19.55b is a dynamic stability chart for a relatively stable geometric condition where $\gamma = 30°$, $\beta = 6.5°$ and for grinding and CWs being of equal diameter. The geometry is close to $\beta = 6.4°$, recommended as optimum by Hashimoto and Lahoti (2004). Odd lobes at $n = 3$ and 5 are more stable for the corresponding high workspeeds, although 7 lobes is less stable than in Figure 19.55a. Workspeeds corresponding to $n = 5, 7, 9, 11, 13, 15, 17, 18, 20, 22, 24, 26$, and 32 lobes should be avoided. The chart suggests aiming for $n = 30$ or even higher. This is achieved by employing low workspeeds n_w rev/s so that $f_0/n_w > 30$.

Geometric stability for $\beta = 6.5°$ is shown in Figure 19.56. This may be compared with Figure 19.49 for $\beta = 7.5°$. The waviness at the integer value $n = 26$ is marginally stable in

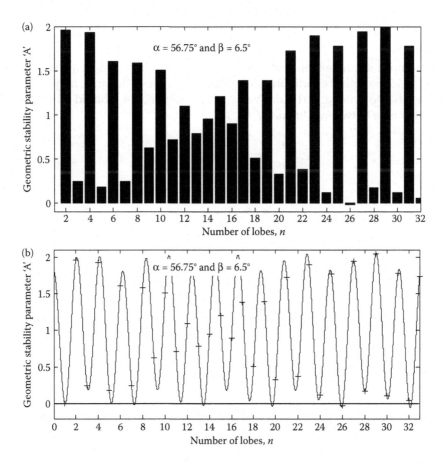

FIGURE 19.56
Geometric stability for 30° workrest and 6.5° tangent angle for (a) integer lobes and (b) noninteger lobes.

Figure 19.49a and marginally unstable in Figure 19.56a. Figures 19.49 and 19.56 strongly indicate that the workspeed corresponding to $f_0/n_w = 26$ should be given a wide berth for both geometries.

Figure 19.56b reveals marginal geometric instability for noninteger lobes. These values correspond with conditions where the dynamic stability charts show red downboundaries in the positive force region or black upboundaries in the negative force region. The conclusion is clear. At these values of lobing, it is possible to experience dynamic instability.

Figure 19.57 shows in greater detail the nature of the classic stability thresholds for numbers of lobes close to the geometrically unstable $n = 26$. Thresholds are plotted against vibration frequency ratio ω/ω_0 for the region covering $n = 25\text{--}27$. At integer lobes, the thresholds approach either plus or minus infinity. Upboundaries are shown as solid black lines and downboundaries as dotted red lines as previously. The main vibration frequencies for a dynamic stability threshold lie just above the natural frequency ω_0 of the main machine vibration mode.

Figure 19.57 has four distinct zones. These are shown as

    A. Absolutely stable: 25.0–25.8 lobes

    B. Absolutely unstable: 25.8–26.0 lobes. Frequency range $1\text{--}1.7 \times \omega_0$

    C. Conditional dynamic instability for low forces: $n = 26.01\text{--}26.03$ lobes

    D. Conditional dynamic instability for high forces: $n = 26.03\text{--}27$ lobes

In this example, it can be seen that care should be taken to avoid frequency ratios f/f_0 in the range 1.0–1.7. For other conditions and number of waves, the range may be even larger. This implies that to avoid a particular value of n, workspeed n_w should be set to avoid the approximate range between f_0/n and $2f_0/n$.

$$\frac{f_0}{n} > n_w \quad \text{or} \quad n_w > \frac{2f_0}{n_w}$$

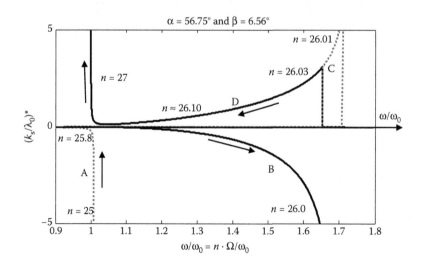

FIGURE 19.57
Frequencies at threshold conditions for 30° workrest and 6.5° tangent angle, $n = 25\text{--}27$ lobes.

This is a very approximate guide since it based on a particular case for a 1 d.o.f. vibration mode.

19.13.3 Workspeed in Relation to Resonant Frequency

The setting of workspeed in relation to the resonant frequency is seen to be of great importance as illustrated by the experimental results in Figure 19.20 and evidence from stability charts. It would therefore be helpful if it were possible to consider particular resonant frequencies and particular workspeeds in combination and then to assess process stability for different numbers of lobes. This is difficult to achieve with classic stability charts because thresholds tend toward negative infinity for integer lobes. Stability thresholds were therefore plotted leaving out integer lobes. Instability for integer lobes can be assessed using geometric stability charts in combination with dynamic stability charts as illustrated in Figure 19.58. Threshold values of k_s/λ_0 are computed so that the real part of the characteristic equation is equal to zero. Upboundaries are filled light gray and downboundaries are filled dark.

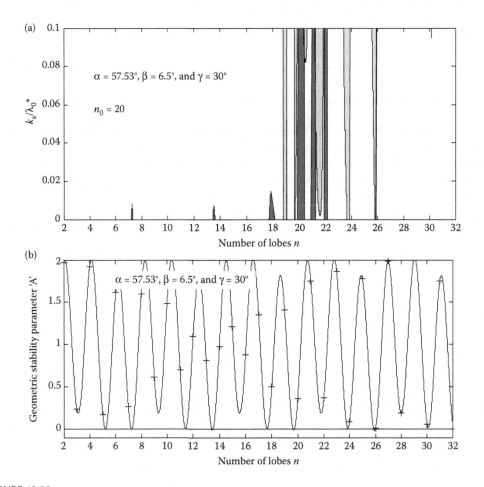

FIGURE 19.58
Threshold stability against number of lobes for a 30° workrest and 6.5° tangent angle: (a) threshold grinding force/machine stiffness with resonant frequency n_0 at $n = 20$ and (b) geometric stability parameter A.

A stability chart for k_s/λ_0 is shown in Figure 19.58a for a near optimum geometry and a fixed resonant frequency n_0 at $n = 20$. In this figure, the calculation is completed for the slightly more stable situation than in the previous example. The CW is smaller than the grinding wheel so that $\alpha = 57.53°$ rather than 56.75°. This small change improves geometric stability at $n = 7, 13, 20,$ and 26 lobes as revealed by comparing Figure 19.56b with Figure 19.58b.

It is immediately apparent from Figure 19.58a that there are no conventional stability boundaries for frequencies that are sufficiently below the resonant frequency which in this result is set at $n_0 = 20$. Negative downboundaries represent absolute stability, so these are not shown. Negative upboundaries are indicated by threshold conditions that descend to zero. This simplifies the diagram and no important information is lost.

Below the resonant frequency, the system is very stable from a dynamic viewpoint. There are small unstable zones where the threshold value k_s/λ_0^* goes down to zero. These positions close to $n = 7, 13,$ and 18 correspond to troughs in Figure 19.58b for geometric stability parameter A. These three cases are downboundaries in Figure 19.58a where very low-grinding force and high contact stiffness are required to make the system unstable. The frequencies are therefore of small importance although vibrations that directly excite these frequencies should be avoided.

Near the resonant frequency, even lobes are more likely to become dynamically unstable as seen in Figure 19.58a. Thresholds occur close to 18, 20, 21, 22, 24, and 26 lobes, even though the integer values have been left out.

Figure 19.59 shows a stability chart using the same technique but where the workspeed has been halved so that the dimensional natural frequency corresponds to $n_0 = 40$ lobes. Minor instabilities remain close to 7, 13, 18, and 26 lobes. However, the major instabilities seen in Figure 19.58 have been eliminated from the frequency range investigated. The instabilities shown in Figure 19.59 are downboundaries, where the instability occurs for grinding forces lower than the threshold value shown. These instabilities are the very mild kind corresponding closely to conditions of marginal geometric stability.

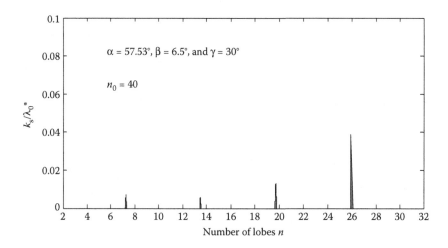

FIGURE 19.59
Threshold stability against number of lobes for a 30° workrest and 6.5° tangent angle: Threshold grinding force/ machine stiffness with resonant frequency n_0 at $n = 40$ lobes.

This technique is not completely satisfactory since the integer lobes have had to be filtered out. The next section employs an easily computed dynamic stability chart that copes with integer lobes.

19.13.4 A New Dynamic Stability Parameter A_{dyn}

It is possible to obtain a clearer picture of stability across the range of frequencies from a new dynamic stability parameter A_{dyn} derived as follows (Rowe 2014). The new parameter has the advantage of simplicity to plot and the ability to reveal stability at integer and noninteger lobe values. From Equation 19.77, the characteristic equation expressed in real and quadrature parts yields the stability test:

$$\frac{k_s}{\lambda_0} \cdot \left[\frac{1}{1-(n/n_0)^2 + j \cdot n/n_0} + \frac{\lambda_0}{\lambda_2} \right] \cdot (1 - e^{j \cdot 2\pi n}) + 1 + K_2 e^{-j \cdot n(\pi - \beta)} - K_1 e^{-j \cdot n\varepsilon} = 0$$

The characteristic equation may be represented as $A_{dyn} - j \cdot B_{dyn}$. The real and imaginary parts must both equal zero for limiting stability. Instability can only occur if the real part becomes negative. The dynamic stability parameter A_{dyn} is defined as the real part of the characteristic equation.

$$A_{dyn} = \frac{k_s}{\lambda_0} \left[\left(C + \frac{\lambda_0}{\lambda_2} \right) \cdot (1 - \cos 2\pi n) + D \cdot \sin 2\pi n \right] + 1 + K_2 \cos n(\pi - \beta) - K_1 \cos n\alpha \quad (19.86)$$

$$B_{dyn} = \frac{k_s}{\lambda_0} \left[\left(C + \frac{\lambda_0}{\lambda_2} \right) \cdot \sin 2\pi n - D \cdot (1 - \cos 2\pi n) \right] + K_2 \sin n(\pi - \beta) - K_1 \sin n\alpha \quad (19.87)$$

where

$$C = \frac{1 - (n/n_0)^2}{[1 - (n/n_0)^2]^2 + [(n/n_0 Q)]^2}$$

$$D = \frac{n/n_0 Q}{[1 - (n/n_0)^2]^2 + [(n/n_0 Q)]^2}$$

The dynamic stability parameter A_{dyn} is shown in Figure 19.60 for the same geometry as previously. The workspeed and resonant frequency in this example have been again set so that $n_0 = 20$ lobes. Instability can only occur for negative values of A_{dyn}. Integer lobes are indicated by a " + " sign. To make the visualization more realistic, stability plots are presented for (a) high-grinding force and low stiffness, $K_s/\lambda_0 = 0.5$, and (b) low-grinding force with very high stiffness, $K_s/\lambda_0 = 0.001$. With high-grinding force, the system is stable up to a frequency slightly below the resonance at $n_0 = 20$, where the first instability is experienced. Above the resonant frequency several orders of waviness are unstable. Low-grinding force and very high stiffness are best. There are no significant instabilities throughout the range. Surprisingly perhaps, but justified by geometrical considerations, high-grinding force and low stiffness gives a better margin of stability at some low frequencies but has the highest growth rates for the high frequency dynamic instabilities.

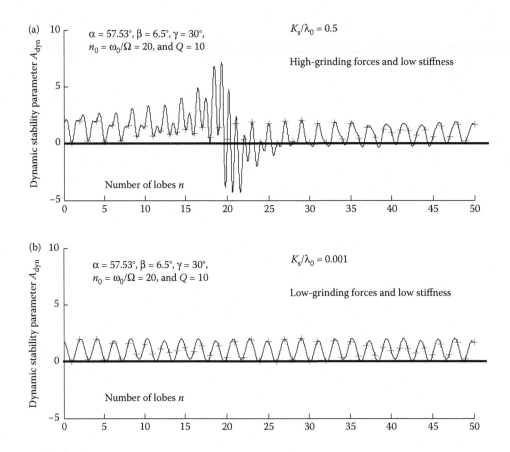

FIGURE 19.60
Dynamic stability parameter A_{dyn} against number of lobes for a 30° workrest and 6.5° tangent angle. Resonant frequency n_0 at $n = 20$. Negative values indicate unstable conditions. (a) High-grinding forces and low-stiffness machine and (b) low-grinding forces and very high-stiffness machine.

At very high-machine stiffness, the dynamic stability parameter becomes equal to the geometric stability parameter. This means $A_{dyn} = A$, giving stability for a completely rigid machine. The advantages of the new dynamic stability parameter A_{dyn} are clear. The easily plotted chart vividly demonstrates the relationship between the combined effects of rounding geometry, workspeed, and resonant frequency for a range of possible lobes.

The advantage of a high-resonant frequency is confirmed in Figure 19.61, where the resonance corresponds to $n = 40$. Figure 19.61 may be compared with Figure 19.60. Dynamic instability is not apparent until just below the resonance at $n = 40$ even for a low-stiffness machine. Since $n = f/n_w$, it is seen that reducing workspeed to half has the same benefit for stability as doubling natural frequency.

Many machines will have a critical resonant frequency in the region of 75 Hz so that with a workspeed of 2 rev/s, dynamic chatter vibrations should be a relatively low risk for numbers of waves up to $n = 30$ even for a machine of low stiffness. However, there will be definite advantage in a machine having a much higher resonant frequency. In the ultraprecision machine described in Section 19.8.3, a resonant frequency of approximately 500 Hz was achieved.

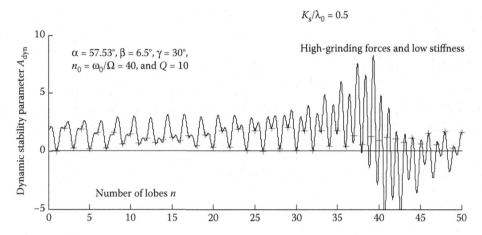

FIGURE 19.61
Dynamic stability parameter A_{dyn} against number of lobes for a 30° workrest and 6.5° tangent angle. Resonant frequency n_0 at $n = 40$.

19.14 Avoiding Critical Frequencies

19.14.1 Selection of Work Rotational Speed

Figure 19.20 shows there is a limiting work rotational speed when dynamic instability becomes a process limit. A compliant machine with a natural frequency of approximately $f_0 = 76$ Hz gave instability at work rotational speeds above $n_w = 4.5$ rev/s. At this workspeed, the number of waves would be expected in the region close to $n = 76/4.5 = 17$ waves.

The results for the very stiff machine presented in Figure 19.20 were for a natural frequency $f_0 = 500$ Hz. Instability was experienced at work rotational speeds above 16 rev/s. At this workspeed, the number of lobes experienced would be expected to exceed 31. However, 31 lobes for a 25-mm diameter workpiece will be attenuated by contact filtering for more than 17 lobes. These results explain why a very stiff machine allowed higher workspeeds to be employed and the smaller amplitude of roundness errors when the process became unstable.

Table 19.6 shows experimental data for strong chatter experienced when grinding 40 mm cast iron workpieces and 50 mm EN9 workpieces on a heavy-duty centerless grinding

TABLE 19.6

Examples of Experimental Data for Strong Chatter

No. of Lobes	Work Speed (rev/s)	Resonant Frequency (Lobe Number)	Resonant Frequency (Hz)	Chatter Frequency (Hz)	
N	n_w	$n_0 = f_0/n_w$	f_0	$f_c = n \cdot n_w$	f_c/f_0
16	6.92	111	108	110.7	1.025
18	13.15	6.77	189	236	1.25
20	10.37	18.22	189	207	1.095
22	3.89	18.5	72	86	1.194
24	9.22	20.5	189	184	0.974
26	8.13	23.24	189	211	1.12
26	6.73	28.08	189	175	0.926

machine. The grinding machine demonstrated several resonant frequencies as listed in the table. Audible chatter was experienced for 16, 18, 20, 22, 24, and 26 lobes. In all cases, lobing corresponded closely with troughs on the stability charts and with nearby resonant frequencies providing validation of the dynamic stability charts.

If roundness problems are experienced, it is necessary to determine the cause of the vibration. Frequency of vibration can be evaluated from the number of lobes on the workpiece multiplied by the work rotational speed. This will assist in identifying the source of vibration and workspeed can be chosen to avoid a convenient waviness. If the frequency is identified with a machine resonance, the usual adjustment is to reduce workspeed.

Figure 19.20 shows that at low workspeed, there is increased risk of thermal damage. If vibration is associated with a thermal damage condition at very low workspeeds, it is necessary to increase workspeed.

19.14.2 Selection of Wheel Rotational Speed

Problems can arise in any grinding process when the wheel rotational speed is an integer multiple of the work rotational speed. This is because any wheel run-out due to wheel unbalance or other cause will be imposed at the same positions on the workpiece in a repetitive action. For example, if the grinding wheelspeed $n_s = 30$ rev/s and the work rotational speed $n_w = 3$ rev/s, it is highly probable that $n = 30/3 = 10$ lobes will be detectable in the work roundness.

It can also be seen that wheelspeed should not be allowed to lie at the same speed as the dominant natural frequency. Any wheel unbalance will lead to large vibration amplitudes. Even direct multiples and submultiples are best avoided.

19.14.3 Selection of Dresser Speed

The modern practice in precision grinding particularly when using superabrasive GWs is to employ a motorized rotary dressing tool. The tool usually consists of a diamond-faced disk with a narrow cutting edge. A rotary disk is far more durable than a single-point diamond and a precision disk allows excellent rotational accuracy for precision dressing. The runout of the dressing disk is likely to be approximately 2 μm.

Typically, the disk is arranged so that when dressing, the dresser surface speed $v_d = 0.4$–$0.8\ v_s$. This means the dressing tool has to rotate at high speed for high-speed grinding. Alternatively, wheelspeed may be reduced for dressing. The dressing tool assembly must be capable of low vibration operation at the wheelspeed employed. The dressing tool rotational speed n_d should not be a multiple of work rotational speed n_w. Neither should the dresser rotational speed be a simple multiple of the GW rotational speed n_s.

If it is decided to perform dressing at reduced grinding wheelspeed, it is important not to excite machine resonance from the dressing tool system or the GW drive system.

19.14.4 Speed Rules

If chatter is experienced, a reduction of workspeed will usually cure the problem.

Summarizing the further requirements outlined in the discussion, the following general rules may be concluded:

- f_0/n_w should not be an integer
- n_s/n_w should not be an integer

- n_d/n_w should not be an integer
- n_d/n_s should not be an integer

The same principle of avoiding multiple speed relationships that directly affect workpiece roundness can be extended to other machine elements in the system, such as the CW, the drive motors and gears.

19.15 Summary and Recommendations for Rounding

- Setup geometry should be set for rapid rounding and to avoid convenient waviness.
- Convenient waviness and marginal stability cannot be completely avoided. However, convenient waviness is unlikely to be a problem unless waviness is forced by vibration at the same frequency.
- Static compliance at the workpiece contact point softens the imposition of errors into the workpiece and hence makes it easier to achieve low roundness errors, although a longer cycle time may be required.
- Resonant frequencies should be avoided since these amplify vibrations and cause large roundness errors. Resonant vibrations are minimized by well-designed stiff structures and high damping.
- Forced vibrations close to a resonant frequency should be avoided. This can be achieved by checking that excitation due to motors, drives, and wheel unbalance is not at a frequency coincident with a dominant machine resonant frequency.
- Sources of forced vibration such as wheel unbalance should not be allowed to excite a frequency of marginal geometric stability indicated by a trough on the stability chart.
- Many problems can be overcome by adjusting workspeed. Ideally, a machine should provide a continuously variable range of workspeeds. Low workspeeds are more stable than high workspeeds.
- A chart of A_{dyn} for a grinding geometry setup can be easily constructed and used diagnostically to see where potential lobing problems may occur for a particular workpiece diameter, workspeed, and resonant frequency.

19.16 Process Control

For the manufacturer producing components by the thousand and the millions, process control assumes great importance. In this situation, it is essential to produce components rapidly and be assured of good quality. It can be wasteful if the process goes out of tolerance, particularly, if the situation is not quickly detected. The process control system should therefore be able to detect changes that may lead to rejection of parts and take appropriate action.

Basically two different strategies are possible in response to detecting an unacceptable process condition:

- Stop the process and give an alarm
- Make an adjustment to the operating conditions and continue

For flexible manufacture of small quantities, it is difficult to do better than a human operator in control of the process. However, as the drive to reduce costs becomes more important, there is a tendency to automate machines and place a number of machines under the control of one operator.

The relative ease of introducing low-cost automation for larger batches was mentioned in Section 19.1. There are various levels of automation, and the manufacturer will want to base the level of automation on such factors as the need to make frequent changes to produce different components, the frequency of repeat batches, the long-term future for production of such components and so forth.

The basic levels of automation are likely to include

- Automated part feeding with bowl feeders, air cylinders, electric drives, and so on
- Automatic size gauging with air gauging, electronic transducers, and so on
- Automated checking of door closure or other safety features
- Automated size compensation using servo drives
- Automated GW balancing and wheel dressing

At the higher levels, automation may be implemented, using a CNC machine having appropriate servo-drives and measuring scales. A CNC is a system capable of being programmed to produce parts on particular machines. A CNC is given information about the specific machining operation through a part program. The part program allows an operator to feed data about the workpiece shape and the required sequence of operations into the CNC. The CNC gives continuous instructions to servo controls to position machine axes following the programmed cycle and reads signals from servo-drives and measuring devices to ensure the correct positions are achieved. The CNC may also send information about the process to a display screen or to a higher-level computer.

The cost of a CNC machine is much greater than the cost of a manually controlled machine. The decision to change to CNC control will not therefore be taken lightly. A CNC that becomes unreliable is a liability and can lead to a machine becoming unusable for long periods.

Before deciding to adopt or reject CNC control for centerless grinding, consideration needs to be given to potential advantages and disadvantages. Potential advantages of a CNC grinding machine are

- Can operate unsupervised for long periods
- Can incorporate control of automation devices and features
- Can incorporate intelligent control for process optimization
- Can in some cases store production data and process data
- Can in some cases communicate with a higher-level computer

It should be noted, however, that a low-cost CNC is unlikely to incorporate more than the first of the advantages listed of operating unsupervised for long periods. A more expensive CNC will probably be required to allow integration with a full range of sizing and other equipment.

Some disadvantages of a CNC grinding machine are

- High initial machine cost
- The need for greater reliability of electronic devices over a long working life
- Dependence on a specialized repair service in the event of CNC breakdown

Very little has been published on the application of CNC to centerless grinding machines, although several papers have considered optimization and CNC in relation to other grinding processes including Peters (1984) and Tönshoff et al. (1986). More recently, attention has turned to intelligent process control based on modern CNC technology (Rowe et al. 1994). In addition to the direct benefits of reduced manning, intelligent process control offers the potential for

- Sustained removal at maximum rate (Rowe et al. 1986)
- Avoidance of thermal damage (Rowe et al. 1988)
- Accelerated spark out by overshoot (Malkin 1981)
- Accelerated infeed through digital closed loop control (Tönshoff et al. 1986)
- Consistent quality and production rate
- Learning strategies for new materials, wheels, dressing conditions, etc.
- Sensing and/or avoidance of chatter
- Feed-forward control of lobing (Frost et al. 1988)
- Combined grinding and measuring cycles

Figure 19.62 shows a basic scheme for intelligent process control of a grinding system. The figure is not intended to describe a system currently in existence. It illustrates a framework into which different system developments can be classified. Various aspects that contribute to such a scheme have been developed and applied in particular systems as illustrated by some of the references listed above. For further details, the reader may refer to Rowe et al. (1994).

The full expert system level illustrated has not yet been commercially developed for grinding, although aspects of such a system may already be incorporated into CNC grinding systems. A prototype expert system was developed and demonstrated that such an approach could be feasible. The operator wishing to machine a new product queried a database for recommendations on feed rate and material removal rate. The system gave advice based on previous products of a similar type or used rules to generate recommendations.

Having decided the basic values to be used for machining the part, these values are entered into the part program using the part-program generator of the CNC. The process controller then initiates a grinding cycle. The first part is usually machined under close operator supervision to ensure sizing devices are correctly adjusted and change points for the grinding cycle feed rate changes take place at suitable positions.

Size gauging devices are a normal feature of process monitoring at the monitoring level indicated. A size reading may be used to trigger a fine feed rate for finish grinding and

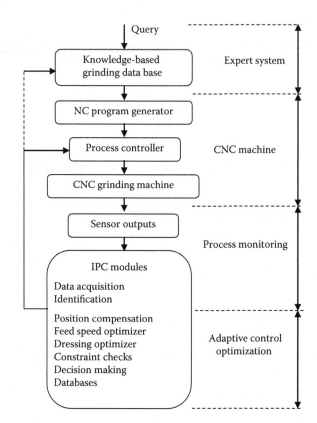

FIGURE 19.62
Basic scheme for intelligent process control of a grinding system.

to trigger the commencement of dwell for spark out. If the size reading takes place after the workpiece leaves the grinding contact, the size reading may be used to trigger a wear compensation for the infeed movement.

A power-monitoring sensor may be incorporated into the system and can be used in several different ways. If the maximum power reading increases after a number of parts have been produced, it is likely to mean that the GW has become blunt. This can be used to signal the need to redress the GW.

If the power reading drops after a number of parts have been produced, it means that the GW wear has had a self-sharpening effect. This may imply that the GW is too soft for the application. However, it may also imply that it is safe to increase feed rate and reduce cycle time. Adaptive control optimization is a process where the system varies the feed rate or other parameters to bring the process closer to optimum operation.

Another possibility is to detect potential thermal damage to the workpieces based on the measured power. In this case, feed rate can be limited if the measured power approaches the maximum permissible for avoidance of thermal damage. The application of such a system relies on a process model that allows the onset of thermal damage to be predicted. Several researchers have proposed such systems but the practice lags well behind the theory.

Over the last two decades since the introduction of CNC, grinding process controllers have become increasingly sophisticated. Basic technology for grinding control may be incorporated by the CNC manufacturer. Further specialized technology for grinding may be added by the manufacturer of the grinding machine. In some cases, the machine user may add further specialized technology.

Where centerless grinding is employed for sustained production of batches of components of high-precision parts, the greatest priority is likely to be for systematic trials to ensure stable grinding conditions together with high accuracy and high production rates.

By studying the information given in this chapter, the reader will achieve a sound grasp of the fundamental principles required to undertake the development of such a system.

References

Becker, E. A., 1965, Krafte und Kreisformfehler beim Spitzenlosen Einstechschleifen. Dissertation, T.H. Aachen.

Cai, R., 2002, Assessment of vitrified CBN wheels for precision grinding. PhD thesis, Liverpool John Moores University, UK.

Cui, Q., Ding, H., Cheng, K., 2015, An analytical investigation on the workpiece roundness generation and Its perfection strategies in centreless grinding. *Proceedings of the Institution of Mechanical Engineers Part B, Journal of Engineering Manufacture*, 229(3), 409–420.

Dall, A. H., 1946, Rounding effect in centreless grinding. *Mechanical Engineering*, April, 325.

Ebbrell, S., 2003, Process requirements for precision engineering. PhD thesis, Liverpool John Moores University, UK.

Frost, M., Horton, B. J., Tidd, J. L., 1988, Lobing Control in Centreless Grinding. SME Paper MR88-610.

Furukawa, Y., Miyashita, M., Shiozaki, S., 1971, Vibrational analysis and work-rounding effect in centreless grinding. *International of Journal of Machine Tool Design and Research*, 11, 145–175.

Gallego, I., 2007, Intelligent centreless grinding: Global solution for process instabilities and optimal cycle design. *Annals of the CIRP*, 56(1), 347–352.

Goodall, C., 1990, A study of the centreless grinding process with particular reference to size accuracy. CNAA PhD thesis, Liverpool Polytechnic in collaboration with the Torrington Company, USA.

Gurney, J. P., 1964, An analysis of centreless grinding. *ASME Journal of Engineering for Industry*, Paper 63-wa-26, 87, 163–174.

Gviniashvili, V. K., Woolley. N. H., Rowe. W. B., 2004, Useful flow-rate in grinding. *International Journal of Machine Tools and Manufacture (Elsevier)*, 44, 629–636.

Harrison, A. J. L., Pearce, T. R. A., 2004, *Reduction of Lobing in Centreless Grinding via Variation of Set-up Angles*, vol. 257–258, Key Engineering Materials, Trans Tech Publications, Switzerland, 159–164.

Hashimoto, F., Kanai, A., Miyashita, M., 1983, High precision method of regulating wheel dressing and effect on grinding accuracy. *CIRP Annals*, 32(1), 237–239.

Hashimoto, F., Lahoti, G. D., 2004, Optimisation of set-up conditions of the centreless grinding process. *CIRP Annals*, 53(1), 271–274.

Hashimoto, F., Lahoti, G. D., Miyashita, M., 1998, Safe operation and friction characteristics of regulating wheel in centreless grinding. *Annals of the CIRP*, 47(1), 281–286.

Jameson, J. R., Farris, T. N., Chandrasekaran, S., 2008, Equilibrium and compatibility simulation of centreless grinding. *Proceedings of the Institution of Mechanical Engineers Part B, Journal of Engineering Manufacture*, 222, 747–757.

Johnson, S. P., 1989, Below-centre centreless grinding. PhD thesis, Coventry Polytechnic, UK and CNAA, April 1989.

King, R. I., Hahn, R. S., 1986, *Handbook of Modern Grinding Technology*. Chapman and Hall Advanced Industrial Technology Series, New York, NY and London, UK.

Klocke, F, Friedrich, D., Linke, B., Nachmani, Z., 2004, Basics for in-process roundness error improvement by a functional workrest blade. *CIRP Annals*, 53(1), 275–280.

König, W., Henn, K., 1982, Spitzenloses Durchlaufschleifen (Centreless through-grinding), *Metalbearbeitung*, Part 1, 9, 64–70 and Part 2, 10, 14–16.

Malkin, S., 1981, Grinding cycle optimization. *CIRP Annals*, 30(1), 213–217.

Marinescu, I. D., Rowe, W. B., Dimitrov, B., Inasaki, I., 2004, *Tribology of Abrasive Machining Processes*. William Andrew Publishing, NY13815.

Meis, F. U., 1980, Geometrie und kinematische Grundlagen fur das spitzenlose Durchlaufschleifen (Geometry and kinematic principles for through-feed centreless grinding). Dissertation, T.H. Aachen.

Miyashita, M., 1965, *Influence of Vibrational Displacements of Machine Elements on Out-of-Roundness of Workpiece in Centreless Grinding*. Memoirs of Faculty of Technology, Tokyo Metropolitan University, 15, 15–27.

Moriya, T., Kanai, A., Miyashita, M., 1994, Theoretical analysis of the rounding effect in generalized centreless grinding. *ASME, Materials Issues in Machining II*, 303–319.

Opitz, H., Gühring, K., 1968, High-speed grinding. *Annals of the CIRP*, 16, 61–73.

Pande, S. S., Lanke, B. R., 1989, Investigation on the through-feed centreless grinding process. *International Journal of the Production Research*, 27(7), 1195–1208.

Pearce, T. R., Stone, B. J., 2011, Unstable vibration in centreless grinding Part I, Geometric instability or chatter? Part 2, Graphical method. *Proceedings of the Institution of Mechanical Engineers Part B, Journal of Engineering Manufacture*, 225, 1227–1254.

Peters, J., 1984, Contributions of CIRP research to industrial problems in grinding, *CIRP Annals*, 33(2), 1–18.

Reeka, D., 1967, Uber den Zusammenhang zwischen Schleifspaltgeometrie und Rundheitsfehler beim Spitzenlosen Schleifen (On the relationship between grinding support geometry and roundness errors in centreless grinding). Dissertation, T.H. Aachen.

Rowe, W. B., 1964, Some studies of the centreless grinding process with particular reference to the roundness accuracy. PhD thesis, Faculty of Science, Manchester University, UK.

Rowe, W. B., 1974, An experimental investigation of grinding machine compliance and improvements in productivity. *Proceedings of the 14th Internationl Machine Tool Design and Research Conference,* Macmillan Press, 479–486.

Rowe, W. B., 1979, Research into the mechanics of centreless grinding, *Precision Engineering*. IPC Press, UK. 1, 75–84.

Rowe, W. B., 2013, *Principles of Modern Grinding Technology*, 2nd edition. Elsevier (William Andrew imprint), Oxford, UK.

Rowe, W. B., 2014, Rounding and stability in centreless grinding. *International of Journal of Machine Tools and Manufacture*, 82/83, 1–10.

Rowe, W. B., Barash, M. M., 1964, Computer method for investigating centreless grinding. *International Journal of Machine Tool Design and Research* 4, 91–116.

Rowe, W. B., Barash, M. M., Koenigsberger, F., 1965, Some roundness characteristics of centreless grinding. *International Journal of Machine Tool Design and Research*, 5, 203–215.

Rowe, W. B., Bell, W. F., Brough, D., 1986, Optimisation studies in high-removal rate centreless grinding. *CIRP Annals*, 35(1), 235–238.

Rowe, W. B., Ebbrell, S., 2004, Process requirements for cost-effective precision grinding. *CIRP Annals*, 53(1), 255–258.

Rowe, W. B., Koenigsberger, F., 1965, The work-regenerative effect in centreless grinding. *International Journal of Machine Tool Design and Research*, 4, 175–187.

Rowe, W, B., Miyashita M, König W, 1989, Centreless grinding research and its application in advanced manufacturing technology. Keynote Paper, *CIRP Annals*, 38(2), 617–625.

Rowe, W. B., Morgan, M. N., Qi. H. S., 1993. The effect of deformation on contact length in grinding. *CIRP Annals*, 42(1), 409–412.

Rowe, W. B., Pettit, J. A., Boyle, A., Moruzzi, J. L., 1988, Avoidance of thermal damage in grinding and prediction of the damage threshold. *CIRP Annals*, 37(1), 327–330.

Rowe, W. B., Richards, D. L. 1972, Geometric stability charts for centreless grinding. *Journal of Mechanical Engineering Science*, 14(2), 155–158.

Rowe, W. B., Spraggett, S., Gill, R., 1987, Improvements in centreless grinding machine design. *CIRP Annals*, 36(1), 207–210.

Rowe, W. B., Yan, L., Inasaki, I., Malkin, S., 1994, Applications of artificial intelligence in grinding. *CIRP Annals*, Keynote Paper, 43(2), 1–11.

Schreitmuller, H., 1971, Kinematische Grundlagen fur die practische Anwendung des spitzenlosen Hochleistungsschleifens. Dissertation, T.H. Aachen.

Tönshoff, H. K., Zinngrebe, M., Kemmerling, M., 1986, Optimization of internal grinding by microcomputer based force control. *CIRP Annals*, 35(1), 293–296.

Yonetsu, S., 1959a, Consideration of centreless grinding characteristics through harmonic analysis of out-of-roundness curves. *Proceedings of the Fujihara Memorial Faculty of Engineering, Keio University*, 12(47), 8–26.

Yonetsu, S., 1959b, Forming mechanism of cylindrical work in centreless grinding. *Proceedings of the Fujihara Memorial Faculty of Engineering, Keio University*, 12(47), pp. 27–45.

Yoshioka, J., Hashimoto, F., Miyashita, M., Kanai, A., Abo, T., Daito, M., 1985, Ultraprecision grinding technology for brittle materials: Application to surface and centreless grinding processes. *ASME, M C Shaw Grinding Symposium*, PED-Vol 16, 209–227.

20

Ultrasonic-Assisted Grinding

20.1 Introduction

A wide variety of applications verify the great potential of high-performance ceramics for components with special requirements. For example, hip-joint endoprostheses on aluminum oxide or zirconium oxide bases, components for slide bearings and burners of silicon carbide, as well as ceramic components for roller bearings or valves are made of silicon nitride (Spur 1989; Pattimore 1998; Popp 1998). Extension of the market share for ceramic components is often opposed by the difficulties of manufacture with respect to achievable component quality and economic efficiency. Manufacturing costs arise mainly in grinding, honing, lapping, and polishing. High costs result from relatively inefficient technologies for machining of brittle–hard materials (Uhlmann 1998). This demonstrates the need to provide economically efficient machining methods for ceramic workpieces. In addition, there is a lack of suitable strategies for economic manufacture of complex geometries such as bores, holes, grooves, spherical surfaces, and sculptured surfaces (Uhlmann and Holl 1998a).

Although ultrasonic lapping and electrical discharge machining (EDM) processes are suitable for manufacture of these geometries, there are significant disadvantages. Only electrically conductive ceramics such as SiSiC can be machined with EDM methods, and there are technological limits to ultrasonic lapping due to the small material-removal rate, the high wear of the forming tools, and unsatisfactory accuracy. Therefore, suitable manufacturing methods for highly accurate and economic machining of ceramic materials have been developed over the past few years. The development of hybrid manufacturing processes on the basis of existing methods opens up new avenues.

In ultrasonic lapping, a forming tool oscillates with ultrasonic frequency and thrusts loose abrasive lapping grains into the surface of the workpiece, thus removing material. Based on this process, ultrasonic action has been superimposed onto conventional machining kinematics in several manufacturing processes over the past few years. In nearly all cases, process results have been improved (Drozda 1983; Nankov 1989; Prabhakar et al. 1992; Pei et al. 1993; Suzuki et al. 1993; Westkämper and Kappmeyer 1994; Pei and Ferreira 1999).

In industry, ultrasonic lapping and ultrasonic-assisted grinding have been applied so far for finishing brittle–hard materials. Due to high material-removal rates and the freedom of geometrical configuration, this method has a wide range of application possibilities. Figure 20.1 shows the advantage of ultrasonic-assisted grinding in comparison to conventional finishing methods with respect to an increase of the material-removal rate in the machining of aluminum oxide through the superposition of grinding kinematics and an ultrasonic frequency.

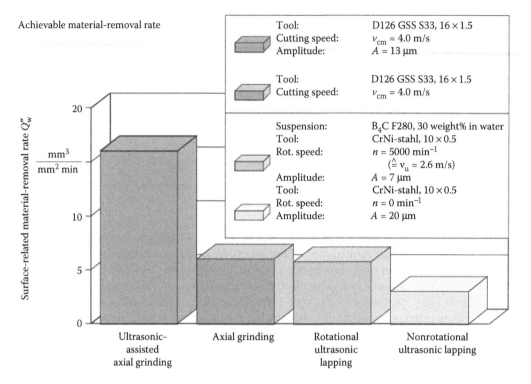

Achievable material-removal rate

Tool:	D126 GSS S33, 16×1.5
Cutting speed:	$v_{cm} = 4.0$ m/s
Amplitude:	$A = 13$ μm

| Tool: | D126 GSS S33, 16×1.5 |
| Cutting speed: | $v_{cm} = 4.0$ m/s |

Suspension:	B_4C F280, 30 weight% in water
Tool:	CrNi-stahl, 10×0.5
Rot. speed:	$n = 5000$ min^{-1}
	($\hat{=} v_u = 2.6$ m/s)
Amplitude:	$A = 7$ μm
Tool:	CrNi-stahl, 10×0.5
Rot. speed:	$n = 0$ min^{-1}
Amplitude:	$A = 20$ μm

Surface-related material-removal rate Q''_w

$\dfrac{mm^3}{mm^2 \, min}$

Ultrasonic-assisted axial grinding — Axial grinding — Rotational ultrasonic lapping — Nonrotational ultrasonic lapping

FIGURE 20.1
Achievable material-removal rates during the machining of aluminum oxide depending on the applied manufacturing method. (From Cartsburg, H., 1993, Hartbearbeitung keramischer Verbundwerkstoffe. Dissertation Technische Universität Berlin. With permission.)

20.2 Ultrasonic Technology and Process Variants

Elastomechanical ultrasonic vibration is generated by the transformation of electric energy in piezoceramic or magnetostrictive sonic converters. A voltage generator converts a low-frequency main voltage into a high-frequency alternating-current voltage. The generated longitudinal vibrations are periodic elastic deformations of the mechanical vibration system in the micrometer range at supersonic frequencies, that is higher than $f = 16$ kHz. The sound-generating unit preceding the actual ultrasonic tool or shape-generating counterpart consists of a sonic converter, an amplitude transformer (transforming sections), and sonotrodes (Spur and Holl 1995). Figure 20.2 shows the design of a vibration system.

The process of sound generation and transformation should be largely free of losses in order to obtain a high total efficiency of the vibration system. At the same time, it is required to produce maximum vibration amplitude at the sonic converter to reach sufficient amplitude at the effective surface for the machining task. A mainly loss-free increase in amplitude is guaranteed by means of resonance that is a vibration with a frequency that corresponds to that of the eigenfrequency of the system. This requires that the geometrical lengths of the single elements must correspond to half the wavelength of the vibration or an integral multiple. Additionally, changes of the resonance frequency, caused by changes in the mechanical system, for example, variation of engagement conditions or

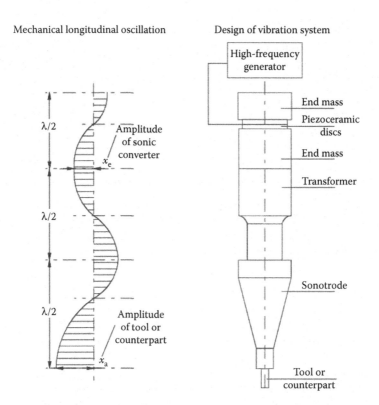

Mechanical longitudinal oscillation

Design of vibration system

FIGURE 20.2
Design of vibration system and vibration course for the example of ultrasonic lapping. (From Haas, R., 1991, Technologie zur Leistungssteigerung beim Ultraschall-schwingläppen. PhD dissertation, RWTH Aachen. With permission.)

weight losses at tools and workpieces, means the resonance frequency has to be continuously readjusted (Conrath 2005).

The amplitude of the converter is generally too low for machining. It can be raised to a value sufficient for machining by a subsequent transformer. The sonotrode serves as a tool holder as well as an adaptation to the resonance of the entire vibration system. In addition, there is the possibility to design the sonotrode in a way that allows extra amplitude to be achieved (Haas 1991). The ultrasonic vibration unit is clamped at the points where vibration nodes occur, because acceleration and amplitude are zero at these points and, thus, normal force freedom predominates.

Electric energy is converted into mechanical vibrations in modern machinery using the piezoelectric effect. It is related to the reversible property of special ceramic materials to deliver electric voltage when affected by external forces. This property is used for the generation of ultrasonic vibration by converting applied voltage into mechanical vibrations. Modern sonic converters usually contain several piezoceramic discs of lead zirconate titanate restricted by two final masses that are mechanically prestressed by a center screwing.

In the course of grinding, ultrasonic frequency can be introduced into the contact zone by the tool as well as by the workpiece. An excitation of oscillation of the workpiece takes place when the dimensions and the weight of the tool do not allow a high-frequency introduction of oscillations. Depending on the position of the active partners relative to each

Process variants of ultrasonic-assisted grinding

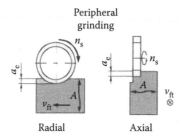
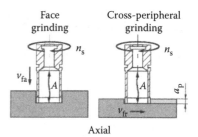

FIGURE 20.3

Process modifications of ultrasonic-assisted grinding. (From Uhlmann, E., Holl, S. -E., 1998a, Entwicklungen beim Schleifen keramischer Werkstoffe. *Vortrag im Rahmen des Seminars, Moderne Schleiftechnologie, am 14 Mai 1998 in Furtwangen,* Deutschland. With permission.)

other and on the direction of oscillations, there are different process modifications that allow the realization of machining tasks (Figure 20.3).

It has been proved that process improvement through ultrasonic superposition can be achieved, particularly where grinding is characterized by constant workpiece–wheel contact. These contact conditions particularly apply to face and cross-peripheral grinding. Moreover, investigations on peripheral grinding showed that the effects resulting from ultrasonic assistance have a more pronounced effect on the work result with increasing contact length (Spur and Holl 1995). Thus, contrary to creep-feed grinding, hardly any process improvement could be observed in the case of reciprocating grinding.

20.3 Ultrasonic-Assisted Grinding with Workpiece Excitation

According to the process variant, a supersonic oscillation is superimposed in the contact zone either vertical or parallel to the workpiece surface in addition to the conventional cutting movement. This change in the speed ratios and in the resultant cutting speed leads to functional and wear mechanisms, which are basically different from those of conventional grinding.

20.4 Peripheral Grinding with Radial Ultrasonic Assistance

Material removal and tool wear mechanisms of peripheral longitudinal grinding with radial ultrasonic excitation of the workpiece can be described through the simulation of the engagement of a single grain (Figure 20.4) and through scratch tests (Uhlmann and Holl 1998b).

If the feed speed is ignored, the path of a grain making a scratch, without ultrasonics, is described by the segment of a circle. The grain penetrates the material up to the depth

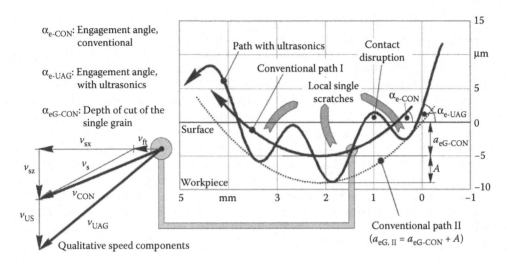

FIGURE 20.4
Parameters and simulation of the individual grain engagement during ultrasonic-assisted grinding. (From Uhlmann, E., Holl, S. -E., 1998b, *Maschinenmarkt.*, 104(48), S34–S37. With permission.)

of cut of the single-grain $a_{eG\text{-}CON}$, with a constant wheel speed v_s, and a defined engagement angle $\alpha_{e\text{-}CON}$. The maximum depth of scratch is reached at the lowest point of the curve. After leaving the surface, the grain has marked a trace of the length $l_{Rt\text{-}CON}$.

Additional longitudinal workpiece vibration causes significant deviations in a radial direction from the path described before. Depending on how many workpiece vibrations are realized per contact phase, a number of single scratches locally strung together with different scratch depths and lengths emerge instead of a circular scratch. The maximum depth of scratch increases by the value of the amplitude at nominally equal single-grain working engagements. As long as $a_{eG\text{-}CON}$ is smaller than the amplitude, complete contact disruptions occur in each case between the local single-grain scratches. Due to the additional speed component in the radial direction, the grain hits the surface at a larger engagement angle with higher active speed. Each local single-grain scratch is characterized by distinctly shorter contact times and single-grain lengths of scratch, as well as higher single-grain depths of scratch.

Figure 20.5 displays the surfaces of the materials, silicon nitride (SSN), and zirconium oxide (ZrO_2) scratched with and without ultrasonic assistance at equal maximum depths of scratch.

Without ultrasonics	With ultrasonics	100 μm
		SSN
		ZrO_2

FIGURE 20.5
Scratches on different ceramics with and without ultrasonics.

The wheel speed of the single-grain diamond was set to $v_s = 5$ m/s. Scratching without ultrasonic assistance leads to continuous traces, which mainly display areas of plastic deformation on the bottom of the scratch, partly showing traces of other single cutting edges. Material-removal processes occur while the diamond grain moves from right to left, which, due to critical stress conditions, causes radial cracks on the scratch borders that run vertically to the direction of motion. Above this are a number of lateral cracks depending on the K_{Ic} value, leading to conchoidal chips of material particles.

Observing the traces generated with ultrasonic assistance, it becomes clear that the entire trace is divided into local single-grain scratches. Repeated contact disruptions are due to periodic oscillation of the workpiece as well as superimposed circular movement of the single-grain diamonds. The effect of complete liftoff and reentry into the workpiece surface was similarly noticed for all tested materials. The scratches are primarily characterized by plastic deformations. Lateral crack systems are formed on the right and left borders of the scratches in relation to the depth of indentation, resulting in conchoidal chips.

The effects of ultrasonic-assisted grinding cause a higher mechanical load on material and diamond grits. In addition, the time of contact and, hence, the friction effects are distinctly reduced, thus decreasing the thermal loads. However, the degree that mechanical stresses appearing due to material displacement processes in front of the cutting edge are influenced remains unclear.

The theoretical center distance of two local single-grain scratches can be determined by approximation directly from the relation of wheel speed v_s, and oscillation frequency f. A comparison of theoretical setting conditions with actually generated structures contributed to the verification of the formation of scratch geometries typical for ultrasonics (Spur and Holl 1995).

Figure 20.6 shows the resulting normal and tangential forces during creep-feed grinding of silicon carbide, zirconium oxide, and aluminum oxide with and without ultrasonics. Ultrasonic assistance leads to a significant reduction of the process forces. Contrary to conventional creep-feed grinding, the course of the process is quasistationary. The ultrasonic superposition leads to higher stresses in the workpiece subsurface and the diamond cutting edges. Without ultrasonic assistance, the diamond cutting edges get flatter in the course of the process, starting from the sharpened state. Thus, the grain strain increases. Above a certain marginal load, the grain parts splinter in big segments or break off. The grains have not flattened in any of the cases with ultrasonic assistance. Rather, an increase in microsplitting could be observed. If transferred to the grinding process, the microsplitting leads to a continuous generation of sharp single cutting edges, resulting in the quasistationary state.

Figure 20.7 shows the integral temperature values in the subsurface of different ceramics during creep-feed grinding with and without ultrasonics. The measurement of temperature values was carried out with thermal elements, which were positioned in the subsurface of the workpieces. It becomes clear that the superposition of ultrasonics leads to a significant decrease in temperature.

The contact interruption between wheel and workpiece as a result of ultrasonic oscillation leads to a reduction of engagement times as described above. Moreover, infeed of coolant at the contact zone is improved by temporary lifting of one of the active partners and by the removal of chips. The reduction of the friction effects reduce grain flattening and the preferred microsplitting of the diamond grains can be considered further reasons for the decrease of process temperature.

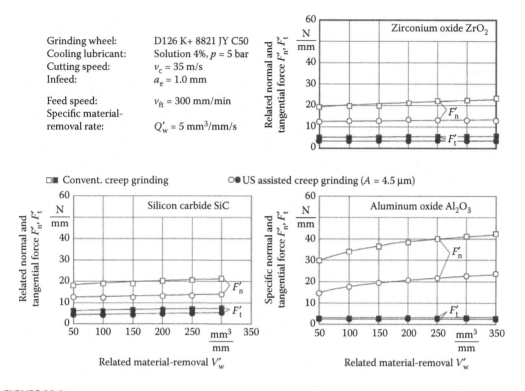

FIGURE 20.6
Process forces during conventional and ultrasonic-assisted creep-feed grinding of different ceramic materials.

Figure 20.8 shows the radial wear of a grinding wheel and the surface quality of a silicon nitride ceramic. Just as in conventional grinding, radial wear of the grinding wheel increases in grinding with ultrasonics as the specific material-removal rate increases. In the case of ultrasonic-assisted grinding, radial wear of the grinding wheel is higher than in conventional grinding. In the case of a material-removal rate of $Q'_w = 20$ mm³/mms, machining was not possible with a conventional process method due to inadmissible process forces.

Surface roughness also increases with specific material-removal rate in both process variants. In ultrasonic-assisted grinding, surface roughness is higher than in conventional grinding.

20.5 Peripheral Grinding with Axial Ultrasonic Assistance

Peripheral longitudinal grinding with axial ultrasonic assistance is characterized by a periodically changing working direction with continuous cutting-edge engagement. The newly created workpiece surfaces have sinusoidal machining marks (Warnecke and Zapp 1995).

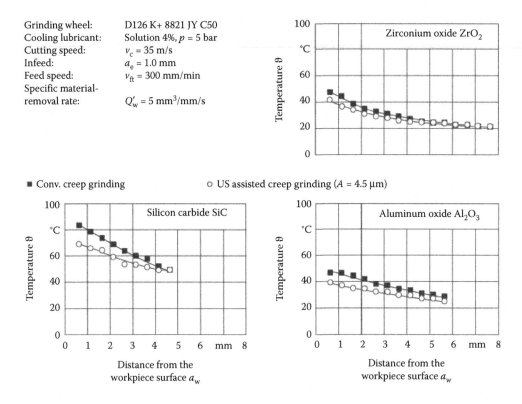

Grinding wheel: D126 K+ 8821 JY C50
Cooling lubricant: Solution 4%, $p = 5$ bar
Cutting speed: $v_c = 35$ m/s
Infeed: $a_e = 1.0$ mm
Feed speed: $v_{ft} = 300$ mm/min
Specific material-
removal rate: $Q'_w = 5$ mm^3/mm/s

■ Conv. creep grinding   ○ US assisted creep grinding ($A = 4.5$ μm)

FIGURE 20.7
Workpiece subsurface temperatures during conventional and ultrasonic-assisted creep-feed grinding of different ceramic materials.

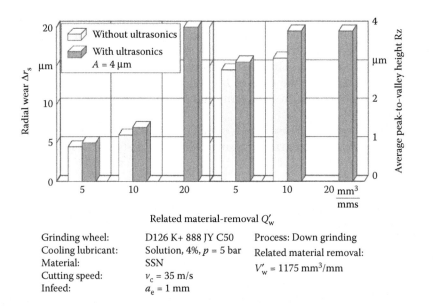

Grinding wheel: D126 K+ 888 JY C50   Process: Down grinding
Cooling lubricant: Solution, 4%, $p = 5$ bar   Related material removal:
Material: SSN $V'_w = 1175$ mm^3/mm
Cutting speed: $v_c = 35$ m/s
Infeed: $a_e = 1$ mm

FIGURE 20.8
Radial wear of the grinding wheel and the surface quality of silicon nitride during conventional and ultrasonic-assisted down grinding.

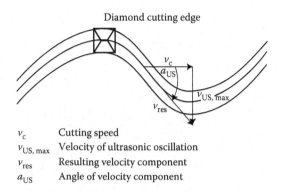

v_c Cutting speed
$v_{US,\,max}$ Velocity of ultrasonic oscillation
v_{res} Resulting velocity component
a_{US} Angle of velocity component

FIGURE 20.9
Principle of velocity superposition during grinding with axial ultrasonic assistance.

Through the superposition of the conventional grinding process with ultrasonic oscillation, there is an additionally arising and constantly changing velocity that is transverse to the cutting speed set on the machine. Since ultrasonic vibration mathematically follows a sinusoidal function, the ultrasonic-related speed fluctuates between zero and a maximum value, determined by the set frequency and amplitude. Hence, the resulting working velocity does not solely occur in the feed direction but, additionally, in the transverse direction. Also, the velocity changes within half a period of ultrasonic oscillation, that is, within 1/44,000th of a second from zero to the maximum value.

The resultant cutting speed leads to a constant change of the stress direction of the diamond cutting edges. Contrary to conventional grinding, the area engaged of the diamond edges is changing. Figure 20.9 shows the principle of the change of the resultant cutting direction as a result of ultrasonic superposition.

In conventional grinding, the machining marks run in the cutting direction. In ultrasonic-assisted grinding with radial excitation, the machining marks are also parallel to the direction of cutting, consisting of individual impact. The vibration in the axial direction causes machining marks transverse to the direction of cutting. Figure 20.10 shows the surface of a silicon nitride workpiece after grinding with axial ultrasonic assistance.

At a cutting speed of $v_c = 1$ m/s, an ultrasonic frequency of $f = 21.27$ kHz, and an amplitude of $A = 4.5$ μm, the ratio of the amplitude of oscillations and the wavelength is so high that the course of the generated oscillations can be clearly recognized. The infeed was defined at $a_e = 2$ μm during these tests.

The creation of surface structures with axial ultrasonic assistance is possible if the tool, the material, and the process parameters are adapted to the requirements. The shape of the formed marks can be specifically set with the choice of the parameters' cutting speed, ultrasonic frequency, and amplitude.

During the machining with axial superposition of oscillations, process forces are reduced in comparison to conventional grinding. Due to the constantly changing working direction, there will be less wear on the diamonds. Thus, sharp edge areas of a cutting grain consistently engage in the material, which leads to the reduction of the process forces. As a result of the reduced proportion of friction through the sharper cutting edges, the thermal stresses are reduced as a whole. The contact zone temperature can be reduced, although additional energy is imported to the process by the high-frequency superposition of oscillations (Warnecke and Zapp 1995).

Material:	SSN	Frequency:	$f = 21.27$ kHz
Cutting speed:	$v_c = 1$ m/s	Amplitude:	$A = 4.5$ μm
Feed:	$v_{ft} = 86$ mm/min	Infeed:	$a_e = 2$ μm
Cooling lubricant:	Solution 4%	Grinding wheel:	D46 K+ 888JY C50

Direction of grinding

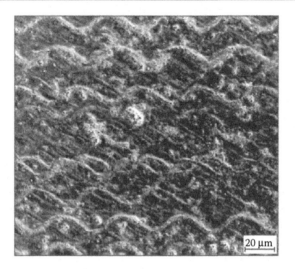

FIGURE 20.10
Structured surface of silicon nitride.

20.6 Ultrasonic-Assisted Grinding with Excitation of the Wheel

Ultrasonic-assisted grinding with an oscillating wheel is well established and commonly applied in industry and can be divided into different process modifications. In the case of ultrasonic-assisted cross-peripheral grinding, the feed movement takes place vertically to the tool spindle axis and thus to the generated oscillation. With this process, grooves, gaps, and radii can be machined in brittle–hard materials. The feed movement in the case of ultrasonic-assisted face grinding is, however, parallel to the tool axis. Bores may be machined with this process.

20.6.1 Ultrasonic-Assisted Cross-Peripheral Grinding

During cross-peripheral grinding, the cutting-edge engagement is not interrupted because the peripheral side of the wheel is in contact with the workpiece. Rather, a sinusoidal grain engagement can be observed as a result of the axial oscillation. The engagement at the front face of the wheel takes place according to the movement conditions described for the peripheral longitudinal grinding with radial excitation, leading to local individual engagements.

Figure 20.11 shows the achievable surface topographies and roughnesses after ultrasonic-assisted cross-peripheral grinding of different ceramics. Surface roughness achieved corresponds to values after conventional grinding.

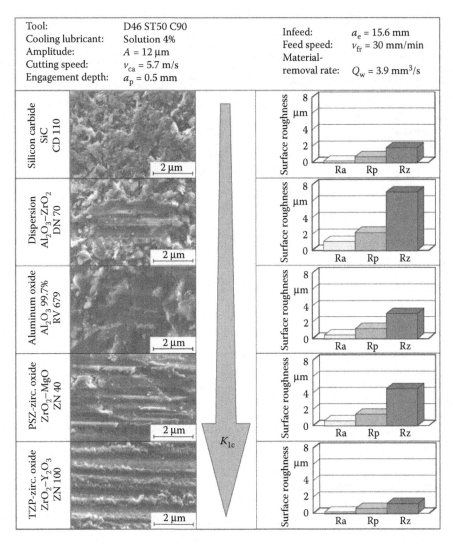

FIGURE 20.11
Surface topography and roughness of ceramic materials after ultrasonic-assisted cross-peripheral grinding.

Under certain kinematic conditions, the result of ultrasonic movement vertically to the workpiece surface is a complete interruption of the workpiece–tool contact. On all materials, typical pocket-type surface structures can be observed as an effect of the axial ultrasonic movement. The topography depends on the fracture toughness of the machined material. Bending strengths of components ground conventionally or with ultrasonic assistance are similar, although higher residual compression is imparted to the subsurface of the material through the changed stress in ultrasonic-assisted grinding (Engel and Daus 1999).

Figure 20.12 shows the achievable material-removal rate during ultrasonic-assisted cross-peripheral grinding of different ceramic materials. The feed speed was controlled in the tests by the determination of a maximum process force in the direction of feed. Smaller process forces lead to higher feed speeds and thus to growing material-removal rates.

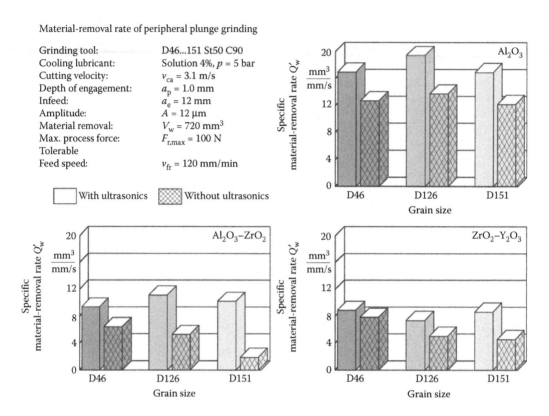

Material-removal rate of peripheral plunge grinding

Grinding tool: D46...151 St50 C90
Cooling lubricant: Solution 4%, p = 5 bar
Cutting velocity: v_{ca} = 3.1 m/s
Depth of engagement: a_p = 1.0 mm
Infeed: a_e = 12 mm
Amplitude: A = 12 µm
Material removal: V_w = 720 mm^3
Max. process force: $F_{r,max}$ = 100 N
Tolerable
Feed speed: v_{fr} = 120 mm/min

☐ With ultrasonics   ▨ Without ultrasonics

FIGURE 20.12
Material-removal rates during conventional and ultrasonic-assisted grinding of ceramics in relation to the grain size of the tool.

In contrast to ultrasonic-assisted grinding with constant feed speed, this kind of process control prevents inadmissibly high stresses on the tool.

As a result of the ultrasonic superposition of the grinding process, higher material-removal rates can be stated for all investigated materials and used diamond grain sizes of the grinding tools. It can be seen that the diamond grain size is not decisive for the achievable material-removal rates during ultrasonic-assisted grinding. There is a correlation between this statement and investigations by Pei and Ferreira (1999), who observed this behavior during the machining of zirconium oxide. It becomes clear that the mechanical properties of the machined material have a significant influence on the machining result.

The surface qualities obtainable are greatly dependent on the characteristics of the machined materials and of the machining process. This can be proved by comparing the surface qualities of ceramics, which were machined with ultrasonic-assisted grinding and plane parallel lapping (Figure 20.13). Excluding the material silicon carbide, the arithmetical mean deviation of samples ground with ultrasonic assistance is slightly above the value established for that of plane parallel lapping.

For the materials silicon carbide and zirconium oxide (ZN 40), the best surface values were measured during ultrasonic-assisted grinding. Summarizing, the surface qualities obtainable for ceramic materials machined with ultrasonic-assisted grinding can be established at the level of conventional finishing procedures. Despite higher mechanical loads, surface qualities comparable to those of lapping can be achieved.

Ultrasonic-assisted cross-peripheral grinding

Grinding tool:	D46 ST50 C90
Cooling lubricant:	Solution 4%
Amplitude:	$A = 12\,\mu m$
Cutting speed:	$v_{ca} = 5.7$ m/s
Depth of engagement:	$a_p = 0.5$ mm
Working engagement:	$a_e = 15.6$ mm
Feed speed:	$v_{fr} = 30$ mm/min
Material-removal rate:	$Q_w = 3.9$ mm^3/s

Plane parallel lapping:

Lapping abrasive	B$_4$C F400
Rotational speed of the inner pin circle:	$n_i = 64$ rpm ni
Rotational speed of the lower wheel:	$n_u = -56$ rpm
Rotational speed of the upper wheel:	$n_o = 33$ rpm
Lapping pressure:	$p_L = 0.8$ Mpa
Lapping wheels:	Perlitic gray cast iron

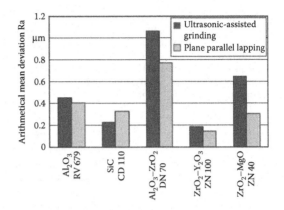

FIGURE 20.13
Arithmetical mean deviation depending on the material and the machining process. (From Spur, G. et al., 1999, *Machining of Ceramics—2000 and Beyond*. With permission.)

Bending strengths of ceramic workpieces were investigated for the machining parameters in Figure 20.13. Ultrasonic-assisted grinding gives bending strengths similar to or higher than those of plan parallel lapping. Moreover, ultrasonic-assisted grinding leads to insignificantly higher deviations of bending strengths. These are caused by the alternating load on the workpiece subsurface (Spur et al. 1999).

20.6.2 Ultrasonic-Assisted Face Grinding

To evaluate the machining process, the development of process forces was analyzed for a path-controlled feed speed. Figure 20.14 illustrates that a grinding operation without ultrasonic assistance produces a poorer process because rapidly increasing axial forces occur. After a certain grinding time, the force already reached a value of $F_z = 240$ N. This

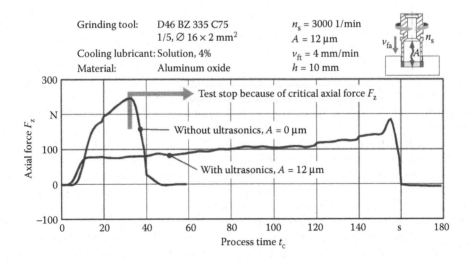

FIGURE 20.14
Comparison between axial forces during tool-path controlled face grinding with and without ultrasonics.

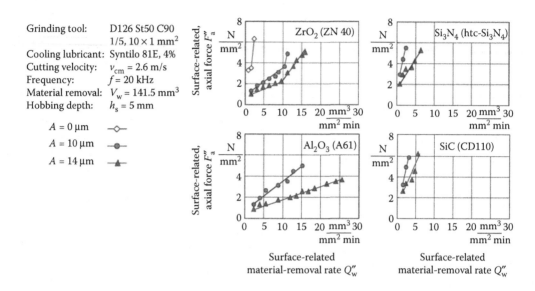

FIGURE 20.15
Surface-related axial forces during face grinding in relation of the surface-related material-removal rate for ceramics. (From Uhlmann, E., 1998, *Annals of the CIRP*, 47(1), 249–252. With permission.)

level could no longer be tolerated, so the process had to be stopped. A permanent wheel–workpiece contact is responsible for faster blunting of the grinding coating, leading to a strongly reduced cutting ability. This results in an enormous increase in force if feed speeds are constant. Therefore, economic production of such contours with conventional methods (grinding without ultrasonics) is not possible.

Figure 20.15 shows the influence of ultrasonic oscillations on the process forces during face grinding. The height of the surface-related axial forces depends on the machined material. Alongside fracture toughness, stability, and hardness of the material play an important role here.

While the highest surface-related material-removal rates of $Q''_w = 25$ mm^3/mm^2/min are achieved for aluminum oxide, they decrease to $Q''_w = 5$ mm^3/mm^2/min during the machining of silicon nitride. A maximum surface-related axial force of 6 N/mm^2 was chosen as a critical value. It can be observed in the case of zirconium oxide how the process forces behave in contrast to conventional face grinding. There are similar surface-related axial forces at $Q''_w = 2.5$ mm^3/mm^2/min during conventional and at $Q''_w = 15$ mm^3/mm^2/min for ultrasonic-assisted grinding with $A_{US} = 14$ μm. This is a sixfold increase of the material-removal rate.

The influence of the engagement angle $\alpha_{eUS,max}$ in ultrasonic-assisted grinding described within the scope of the kinematical investigations of the process becomes clear for all four tested materials. The comparison of the effect of two different ultrasonic amplitudes of $A_{US} = 10$ and 14 μm on the process forces shows the influence of the engagement angle $\alpha_{eUS,max}$. The reduction of the ultrasonic amplitude leads to a decrease of $\alpha_{eUS,max}$ ($A_{US} = 14$ μm) = 32° to $\alpha_{eUS,max}$ ($A_{US} = 10$ μm) = 24°. The result of these kinematical changes is smaller grain acceleration due to the ultrasonics and a reduction of the mechanical, pulsed stresses of the single grains. Basically, these mechanical stresses are the cause of microsplitting of the abrasive grains. In turn, this microsplitting leads to the generation of new sharp cutting edges (Uhlmann 1998).

In recent years, fundamental research has been undertaken, in order to simulate and model the engagement conditions at ultrasonic-assisted face grinding (Uhlmann and Hübert 2006a,b, 2007; Uhlmann and Sammler 2010). The engagement conditions at conventional and ultrasonic-assisted face grinding differ significantly (Figure 20.16). The single-grain path and engagement angel of both processes can be calculated with the represented equations.

From experience, the ultrasonic amplitude at excitation of the tool is limited due to the ultrasonic system and thus the limiting factor of achievable removal rates (Lauwers et al. 2010; Uhlmann and Sammler 2010). As a major requirement for stable ultrasonic-assisted face grinding processes, a contact disruption of grain and workpiece has to take place, in order to achieve microsplitting and avoid single-grain blunting, caused by continuous grain contact (Figure 20.16). Therefore, the engagement conditions, chipping parameters, and resulting surface were calculated through kinematical three-dimensional voxel-based simulation (Figures 20.17 and 20.18).

The conchoidal chip shape was confirmed through the simulation as well as increasing ultrasonic amplitudes must not lead to a higher engagement angle and a higher single- grain chip thickness, when disruption is ensured. Hence, if unstable process behavior under ultrasonic conditions and maximum amplitude occurs, the grain size and concentration of the tool can be used in order to reduce the maximum chip thickness to be below the ultrasonic amplitude (Uhlmann and Sammler 2010). By using computational fluid dynamics simulation, the positive influence on coolant volume flow passing the engagement zone at internal coolant supply could be proven as well as the cavitation effect of local low-pressure zones, caused by the high-frequency movement of the tool (Uhlmann and Hübert 2007). A sharpening effect of the bond and an increased material transport of the removed material are supported through this.

$$\vec{I}_e(t) = \begin{Bmatrix} I_{ex}(t) \\ I_{ey}(t) \\ I_{ez}(t) \end{Bmatrix} = \begin{Bmatrix} r_s \cdot \sin(2 \cdot \pi \cdot n_s \cdot t) \\ r_s \cdot \cos(2 \cdot \pi \cdot n_s \cdot t) \\ V_{fa,CON} \cdot t - A \cdot \sin(2 \cdot \pi \cdot f \cdot t) \end{Bmatrix}$$

$$\alpha_{e,US} = \arctan \left(\frac{V_{fa,CON} - \frac{1}{2} \cdot A \cdot \pi \cdot f \cdot \cos\left(\pi \cdot \left(1 - 2 \cdot \frac{I_{ch,US}}{I_{US}}\right)\right)}{V_c} \right)$$

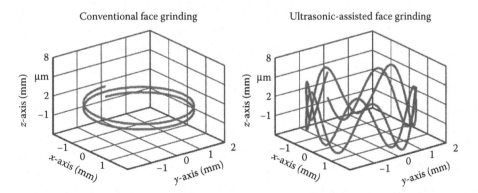

Conventional face grinding   Ultrasonic-assisted face grinding

FIGURE 20.16
Single-grain path at conventional and ultrasonic-assisted face grinding.

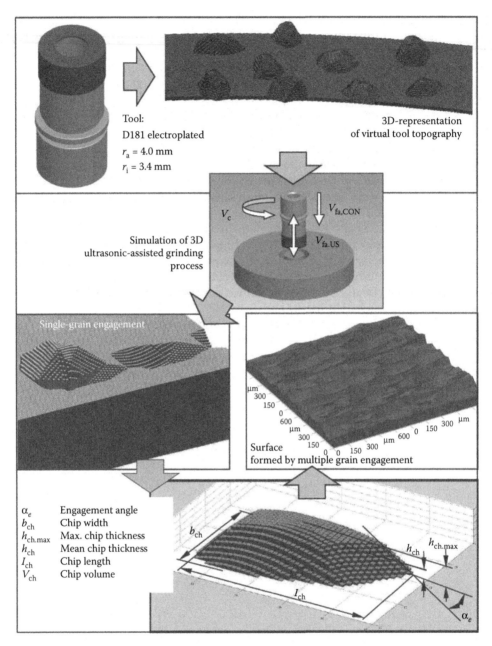

FIGURE 20.17
Three-dimensional voxel-based simulation of ultrasonic-assisted grinding.

20.7 Summary

Ultrasonic-assisted grinding is a finishing process for economic machining of brittle–hard materials. The superposition of the kinematics of the grinding process with ultrasonic oscillations leads to removal and wear mechanisms different from those during

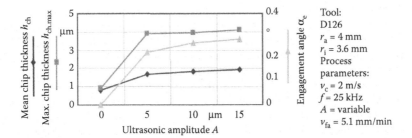

FIGURE 20.18
Engagement parameters calculated through simulation in dependency of ultrasonic amplitude.

conventional grinding. There is microsplitting due to the high stress of the abrasive grains, provoking the constant formation of new sharp edges. The processes are characterized by small process forces with a quasistationary process during the machining allowing an increase of the material-removal rates.

In peripheral grinding with radial ultrasonic superposition, a reduction of process forces of up to 90% could be observed in contrast to conventional grinding with the same specific material-removal rate. At the same time, wear of the grinding wheel as well as surface quality of the machined workpieces increase slightly. In peripheral longitudinal grinding with axial superposition, structured surfaces could be formed and directional machining marks avoided with the help of the kinematics.

In cross-peripheral grinding with axial ultrasonic excitation, complex contours with high specific material-removal rates can be achieved despite the low-cutting velocities. At the same time, high surface qualities and shape accuracies can be realized.

These results show that ultrasonic-assisted grinding ensures the machining of ceramics in terms of high economic efficiency and component quality. Through the variety of kinematic process variants, it is possible to machine different geometrical elements on brittle–hard materials (Figure 20.19).

FIGURE 20.19
Process of ultrasonic-assisted cross-peripheral grinding and possible geometrical elements. (From Uhlmann, E., Holl, S. -E., 1998a, Entwicklungen beim Schleifen keramischer Werkstoffe. *Vortrag im Rahmen des Seminars, Moderne Schleiftechnologie, am 14 Mai 1998 in Furtwangen*, Deutschland. With permission.)

References

Cartsburg, H., 1993, Hartbearbeitung keramischer Verbundwerkstoffe. PhD dissertation Technische Universität Berlin.

Conrath, M., 2005, Systematische Gestaltung von frequenzadaptierbaren Ultraschall-Werkzeugsystemen zum Einsatz in fertigungstechnischen Prozessen. PhD dissertation Universität Karlsruhe.

Drozda, T. J., 1983, Mechanical nontraditional machining processes. *Manufacturing Engineering*, 91(1), 61–64.

Engel, H., Daus, N. -A., 1999, Veränderung des Wirkmechanismus beim Schleifen durch Ultraschallunterstützung und daraus resultierende Prozeßverbesserungen. Vortrag zum Seminar "Hybride Prozesse der Zerspan- und Abtragtechnik." TU Dresden.

Haas, R., 1991, Technologie zur Leistungssteigerung beim Ultraschall-schwingläppen. PhD dissertation, RWTH Aachen.

Lauwers, B., Bleicher, F., Ten Haaf, P., Vanparys, M., Bernreiter, J., Jacobs, T., Loenders, J. 2010, Investigation of the process-material interaction in ultrasonic assisted grinding of ZrO_2 based ceramic materials. In *Proceedings of 4th CIRP International Conference on High Performance Cutting*.

Nankov, M. M., 1989, Supersonic activation or cup-shape diamond disk grinding. *ISEM-9, Proceedings of the Symposium for Electro Machining*. The Japan Society of Electro-Machining Engineers, Nagoya, Japan.

Pattimore, J., 1998, Optimisation of grinding for cylindrical silicon nitride components for mass production. Lecture: *Informativer Arbeitskreis Keramikbearbeitung*. IWF Berlin.

Pei, Z. J., Ferreira, P. M., 1999, An experimental investigation of rotary ultrasonic face milling. *International Journal of Machine Tools and Manufacture*, 39, 1327–1344.

Pei, Z. J., Prabhakar, D., Haselkorn, M., 1993, *Mechanistic Approach to the Prediction of Material Removal Rates in Rotary Ultrasonic Machining*. Manufacturing Science and Engineering American Society of Manufacturing Engineers, Production Engineering Division (Publication) PED. Volume 64. ASME, New York.

Popp, M., 1998, Seriengerechte Herstellung von Bauteilen aus Hochleistungskeramik am Beispiel des Keramikwälzlagers. *Proceedings: "Karlsruher Arbeitsgespräche 1998,"* BMBF/PFT, Karlsruhe, 12 und 13.03.

Prabhakar, D., Ferreira, P. M., Haselkorn, M., 1992, An experimental investigation on material removal rates in rotary ultrasonic machining. *Konferenz-Einzelbericht: Transactions of the North American Manufacturing Research Institution of SME 1992, the 20th NAMRC Conference*, Washington State University, Pullman, WA.

Spur, G., 1989, *Keramikbearbeitung*. Hanser Verlag, München, Wien.

Spur, G., Holl, S. -E., 1995, Ultraschallunterstütztes Schleifen von Hochleistungskeramik. In *Jahrestagung 1995, Kurzreferate. Deutsche Keramische Gesellschaft*, Aachen, Germany.

Spur, G., Uhlmann, E., Holl, S. -E., Daus, N. -A., 1999, Ultrasonic machining of ceramics. In I. D. Marinescu (Ed.), *Machining of Ceramics—2000 and Beyond*.

Suzuki, K., Tochinai, H., Uematsu, T., Nakagawa, T., 1993, New grinding method for ceramics using a biaxially vibrated nonrotational ultrasonic tool. *Annals of the CIRP*, 42(1), S375–S378.

Uhlmann, E., 1998, Surface formation in creep feed grinding of advanced ceramics with and without ultrasonic assistance. *Annals of the CIRP*, 47(1), 249–252.

Uhlmann, E., Holl, S. -E., 1998a, Entwicklungen beim Schleifen keramischer Werkstoffe. *Vortrag im Rahmen des Seminars, Moderne Schleiftechnologie, am 14 Mai 1998 in Furtwangen*, Deutschland.

Uhlmann, E., Holl, S. -E., 1998b, Schwer zerspanbare Werkstoffe ultraschallunterstützt schleifen. *Maschinenmarkt.*, 104(48), S34–S37.

Uhlmann, E., Hübert, C., 2006a, Advances in ultrasonic assisted grinding of ceramic materials. *Tagungsband 11th International Ceramics Congress. International Congress on Modern Materials & Technologies, Acireale, Sicily,* Italy.

Uhlmann, E., Hübert, C., 2006b, Modellbildung und Simulation innovativer Verfahren der Feinbearbeitung. In Hrsg.: C. Becher, G. Pritschow (Eds.), *Simulationstechnik in der Produktion.* VDI (Fortschritt-Berichte VDI, Reihe 2: Fertigungstechnik 655), Düsseldorf.

Uhlmann, E., Hübert, C., 2007, Ultrasonic assisted grinding of advanced ceramics. *Proceedings of the ASPE Spring Topical Meeting.*

Uhlmann, E., Sammler, C., 2010, Influence of coolant conditions in ultrasonic assisted grinding of high performance ceramics. *Production Engineering,* 4(6), S581–S587.

Warnecke, G., Zapp, M., 1995, Ultrasonic superimposed grinding of advanced ceramics. *Proceedings of the 1st International Machining and Grinding Conference,* Dearborn.

Westkämper, E., Kappmeyer, G., 1994, Feiner Abtrag—Anwendungsgerechtes Auslegen von Werkzeugen zum Ultraschallhonen. *Maschinenmarkt,* 100(30), S28–S33.

Glossary

Symbols

a, a_p, A	Set depth of cut = Programmed depth of cut = Apparent depth of cut
a_d	Dressing depth
a_e	(Real) depth of cut
$a_e(\theta)$	Depth of cut at position θ
a_{ed}	Real depth of cut while dressing
a_{sw}	Depth of wheel wear
b_{cu}	Uncut chip width
b_d	Active width of dressing tool
b_s	Contact width on grinding wheel
b_w	Contact width on workpiece
c	Damping
c	Specific heat capacity
c, c_l, c_s	Costs, labor rate, wheel cost
c_p	Specific heat at constant pressure
c_v	Specific heat at constant volume
d_{cap}	Diameter of capillary
d_e, d_{eq}	Equivalent wheel diameter
d_g	Grain diameter or mean grain diameter
d_{jet}	Fluid jet diameter
d_s	Diameter of grinding wheel
d_{ss}	Diametral standoff
d_w	Diameter of workpiece
d_{ww}	Diametral stock removal
e_c or U	Specific cutting energy or energy per unit volume
e_{ch}	Energy per unit volume carried away by chips
f	Frequency in cycles per second
f	Ultrasonic frequency
f_d	Dressing feed per wheel revolution
f_n	Normal force on a grain
f_t	Tangential force on a grain
h	Heat convection coefficient for moving surface. Heat per unit area per degree temperature difference
h_{air}	Thickness of the boundary layer of air surrounding a wheel
h_{cu}	(Uncut) chip thickness
V'_w	Mean thickness of uncut chips
h_{cueff}	Effective chip thickness
h_{eq}	Equivalent chip thickness

h_f	Convection coefficient for fluid cooling in contact zone
h_{fu}	Mean thickness of the useful process fluid layer that passes through the contact
h_g	Convection coefficient for an abrasive grain
h_{jet}	Thickness of a uniform jet of process fluid
h_p	Depth of fluid penetration into the wheel
h_{pores}	Mean depth of pores at surface of wheel
h_{slot}	Slot nozzle gap thickness
h_w	Work height in centerless grinding
h_w	Convection coefficient for workpiece
h_{wq}	Convection coefficient for the workpiece at a grain contact
k	Shear flow stress
k	Thermal conductivity
k_m	Stiffness of the machine–tool–workpiece system
k_s	Normal grinding force per unit depth of cut (sometimes termed grinding stiffness)
k_w	Thermal conductivity of work material
l_c	Contact length of tool and workpiece
l_c	Uncut chip length
l_{cap}	Capillary length
l_{cu}	Chip length
l_f	Contact length due to force arising from deflection
l_{fr}	Contact length due to force for a real surface with roughness
l_{fs}	Contact length due to force for a smooth body
l_g	Geometric contact length
l_{mc}	Length of median crack
n	Number of waves
n_b	Number of parts per batch
n_d	Number of dressing passes
n_s	Wheel rotational speed
n_w	Workpiece rotational speed
p	Angular pitch of waviness
p	Pressure
p	Normal stress
p_p	Pumped pressure
q	Tangential stress
q	Speed ratio: wheel speed over work speed
q	Speed quotient
q	Heat flux. Heat per unit area per unit time
q_{ch}	Heat flux to chips
q_d	Dressing speed ratio (crush ratio)
q_f	Heat flux to process fluid
q_o, q_t	Mean heat flux. Total heat flux
q_s	Heat flux to abrasive
q_w	Heat flux to workpiece
r	Ratio of chip groove width to groove thickness
r_{cu}	Uncut chip aspect ratio
r_o	Effective radius of wear flat on tip of grain
r_g	Effective radius of grain
r_s	Radial wheel wear per dress

$r(\theta)$	Reduction in radius at position θ
s	Feed of workpiece per cutting edge
s	Laplace operator
t	Time
t_a	Spark-out time
t_c	Time of contact of point on workpiece based on real contact length
t_c, t_d, t_t	Cycle time per part, dressing time per part, total cycle time per part
t_g	Time of contact of point on workpiece based on geometric contact length
t_{gc}	Time for passage of one grain through the contact zone
t_p	Proportion of bearing area at a depth p below highest peak
t_s	Machining time
t_{so}	Dwell time for spark out
t_w	Workpiece temperature
v	Speed of a moving heat source
v_c	Cutting speed
v_{cap}	Mean fluid velocity through capillary
v_{cds}	Sharpening velocity
v_{cm}	Mean cutting speed while ultrasonic grinding
v_d	Velocity of roll dresser
v_f	Feed rate or infeed rate or through-feed rate of workpiece
v_{fa}	Feed rate in axial direction
v_{fd} or v_d	Dressing traverse feed rate
v_{fe}	Effective feed rate in centerless grinding
v_{fr}	Feed rate in radial direction
v_{ft}	Feed rate in tangential direction
$V_{f.wear}$	Rate of reduction of wheel radius due to wear
v_{jet}	Fluid jet velocity
$v_{orifice}$	Fluid velocity from orifice
v_r	Resultant control wheel speed in centerless grinding
v_s	Speed of the wheel
v_w	Speed of the workpiece tangential to grinding wheel speed
v_w	Table speed
y_b	Thermal boundary layer thickness of the process fluid in the contact zone
y_t	Payback time for machine
w_{slot}	Width of slot nozzle
$x(\theta)$	Dynamic deflection at position θ
z_s	Cutting-edge depth
A	Apparent = set depth of cut
A	Geometric stability parameter for centerless grinding
A	Ultrasonic amplitude
A_c	Contact area. Overall contact area
A_{cu}	Mean cross-sectional area of the uncut chips
A_r	Real area of contact
B	Brittleness index
B	Cutting-edge spacing in lateral direction. Mean grain spacing
B	Peclet equivalent for inclined band source
C, C_a	Cutting-edge density. Number of active cutting edges per unit area
C_{stat}, C_{dyn}	Static and dynamic cutting-edge density

C	Factor for temperature solution taking account of Peclet number, flux distribution, and geometry
C, C_p, C_v	Specific heat capacity. For gases, specific heat at constant pressure or at constant volume
C_a	Area contraction coefficient for fluid jet
C_d	Orifice discharge coefficient
C_l	Labor cost per part
C_m	Machine cost per part
C_{mc}	Machine cost
C_s	Wheel cost per part
C_v	Velocity coefficient for orifice or nozzle jet
E	Young modulus of elasticity
E	Energy
E^*	Equivalent elastic modulus for two bodies in contact
E_f	Fracture energy
E_p	Plastic flow energy
F	Force. Resultant of normal, tangential, and axial forces
F^*	Critical load to initiate a crack
F_{ij}	Notation used to represent forces on atom "i" from atoms "j"
F_a	Force parallel to wheel axis
F_a''	Surface-related axial force
F_n	Force normal to wheel surface
F_{nS}	Cutting-edge normal force
F_n'	Normal force per unit width
F_t	Force tangential to wheel surface
F_t'	Tangential force per unit width
F_{tS}	Cutting-edge tangential force
F_z	Axial force
G	Grinding ratio: Volume of work material removed divided by wheel wear volume
H	Hardness. Brinell hardness
H_b	Hardness of material bulk
H_m	Momentum power of fluid at wheel speed
H_p	Fluid pumping power
H_s	Hardness of material at surface
H_t	Total pumping and momentum power required to deliver fluid through grinding contact
H_v	Vickers hardness
H_{RC}	Rockwell hardness
K	Machining-elasticity parameter
K	Archard–Preston wear coefficient
K	Yield shear stress. Flow stress
K	Permeability of a wheel
$K_o[u]$	Bessel function of second kind order zero for an argument of value u
K_{1C}	Fracture toughness
L	Cutting-edge spacing in cutting direction
L	Peclet number for moving band heat source
L_{st}	Static cutting-edge spacing
M	Mesh number. Measure of grain size based on number of wires used in sieve

M'_{air}	Momentum of air boundary layer around wheel per unit wheel width
M_f	Momentum of process fluid
M'_f	Momentum of process fluid per unit wheel width
N_d, N_w	Number of parts per dress. Number of parts per wheel.
N_{mc}	Number of parts produced by the machine within payback period
N_{sact}	Actual number of cutting edges per surface unit
N_{skin}	Kinematic number of cutting edges per surface unit
N_{stat}	Static number of cutting edges per surface unit
P, P_c	Cutting or grinding power
P ratio	Ratio of material volume removed to wheel surface area
P'_c	Cutting power per unit width
$P_c(x)$	Probable number of active cutting edges per unit length in cutting direction
$P_c(z)$	Probable number of active cutting edges per unit length in lateral direction
P_e	Peclet number for moving band source
Q	Dynamic magnification
Q	Rate of work material removal
Q'_{ds}	Rate of work material removal per unit width while sharpening
Q_f	Flow rate of process fluid
Q'_f	Flow rate of cooling fluid per unit nozzle width
Q_{fcz}	Flow rate of cooling fluid through the contact zone
Q'_{fcz}	Flow rate of cooling fluid through the contact zone per unit width
Q_{max}	Mean maximum uncut chip cross-sectional area
Q_{fu}	Useful flowrate. Flowrate that passes through contact zone
Q_s	Rate of tool wear
Q'_s	Rate of tool wear per unit width
Q_w	Rate of work material removal
Q'_w, Q'	Rate of work material removal per unit width of contact
Q''_w	Surface-related material removal rate
R_a	Arithmetic mean deviation of the roughness profile
R_{ch}	Proportion of total heat taken by chips
R_e	Reynolds number
R_f	Proportion of total heat taken by fluid
Rk, Sk	Core roughness depth
R_L	Ratio of real contact length to geometric contact length
Rp, Sp	Maximum peak height of the roughness profile
Rpk, Spk	Reduced peak height
R_r	Roughness factor. Ratio of contact lengths due to deflection for rough and smooth surfaces
R_s	Proportion of total heat taken by abrasive
Rt, Rz	Measures of peak-to-valley roughness
Rt	Total height of the roughness profile
Rvk, Svk	Reduced valley depth
R_w	Partition ratio. Proportion of total heat taken by workpiece
R_{ws}	Partition ratio for the workpiece–wheel rubbing contact
Rz, Sz	Maximum height of the roughness profile
$R(\theta)$	Set reduction at position θ, that is, ignoring deflection
S_{act}	Active cutting-edge number
S_{cu}	Uncut chip surface area
S_o	Sommerfeld number

S_{stat}	Static number of cutting edges (per unit length)
T	Tool life or tool redress life
T	Temperature
T	Time for one work revolution
T_d	Maximum wear spot on diamond dressing tool
T_{max}	Maximum value of workpiece background temperature
T_w	Workpiece background temperature in contact zone
T_{wg}, T_g	Spike (or flash) temperature at a contact between grain and workpiece
T_μ	Critical cutting depth
U_d	Overlap ratio in dressing. Active width of dressing tool divided by dressing feed
V_a	Useful volume of abrasive layer
V_b	Volume percentage of bond material in an abrasive structure
V_{cu}	Uncut chip volume
V_g	Volume percentage of abrasive grain in a wheel structure
V_p	Volume percentage of air in an abrasive structure. Porosity
V_{pw}	Effective porosity ratio at wheel surface. Pore volume/wheel volume at surface
V_s	Total wear volume of wheel or tool material removed
V'_{Sg}	Specific wear volume per unit width
V_w	Total volume of workpiece material removed
V'_w	Total volume of workpiece material removed per unit width
W_d	Diamond size (carats)
$X(\theta)$	Infeed movement at position θ
α	Feed angle in angle-approach grinding
α	Thermal diffusivity
α	Angle between wheel and workrest contact
α	Clearance angle
β	Included tangents angle in centerless grinding
β	Thermal property for transient heat conduction
β	Wheel angle in angled-wheel grinding
β_{cw}	Elevation angle of workpiece from control wheel center in centerless grinding
β_s	Elevation angle of workpiece from grinding wheel center in centerless grinding
γ	Top workrest angle in centerless grinding
γ	Dullness ratio of abrasive grain
γ	Sharpness ratio of dressing tool
γ	Rake angle
δ	Deflection
χ	Phase angle between waviness and deflection
λ	Machine system stiffness
λ_o	Machine system static stiffness
θ_{mp}	Melting temperature of chips
ϕ	Phase angle between force and deflection
ϕ	Angle between plane of motion and inclined plane of heat source
ϕ_{pores}	Porosity of wheel at surface
μ	Friction coefficient
μ	Grinding force ratio
η	Dynamic viscosity
η	Effective direction angle

η_p	Dynamic viscosity at elevated pressure
ν	Kinematic viscosity
ν	Poisson ratio
ρ	Density
σ_{hs}	Hydrostatic stress
σ_n	Direct stress
τ	Shear stress or time constant
ψ	Pressure angle
ω	Frequency in radians per second
ω_o	Undamped natural frequency
ω_s	Grinding wheel angular speed
$\Delta, \delta R$	Roundness error
Δr_s	Radial grinding wheel wear
Λ	Stock removal rate parameter. Removal rate divided by normal force
Φ	Control wheel tilt angle
Φ_d	Dressing angle
Ω	Work rotational frequency in radians per second

A

AE sensor: Acoustic emission sensor used to detect wheel contact

AFM: Atomic force microscopy

Absorption: Assimilation of matter from surround into a surface and into the body of the material

Abrasion: Wearing or machining by rubbing, scratching, or friction

Abrasion, two-body: Wear of one body abraded by another

Abrasion, three-body: Wear of one body acted on by another with free abrasive in the contact

Abrasive: Tendency to cause wear. A hard and sharp material that wears a softer material

Abrasive composite: Material consisting of abrasive grains and bond structure

Abrasive, conventional: Hard abrasive mineral grit such as alumina and silicon carbide

Abrasive grains: Hard particles in an abrasive tool that provide the cutting edges

Abrasive machining: Machining by an abrasive process such as grinding, honing, lapping, or polishing

Abrasive medium: Suspension of abrasive grit in a liquid as used in lapping and polishing

Abrasive tool: Tool used for abrasive grinding, lapping, honing, or polishing

Abusive grinding: Grinding under conditions which produce severe damage, cracks, or burn

AC additive: Anticorrosion additive

Accuracy: Difference between intended value and achieved value

Acoustic emission: Process noise produced due to mechanical and metallurgical interactions

Active cutting depth: Depth of abrasive surface that actively engages the workpiece surface

Active layer: The layer of the abrasive tool that contains the abrasive

Active surface: The surface of the active layer that makes abrasive contact

Additivation: The adding of additives to a lubricant or process fluid

Additive: Substances added in small quantities to modify properties of a lubricant or process fluid

Adhesive loading: Work material adhering to tips of abrasive grains of active layer

Adhesion: Sticking together of two surfaces

Adiabatic: Process takes place without heat loss or gain within an immediate volume

Adsorption: Assimilation of material into close physical–chemical contact with a surface

Aerostatic bearing: See air bearing

Affinity: Measure of possibility of chemical reaction between two materials

Agglomeration: Tendency of particles to group together in clumps within a mass

Air barrier: Boundary layer of air around a high-speed wheel tending to deflect grinding fluid

Air bearing: Aerodynamic or pressure-controlled gas bearing (see also hydrostatic bearing)

Algorithm: Series of statements constituting a calculation routine

Aloxide, Alox: Terms sometimes used for aluminum oxide abrasive

Alumina, Al_2O_3: Aluminum oxide abrasive

Ambient: Conditions in the atmosphere surrounding the process

Amorphous layer: A thin layer lacking organized structure due to abrasive deformation

Analytical: Based on mathematical or logical consideration

Angle grinding: The axis of the wheel or the feed motion is angled to the workpiece axis

Anisotropic: Material having properties dependent on direction of measurement

Annulus: Space or shape between two concentric diameters

AO additive: Antioxidation additive used to increase service life of a process fluid

Antifog additive: Encourages aerosol particles to combine as droplets

Apparent contact area: Total area of contact including spaces between contacts

Apparent contact pressure: Normal force divided by apparent contact area

AR additive: Antirust additive

Archard constant: Coefficient used in Archard–Preston law for wear rate

Archard law: Relationship between wear, normal force, hardness, and sliding distance

Aromatic oils: Cycloaromatic and mixed structure nonsaturated hydrocarbons

Arrhenius law: Relationship for rate of chemical action based on temperature and time

Asperity: Sharp edge

Atmosphere: Air surrounding a process

Austenite: Gamma phase of low-carbon iron

Atp: Standard atmospheric temperature and pressure

Auxiliary processes: Processes additional to the main abrasive process. For example, fluid delivery, dressing

AW additive: Antiwear additive

Axial force: Component of force axial to the grinding wheel surface

B

Background temperature: Mean temperature at a point in a region of many flash contacts

Backlash: Lost movement on reversal of motion due to clearance in a feed drive

Bactericide: Additive that kills bacteria

Balancing: Adding or removing weight to improve wheel balance

Ballscrew: Leadscrew with recirculating rolling ball elements

Band heat source: Wide heat source having finite length

Barite, barite: Barium sulfate, sulfide mineral used for polishing

Barkhausen noise: Electromagnetic technique for detecting subsurface structural changes

Bauxite: Impure ore of aluminum and alumina

Bcc: Body-centered cubic lattice structure

Bearing area curve: Area of solid phase increasing with depth into surface expressed as a fraction

Bearing steel: Group of steels used for manufacture of rolling bearing elements

Bed: Machine base

Bell mouthing: Internal bore where the diameter at the end is larger than the middle

Bessel functions: Mathematical functions used to solve moving heat source problems

Beta (β): Thermal property for transient heat conduction in a material

Binder: Medium for suspension of abrasive particles

Biocide: Additive to kill bacteria and improve life of an oil

Black box: Unknown system characterized by measuring inputs and outputs

Blunt cutting edge: Cutting edge worn to a flat

Boiling temperature: Temperature at which grinding fluid rapidly vaporizes producing bubbles

Bond material: Material that bonds abrasive grains in a tool

Bond post: Bond material joining one grain to another

Bore grinding: Grinding an internal cylindrical surface

Boundary layer: Region close to a surface. For example, ϵ as in fluid

Boundary lubrication: Sliding contact with molecular layer of lubricant

Brake dresser: A dressing tool driven by the wheel; speed is controlled by a brake

Brinell hardness: Measure of hardness using a standard ball indenter

Brinelling: Indentation due to compressive action

Brittle: Tendency to fail by cracking fracture

Brittleness index: Measure of brittleness based on fracture toughness and Vickers hardness

BUE: Builtup edge. Material piled up at leading edge of tool while cutting

Bulk temperature: Mean temperature of the whole workpiece

Burn: Action of oxidation or material damage due to high machining temperature

Burnout, fluid: The complete drying out of grinding fluid in the contact zone at high temperature

Burnishing: Smoothing by action of friction

Burr: Lip of deformed material extending from edge of cut surface. Deburr is removal of burr

Bursting speed: Speed at which a wheel fails due to hoop stresses

C

Cam grinder: Machine for grinding engine camshafts

Carat: Measure of diamond grain size

Carborundum: Early name for silicon carbide abrasive

Carcinogenic: Tending to lead to cancer in humans

Cation: Positively charged ion

CBN, cBN: Cubic boron nitride. An extra-hard allotropic form of boron nitride

CD: Continuous dressing

CFD: Computational fluid dynamics. Type of computer software for fluid flow

CFRP: Carbon fiber-reinforced plastic

CVD: Chemical vapor deposition

Center: A conical pin that locates in the center hole to hold a cylindrical workpiece

C factor: Temperature factor for heat conduction from a moving source into a workpiece

Capillary tube: A tube of large length to diameter ratio

CDCF: Continuous dress creep feed grinding

Center grinding: Grinding between centers

Centerless grinding: Process for grinding without center holes for workpiece support

Cementite: Iron carbide

Ceramic: Materials made by firing clays or similar materials. For example, silicon nitride ceramic

Chalk: Fine white calcium carbonate powder used for polishing

Characteristic equation: Denominator of a transfer function set to zero

Chatter vibration: Regenerative vibration arising from an unstable machining process

Chemisorption: Assimilation of a material on a surface by chemical action

Chip aspect ratio: Ratio of length/width of uncut chip

Chip, uncut chip: Undeformed workpiece material in the path of an oncoming abrasive grain

Chip, cross-sectional area: Cross-sectional area of the uncut chip

Chip length: Length of the uncut chip

Chip thickness: Maximum or mean thickness of uncut chip

Chip volume, mean: Volume of material removed divided by number of chips

Chip width: Width of the uncut chip

Chips: Pieces of material or swarf cut from workpiece

Chuck: A type of workholding device for cylindrical parts

CIRP: International College/Institution of Production Engineering Research

Cleaning up: Removing a layer of abrasive to remove loading and restore unworn surface

Climb grinding: The grinding motion and the workpiece motion are in same direction

CMC: Ceramic matrix composite

CMM[1]: Chemomechanical machining

CMM[2]: Coordinate measuring machine

CMP: Chemomechanical polishing

CNC: Computer numerical control of a machining system

Coated abrasive: Abrasive applied as a coating to a belt

Compatibility: Suitability of two materials to form rubbing couple without severe damage

Compliance: Inverse of stiffness. Movement per unit force

Composite: Body formed from mixture of two or more materials

Coarse dressing: Dressing with large dressing depth and large dressing feed rate

Concentration: Proportion of a material in a mixture or in a solution. Measure of abrasive in a wheel

Conditioning: Process to prepare an abrasive surface for machining. See also dressing

Conduction: Transfer of energy through a body. For example, heat conduction

Conformal contact: Convex surface contacting within a concave surface

Constant force process: Abrasive process controlled by application of constant force

Contact: Touching between one body and another

Contact angle: Half-angle subtended by contact arc at the wheel center

Contact area: Apparent area of contact between abrasive tool and workpiece

Contact area, real: Sum of grain contact areas with workpiece

Contact, grain: Contact between grain and workpiece

Contact, wheel: Contact between wheel and workpiece

Contact length: Length of tool and workpiece contact parallel to grinding direction

Contact length, force: Contact length due to effect of force and deformation

Contact length, geometric: Contact length predicted ignoring roughness, speeds, and deformation

Contact length, kinematic: Geometric contact length modified for speed of wheel and workpiece

Contact length ratio: Ratio of real contact length/geometric contact length

Contact length, real: Contact length including all influences. Also effective contact length

Contact mechanics: Analysis of contact, particular due to elastic/plastic deflections

Contact pressure: Normal force divided by contact area. See real and apparent contact pressure

Contact radius: Mean radius of the wear flat on an abrasive grain

Contact surface: Surface in (apparent) contact area during abrasive machining

Continuous dressing: Process of dressing concurrent with grinding

Contour grinding: Profile grinding by generation of the required contour

Controlled feed process: Abrasive process controlled by application of feed motion

Control wheel: Wheel used to control workpiece motion in centerless grinding

Control wheelhead: Powered assembly to carry and rotate control wheel

Convection: Process of carrying by physical movement. For example, heat carried by motion of fluid

Convection coefficient: Convective heat transfer per unit contact area per degree temperature difference

Convenient waviness: Nonround shapes that are not corrected in incorrect centerless grinding

Conventional: In accordance with traditional or widest practice

Conventional speed: Wheel speeds from 20 to 45 m/s

Coolant: Grinding fluid

Corundum: Traditional name for aluminum oxide mineral

Corrosion: Conversion of work material surface by chemical or electrochemical action

Corrosion wear: Removal of material from a surface by corrosion

Cost: Price to be paid

Couette flow: Flow induced by proximity to a sliding surface. Entrained flow

Coulomb friction: Friction force is proportional to normal force

Covalent: Electrons are shared by neighboring atoms

Crankshaft grinder: Machine for grinding engine crankshafts

Creep grinding: Grinding at very low workspeeds and usually with large depth of cut

Criterion: Standard for making a judgement. Basis for making a decision

Critical: Point at which an abrupt change takes place. For example, critical temperature

Critical temperature: Temperature for a change of physical condition. For example, softening

Cross-axis dressing: Rotary dressing tool with axis at 90° to the grinding wheel axis

Cross section: View of section of body cut through to reveal internal structure

Cross-sectional area: Area of a cross section

Crushing: Removing a layer of abrasive by applying pressure with a block of softer material

Crushing roll: Roller or disc used for crush dressing

Crystal: Solid of regular atomic structure, such as quartz

Crystalline: Having an ordered atomic lattice structure

Cup dresser: Cup-shaped rotary dressing tool

Curvature: Inverse of radius

Custom: Designed to suit a particular customer's requirements

Cut, cutting: Process of shearing a material. Abbreviation for depth of cut

Cutoff grinding: Cutting through material with a thin wheel to remove a slice or slab

Cutting edge: That part of an abrasive particle that engages the workpiece

Cutting-edge depth: Depth of penetration of the grains into the workpiece

Cutting-edge density: Number of cutting edges per unit area. Varies with depth

Cutting-edge spacing: Measure of spacing between cutting edges on an abrasive surface
Cutting-edge width: Measure of the width of the cutting edge in contact with the workpiece
Cutting force: Resultant force in cutting process. Vector sum of component forces
Cutting speed: Wheel surface speed
CVD: Chemical vapor deposition process used to form diamond layer
Cycle time: Mean time for the machining of a part
Cylindricity: Measure of deviations from a cylindrical shape
Cylindrical grinding: Grinding a cylindrical surface by rotating a workpiece

D
d.o.c.: Depth of cut
D'Alembert force: The reaction experienced when accelerating a mass
Damping force: Reaction force proportional to speed
DBDS: Dibenzyl disulfide
Dead center: Nonrotating work location pin
Debris: Small particles of abrasive and workpiece. For example, Swarf
Deep grinding: Grinding depths of cut much in excess of 0.1 mm
Deflection: Movement due to pressure, force, or temperature
Delamination: Plate-shaped particles breaking out of surface
Density: Mass per unit volume
Depth of cut: Instantaneous normal thickness of layer of material to be removed
Detergent: Cleansing agent. Additive to oil for cleansing surfaces
Diamond tool: Cutting tool using diamond cutting edges
Dicing: Cutting into slices
Disc or disk grinding: Grinding across the face of a disc-shaped wheel
Diffusion: Migration of energy or atoms through a material
Diffusivity: Thermal property: Conductivity divided by density and specific heat
DIP slide: Used for measurement of bacterial concentration
Direct stresses: Tensile or compressive stresses
Disc dresser: Disc-shaped dressing tool, usually rotary
Dislocation: Vacancy in a structure. For example, within structure of atoms
Dispersing additive: Additive to keep solid particles in suspension in the fluid
Disturbance: Change to input value which causes system output to change
DN value: Diameter times speed in revolutions per minute
Dog drive: An eccentric pin used to drive a rotating cylindrical workpiece
Double-side grinding: Process for simultaneously grinding two sides of a workpiece. Duplex grinding
Down grinding or down cut: The wheel motion and workpiece motion are in the same direction
Dressing: Process to prepare a wheel surface for machining. See also truing and conditioning
Dressing, continuous: Process of dressing at the same time as grinding
Dressing depth: Depth of cut in dressing operation
Dressing feed rate: Traverse/feed rate of dressing tool in dressing operation
Dresser head: Powered assembly to carry and drive rotary dressing tool
Dressing height: Offset height of dressing tool relative to grinding wheel axis
Dressing increment: Dressing depth
Dressing lead: Traverse/feed distance of dressing tool per wheel revolution
Dressing plate: Plate dressing tool with diamond grit in the tool cutting surface

Dressing, roll: Friction or power driven roll-shaped dressing tool
Dressing sharpness ratio: Dressing depth/dressing width of wedge- or cone-shaped tool
Dressing speed: Peripheral speed of a dressing roll
Dressing stick: A stick-shaped abrasive tool used to dress a wheel
Dressing tool: Tool used for dressing. For example, diamond, fliese, or rotary dressing tool
Dressing tool life: Tool life of the dressing tool
Dry machining: Machining without use of a process fluid
Ductile: Tendency to deform by shear without cracking fracture
Dull: Blunt. Opposite of sharp
Dynamics: Analysis of motions including accelerations due to forces
Dynamic balancing: Adding or removing weight in two planes to improve couple balance.
Dynamic stiffness: Usually, stiffness at a particular frequency or less usually, at resonance

E
E: Young's modulus
Eccentricity: Displacement of a center of rotation relative to another
Effective porosity ratio: Pore volume in active layer divided by total layer volume
Elastic deformation: Recoverable linear deformation under load
Elastic modulus: Material property. Rate of increase of tensile stress with strain in tensile test
Electrolysis: Material transfer from an electrode due to electric current and an ionized electrolyte
Electrolyte: Electrically conducting liquid used in electrolysis
Electroplated abrasive: Abrasive grains attached to tool by electroplated metal
Element: A part of a system
ECG: Electrochemical grinding
ECM: Electrochemical machining
EDM: Electrical discharge machining
EHL: Elastohydrodynamic lubrication
Eigenfrequency: A particular frequency at which a machine resonates
ELID: Electrolytic in-process dressing. Process for metal-bonded wheels
EP Wheel: Electroplated wheel
EPHL: Elastoplasto hydrodynamic lubrication
Emery: Black abrasive based on corundum with magnetite or hematite
Empirical: Deriving from a limited range of measurements. Not based on physics
Emulsion: Grinding fluid consisting of a well-dispersed suspension of oil in water
Emulsion stabilizer: Surfactant added to resist separating out of the dispersed phase
Encoder: Position measuring device based on reading pulses from a rotary or linear scale
Energy: Capacity to do work. Measure of work expended
Energy dissipation: Transformation of work energy into heat
Energy partition: Analysis of energy dissipation to particular heat sinks
Entrained flow: Fluid flow induced by parallel sliding of surface. Couette flow
Environment: Conditions in surround. For example, atmosphere, noise, and temperature
E.P.: Extreme pressure conditions as in extreme pressure lubrication
EP additive: Extreme pressure additive for process fluid
Erosion: Wear by series of small impacts from gas, liquid, or solid particles
Equivalent diameter: Grinding wheel diameter modified to allow for workpiece diameter
Equivalent chip thickness: Thickness of the layer of chips emerging from grinding action
Error: Difference between measured value and ideal value. For example, size error

Error function: An integral function used in heat transfer calculations

Esters: Compounds produced by acid–alcohol reactions with elimination of water

Exoemission: Radiation of photons or electrons

External grinding: Grinding an external surface particularly for cylindrical grinding

Extreme pressure (EP): Extreme pressures as in extreme pressure lubrication

F

Face grinding: Grinding a flat surface using the side face of a grinding wheel

Fatigue life: Expected life under a cyclic loading condition

FCC: Face-centered cubic lattice structure

Feed: An increment of grinding wheel position relative to machined surface, either tangential or normal

Feed cycle: A series of feed movements and retractions required to complete a grinding cycle

Feed per cutting edge: Distance moved by workpiece in interval between succeeding cutting edges

Feed rate: Speed of grinding wheel movement normal or tangential to machined surface

Ferrous material: Material containing mainly iron

Ferrite: Ductile alpha phase of low-carbon iron

Firing temperature: Temperature of vitrification for a vitrified wheel

Fine dressing: Dressing with small dressing depth and small dressing feed rate

Finish surface: Workpiece surface after grinding

Fixture: Fixed (holding) device

FEM: Finite element modeling

Flash point: Lowest temperature at which vapor above a liquid may be ignited in air

Flash temperature: Peak temperature at an individual cutting edge/grain

Flatness: Measure of deviations from a flat plane

Fliese: Diamond-coated wedge-shape tool used for dressing

Flood nozzle: Nozzle delivering large volume of fluid at low velocity

Flow utilization: Useful flow divided by total flow

Fluid: Liquid or gas that can flow

Flushing: Displacing swarf by the action of a fluid jet

Flux[1]: Flow

Flux[2] heat: Rate of heat flow. For example, heat per unit area per second

Flux distribution: Distribution of flux in contact zone. For example, triangular distribution

FM additive: Friction modifier additive for process fluids

Foam depressant: Additive to promote bubble coalescence

Fog: Mist

Forced vibration: Vibration of constant amplitude due to application of a harmonic force

Force loop: The interacting machine elements and workpiece that resist the grinding force

Form or profile: Shape of a section that is not a straight line. Shape that is not a flat surface

Form dressing: Dressing a form profile on a wheel with a form tool or by form generation

Form grinding: Grinding a form profile with a form tool or by form generation. See generation

Foundation: Floor or structure within floor on which a machine is mounted

Fracture toughness: Measure of resistance to fracture under impact loading

Free radical: Reactive atom or molecule containing an unpaired electron

Free abrasive: Loose abrasive grains used in free abrasive processes

Free surface: Surface open to the environment

Friable: Tending to fracture under compression
Friction: Resistance to slip between two sliding surfaces
Friction coefficient: Ratio of tangential force to normal force between two sliding bodies
Friction couple: The pair of interacting body materials in abrasive contact
Friction pair: Friction couple
Friction polymer film: Polymer film on work surface formed by friction process
Friction power: Power required to overcome frictional drag
Friction ratio: Ratio of interface shear stress/shear flow stress of the softer bulk material

G
Garnet: Crystalline silicate abrasive
Gauging: Use of sensors as for in-process diameter measurement
Gaussian distribution: Normal distribution
Gear grinding: Grinding a gear surface
Generation: Producing a shape by compound motions
Geometrics: Analysis of points, lengths, lines, curves, and shapes
Gib: Tapered strip used to adjust clearance in a slideway assembly
Glassy bond: Noncrystalline vitreous bond
Glazing: Condition of large wear flats on the abrasive grains
Gleichdicke shape: Equal diameter shapes that deviate from round
GMC: Glass matrix composite
Grade, grade letter: System used to classify hardness of abrasive layer
Grain boundary: Boundary of grains within a granular structure such as soft steels
Grain, grit: Abrasive particle
Grain contact: Touching between a grain and the workpiece
Grain contact time[1]: Time a grain is in contact with workpiece
Grain contact time[2]: Time a point on workpiece is in contact with a grain
Grain force: Resultant force on an abrasive grain
Grain depth: Depth of penetration of a grain into the workpiece
Grain penetration: Depth of penetration of a grain into the workpiece
Grain protrusion: Measure of height of grain tips above surrounding bond
Grain size: Measure of the sizes of abrasive grains
Grating: Series of parallel wires or lines
G **ratio:** Volume of material removed divided by volume of grinding wheel removed
Grindability: Qualitative or other measure for a material of ease of grinding
Grinder: Shop-floor term for grinding machine or operative
Grinding: Removal of material from a surface by abrading with a hard rough surface
Grind hardening: Process proposed for hardening steels from soft state by grinding at high temperatures
Grinding fluid: Fluid used to lubricate, cool and flush in abrasive processes. See also process fluid
Grinding force: Resultant force in grinding. Vector sum of component forces
Grinding force ratio: Ratio of tangential force to normal force
Grinding power: Product of tangential force and tangential grinding wheel speed
Grinding ratio: Workpiece volume removed by grinding divided by wheel wear volume
Grinding time: Part of time spent in grinding
Grinding wheel: Cylindrical abrasive tool rotated at high speed in machining
Grit: Small hard particle of abrasive. Also grain

H

Hardness: Measure of resistance to penetration

Hardness, hot: Hardness of a material at elevated temperature

Harmonic: Integer multiple of basic frequency

Headstock: Machine structure that carries a spindle drive and/or center usually for the work drive

Heat: Thermal form of energy

Heat conduction: Transfer of heat through a body due to temperature gradient

Heat sink: Where heat goes: Heat flows from a hot source to a cool sink

HEDG: High efficiency deep grinding

Hertz: Unit of frequency: Cycles per second

Hertzian: Smooth elastic contact between a sphere and a plane or two spheres

High speed: Wheel speeds in excess of 45 m/s

HLB: Hydrophile–lipophile balance

Homogeneous, homogenous: Constant material composition throughout

Honing: A cylindrical process using abrasive stones

Honing stones: Abrasive blocks inserted in honing tools

Horizontal grinder: Grinding wheel axis is horizontal

Horsepower: Old unit of power. 746 W

HPSN: Hot-pressed silicon nitride

HSS: High-speed steels used for cutting tools

h **value:** Heat convection value for oil under a standardized condition

Hv, HV: Vickers hardness measure

Hybrid: Combining two materials, for example, ceramic/steel or modes, for example, hydrostatic and hydrodynamic

Hydrodynamic[1]: Usually refers to action of fluid due to surface movements

Hydrodynamic[2]: Strict meaning is action due to fluid motion

Hydrogenation: Hydrogen reaction used typically to saturate and stabilize a fatty oil

Hydrostatic bearing: Load support arranged by pressure-controlled liquid bearing

Hydrostatic stress: Contribution to stress system for equal direct stresses, usually compressive

I

ID: Internal diameter

Image processing: Techniques for extracting information from surface data

Impregnated truer: Dressing tool with small diamonds held in a metal matrix

Inclined heat source: Heat source moves in a plane inclined to the workpiece surface

Infeed: Feed of the grinding wheel normal to the ground surface

Internal grinding: Grinding an internal surface particularly bore grinding

Interrupted cut: The wheel repeatedly engages and disengages contact with the workpiece

Ion: Electrically charged atom or group of atoms

Ionic: Having electrically charged atoms or groups of atoms

ISO: International Standards Organisation

Isotropic: Material with properties constant in all directions

J

Jet: High-velocity fluid stream. Orifice for high-velocity fluid

Journal: Rotational shaft as in a journal bearing

Junction growth: Growth of an area of sticking contact

Jewellers rouge: Red iron oxide powder used for polishing

K

K₁c: Measure of fracture toughness
Kaolin: Fine white clay made into a paste for polishing
Kinematics: Analysis of motions, ignoring forces
Kinematic similarity: Uncut chip dimensions remain unchanged
Kinetics: Effects or study of rates of action
Kramer effect: Exoemission under abrasive conditions

L

LDA: Laser–Doppler anemometry.
Laminar: Fluid tending to move along steady parallel paths. Nonturbulent
Lap: An abrasive tool used for lapping
Lapping: Process of improving form using a lap and abrasive paste/fluid
Lattice: Structure of directional bonding of atoms
Leadscrew: Nut and screw device for transforming rotary motion into linear motion
Limit: Maximum or minimum permissible value
Limit chart: Chart showing maximum achievable values of process control variables
Linear motor: Usual meaning: An electric motor that produces linear motion directly
Liquid: Intermediate phase on cooling between gas and solid. For example, water or oil
Live center: Rotating work location center pin
Log, diamond: Synthetic diamond coated on a prismatic log shape
Longitudinal grinding: Grinding with a wheel traversing the workpiece length
Losses: Difference between input and useful output. For example, power losses
Lower bound: Estimate known to be lower than real value
Lubricant: Medium such as oil or graphite which eases sliding between surfaces
Lubricity: Ability of a fluid to reduce friction other than by its viscosity

M

Machinability: Qualitative or other measure for a material of machining ease
Machine tool: A powered machine or system used in part production
Machining: Production of shape by removal of material from a part
Machining center: A multi-tool-head machine, for example, milling, drilling turning, and grinding
Machining conditions: Process conditions. Values of parameters employed for machining
Machining-elasticity parameter: Ratio of real depth of cut to set depth of cut
Magma plasma: Theory of material in energetic state due to intense abrasive deformation
Magnetic abrasive machining: Process where magnetism applies force on the abrasive
Malleable: Capable of large plastic deformations by pressing and hammering
Martensite: Hard brittle phase of carbon dissolved in iron produced by quenching. White phase
Material removal: Volume of material removed from workpiece
Metal-bond wheels: Superabrasive wheels bonded with cast iron or other metal compositions
Mean: Arithmetic average of a series of readings
Mesh number: Measure of grit size based on number of wires in a sieve. High mesh number yields small grit size
MIC value: Minimum inhibitory concentration of fluid
Microscopy: Magnified visualization of a surface using one of several physical principles
Micro-hardness: Hardness measured with an extremely small indenter

Mineral oil: Natural hydrocarbon oils
Misalignment: Deviation from parallelism between two elements
Mist: An aerosol dispersion of particles/fluid in the atmosphere
Mixed lubrication: Transition between hydrodynamic lubrication and boundary lubrication
MMC: Metal matrix composite
MNIR: Maximum normal infeed rate
Model: Mathematical, physical, or conceptual representation of a structure or process
Momentum power: Rate of kinetic energy
Monocrystal: Grains constitute a single crystal
Morphology: Shape, form, and structure of a body
MQL: Minimum quantity lubrication

N
Nano-: Refers to the nanometer order of magnitude. Less than 0.02 microns.
Nanoadditive: Nanosize powders used as additives
Nanogrinding: Grinding with nanosize grain penetration
Nanometer, nm: 1 m divided by 10 raised to the power 9
Nanotechnology: Technology involving machines and processes at the scale of a few nanometers
Napthenes: Cycloparaffinic hydrocarbon oils
Natural fatty oil: Animal, vegetable, or fish oil
ND: Natural diamond
Neat oil: Oil not mixed with water. Immiscible oil
Nip: Convergent gap at entry to grinding contact zone that creates coolant wedge
Nodular iron: Iron containing spherical graphite particles
Noise: Unwanted vibration
No-load power: Power with grinding wheel rotating but not grinding
Normal: 1. Perpendicular. 2. Usual
Normal distribution: Continuous random distribution with same mean, median, mode. For example, Bell curve
Normal force: Component of force perpendicular to grinding wheel surface
Nozzle: The end of a pipe or hose shaped to direct grinding fluid
Nozzle, jet: Nozzle shaped to intensify exit velocity of the grinding fluid
Nozzle, shoe: Fluid delivery nozzle shaped to fit snugly around a wheel
Nyquist criterion: A statement of a necessary condition for stability
Nyquist plot: A graph of amplitude and phase with frequency for an open-loop transfer function

O
OD: Outside diameter
OEM: Original equipment manufacturer
OOR: Out of roundness. Roundness error
Orbital grinding: The grinding wheel orbits in addition to rotating
Organic[1]: Having biological origins
Organic[2]: Based on or related to carbon compounds
Operator, operative: Person who operates a machine
Oscillating: Motion to and fro, usually for short distances
Overlap ratio: Contact width of dressing tool/feed per revolution of wheel

P

Padding: Solids added to bond mixture to modify the effective hardness of an abrasive layer

Paraffinic oil: Linear or ramified saturated hydrocarbons

Particle: Small part, fragment, grit, or grain

Partition ratio: Energy to a heat sink divided by total heat energy

Passivation: Slowing corrosion due to inhibitor or protective layer

Passivator: Additive to prevent catalytic reaction

Paste: Thick mixture of abrasive particles in liquid or wax

PCBN, PcBN: Polycrystalline cubic boron nitride

PCD: Polycrystalline diamond

Pearlite: Eutectoid phase of carbon in iron

Peclet Number: Dimensionless speed parameter for moving heat sources

Peel grinding: Deep cylindrical traverse grinding with high workspeeds and wheelspeeds

Pendulum grinding: Shop-floor term for forward and back traverse grinding. Reciprocating grinding

Peripheral grinding: Grinding with the periphery of the wheel. C.f. face grinding

Permeability: Measure of the ability of fluid to diffuse through a material

pH: Measure of acidity/alkalinity. Acids have pH less than 7

pH meter: Alkalinity meter

Physisorption: Layers bonded by Van der Waals weak forces

Pickup: 1. Scuffing. 2. A sensor

Pitch error: Error in spacing of gear teeth at the pitch circle diameter

Plain: Smooth, lacking in features, nonrecessed

Plane: A flat section of space

Plasma: Hot ionized gas

Plastic deformation: Nonrecoverable deformation by ductile shear within a material

Plateau honing: Removing the peaks of the machining marks to form plateaux

Plated wheel: A grinding wheel having a single layer of abrasive bonded to the core by plating

Ploughing, plowing: Grooving a surface without loss of material

Ploughing energy: Cutting energy less chip energy and sliding energy

Plunge dressing: The dressing tool is fed directly into the grinding wheel

Plunge grinding: The grinding wheel is fed directly into the workpiece

PMC: Polymer matrix composite

Pneumatic: Using air as the power source

Poisson distribution: A discrete distribution. Alternative to binomial distribution

Poiseuille flow: Flow induced by a pressure gradient

Polishing: Use of a conformable pad and an abrasive to smooth a surface

Polycrystalline: Many small closely packed crystals form the grains

Pore: A small hole or channel in a structure. Space between grains in a wheel

Pore loading: Abrasive layer with work material loaded into pores

Porosity: Measure of air or pores in a structure

Potential energy: Energy having the potential to do work. For example, pressure or height energy

Potential function: Model of variation of potential energy. For example, between atoms

Preston equation: Alternative source of Archard equation

Primary shear zone: Zone of primary shear between workpiece and chip

Process: Action or sequence of actions or operations

Process conditions: Specification of speeds, feeds, tools, fluid, and all conditions of the process

Process fluid: Liquid or gas used to lubricate, flush, and cool an abrasive process

Profile: See form

Profile grinding: Grinding a form on a workpiece with a form tool or by generation

Process limits: The maximum permissable values of operating speeds for a process such as grinding

Process monitoring: Continuous measurement of any characteristic such as size or power

Profilometry: Techniques for measuring shape or surface texture

P-type corrosion: Corrosion with passivated progression

Pumice stone: Porous volcanic abrasive used for scouring and polishing

Pumping power: Power required to deliver a flow from a pump

Punch grinding: Grinding of punches

PVD: Physical vapor deposition

Q

Quadrature: At 90° orientation

Qualification: Measurement/proof of quality

Quality: Term used for attributes such as fitness for purpose, roughness, and durability

Quartz: Hard crystalline silicon dioxide colorless rock

Quasi-: Almost but not exactly the real situation

Quill: Internal grinding spindle and wheel mount

R

R_a, R_z, R_t: 2D measures of surface roughness

Rake angle: Rake angle is between workpiece normal and leading cutting-tool face

Range: Difference between minimum and maximum values. For example, cutting-edge depth range

Rare earth element: Elements of the lanthanide series

Real contact pressure: Normal force divided by real contact area

Real depth of cut: Actual depth of cut taking account of wheel wear and deflections

Recess: Well or pool within a bearing land to spread fluid pressure

Reciprocal grinding: Successive forward and back traverse grinding

Redundant energy: Energy consumed in nonuseful deformation

Redress life: Machining time between dress and redress of a wheel

Rehardening: Heating of hardened surface to transformation followed by quenching

Rehbinder effect: Strength reduction due to adsorption of fluid molecules onto material surface

Removal rate: Volume rate of material removal from a workpiece

Repeatability: Range for series of repeated measurements. For example, sizes or positions

Replica, surface: A molded impression of a surface

Residual stress: Stress that remains in a material after load is removed

Resinoid wheel: Wheel having abrasive grains in resin-based bond

Resonance: Large vibrations experienced at a resonant frequency

Retrofit: Equipment added subsequently after a machine has been supplied.

Reynolds number, Re: Dimensionless parameter increases with ratio of inertia forces to viscous forces

Ringing: Acoustic test for flaws in a wheel from sound of a sharp tap

RMS, rms: Square root of the mean of the sum of the squares of the differences from the mean

Rolling: Motion without sliding, as of roller on plane

Rolling bearing: Load support on ball or roller elements

Roll dresser: Dressing tool in form of roller or disk

Roll grinding: Grinding of large rolls

Roots: Conditions for a maximum amplitude of a dynamic system

Roughness factor: Ratio of contact lengths due to normal force for rough and smooth contact

Rounding: Process of improving roundness

Roundness, error: Measure of deviations from a circle. Out of roundness

Roundway: Round guideway

Roughness: Measure of microdeviations in height of a surface

Rubbing: Sliding of one surface on another with frictional contact

Rules of mixture: Rules for properties of a mixture related to element properties

Run out: Error of circular motion

S

Sanding: Sandpaper smoothing of wood or similar material

SAE52100: A bearing steel specification

Scratch: A groove made by dragging a sharp hard tool along a surface

Screw grinding: Grinding a screw surface by rotation and axial feed of a workpiece

Scuffing: Welded workpiece material on tool pulls material out of the workpiece surface

SD: Synthetic diamond

SEM: Scanning electron microscopy

Secondary shear zone: Zone of shear between chip and cutting face of tool

Seeded gel, SG: A tough alumina abrasive produced from alpha-phase chemically precipitated crystals

Segmented wheel: Wheel with a number of pieces or segments of abrasive layer attached to a holder

Seizure: General meaning is the welding together of two rubbing elements

Self-dressing: Self-sharpening. Grinding process wear that provides new sharp cutting edges

Self-excited vibration: Vibration that builds up when forced

Self-sharpening: Tendency of grains to produce sharp edges when worn edges fracture

Sensor: Measuring device or transducer used to sense a process variable

Servomotor: Variable speed motor with feedback control for speed or position

Set depth of cut: Depth of cut ignoring wheel wear and deflection errors

Setup: Geometry, speeds, and conditions for grinding

Shallow grinding: Grinding depths of cut less than 0.1 mm

Sharp cutting edge: Pointed cutting edge

Sharpness: Measure of angularity or narrowness of cutting tip

Shear: Plastic sliding within a material acting like a pack of cards

Shear strain rate: Rate of plastic sliding/thickness of sheared zone

Shellac abrasive: Abrasive grains held together in a shellac resin bond

Shoe centerless grinding: Blocks are used for radial location of the workpiece

Shoe nozzle: Nozzle that fits around the wheel periphery

Side grinding: Grinding with the side of the wheel. Face grinding

SI: Systeme International d'Unités. International system of units

Siccative, siccativator: Drying agent

Silica, SiO$_2$: Silicon dioxide. Occurs naturally as quartz

Silicon carbide, SiC: A hard bluish-black abrasive

Simulation: Step-by-step imitation of a process over a short-time period

Single-point dressing: Dressing with a single-point tool such as a single diamond

Sintering: Process of forming solid by partial fusing of compacted powder

Size effect: Tendency for reduced specific energy with larger chip size

Slide, slideway: Machine element for guiding sliding motion

Slideway grinder: Long-bed machine for grinding slideways

Sliding: Relative movement of two surfaces in tangential direction

Sliding energy: Component of energy proportional to sliding area of grains

Sliding heat source: Heat source that moves parallel to the workpiece surface

Slip line field: Lines of maximum shear stress in a plastic field

Slot nozzle: Nozzle having a large gap width-to-thickness ratio

Slurry: Dense suspension of powder in a liquid

Soap[1]: Product of hydrated oxides and fatty acids

Soap[2]: Metallic salt of a fatty acid

Sol gel: An alumina abrasive produced from composite-phase chemical precipitation

Solid lubricant: Lubricant in the form of a lamellar solid such as graphite

Solution: Mixture formed of material dissolved in a liquid or solid

Spacing length: Cutting-edge spacing in direction of grinding

Spacing width: Cutting-edge spacing in lateral direction

Spark out: Period of dwell while depth of cut decreases

SPC: Statistical process control methods

Specific: Particular form of a parameter. For example, Sp. Force = force/unit width

Specific energy: Energy per unit volume of material removed

Specific energy in chips: Energy per unit volume of material removed carried within chips

Specific energy in fluid: Energy per unit volume of material removed carried within fluid

Specific energy in wheel: Energy per unit volume of material removed carried within wheel

Specific energy in workpiece: Energy per unit volume of material removed carried within workpiece

Specific force: Force per unit width of grinding wheel contact with workpiece

Specific heat capacity: Specific heat: Heat per degree temperature rise

Specific power[1]: Power per unit width of grinding wheel contact with workpiece

Specific power[2]: Power per unit area of contact

Specific power[3]: Power per unit volume removed

Specific removal rate: Removal rate per unit width of contact

Specific wear rate: Wear rate per unit width

Specific wear resistance: Inverse of specific wear rate

Speed ratio: Ratio of surface speed of wheel to surface speed of workpiece

Speed-stroke grinding: Grinding with high work speeds of the order of 1 m/s

Spike temperature: Temperature at a grain contact. Also flash temperature

Spindle power: Power required to drive the main wheel spindle

Spinel: Any of a group of hard glassy minerals

Spinner: A workpiece that accelerates and spins out of control in centerless grinding

Spray: Application of jet or shower of particles, usually liquid

SSD: Single synthetic diamond

Stability limit: The condition when a small disturbance neither builds up nor decays

Stainless steels: A group of corrosion steels containing a high chromium content

Static deflection: Movement due to a steady applied force

Statics: Analysis with steadily applied forces. C.f. dynamics

Stepper motor: Motor that controls position in small steps

Stick dressing, sticking: Use of a "soft" abrasive stick to open/clean a superabrasive wheel surface

Sticking friction: Friction due to tangential shearing of a material

Stiffness: Resistance to movement: Rate of force divided by movement

Stochastic: Acts randomly according to some statistical distribution

Stock removal: Normal, radial, or diametral reduction of workpiece dimension

Stone: 1. A gem. 2. An abrasive tool

Stress: Local ratio of force per unit area. For example, shear stress, tensile stress.

Stribeck curve: Graph of friction coefficient against Sommerfeld/Hersey number.

Structure number: Number used to indicate proportions of grit and bond volumes in an abrasive layer

Stylus measurement: Profilometry using a stylus contact to sense shape or texture deviations

Surface finish: Nonscientific term used to describe surface roughness

Surface grinding: Peripheral grinding of a flat surface or a profiled surface. C.f. face grinding

Surface texture: Measure or nature of surface topography. See roughness

Superabrasive: Extra-hard abrasive grit such as diamond or cubic boron nitride

Superfinishing: Process for production of very low roughness

Suspension: Dispersed particles in a liquid medium

Swarf: Material debris and chips machined from workpiece

Synthetic oil: Oil produced chemically. For example, silicone oils

Synthetic emulsion: Emulsion of synthetic oil in water

System: Process that transforms inputs to outputs

T

Table: Machine element on which a workpiece or work fixture is mounted

Tailstock: Machine subassembly to support a rotating workpiece at opposite end to headstock

Talc: Fine magnesium silicate powder used for polishing

Tangential force: Component of force tangential to grinding wheel surface

TDA: Thermographic differential analysis

Temper: Diffusion process leading to softening and temper colors

Temper colors: Colors produced on a surface by oxidation at high temperatures

Temperature gradient: Rate of increase or reduction in temperature with increasing distance

Thermal boundary layer: A thermally affected layer of fluid near a surface

Thermal conductivity: Heat transmitted per degree per unit length of transmission

Thermal stress: Stress in a structure arising due to change of temperature

Thermocouple: Temperature measurement device based on junction of dissimilar metals

Thermal stability (TS): Temperature above which fluid chemical composition breaks down

Thread grinding: Screw grinding of screw thread forms

Threshold grinding force: Minimum force to achieve chip removal

Through-feed grinding: The workpiece is fed between centerless grinding wheel and control wheel

TiC: Titanium carbide

Tilt: Deviation from vertical/horizontal plane. For example, Control wheelhead tilt

Tolerance: Defined as difference between maximum and minimum permissible limits

Topography: Description of surface shape at macro or microlevel

Tool: A cutting part or implement

Tool life: Machining time between dress and redress of tool

Tool steel: A group of steels used for manufacturing cutting tools

Touch dressing: Dressing process using 1–10 μm dressing depth

Transfer function: Ratio of output to input for a system usually in Laplace form

Transitional flow: The domain between fully laminar and fully turbulent flow

Traverse: Linear movement. Fast movement on a slideway

Traverse grinding: Grinding with the wheel traversing the workpiece length

Trial: Series of grinding operations to evaluate a set of grinding conditions

Triangular heat flux: Band heat source: Flux varies linearly from maximum to zero

Tribo-, tribocontact: Relating to conditions of abrasive deformation and surface generation in sliding contacts

Tribocatalytic: Catalytic effects due to tribocontact conditions

Tribochemistry: Tribological phenomena particularly related to chemical action

Tribochemical wear: Chemical wear accelerated by abrasive action

Tribology: The science of friction, lubrication, and wear of sliding contacts

Tribomechanical: Quasi-tribophysical. Mechanical aspects of tribology

Tribometer: Machine to test friction and rate of wear for particular speed, force, and contact condition

Tribooxidation: Oxidation mechanism under tribocontact conditions

Triboplasma: Material in energetic state due to intensive tribocontact

Triboreaction: Reaction under tribomechanical activation

Tribosimulation: Testing on a rig replicating the real tribological contact conditions

Tribosorption: Assimilation of a material on a surface under tribomechanical action

Tripoli: Soft powder of siliceous rock (originally from Tripoli, Libya)

Truing: Process to produce accurate form required on an abrasive tool

Turbulent flow: Moving forward with agitated sideways buffeting motion

Turret: Indexable tool head

U

Ultrasonic machining: Machining by ultrasonic vibration of a tool against a workpiece

Unbalance: Eccentric mass on rotating disk causing a harmonic force

Uncut chip thickness: A notional thickness of the chips based on grain kinematics

Uniform heat flux: Band heat source of constant flux magnitude

Universal machine: Machine capable of internal and external grinding

Upgrinding or upcut: The wheel motion and workpiece motion are in opposite directions

UPM: Ultraprecision machining

Upper bound: Estimate known to be higher than the real value

Useful flowrate: Flowrate of grinding fluid that passes into the contact zone

V

Vertical grinder: Grinding wheel axis is vertical

VI improver: Viscosity index additive used to reduce temperature dependence

Vickers hardness: Hardness measured using standard diamond shape indentor

Vienna lime: Powder of calcium and magnesium oxides used for polishing

VIPER process: A high-rate process developed by Rolls-Royce (vitrified improved performance)

Viscous friction: Friction force is proportional to sliding speed

Vitreous, vitrified: Glassy phase produced by heating

Vitrification: Conversion into a glassy form by heating

Vitrified wheel: Wheel having abrasive grains held together by vitreous bond

W

Waviness: Surface shape errors having a pattern of undulation

Way: Slideway or rolling guideway

Wear: Loss of material from a body by abrasive rubbing and sliding processes

Wear, attritious: Slow wear evidenced by a polished surface

Wear flat: Plane worn area on the tip of an abrasive grain formed by rubbing

Wear, fracture: Removal of parts of a body by cracking and breakage

Wear, microfracture: Removal of very small parts of a body by small localized cracks

Wear rate[1]: Wear volume per unit time

Wear rate[2]: Wear depth per unit time

Wear ratio: Inverse of grinding ratio

Workpiece: The part or body to be machined

Wheel: Disc that rotates on a spindle. Abbreviation for grinding wheel

Wheelhead: Powered spindle assembly and headstock to carry and drive grinding wheel

Wheel flange: Plate used to clamp wheel on hub

Wheel hub: Adaptor, arbor used to mount wheel on a spindle

Wheel loading: Tendency for softened workpiece material to adhere to the wheel surface

Wheel-regenerative chatter: Self-excited vibration caused by build up of waves on the wheel

Wheel speed: Usual: tangential surface speed of wheel. Sometimes: rotation speed

Wheel-work partition: Division of energy at workpiece and grain contacts

White layer: White layer of workpiece at surface due to mechanical or thermal effects

Width of cut: Width of grinding wheel contact with workpiece

Workblade, workplate, workrest: Plate to support workpiece in centerless grinding

Workblade angle: Angle of top surface of workblade

Work hardening: Tendency for shear stress to increase with plastic shear strain

Workhead: Powered spindle assembly and headstock to locate and drive workpiece

Work height: Height of the workpiece axis above the grinding wheel axis in centerless grinding

Work material: Workpiece material

Work partition ratio: Proportion of heat conducted into workpiece in contact zone

Workpiece: The part to be machined

Work regenerative chatter: Self-excited vibration caused by build up of waves on the wokpiece

Workrest: Plate used to support workpiece in centerless grinding. Also workplate or workblade

Workspeed: Tangential speed of workpiece parallel to wheel speed
Wurtzite: A form of crystal structure as in wurtzitic diamond or boron nitride

X
XRD: X-ray diffraction technique used for measuring residual stresses

Z
ZDTP: Zinc dialkyl dithiophosphates. AW additive for lubricating and process fluids
ZrO$_2$: Zirconium dioxide
ZTA: Zirconia-toughened alumina

Appendix: Notation and Use of SI Units

Quantity	Notation	Unit	Notation
Base units			
Mass	[M]	Kilogram	kg
Length	[L]	Meter	m
Time	[T]	Second	s
Electric current		Ampere	a
Thermodynamic temperature	[K]	Kelvin	k
Amount of substance		Mole	mol

	Unit	Notation	Equivalence
Supplementary units			
Celsius temperature	Celsius	C	
Angle	Radian	rad	–
Frequency	Hertz	Hz	$[T]^{-1}$ (or cycles s^{-1})
Force	Newton	N	$[M][L][T]^{-1}$
Pressure and stress	Pascal	Pa	$[M][L]^{-1}[T]^{-2}$ (or N m^{-2})
Energy, work, heat	Joule	J	$[M][L]^2[T]^{-2}$
Power	Watt	W	$[M][L]^2[T]^{-3}$ (or J s^{-1})
Kinematic viscosity	Stoke	St	$[L]^2[T]^{-1}$ (or 10^{-4} m^2 s^{-1})
Dynamic viscosity	Poise	P	$[M][L]^{-1}[T]^{-1}$ (or 1 Pa s = 10 P)
Atomic distances	Ångström	Å	$[L]$ (or 10^{-10} m)
Thermal conductivity	–	–	$[M][L][T]^{-3}[K]^{-1}$ (or W $m^{-1}K^{-1}$)

A.1 Use of Units

- Units should be associated with numbers
- Units should not be associated with symbols
- Equations should be stated without units since any set of consistent units is equally valid
- Equations should not contain conversion factors for inconsistent units

A.1.1 Examples of Correct and Incorrect Practice

$$F = m \cdot a \quad \textbf{CORRECT}$$

The above equation is simple and elegant. It works with any set of consistent units such as SI or British Engineering units.

$$F = m \cdot a / 32.2 \quad \textbf{WRONG}$$

The second equation is incorrect and messy. It does not work with consistent SI or British Engineering units. To make the equation work requires that each symbol must be associated with a particular inconsistent unit. This is poor practice and leads to mistakes.

It is equally important that numbers must be associated with units. 1 lb is not the same as 1 slug. Failure to specify units with numbers can lead to extremely expensive disasters as discovered after a failed Mars mission.

When values are inserted into an equation in order to calculate the value of an unknown, it is important that a consistent set of units is employed. The SI system ensures consistency when using base force, length, mass, and time units.

Some examples of consistent sets used in engineering are listed below. Other engineering quantities can be expressed in terms of these units, thus ensuring consistency. Some examples of equivalence in SI units are given above. Some conversion factors between SI and British units are given below.

System	Force	Length	Mass	Time
SI	N	m	kg	s
British Engineering Units	lbf	ft	slug	s
Alternative British Units	lbf	in.	lbf in.$^{-1}$ s^2	s
British Physical Units	pdl	ft	lb	s

A.1.2 Factors for Conversion between SI Units and British Units (Values Rounded)

Mass: 1 kg = 2.205 lb = 0.06848 slug = 0.8218 lbf in.$^{-1}$ s^2

Length: 1 m = 3.281 ft = 39.37 in.

Temperature rise: 1 K = 1°C = 1.8 F = 1.8 R

Force: 1 N = 7.233 pdl = 0.2248 lbf

Volume: 1 m^3 = 1000 L = 61,020 in.^{3} = 35.32 ft^3

Pressure: 1 bar = 0.1 MPa = 14.5 lbf in.$^{-2}$. 1 N m^{-2} = 0.000145 lbf in.$^{-2}$

Density: 1 kg m^{-3} = 0.06243 lb ft^{-3} = 0.001939 slug ft^{-3} = 0.00001347 lbf in.$^{-4}$ s^2

Energy: 1 J = 0.7376 ft lbf = 8.851 in. lbf

Power: 1 W = 0.7376 ft lbf s^{-1} = 8.851 in. lbf s^{-1}

Thermal conductivity: 1 W m^{-1} K^{-1} = 0.0141 lbf s^{-1} R^{-1}

Heat transfer coefficient: 1 W m^{-2} K^{-1} = 0.003172 lbf in.$^{-1}$ s^{-1} R^{-1}

Kinematic viscosity: 1 m^2 s^{-1} = 10^4 St = 10.76 ft^2 s^{-1} = 1550 in.^{2} s^{-1}

Dynamic viscosity: 1 N s m^{-2} = 10 P = 1000 cP = 0.000145 lbf s in.$^{-2}$ = 0.000145 reyns

Gem size: 1 metric carat = 2 × 10^{-4} kg

Index

Printed in the United States
by Baker & Taylor Publisher Services